D1160003

LABORATORY SAFETY: PRINCIPLES AND PRACTICES

QR
54.7
.L33
1986

LABORATORY SAFETY:
PRINCIPLES
AND PRACTICES

EDITOR IN CHIEF: BRINTON M. MILLER
EDITORS: DIETER H. M. GRÖSCHEL
JOHN H. RICHARDSON
DONALD VESLEY
JOSEPH R. SONGER
RILEY D. HOUSEWRIGHT
W. EMMETT BARKLEY

AMERICAN SOCIETY FOR MICROBIOLOGY
WASHINGTON, D.C. 1986

Tennessee Tech. Library
Cookeville. Tenn.

363517

Copyright © 1986 American Society for Microbiology
1913 I St., N.W.
Washington, DC 20006

Library of Congress Cataloging in Publication Data

Main entry under title:

Laboratory safety.

Includes indexes.
1. Microbiological laboratories—Safety measures.
I. Miller, Brinton M. (Brinton Marshall), 1926–
II. Gröschel, Dieter H. M. (Dieter Hans Max), 1931–
III. American Society for Microbiology.
QR647.L33 1986   576′.028′9   85-20048

ISBN 0-914826-77-8

All Rights Reserved
Printed in the United States of America

Laboratory Safety: Principles and Practices

ERRATA

Preface, p. ix, col. 2. The past affiliations of Riley D. Housewright should be correctly described as Chief of the Microbial Physiology and Chemotherapy Branch in the Safety Division, Fort Detrick, Maryland, and subsequently Scientific Director of Fort Detrick.

Page 80. Table 2 should appear as follows.

TABLE 2. Poxviruses with zoonotic potential

Genus	Virus species	Host range	Potential reservoir hosts
Orthopoxvirus	Variola, alastrim	Humans, monkeys, mice	Humans, monkeys, rodents
	Whitepox	Monkeys, rodents	Monkeys, rodents
	Monkeypox	Monkeys, humans, rodents	Monkeys, rodents
	Cowpox	Cattle, humans, other mammals	Unknown
	Cowpoxlike viruses	Rodents, carnivores, elephants, humans	Rodents
	Poxviruses of buffalo, camel, mouse, other mammals	Species of association, possibly humans	Species of association
	Vaccinia	Humans, other mammals, birds	Unknown
Parapoxvirus	Pseudocowpox (milker's node, paravaccinia)	Cattle, humans	Cattle
	Orf (contagious ecthyma)	Sheep, goats, cattle, humans	Sheep, goats
	Bovine papular stomatitis	Cattle, humans	Cattle
Capripoxvirus	Goatpox	Goats, humans	Goats
Unclassified	Molluscum contagiosum	Humans, chimpanzees	Humans
	Tanapox	Monkeys, humans	Monkeys
	Yaba monkey tumor pox	Monkeys, humans	Monkeys

Page 84. Table 3 should appear as follows.

TABLE 3. Some togaviruses of zoonotic importance

Genus	Subgroup	Vector	Examples of viruses	Host Reservoir	Host Diseased
Alphavirus	VEE	Mosquito	VEE	Rodents, equines	Humans, equines
			Semliki Forest	Chimpanzees, rodents	Humans
			Chikungunya	Humans, monkeys	Humans
	WEE	Mosquito	WEE	Birds	Humans, equines
			Sindbis	Mammals, birds	Humans
	EEE	Mosquito	EEE	Birds, turtles	Humans, equines, pheasants
Flavivirus	Tick-borne	Tick	Kyasanur Forest	Monkeys, rodents, ticks	Humans, monkeys
			Louping ill	Sheep, rodents	Humans, sheep
			Hypr	Rodents, birds	Humans
			Russian spring-summer encephalitis	Rodents, birds	Humans
	Mosquito-borne	Mosquito	Yellow fever	Humans, monkeys	Humans
	West Nile		West Nile	Birds	Humans
			Japanese (B) encephalitis	Birds, pigs	Humans
			Wesselsbron	Humans, sheep	Humans, sheep
	Dengue		Dengue, types 1 through 4	Humans, monkeys	Humans
	Spondweni		Spondweni	Cattle, sheep, goats	Humans
			Zika	Humans, monkeys	Humans

Contents

Section V. Accidents and Medical Emergencies

Contributors

Hans K. Adldinger
Department of Microbiology, College of Veterinary Medicine, University of Missouri, Columbia, Missouri 65211

David M. Asher
Laboratory of Central Nervous System Studies, Intramural Research Program, National Institute of Neurological and Communicative Disorders and Stroke, Bethesda, Maryland 20892

Donald C. Blenden
Department of Microbiology, College of Veterinary Medicine, and Department of Medicine-Infectious Diseases, School of Medicine, University of Missouri, Columbia, Missouri 65211

Herbert M. Bloom
16 Fairview Avenue, Frederick, Maryland 21701

Walter W. Bond
Hospital Infections Program, Centers for Disease Control, Atlanta, Georgia 30333

John M. Boyce
The Miriam Hospital, Providence, Rhode Island 02906

G. Alexander Carden
Tropical Medicine Institute, North Miami, Florida 33162

Mark A. Chatigny
Naval Biosciences Laboratory, University of California, San Lorenzo, California 94580

Kermit G. Dwork
Department of Medicine, State University of New York, Stony Brook, New York 11794, and Queens Hospital Center, Jamaica, New York 11432

Richard F. Edlich
Advanced Life Support Learning Center, Department of Plastic Surgery, University of Virginia School of Medicine, Charlottesville, Virginia 22908

Martin S. Favero
Hospital Infections Program, Centers for Disease Control, Atlanta, Georgia 30333

John E. Forney
Laboratory Improvement Program, Centers for Disease Control, Atlanta, Georgia 30333 (retired)

D. Carleton Gajdusek
Laboratory of Central Nervous System Studies, Intramural Research Program, National Institute of Neurological and Communicative Disorders and Stroke, Bethesda, Maryland 20892

Clarence J. Gibbs, Jr.
Laboratory of Central Nervous System Studies, Intramural Research Program, National Institute of Neurological and Communicative Disorders and Stroke, Bethesda, Maryland 20892

Dieter H. M. Gröschel
Department of Pathology, University of Virginia School of Medicine, Charlottesville, Virginia 22908

Mary M. Halbert
Department of Environmental Health, School of Public Health, University of Minnesota, Minneapolis, Minnesota 55455

Everett Hanel, Jr.
Program Resources, Inc., National Cancer Institute-Frederick Cancer Research Facility, Frederick, Maryland 21701

John E. Jaugstetter
Office of Biosafety, Centers for Disease Control, Atlanta, Georgia 30333

Dennis D. Juranek
Parasitic Diseases Division, Center for Infectious Diseases, Centers for Disease Control, Atlanta, Georgia 30333

Arnold F. Kaufmann
Division of Bacterial Diseases, Center for Infectious Diseases, Centers for Disease Control, Atlanta, Georgia 30333

James N. Keith
Chemistry and Chemical Engineering Research Division, IIT Research Institute, Chicago, Illinois 60616

John G. Kenney
Department of Plastic Surgery, University of Virginia School of Medicine, Charlottesville, Virginia 22908

James Lauer
Department of Environmental Health and Safety, Boynton Health Service, University of Minnesota, Minneapolis, Minnesota 55455

Elaine Levesque
Advanced Life Support Learning Center, University of Virginia Medical Center, Charlottesville, Virginia 22908

Don C. Mackel
Nosocomial Infections Laboratory Branch, Hospital Infections Program, Center for Infectious Diseases, Centers for Disease Control, Atlanta, Georgia 30333

Dorothy M. Melvin
Parasitology Training Section, Centers for Disease Control, Atlanta, Georgia 30333

Brinton M. Miller
Neogen Corp., Lansing, Michigan 48912

Raymond F. Morgan
Department of Plastic Surgery, University of Virginia School of Medicine, Charlottesville, Virginia 22908

Norman J. Petersen
Bureau of Disease Control, Arizona State Department of Health Services, Phoenix, Arizona 85008

G. Briggs Phillips
Becton Dickinson & Co., Paramus, New Jersey 07653

Robert S. Runkle
Pharmaplastics Closures Inc., Baltimore, Maryland 21230

Monica H. Schaeffer
Division of Safety, Office of Research Services, National Institutes of Health, Bethesda, Maryland 20892; presently with Department of Medical Psychology, Uniformed Services University of the Health Sciences, Bethesda, Maryland 20814

Leonard W. Scheibel
Department of Preventive Medicine, School of Medicine, Uniformed Services University of the Health Sciences, Bethesda, Maryland 20814

Edward A. Schmidt
ServiceMaster Industries Inc., Downers Grove, Illinois 60515

Kimberly A. Silloway
Department of Plastic Surgery, University of Virginia School of Medicine, Charlottesville, Virginia 22908

Joseph R. Songer
National Animal Disease Center, U.S. Department of Agriculture, Ames, Iowa 50010

Daniel A. Spyker
Department of Medicine, University of Virginia School of Medicine, Charlottesville, Virginia 22908

Jerome W. Staiger
Department of Environmental Health and Safety, University of Minnesota, Minneapolis, Minnesota 55455

Bruce W. Stainbrook
BBL Microbiology Systems, Cockeysville, Maryland 21030

James F. Sullivan
National Animal Disease Center, U.S. Department of Agriculture, Ames, Iowa 50010 (retired)

John G. Thacker
Department of Mechanical and Aerospace Engineering, University of Virginia, Charlottesville, Virginia 22901

Donald Vesley
Department of Environmental Health and Safety, Boynton Health Service, and School of Public Health, University of Minnesota, Minneapolis, Minnesota 55455

William M. Wagner
Office of Biosafety, Centers for Disease Control, Atlanta, Georgia 30333

Gailya P. Walter
Center for Environmental Health, Centers for Disease Control, Atlanta, Georgia 30333

David Weathers
Office of Biosafety, Centers for Disease Control, Atlanta, Georgia 30333

Richard P. Wenzel
Department of Internal Medicine, University of Virginia School of Medicine, Charlottesville, Virginia 22908

David L. West
Center for Devices and Radiological Health, Food & Drug Administration, Rockville, Maryland 20857

Preface

Many scientists in both the public and private sectors, especially microbiologists, have long recognized the great variety of pathogenic microorganisms and the hazards of dealing with them. Microbiologists in clinical and research laboratories alike process specimens, cultures, toxins, and countless other types of organic material that present a risk to the handler and the environment. Further, many chemicals, reagents, and other substances regularly encountered in laboratory work can affect human health, often seriously.

With the advent of recombinant DNA and new biotechnologies, there has been a large increase in the use of procaryotic and eucaryotic cells in research, which will spread to manufacturing. Early recognition of the potential hazards of these new organisms led to the well-known Asilomar Conference in California in the mid-1970s, followed shortly by the establishment of the Recombinant DNA Advisory Committee (RAC) of the National Institutes of Health (NIH).

An increasingly important consideration in biotechnology research and applications is that workers in these fields are not necessarily trained in microbiological techniques, including the safe handling of pathogens or potential pathogens. All kinds of engineers, chemists, farmers, food processors, technical representatives, and even executives now find themselves, by choice or circumstance, in a position of exposure to hazardous organisms, toxins, or toxic materials used in research and development.

In 1978 Robert A. Day, then Managing Editor of the American Society for Microbiology (ASM), proposed to the ASM Publications Board that ASM publish a manual of laboratory safety. At that time there was no thorough and up-to-date text available devoted to safety procedures designed for the field of microbiology. What is now called modern biotechnology was beginning its present rapid growth, involving increasing numbers of nonmicrobiologists. It was suggested that the Society had a responsibility to provide this book for the field, much as it had provided the highly regarded *Manual of Clinical Microbiology*. The Publications Board and both of its chairmen during the following years, Leon Campbell and Helen Whiteley, enthusiastically endorsed the project.

A top-flight board of editors was enlisted, each with experience within the wide spectrum of biosafety. They included:

—Dieter H. M. Gröschel, director of the Microbiology Laboratory at the University of Virginia Medical Center, Charlottesville.

—John H. Richardson, biological safety consultant, Division of Laboratory Training and Consultation at the Centers for Disease Control (CDC), who with W. Emmett Barkley edited the 1984 CDC/NIH publication "Biosafety in Microbiological and Biomedical Laboratories."

—Donald Vesley, Director of Environmental Health and Safety and with the School of Public Health at the University of Minnesota, Minneapolis, a well-known writer-researcher of programs to control pathogens in the laboratory workplace.

—Joseph R. Songer, Biohazard Control Officer and now Chief, Environmental Health and Safety, at the National Animal Disease Center in Ames, Iowa.

—Riley D. Housewright, an early and active worker in laboratory biosafety, Safety Director and subsequently Scientific Director in the Safety Division at Fort Detrick, Maryland, and later extensively concerned with microbiological problems of water systems while with the National Academy of Sciences.

—W. Emmett Barkley, Director of the Division of Safety at NIH, a past principal on the RAC, and co-editor of "Biosafety in Microbiological and Biomedical Laboratories."

—Brinton M. Miller, Editor in Chief, former Director of Animal Infections and Biosafety Officer for the Merck Sharp & Dohme Research Laboratories in Rahway, N.J.

Initially the editorial board tried to design a basic manual providing "how-to" information relative to safety practices. However, it was soon apparent that before practices of safety could be put to paper, the concept of and the reasons for safety programs must be addressed. We believe that the laboratory supervisor, the person in charge of the individual physical working unit (the laboratory), is the keystone of good safety practice. We also believe, especially as several of us have been or are institutional biosafety officers, that it is necessary for laboratory management to support by both word and deed the safety practices of their workers. Consequently, we decided that the principles of safety management and assessment of hazard should be covered before the methods of safety practice could be detailed.

During these early deliberations we learned that the CDC and NIH were jointly to revise the older CDC publication "Classification of Etiologic Agents on the Basis of Hazard." The newer document would deal with historical perspectives plus detailed descriptions and classification of laboratories for safe handling of microorganisms, and especially would provide a new and reasonably extensive description of pathogenic agent groups. The *Laboratory Safety* editors therefore decided to defer publication of the ASM book so that this valuable information could be included in the text. Meanwhile, greater attention could be given in the book's chapters to the principles and management of safety programs, which the CDC/NIH publication could not provide.

Objectives. The objectives of this book are to

describe safety management, assessment of hazard, design of laboratory and equipment for safe practice, special containment techniques used to handle potential hazards, and what to do in case of accidental exposure to a pathogen or hazardous materials used in a microbiological laboratory.

In Section I the responsibilities of management are described. Institutions must be committed to responsible management practices as a foundation for safety programs in the laboratory. In Section II, assessment of hazard is taken up in considerable detail, beginning with an overview of laboratory-acquired infections and their epidemiology. Following is an excellent discussion of human factors, which are all-important to any sound, workable program. The individual and his or her collaboration with others are the starting point for the common-sense precautions essential to any worthwhile safety program. Next, we selected a spectrum of infectious disease areas as examples for assessing hazards. These are by no means limiting. Obviously, every laboratory situation will have its own priorities. This array of topics, however, presents various successful approaches adapted to unique situations, and the reader can use any of them as a model for his or her own circumstances. Included, we believe for the first time, is a thorough discussion of parasitic infections. To the purist these may not concern either microbiology or molecular biology. However, one glance through current scientific literature will show that these well-known higher organisms are being widely studied nowadays and even being used in biotechnology.

In Sections III and IV, physical equipment and actual techniques are described. Safety procedures must consider the entire laboratory, its parts, and everything that enters or leaves the facility. The "parts" include a wide variety of devices and equipment for protection of the worker, the experiment, and the environment. The techniques presented cover methods for containment of many of the chemical agents and especially radioactive substances used so often in work at the cellular and molecular levels.

Section V contains three chapters that deal with "on-the-spot" accidents and medical emergencies which can occur in the laboratory. Authors for these chapters include practicing physicians who have been called upon many times to treat such accidents. The editors respect their opinions and their practice; the editors also realize that each accident will have its own set of circumstances and that the attending physician will be guided by his or her own training and expertise.

Of the four appendixes, the first reprints in its entirety the CDC/NIH publication "Biosafety in Microbiological and Biomedical Laboratories" discussed above. As well as other valuable information, this publication includes an excellent, if brief, historical perspective of reported laboratory-acquired infections. (We emphasize "reported" because we believe the number of actual accidents to be far greater than suggested by the incidental reporting and occasional surveys cited.)

Appendixes 2 and 3 list the offices of U.S. State Epidemiologists and Health Officers and the addresses and telephone numbers of Poison Control Centers across the country. We recommend that these specialists be consulted on any question of assessing the hazards of working with a particular agent. More importantly, in case of accidental exposure to a suspected infectious or toxic agent, these offices should be called upon as soon as possible for guidance, treatment, and follow-up. Finally, Appendix 4 provides an illustrated first aid guide. The illustrations may be recognized by readers as the same as those appearing in some telephone directories; we are indebted to Richard Edlich for obtaining permission to use them.

Conclusions. This book was conceived in late 1978 and early 1979. The years since then have been a significant period in science during which much new research has been completed. Changes in safety programs have also taken place, albeit a bit more moderately. Consequently, some chapters written early had to be revised and updated, while others were virtually rewritten to include important developments in safety awareness, training, and especially equipment. The publication of "Biosafety in Microbiological and Biomedical Laboratories" made possible further revisions.

As might be expected, since there are only so many techniques used for handling cellular populations, there is some overlap between chapters. Rather than being edited out, such reiterations have been retained for emphasis.

We have not included in this book remarks on what to do with laboratory waste. The book's concern is with what happens within the laboratory before, during, and after the research exercise, essentially stopping at the laboratory door. If live organisms still remain in material to be discarded, laboratory management should make sure they are killed; if hazardous toxins or other materials remain, they must be neutralized. However, the whole subject of disposal once waste has left the laboratory has been left to others.

Readers acquainted with good safety programs may recognize throughout this book safety techniques that have not changed over the years. Many safety techniques became standard practice immediately and have not required revision. Improved, more modern tools and equipment, however, call for some adaptation or change. This is true expecially for safety equipment. Personally, I much prefer a good biosafety cabinet to the old "transfer room."

We editors invite critiques of the book. Because it is a "first of its kind" for ASM, we are sure there are shortcomings, subjects missed, subjects under- or overemphasized, etc. We trust the readers will make suggestions for the future so that a second edition can include both new subject matter and improvements on the old.

We acknowledge with thanks all those who contributed to the book's publication, including but not limited to the following.

—Robert Day, our original Managing Editor, who with R. G. E. Murray suggested the book be written.

—Leon Campbell and Helen Whiteley, chairmen of the ASM Publications Board, who encouraged, advised, and urged us both to start and to finish the book. Also, members of the ASM Publications Board are recognized for their constant support.

—Our reviewers of draft chapters, John O. Fish (University of Washington, Seattle), Page S. Morahan (Medical College of Pennsylvania, Philadelphia), G. Briggs Phillips (Health Industry Manufacturers Association, Washington, D.C.), and Gladys Schlanger (Woodhull Medical Center, Brooklyn, N.Y.). They provided good, solid constructive criticism for which we are truly grateful.

—Our authors, without whose dedication there would not be a book. They wrote and rewrote, reviewed, and proofed. Most importantly, they were willing to share with us their wide knowledge of the principles and practices of sound safety programs.

We are also deeply obliged to the ASM Publications Department, especially Ellie Tupper, Susan Birch, and Linda Illig. They have read, copy edited, proofed, and corrected our mistakes. Also praise should be given to Mrs. Tupper for her good humor and positive attitude over the years toward getting the job done.

Brinton M. Miller
Editor in Chief

SAFETY MANAGEMENT

Chapter 1

Laboratory Safety Program Organization

JOSEPH R. SONGER

Properly conceived, the organization of a laboratory safety program permits laboratory management to share and to assign responsibility for accident prevention and to ensure adherence to safety standards among the staff. The safety program is not something that is imposed on the organization; rather, it must be built into every technique and operation within the laboratory. It must be an integral part of each laboratory function. While there are no specific sections of the Occupational Safety and Health Act of 1970 which deal exclusively with laboratories, the provisions of the General Industry Standards (29 CFR, Section 1904.8), which apply to all work sites subject to Occupational Safety and Health Administration (OSHA) regulations, are applicable.

The objective of the Occupational Safety and Health Act, to reduce occupational injuries and illnesses, applies to the laboratory as well as other work environments. Yet mere compliance with OSHA standards, which largely involve equipment and physical conditions, is not enough. OSHA directives serve only as a place to begin; a good safety program will exceed them.

TOP MANAGEMENT RESPONSIBILITY FOR SAFETY LEADERSHIP

Assumed responsibility

The attitude of top management toward safety is reflected in the attitudes of supervisors. Similarly, the worker's attitude usually reflects that of his supervisor. If top management is not genuinely interested in accident prevention, no one else is likely to be.

Details for carrying out a laboratory accident prevention program may be assigned, but responsibility for the basic policy cannot be merely delegated to a safety officer or safety committee. The laboratory director is responsible for overall safe performance and must constantly review this performance, which is best done through lines of supervision.

Declaration of policy

Managing safety in the laboratory must start with a written safety policy. Petersen (9) states that a safety policy should (i) affirm a long-range purpose; (ii) commit management at all levels to reaffirm and reinforce this purpose in daily decisions; and (iii) indicate the scope left for discretion and decision by lower-level management.

Safety is a line responsibility, but the line organization will accept this responsibility only when management assigns it. It is important that the safety policy be written to ensure that there will be no confusion concerning direction and assignment of responsibility.

The safety policy must be put in writing, and each employee should be made aware of it. It should be a part of the laboratory's operational manual. The policy statement should include not only the intent but also the scope of activities covered. Responsibility and accountability should be clearly stated, including the roles of the staff safety officer and any safety committees appointed. The policy statement may be conceived and written by the safety officer, but it should be signed by the laboratory director.

ASSIGNMENT OF RESPONSIBILITY

While the laboratory director has the ultimate responsibility for safety, he delegates authority for safe operation down through all levels of management.

Management officials

Middle levels of management should play an active role in the safety program. Managers of divisions within a laboratory should provide personal leadership in implementing the laboratory safety policy. They should interpret the policy as it applies to their division and give it their full support. A casual or indifferent approach to safety on the part of middle

1

management will be reflected in an indifferent attitude among supervisors and laboratory workers.

Safety officer

Any laboratory with more than a few employees should have a safety officer. It is important that management definitely assign full staff responsibility for safety activities to one responsible individual. The safety officer will advise top management, help in administering the safety policy, and ensure the continuity of the safety program.

The duties of the safety officer usually include:
1. Planning, administering, and making necessary changes in the accident prevention program
2. Reporting to the laboratory director on a regular basis on the status of safety in the laboratory
3. Acting as an advisor on all matters of safety to all levels of management, laboratory supervisors, and other departments such as purchasing, engineering, and personnel
4. Maintaining accident records, investigating accidents, obtaining supervisor's accident reports, and checking corrective action
5. Conducting or coordinating supervisor safety training program; advising and assisting supervisors in employee safety training
6. Developing and coordinating a medical surveillance program
7. Making personal safety inspections and supervising inspections by the safety staff and special employee committees for the purpose of discovering and correcting unsafe conditions or practices in the laboratory
8. Reviewing and approving designs of new equipment and facilities
9. Supervising fire prevention activities
10. Making sure that the laboratory is in compliance with OSHA requirements and state or local laws and ordinances
11. Maintaining outside professional contacts to exchange safety information relevant to laboratory operations

Supervisors

The first-line supervisor is a key person in any safety program because he is in constant contact with the employees. The laboratory supervisor has the following responsibilities:
1. Establish work methods in the laboratory
2. Give job instructions to employees
3. Make job assignments
4. Supervise the work of employees under him
5. Be responsible for maintenance of equipment and laboratory facilities

All of these responsibilities have safety overtones. Establishing work methods that are well understood and consistently followed is essential to an orderly and safe operation. Many laboratory accidents have been reported to result from unsafe methods or procedures.

Giving job instruction, with emphasis on safety aspects of the job, will help eliminate one of the most frequent causes of accidents: lack of knowledge. A technique called job safety analysis is frequently used in industry to emphasize safety in the job assignment (8). This technique can be used in planning new jobs or elucidating hazards associated with established jobs. Job safety analysis is covered in detail later in this chapter.

Assigning people to jobs is closely related to job instruction. Whenever a laboratory supervisor makes a work assignment, safety as well as good job performance requires that he be sure the worker is qualified to do the job and thoroughly understands the work method.

Supervising people in the laboratory is necessary even after a safe work method has been established. People deviate from established safe practices, and injuries result. It is the supervisor's responsibility to make sure that his subordinates are working safely.

Maintaining equipment and laboratory facilities in safe condition is the supervisor's responsibility. Accidents can result from unsafe equipment, a disorderly workplace, or makeshift arrangements. The supervisor who keeps his laboratory and equipment in top condition helps prevent accidents as well as improve efficiency.

Safety committees

A laboratory of 20 or more employees should have a safety committee chaired by the safety officer and representing all levels of employment. Meetings should be held once each month, on laboratory time, to discuss safety problems and programs. Minutes should be taken. The safety committee will review accidents and take corrective action. They will also review safety suggestions and return their favorable or unfavorable response to the originators of the suggestions.

When a safety committee is formed, certain policies and procedures should be set forth in writing, including at least: (i) the scope of the committee's activities; (ii) the extent of the committee's authority; and (iii) points of procedure for meetings, including frequency, time, and place, attendance requirements, the order of business, and records to be kept.

Committee membership should rotate at intervals long enough so that each member has time to make a positive contribution and to gain personal experience and interest, and yet short enough so that the opportunity to serve can be passed around.

Safety inspection committees should understand that their job is essentially a helpful and constructive one. The laboratory supervisor should accompany the committee on its inspection of his area. The supervisor can supply information needed by the committee and should be kept informed of its actions. In making inspections, the committee should watch particularly for unsafe practices and report them immediately to the supervisor. The supervisor should then be allowed the opportunity to correct them.

The inspection committee should comply with all regulations. If laboratory coats, safety goggles, rubber gloves, or respirators are required for workers, committee members should wear them also.

Under no circumstances should the committee or its members interfere with the work of employees or with the conditions of the laboratory nor usurp any of the supervisor's authority.

EMPLOYEES' RESPONSIBILITY FOR THEIR OWN SAFETY

Regardless of how well a laboratory is engineered for safety, the employees' safety depends most on their own conduct. It is impossible to order everything a worker does in a laboratory, so in addition to providing direct supervision, it is necessary to influence the voluntary acts of workers by education and motivation. Management can state in its policy that safe performance from employees is desirable; it cannot, however, force the employees to perform safely. Employees decide for themselves whether they will work, how hard, and how safely. Their decisions are shaped by their attitudes towards themselves, their environment, their supervisor, the laboratory, and the job, as well as by their own knowledge and skills. Group attitudes also affect safety performance. While group attitudes are more obvious in factory settings among assembly line workers than in the usual laboratory setting, they nevertheless do exist in the laboratory. Once an activity becomes an accepted norm among a group of laboratory workers, it is done without question. Essentially, the group sets and keeps to its own safety standards, regardless of what management's standards are. Employees might even abide by the letter of the safety directives but still not perform safely. By building the staff into strong groups of mature and competent people, management is more likely to be able to motivate them to perform safely.

Employee selection is very important in building mature, competent groups. Petersen points out that government today has a large amount of control over this process, whereas some industries have almost abdicated control in this area (9).

Training is also a powerful influence and motivator for safety in the laboratory, just as it is in many other areas. With training, management gives the employee two things—knowledge and skill; if it is successful, three are returned—knowledge, skill, and motivation.

Job safety analysis

A technique widely used in industry which can be adapted for use in the laboratory is job safety analysis. This technique can be used in planning new jobs or for elucidating hazards associated with established jobs. Using a work sheet, the job safety analyst (i) selects the job; (ii) breaks the job down; (iii) identifies hazards and potential accidents; and (iv) develops solutions.

Such jobs as operating laboratory equipment or titrating a virus are appropriate selections for job safety analysis; large, complex jobs as well as simple, one-step jobs should be avoided. In the job breakdown, the analyst should be especially alert for potential accidents. For example, what is the potential for aerosol generation which could result in respiratory exposure? for a puncture injury from a hypodermic needle or an animal bite? for respiratory or contact exposure with toxic chemicals? Methods of control should be sought for each potential hazard.

In the laboratory, job safety analysis should be done by the supervisor, with the technician or other subordinate participating. This process helps the supervisor to understand better the jobs he supervises; the employee's understanding is also increased, improving his safety knowledge and attitude.

A common job in the laboratory which can be used to illustrate job safety analysis is operating a sterilizer. Improper use of sterilizers has resulted in laboratory infections and cross-contamination. The analysis sheet (Fig. 1) is divided into three columns entitled "What To Do," "How To Do It," and "Key Points." In column 1 the steps of the job are listed. Column 2 provides instructions for doing the job, and column 3 lists key points to be kept in mind.

Job safety analysis has been used extensively in industry. Some industrial plants have prepared in excess of 30,000 such analyses. This technique, if applied to laboratory procedures, could help pinpoint potential hazards and enable management to eliminate or protect against them.

Safety promotion programs

It is difficult to maintain a high level of interest and awareness of safety over extended periods of time. Because of this, an ongoing safety awareness program should be instituted. Even if the laboratory has been engineered for safety, laboratory procedures and equipment have been made as safe as possible, and supervisors have trained their workers thoroughly and continue to enforce safe work procedures, it is necessary to maintain interest in safety. Accident prevention basically depends upon the desire of people to work safely.

Many hazardous conditions and unsafe acts cannot be anticipated, so each employee must frequently use his own imagination, common sense, and self-discipline to protect himself. A program which will

WHAT TO DO (Steps in Sequence)	HOW TO DO IT (Instructions)	KEY POINTS (Safety Always a Key Point)
1. Prepare for start-up	Remove plug screen from bottom of chamber and clean.	If not clean, will interfere with free flow of steam.
	Replace chart in controller.	Chart is record to verify sterilization cycle.
2. Heat jacket	With operating handle at "off" position, open steam supply valve.	Do not attempt to sterilize until jacket gauge shows 15 to 17 lb of pressure.
3. Arrange load	Place flat packs of supplies on edge. If several tiers, place alternative tiers crosswise.	To assure adequate steam circulation.
	Loading containers of liquid:	
	—Do not mix loads of liquids with other supplies.	
	—Use only vented closures.	Sealed bottles may explode.
	—Use only type 1 borosilicate (Pyrex) glass bottles.	Stress of pressure and temperature may rupture ordinary glass.
	—Use sterilizer slow exhaust cycle only.	Fast exhaust causes rapid boiling within the bottles with loss of fluids. Do not place flammable chemicals or chemicals which are unstable at high temperatures in the sterilizer.
	Moisten loads of clothing or other fabrics.	Dry fabrics remove moisture from steam, causing superheating which chars the fabric.
4. Close autoclave door	If equipped with "quick throw" handle, move "quick throw" handle clockwise until arms are positioned radially in slots.	
	Continue tightening door by turning hand wheel clockwise.	
5. Sterilize	To sterilize all materials:	
	—Open "steam to chamber valve" and "chamber drain valve."	
	—Close "air filter valve," "vacuum valve," and "condenser exhaust bypass valve."	
	—Turn timer to desired exposure period.	Time required for sterilization varies with load:
	—Turn selector to appropriate position:	—surgical packs, 30 min
	"slow exh" for liquids	—solutions (aqueous) in Pyrex flasks, 30 min
	"fast exh" if drying not required	—chicken eggs in 10-gal GI cans, 6 h
	"fast exh and dry" for surgical packs, wrapped supplies, etc.	
	—Turn operating handle (clockwise only) to "Ster"	
6. Open sterilizer	Turn hand wheel counterclockwise. Move "quick throw" handle counterclockwise and open door.	
When load is completely processed, sterile light will come on and alarm will sound.	At end of a liquid cycle, open sterilizer door no more than 0.5 in. Wait 10 min before unloading sterilizer.	Rapid boiling will occur if door is opened too soon.

FIG. 1. Job safety analysis for operating a sterilizer.

stimulate safety awareness will help prevent accidents.

The safety officer should be a well-informed specialist. His role is to coordinate the safety awareness program and supply the ideas and inspiration, while enlisting the wholehearted support of management, supervisors, and employees.

The laboratory supervisor is the key figure in any program to create and maintain interest in safety because he is responsible for translating management's policies into action and for promoting safety activities directly among the employees. The supervisor's attitude toward safety is a significant force in the success, not only of specific promotional activities, but also of the entire safety program, for his views will generally be reflected by the employees in his laboratory. The supervisor who is sincere and enthusiastic about safety can be more effective than the safety officer in maintaining interest. Supervisors sometimes resist safety promotional ideas; in that case it is the safety officer's job to sell such programs to the supervisor.

Supervisors can motivate by example. Wearing safety glasses and other protective equipment when they are needed is one of the best ways a supervisor can promote safety.

The laboratory supervisor should maintain a high level of safety in the laboratory at all times. He should not have to suddenly adopt a "get tough" approach to enforce safety rules. He should be consistent, firm, and fair. If the workers sense that the supervisor either cannot recognize unsafe conditions and unsafe acts or is indifferent to them, they, too, will become lax.

Safety films with a laboratory orientation are good safety motivators. Films on microbiological hazards, chemical hazards, the hazard of compressed gases, and many other subjects are available.

Regularly scheduled "stand-up" safety meetings are also good safety motivators. Subjects and materials for these regular meetings can be suggested by the safety officer, but the meetings themselves should be conducted by the supervisor. These meetings should be short and timely and oriented toward the work at hand.

Information on safety off the job ought to be distributed to employees. Interestingly, most of the time lost from work due to accidents is from accidents which occur outside work. The National Safety Council publishes an "Off the Job Safety" section in *National Safety News* each month. The council's journal, *Family Safety,* also contains much information on off-the-job subjects.

Safety is always a timely subject for inclusion in the laboratory newsletter or publication. Cartoons in good taste can be very effective (Fig. 2).

Well-planned and properly conducted safety contests can also be effective. One successful contest is "Safety Suggestion of the Month": employees are encouraged to place suggestions on safety in well-

FIG. 2. Cartoons done in good taste are very effective safety communicators.

located suggestion boxes, with the monthly winner receiving, for example, a month's parking in a choice "reserved" parking stall. This has the dual benefit of stimulating employees to think about safety and bringing out safety problems of which supervisors might not have been aware.

Safety bulletin boards by themselves are not very effective; however, in a well-rounded safety program, they can be useful as information centers. Timely safety posters are of value if they are part of a total program. Safety-related displays and exhibits can also be effective, especially if they relate directly to the laboratory. Accidents or "near misses" can be very effectively used to call attention to hazards. Figure 3 presents an example. A CO_2 cylinder scheduled for delivery to a research laboratory was overfilled and fitted with an improper safety device. When it exploded in a storage area, it was hurled through the roof and landed 330 ft from the site. Had the cylinder been delivered on schedule, the explosion would have occurred in a laboratory. A graphic illustration of this type is worth a thousand words.

It is important in all motivational efforts to convey the concept that safety is not something which a worker puts on like his hat, but should be an integral part of his life. There are, of course, no absolutes in safety, and to some extent risk is the essence of life. The objective is to be aware of and to minimize the risks faced daily in the laboratory and elsewhere.

ESTABLISHING RESPONSIBILITY FOR MAINTAINING SAFE WORKING CONDITIONS IN THE LABORATORY

Maintaining safe working conditions in the laboratory is everyone's responsibility; however, some key positions play an important role.

FIG. 3. Near misses such as this exploded cylinder can be used as examples to increase safety awareness.

Inspectors

Inspectors play an important part in maintaining safety in the laboratory. It is, at times, necessary that many different people assume the role of inspector. The inspector may be a safety professional trained to look for specifically hazardous equipment and activity, or a person from any level of management, including the laboratory director who makes a casual inspection to support the safety program. The first-line supervisor is one of the most important inspectors in the laboratory and is a key figure in the maintenance of safe working conditions. Practically all of the supervisor's time in the laboratory is spent in constant contact with the workers. The supervisor should be thoroughly familiar with all hazards which may develop and should be constantly on the alert to correct unsafe conditions and practices.

Maintenance engineers should make frequent trips through the laboratory to assure that proper ventilation and air balance are maintained and that mechanical equipment is functioning properly. At times very sophisticated equipment and systems are installed in laboratories with no provision for their inspection, evaluation, and maintenance. Such equipment as sterilizers, laminar flow biological safety cabinets, and fume hoods need to be monitored on a regular basis.

Employees who are constantly on the alert can be of great value in preventing accidents. Any safety problems should be reported immediately to the laboratory supervisor.

Safety committees often function in an inspection capacity. They may make periodic or intermittent inspections of laboratories which serve to uncover hazards as well as call attention to safety. Safety committees may also investigate accidents.

Paperwork for inspectors should be kept to a minimum. Checklists will speed up the process, but space should be allowed for listing those problems not covered in the checklist.

Purchasing

Purchasing departments should have excellent liaison with the safety department. Safety standards should be developed to be used as guidelines when equipment is purchased for the laboratory. Refrigera-

tors, for example, which are to be used for holding flammable solvents should be explosion-safe. If purchase orders for refrigerators are routed through the safety department, proper equipment can be selected and proper labeling can be done. Many hazardous chemicals can now be purchased in plastic-coated glass bottles. Orders for hazardous chemicals should be reviewed by the safety department to assure that this safety feature is requested. Safety should always be of greater consideration than cost.

Engineering

Engineering design plays an important role in accident prevention. The goal of engineering should be to design laboratory environments and equipment and to set up job procedures so that the laboratory worker's exposure to infection or injury will be eliminated or minimized. Human factors engineering is a scientific approach to matching humans with their work environment to maximize safety and production while minimizing effort and risk. Counter heights, location of equipment, door locations, visual displays, operating knobs, and many other factors influence how efficiently and for how long a laboratory worker can satisfactorily perform a job. Fatigue and strain often contribute to accidents.

Supervisors

Supervisors, as has been previously stated, and laboratory instructors play a key role in maintaining safe working conditions in the laboratory. The supervisor should be satisfied that all equipment is operating safely and that laboratory procedures are outlined and followed in a safe manner. Job safety analysis should be applied to each job performed in the laboratory.

Legal liability

The Resource Conservation and Recovery Act of 1976 was enacted by Congress for the purpose of providing technical and financial assistance for the development of management plans and facilities for the recovery of energy and other resources from discarded materials and for the safe disposal of discarded materials and to regulate the management of hazardous waste. Standards governing the generation, storage, transport, and disposal of hazardous waste were put into effect 19 November 1980. The full impact of these standards on the laboratory has yet to be realized. Compliance with these laws is primarily the responsibility of management.

Legal liability in the laboratory can be incurred in a number of ways; however, for this discussion tort liability will be of primary interest. Most people are only vaguely aware of what lawyers call torts. This is largely due to the fact that no one has satisfactorily defined a tort. A person is far more likely to be the victim of a tort, or to commit a tort, than a crime.

A tort is a civil wrong against an individual. A crime is an offense against the public at large or the state (1). A tort is an act that violates one's private or personal rights. If a tort is committed against you, it is entirely up to you to seek relief by suing the offender in the civil courts. Costs for bringing such suits are borne by you and all damages are paid to you. Probably torts is the most social of all the areas of the law, in that it affects more people in their day-to-day living than do contracts, criminal law, bankruptcy, or any of the other legal fields. Over 75% of all of the litigation in the United States is for torts (5).

To recover damages for a tort one must prove either that the act was committed with deliberate intent or that it was the result of negligence (1). Intent is an essential element in such torts as libel or trespass.

Torts which might occur in the laboratory are for the most part likely to be the result of negligence. Negligence in the eyes of the law may be defined as conduct that falls below a standard of care established by law to protect others against an unreasonable risk of harm (3). If the standard of care has not been specifically established by statute, the actions or inactions of an individual will be measured against what a hypothetical reasonably prudent individual would have done under the same circumstances (2).

One important aspect of the conduct of the hypothetical reasonable person is anticipation. A supervisor must be able to anticipate the common ordinary events and, in some cases, even the extraordinary.

Joyce (6) points out that there are two major sources of any legal standard or requirement: (i) the requirement may be imposed by a statute or government regulation that is ultimately derived from a legislature, whether federal or state, or (ii) it may be contained in the common law, that is, the law established by judges in actual courtroom cases.

The law of negligence comes primarily from the common law. Court opinions will provide guidance about most of the legal requirements placed on teachers and laboratory supervisors. In a sense, the courts are continually recalibrating the yardstick of social values (7).

Legal liability will also differ significantly between academic and nonacademic laboratories. Academic laboratories in institutions of learning are more likely places for torts to occur than are industrial or governmental laboratories. Schmitz and Davies (10) point out that educational institutions undertaking to provide laboratory courses have a duty to provide an outline of reasonably safe experiments. Due care should be exercised to exclude dangerous experiments with limited educational benefits.

Laboratories must be equipped with necessary safety devices such as proper ventilation and fume hoods, operable fire fighting equipment, safe fire escapes, and adequate access for first aid facilities. Laboratory safety rules should be established, and instructors are expected to enforce them.

Schools have a duty to furnish qualified and responsible instructors educated in science and having a knowledge of the specific dangers involved in handling hazardous materials. Laboratory instructors have an active duty to maintain proper supervision commensurate with the potential dangers surrounding the prescribed experiments. To protect their instructors, a number of chemistry departments have resorted to giving tests on safety before beginning the course. One such department required that the students get a perfect score on the examination before laboratory work could be started. The examination could be taken any number of times but had to be signed (11).

Laboratory procedures manuals cannot be relied upon to instruct students sufficiently. Specific warnings describing latent dangers of each experiment must be conveyed by the instructor (10). Corbridge and Clay (4) point out that the duty to warn also applies to the writer of a textbook which sets out laboratory experiments for students. The California Supreme Court has said that a jury may find a writer of a laboratory manual negligent for failing to include a warning that an iron mortar is not to be used when grinding a certain compound. The court also held that the teacher may also be held liable for such a failure to warn if he knew that an iron mortar might be used by the student conducting the experiment and he knew that the use of an iron mortar could cause injury (4). Wyatt and Wright (13) published a paper entitled "How Safe Are Microbiology Texts?" in which they were quite critical of several textbooks. Laboratory techniques and procedures recommended in these books were replete with hazards. They concluded that writers and publishers of textbooks have a responsibility to eliminate hazardous procedures or at least to provide adequate warning.

"Contributory negligence" is sometimes a factor in a liability case. Contributory negligence simply means that the plaintiff was negligent in helping to cause his own injury. The modern trend in the law is for the jury to compare the plaintiff's neglect with the defendant's neglect and reduce the plaintiff's recovery by the percentage of neglect attributed to him (4).

The Sovereign Immunity Doctrine is based on the concept that any governmental operation can do no wrong and therefore cannot be sued without its consent. The United States government and the governments of many states have in recent years passed laws to permit such suits to be brought against them. Some states have laws which permit schools to buy general liability insurance and waive immunity up to the amount of the policy.

The Sovereign Immunity Doctrine varies from state to state. For example, in 1974 a suit charging negligence was brought against a chemistry professor who taught at a state college in Pennsylvania. The courts held that mere negligence is not sufficient to place liability on a public official. Something more, such as malice or recklessness, must be present to overcome the governmental immunity doctrine of Pennsylvania. This governmental immunity is rarely found to extend to professors. In Ohio the Attorney General issued an opinion that governmental immunity may extend as far as the dean of a college in a state university (12).

In some states statutes exist which require indemnification of the teachers by the school board. Where the teacher is found liable for injuries, the board must "save the teacher harmless" by paying the damages. This shifts the financial hardship from the teacher to the board.

A teacher may insure himself against a recovery based on his simple negligence. Some education associations carry insurance which either may cover members as a benefit of membership or may be purchased from the association.

In tort liability cases, damages are established by the jury. In general, when physical injury has occurred, a plaintiff seeks to recover for medical expenses both past and future, for income lost since the injury, for the impairment of earning capacity which will lead to income not gained in the future, and for pain and suffering. These are known as compensatory damages. All juries can do is make a guess at what amount would be fair and reasonable compensation.

Punitive damages may also be sought by the injured party. The purpose of these damages is to punish the wrongdoer and to make him an example to others who might think of acting wrongly (4).

Industrial laboratories. Schmitz and Davies (10) point out that, in contrast to a student, the industrial laboratory supervisor is confronted with a master-servant status; and to recover for injuries related to research employment, master-servant laws must be considered.

Under Workmen's Compensation laws, the employer is strictly liable for injuries irrespective of the employer's negligence and disregarding common-law defenses. All states of the Union, as well as the federal government, have adopted Workmen's Compensation programs, administered by Workmen's Compensation agencies or by the courts.

A number of states limit payments for injuries suffered in certain occupations that are recognized as particularly hazardous. Some states are also restrictive when it comes to occupational diseases. These states list the disorders that entitle a worker to compensation. If a person is not eligible for Workmen's Compensation, it is his right to bring a suit for damages against his employer or any other person who may be responsible for an injury. However, if the worker is entitled to Workmen's Compensation, state laws usually prohibit filing a suit.

Safe equipment and facilities. The employer is charged with a duty to provide safe equipment and instruments and must exercise reasonable care in keeping them in safe operating condition. Liability may arise for failure to provide adequate safety equipment such as safety goggles, respirators, etc.

The employer has a duty to warn of known latent dangers and dangers existing which are chargeable to the employer's knowledge. The duty to warn is particularly applicable to young and inexperienced employees. Thus a technician may rely on his supervisor's assurance that no danger exists, and the employer is liable if, in fact, an unsafe working condition does exist. By the very nature of laboratory work a worker must assume some risk; however, he is not expected to assume the risks of perils arising from his employer's negligence. The principle is well established that the cost of doing business includes providing medical care and compensation for employees injured in the course of business (10).

SUMMARY

In summary, the development of a safety program is a management responsibility. Implementation and enforcement of the safety program are line responsibilities, and the most important person in the system is the first-line supervisor. Ultimately, however, each individual is responsible for his or her own safety. While management desires safe performance, the employees ultimately decide if they will work safely. Employees' decisions are shaped by their attitudes towards themselves, their supervisors, the laboratory, and their entire situation. An effective safety program, including training and other motivational efforts, will help develop proper attitudes toward safety.

Safe working conditions are the responsibility of many disciplines. Persons with inspection responsibilities, purchasing agents, engineers, and first-line supervisors all have the responsibility of ensuring that safety is given due consideration in the selection of equipment and maintenance of work environments.

Legal liability in the laboratory can be incurred in a number of ways; tort liability is the most likely. Torts which occur in the laboratory are, for the most part, likely to be the result of negligence. Supervisors must be informed about the potential hazards in their work environments and must convey this knowledge to workers.

LITERATURE CITED

1. **Anonymous.** 1977. You and the law. The Reader's Digest Association, Inc., Pleasantville, N.Y.
2. **Anonymous.** 1977. Legal aspects of classroom safety, p. 250. *In* Safety in the school science laboratory. National Institute of Occupational Safety and Health, Cincinnati.

3. **Anonymous.** 1981. Legal liability, p. 17. *In* J. A. Gerlovich and G. E. Downs (ed.), Better science through safety. Iowa State University Press, Ames.

4. **Corbridge, J. N., Jr., and A. R. Clay.** 1979. Laboratory teachers and legal liability. Phys. Teacher **17:**449–454.

5. **Evans, J. E., Jr.** 1971. Legal liability for laboratory accidents, p. 64–66. *In* N. V. Steere (ed.), Handbook of laboratory safety. Chemical Rubber Co., Cleveland.

6. **Joyce, E. M.** 1978. Law and the laboratory. Sci. Teacher **45:**23–25.

7. **Lowrance, W. W.** 1976. Of acceptable risk. Wm. Kaufmann, Inc., Los Altos, Calif.

8. **National Safety Council.** 1974. Accident prevention manual for industrial operations, 7th ed. National Safety Council, Chicago.

9. **Petersen, D.** 1978. Techniques of safety management, 2nd ed. McGraw-Hill, Inc., New York.

10. **Schmitz, T. M., and R. K. Davies.** 1967. Laboratory accident liability: academic and industrial. J. Chem. Educ. **44:**A654–A659.

11. **Scott, R. B., Jr., and A. S. Hazari.** 1978. Liability in the academic chemistry laboratory. J. Chem. Educ. **55:**A196–A198.

12. **Sweeney, T. L.** 1977. The personal liability of chemical educators. J. Chem. Educ. **54:**134–138.

13. **Wyatt, H. V., and K. A. Wright.** 1974. How safe are microbiology texts? J. Biol. Educ. **8:**216–218.

Chapter 2

Laboratory Safety Training: Techniques and Resources

MONICA H. SCHAEFFER

Training is generally defined as the acquisition of new knowledge, skills, and attitudes that result in improved performance. Table 1 provides a listing of knowledge and skill areas that are generally included in laboratory safety training programs. The categories of areas included in this list have been compiled after review of a variety of existing books, training programs, and guidelines that address laboratory safety.

The knowledge areas serve as a basic framework that can be tailored to a specific aspect of laboratory safety and to a particular audience (e.g., researchers, engineers, support staff). For example, for a program on biosafety for laboratory personnel, the "recognition of hazards" would need to address specifically the risk factors (e.g., mode of transmission, use characteristics, etc.) associated with the agents being used in the laboratory. The control procedures would then include a description of primary barriers such as biological safety cabinets. Similarly, emergency procedures would address procedures for containing a spill on the open bench or in the cabinet.

The list of skill areas can generally be regarded as the application of basic safety principles (those contained within the "knowledge" column) to continually changing research or clinical procedures performed in the laboratory. An overall skill to be acquired by recipients of laboratory safety training is the ability to select, from the various control procedures, those that will minimize harmful exposures.

While the listing of categories of knowledge and skills in Table 1 is not meant to be exhaustive, it defines major content or information areas that can be arranged or combined in various ways to form the technical portion of the training program. Table 1 also contains a list of attitudes—those that often have been expressed by laboratory personnel and, therefore, must be dealt with effectively in laboratory safety training initiatives, and those that need to be conveyed.

The objective of laboratory safety programs is improved safety performance in the laboratory. The challenge of laboratory safety training is to transmit the necessary information in such a way that it will be accepted, retained, and used by the target audience. This chapter suggests ways of meeting this challenge, describes various training methods, including their advantages and disadvantages, provides training resources that currently exist, and finally describes evaluation and follow-up measures that may be employed both to determine the effectiveness of laboratory safety programs and to ensure that continual attention is given to the area of laboratory safety.

TABLE 1. Areas addressed in laboratory safety training

Knowledge

Recognition of hazards and laboratory-acquired infections

Control of hazards
   Personal practices
   Containment equipment
   Personal protective clothing and equipment
   Facility design

Waste disposal practices

Emergency response measures

Safety information resources

Skills

Manipulation of devices (e.g., mechanical pipetting aids, safety blender, etc.)

Proper use of containment equipment (e.g., biological safety cabinets and fume hoods)

Spill clean-up procedures

Sterilization and decontamination tasks

Animal handling procedures

Use of safety information resources

Attitudes

Those encountered
   Old methods produce consistent results (e.g., mouth pipetting)
   What evidence supports the following of so-called safe practices?

Those to be conveyed
   Safety is everyone's responsibility

   Safe practices can improve integrity of research results

   Safe practices will prevent harmful exposures to others indirectly affected by one's actions

   Safe practices will decrease probability of chronic effects from handling toxic compounds

MEETING THE CHALLENGE:
FRONT-END PLANNING

When faced with the task of developing a training program, one needs to view the development stage within an overall framework. The training process is composed of three phases (Fig. 1). Many training programs fail because the designer of the program only looked at one phase, usually the actual delivery or presentation of the material, without considering the characteristics and needs of the persons who are to receive the training. Others have stressed the importance of the participant while in the training environment, but do not relate this back to performance on the job, in this case in the laboratory.

The first phase consists of front-end planning designed to answer the question, "What will the participants be able to do after the completion of the laboratory safety program?" The key to answering this question is to gather information about the participants' training needs and their background and experience level. Specifically, information needs to be collected with regard to their educational background, working experience, and knowledge of the basic areas listed in Table 1, the current practices being employed to control hazards, the attitudes the employees hold toward safety, and incentives or motivators for performing their job. By collecting such information on the target population, one is in a better position to prepare the materials at a level suitably matched to the participants' educational background and experience. Questionnaires, mailed around 6 months before the program is due to start, can elicit much of this useful information.

In cases where this much lead time is not available, one should still attempt to obtain a profile of the participants. A shortened version of such a questionnaire may be used as part of the introduction to a biosafety workshop or program. The answers to the questions on the form may be compiled on a blackboard and can serve two purposes. The participants

who come from various organizations are able to see from the onset of the program that they share common "needs" for information. The speakers are then able to adapt their presentations to address most of the expectations and concerns raised by the group.

Another very effective means of gathering information about the target training population as well as about the specific topics to be covered is to organize a committee to include representatives from the group that is ultimately to benefit from the training experience. The Division of Safety, National Institutes of Health (NIH), was charged with the task of developing a training program to assist in implementing the "NIH Guidelines for the Laboratory Use of Chemical Carcinogens." Members of the Division of Safety held certain views on the length and content of such a program. However, it was recognized that the views of the ultimate users of the information, namely, laboratory directors, principal investigators, and technicians, also needed to be voiced if the program was to be a success. A committee was organized that consisted of representatives from the Division of Safety, as well as representatives of technicians, principal investigators, and lab supervisors. One of the Institute Scientific Directors chaired the committee. As a result of the committee's discussions, the program's scope was expanded to deal with all chemical hazards rather than only carcinogens. The two-way communication between safety professionals and the target research community also exposed differences of opinion on how safety information should be communicated. The safety professionals' approach to the topic was to describe hazards in terms of general categories (e.g., toxicity, reactivity, flammability, etc.). Members of the research community, on the other hand, were concerned foremost about the specific research procedures they were conducting and thus were interested in learning about the specific hazards associated with the particular procedures. The committee developed a list of common research procedures that could be used as examples to convey the basic safety principles and practices. Illustrations of accidents and injuries were also compiled to be used by presenters to enhance the relevance of the subject matter to the intended audience.

The decision reached by the committee was to provide a half-day program that would provide a general framework that could be applied by laboratory personnel in assessing their particular hazards and selecting control measures based on their risk assessment. Poster sessions that were developed by members from the research community illustrated the application of control strategies to common laboratory procedures, including animal experimentation, cell culture procedures, and liquid scintillation counting. Table 2 is the agenda that resulted from the committee members' input. This initial participation by representatives of the biomedical community (target population) allowed the Division of Safety to

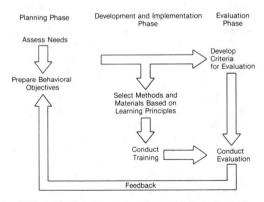

FIG. 1. The training process. Adapted from I. Goldstein, *Training: Program Development and Evaluation* (Brooks/ Cole Publishing Co., Monterrey, Calif., 1974).

TABLE 2. Recognition and control of chemical hazards in the laboratory: agenda

---

Introduction
Factors to consider in assessing risk
Controlling hazards through containment
Controlling hazards through personal practices and protective equipment
(Break)
Procedures and practices to minimize risks to others
Applications of the NIH Chemical Carcinogen Guidelines
Where to go for information and assistance
Poster sessions
    Personal protective equipment
    Dispensing aids
    Safety information resources
    Primary containment
    Chemical storage guidelines
    NIH Chemical Carcinogen Guidelines
    Animal experimentation
    Liquid scintillation counting
    Cell culture procedures

---

present a credible, highly accepted safety training effort.

As mentioned before, the reason for this front-end planning is to be able to describe what the participant should be able to accomplish after completing the training. In other words, one needs to develop a set of objectives, not for what the instructors will accomplish, but rather for what the participants will accomplish. Rather than stating that the participants should "understand" or "appreciate" the information, efforts should be made to specify, or define operationally, how the presented information is to be used back in the laboratory. By taking the time at this point in the training process to formulate specific, practical, and observable or measurable objectives, the tasks of developing the content of the program and developing the evaluation criteria (measures of safety performance) are that much easier.

SELECTING EFFECTIVE
TRAINING METHODS

Once the program's objectives are formulated and agreed to and the major topics to support these objectives are selected, the training process enters the second phase, namely, matching the objectives to the most effective training method. The purpose of this step is to choose those methods that will maximize the participants' retention of the material in the training environment as well as back in the laboratory.

It is important to keep in mind that no one training method will be appropriate for all objectives. In fact, there are many occasions in which a combination of methods is most effective.

Lectures. The lecture is the traditional method of transmitting information to others in a formal class-room setting. It can most effectively be used for transferring information that is factual, descriptive, or explanatory in nature. The primary advantage of the lecture is its economy; namely, the lecture can be used to instruct a relatively large number of people in a rather short period of time. The instructor has primary control over both the amount that is covered and the content. While this method may be thought of as the safest from the instructor's point of view and one that will meet management's desire to minimize the time away from the laboratory, it should be recognized that it is often difficult to stimulate the audience into actively participating in the subjects being presented.

The lecture method might most appropriately be used for new laboratory personnel to lay the foundation for or give an orientation to basic laboratory safety principles and to describe laboratory safety resources that can be called upon to provide future information and assistance. The lectures may be made more effective by: (i) clearly stating the objectives at the outset; (ii) utilizing recognized experts as the presenters; (iii) incorporating examples drawn from the audience's term of reference, such as accidents or injuries that have occurred; (iv) using visual aids to help the audience visualize points being made; and (v) including discussion or review periods to get feedback on whether the objectives have been fullfilled. After one has used the lecture method to set a general framework, other methods described below can be used in conjunction with specific lectures for more detailed disclosure of other kinds of information.

Seminars. To avoid some of the limitations of the lecture and to take advantage of the participants' experience, consideration should be given to developing laboratory safety seminars and workshops. The audience one is trying to attract to laboratory safety programs uses the seminar as a frequent and accepted forum for acquiring and sharing new knowledge in all research areas. By structuring and convening a seminar instead of a "safety lecture" or "safety training" program, one is more likely to appeal to the subject audience. Topics for safety seminars could include a discussion of the research that has been conducted on the effectiveness of various types of glove materials, or the methods that have been developed for safely treating and disposing of laboratory wastes containing carcinogens. Better understanding of the reasons behind safety rules which are otherwise considered unfounded can result in much wider acceptance of good laboratory practices.

Workshops. The workshop, as a training method, offers the very important advantage of allowing participants to become actively involved in the learning process. It has been shown that individuals learn and remember better the things they discover for themselves. Theory and practical applications can be treated concurrently by incorporating case studies after general laboratory safety principles have been discussed. This approach was effectively used in a workshop entitled "Laboratory Biosafety Principles and Practices" that was developed by the Division of Safety, NIH, for the American Society for Microbiology Workshop Series at the 1983 Annual Meeting. Case studies were prepared that were based on actual laboratory-acquired infections that had occurred in various types of research/clinical settings. Workshop participants were divided into groups of five and asked the following questions. (i) What was the probable source of the infectious agent? (ii) What was the probable mechanism of exposure? (iii) What safeguards (i.e., practices, equipment, facilities) might have reduced or prevented the laboratory-acquired infection? (iv) What steps need to be taken to put the identified safeguards into practice?

The case studies required the participants to use their knowledge about infectious agents and to share information with others about relevant experiences that had a bearing on the case. The case studies also served to reinforce the principles and practices that had been discussed earlier in the day. After a general discussion was held on each case study, participants were given a summary sheet that answered the particular questions and a set of references that provided more detailed information. This particular workshop was composed of researchers and safety professionals from a variety of different institutions, but the same format could be conducted effectively by gathering individuals from a variety of laboratories and the safety office within one institution. Case studies offer the advantage of demonstrating to the assembled group that there is often more than one solution to a problem.

Other exercises for group discussion and problem solving might involve asking participants to devise methods for promoting safety as an integral part of conducting work in their laboratories, or asking participants to observe their own laboratory procedures and operations, assess the hazards that exist, and suggest means for controlling or eliminating them. The probability of the participants' implementing these methods after this type of exercise is considerably higher than if they are required to listen to a lecture on either the need for safety or the recognition of hazards in the laboratory.

Laboratory exercises. This discussion of training methods has so far emphasized those that address gaps in the knowledge areas listed in Table 1. If the front-end needs assessment indicates that certain skills are lacking, such as manipulating mechanical pipetting aids or working in a biological safety cabinet, then methods need to be chosen to allow the participant to practice the skills and to receive feedback on his or her efforts. The development of laboratory exercises is one method that offers several advantages. The obvious benefit is that the training environment is either the same as or similar to the job so that the instructor does not need to worry about the transfer of learning from a classroom-type environ-

ment back to the job. In addition, in performing a laboratory exercise, the participant has an opportunity to use the equipment and practice the "correct" response. The instructor can, in turn, provide immediate feedback on the participant's performance. For example, if there is a resistance to using mechanical pipetting devices, one might consider developing an exercise in which participants handle a number of devices for various procedures currently performed in the laboratory. As a result, the participant would learn the required skill and at the same time select a device that suited his or her personal preference. The primary disadvantages to this training method are the limited number of persons that can be accommodated and the amount of time needed.

A method that holds considerable promise is integrating safety into new or existing technique or procedural courses. Traditionally, laboratory safety has been treated either as a separate entity that is usually "tacked on" to the end of a program or as the principal subject of a specific course. Great strides in laboratory safety could be made if the requisite knowledge, skills, and attitudes were taught as part of the procedure or research technique. Such a method would capitalize on the laboratory worker's motivation to conduct quality research.

Ideally, this concept would be incorporated into the teaching of all basic laboratory courses in microbiology. For example, when a student is first instructed to use a pipette for a particular procedure, this student should also be provided with the reasons for, and the techniques of, using mechanical pipetting aids. The individual learning the technique would thus view the use of safety equipment as part of good laboratory technique. The difficult task of *unlearning* a hazardous technique would no longer be an objective of laboratory safety training programs.

Similar efforts toward integrating safety into laboratory techniques should be included in instruction for advanced students and investigators in new experimental methods such as the splicing of genes by recombinant DNA technology. Course developers should take special effort to ensure that all techniques are performed in accordance with established safety guidelines. Thus the "safe way" becomes the learned way, and, even more critically, the safe way becomes the only way that the technique is performed.

Job aids. In certain cases, rather than a formal program, written instructions or job aids might be the method of choice. Job aids are especially beneficial for those procedures where the laboratory worker can refer to a chart or checklist rather than depend on recalling the procedures from memory. The Division of Safety, NIH, in conjunction with the NIH research community, has developed two such job aids. One of these deals with the "Recommendations for the Safe Handling of Parenteral Antineoplastic Drugs." Members of the Division of Safety, in conjunction

TABLE 3. Selection and use of training aids

Type of training aid	Purpose	Advantages	Disadvantages	Comments
Printed material				
Information handout or job aid	Present detailed information	Provides posttraining reference material	Passive medium—cannot evaluate whether the message was received or understood	Material must be well organized and well written
	Summarize key concepts	Provides detail inappropriate for oral or visual presentation	Static—cannot portray a dynamic concept	Relevance of material must be explained
		Easily developed, maintained, and distributed		
		Tool for group exercises		
		Economical		
Flip charts	Record results of group discussions or small group exercises	Requires synthesis of ideas	Limited visibility	A person should be appointed to record information
		Allows comparison of results of several groups	Some groups do not like to use them	
		Flexibility		
Display posters (poster sessions)	Portray a central theme or concept	Can remain in place throughout the session	Motion can only be implied	Should be placed near coffee table or another gathering place for easy access
		Participants can view at their convenience	Limited control over presentation	
		Provides a focal point for discussions	Can be expensive and difficult to transport	
Visual stills				
Slides, viewgraphs (transparencies)	Highlight key points	Clear display of visual concepts	Projector and screen required	Equipment must be checked out beforehand
	Portray visual concepts	Good visibility	Limited data can be shown at one time	Extra bulb for the projector should be handy
	Focus group discussion	Flexibility	Effectiveness depends on presenter's use	Only one main point should be covered per slide
		Focal point for discussion		

		Simple to use Economical	Difficult to see if complicated	Slides, and especially viewgraphs, should be kept out of sight until used and turned off as soon as one is finished Presenter should not stand in front of the display
Slide/tape "canned programs"	Convey audio and visual impressions	Assures consistency in presentation Slides provide increased trainer flexibility to rearrange, delete, etc. Fairly economical Good for small groups	Motion can only be implied Limited data displayed at one time Usually requires more lead time for preparation	Equipment must be checked out beforehand Visibility and adequacy of volume must be checked Relevance should be explained to audience
Visual motion Films, video cassettes (audio/visual)	Portrays situations involving human interaction Conveys attitudes, emotions, values	Consistency in presentation Maintains audience attention Effective with a large group Dynamic Stimulates group discussion	Lack of flexibility to specific audiences' needs Requires 16mm projector or video tape equipment Fairly expensive, and high quality expected	Equipment must be checked out beforehand Material must be relevant and of adequate quality Everyone must be able to see and hear it Introduction and summary should be provided to audience to stimulate discussion
Real objects	Visualize concepts or points too complex to explain with pictures or words alone	Use in demonstrations of specific procedures Participants can view at their convenience	Difficult to transport to remote training site Restricted mainly to small items Expensive	Materials should be displayed with guidelines for proper handling Adequate time should be provided for participants to handle equipment

with the Clinical Center Pharmacy Department and the Cancer Nursing Service, observed the steps that were being performed by pharmacists and nurses in the preparation and administration of antineoplastic drugs. A brochure which opened up to a poster was developed that linked photographs or illustrations depicting proper and safe techniques with a narrative describing each step.

A second job aid that was developed for the NIH community covered waste disposal practices. A calendar-type chart was developed which allows the user to flip to one of four waste disposal routes currently operating at NIH. For each waste route, packaging guidelines and disposal information are presented in a decision table. Rather than wading through paragraphs of narrative, the user can select the type of material that needs to be discarded and follow the proper disposal guidelines which are listed in sequential order.

TRAINING AIDS

A critical component of any training method is the proper selection and use of training aids. Basically, a training aid is anything that helps the participant achieve the objectives and enhances the presenter's effectiveness in communicating the message. Studies have shown that people remember only about 10% of what is heard and forget about 75% over a period of 2 days. On the other hand, about 75% of what is learned is through the sense of sight. Thus, the visual sense needs to be considered when developing a training strategy.

With regard to laboratory safety information, training aids can most effectively be used:

1. To describe an object or piece of equipment. For example, when new pipetting aids are being described, slides or, even better, the actual devices should be shown. Afterward, the participant should be allowed to practice with them.
2. To present an issue that is too complex for the spoken or written word alone. For example, if one is describing the inner workings of a biological safety cabinet or steam autoclave, a photograph coupled with a graphic or schematic diagram can cut down drastically on the time that it would take to present a verbal description.
3. To describe the steps to follow in performing a specific technique. The participant's attention can be fixed on these steps with the use of a visual aid in which the steps are numbered and sequentially highlighted. These steps can then be reinforced by having the participants perform the technique in the same order.
4. To summarize and reinforce the main points discussed.

There are a wide range of training aids available.

Table 3 categorizes these into four major classes: (i) printed material, (ii) visual stills, (iii) visual motion, and (iv) real objects. The table also provides a brief description of the main purpose, advantages, and disadvantages of each type of aid and techniques for effectively using these training aids. A more detailed and comprehensive explanation on selecting and developing different types of training aids can be found in an excellent resource by R. H. Anderson (1).

Before developing original material, it is always prudent to review existing training materials. Table 4 provides a listing of those organizations which maintain a collection of films, slide/cassette programs, and video tapes.

Recently, a new set of training resource materials on microbiological safety was developed under contract by the University of Minnesota's School of Public Health for the National Institute of Allergy and Infectious Diseases and the Division of Safety, NIH. The five-unit resource, entitled "Fundamentals for Safe Microbiological Research," consists of 32 modules, each including an annotated outline accompanied by a set of 35mm slides keyed to the outline. Each outline is preceded by a cover sheet noting both the qualifications that are needed by the individual who will be presenting the lecture or leading the discussion and the objectives to be fulfilled by the participants receiving the information. A set of references and a set of multiple-choice questions, designed to evaluate whether or not the objectives were achieved by the participants, are also supplied with each module. These training materials are intended to provide a minimum base of knowledge and skills that should be demonstrated by individuals working with potentially biohazardous agents.

The University of Minnesota materials were designed to give biomedical institutions maximum flexibility in presenting the information. For example, a university may use the material to structure a new semester course or may incorporate certain modules into an existing curriculum on microbiology. The materials might also be used by government, academic, and private institutions to develop special seminars on selected topics. The laboratory exercises contained in the fifth unit can be used to reinforce and apply the principles and practices included in the other four units.

EVALUATING TRAINING EFFORTS

Although most training programs conclude after the participants leave, the organizers at this point cannot answer the critical question: did the training program accomplish what it was supposed to? The third phase of the training process, that of evaluation, can provide answers to this question. Specifically, the evaluation phase involves determining the level of acceptance of the material presented in the training program, the amount of learning that has taken place,

TABLE 4. Sources for training aids in microbiological safety

National Institutes of Health, Division of Safety

Division of Safety
Bldg. 31, Room 1C02
Bethesda, MD 20892
(301) 496-2801

Contact the Division to borrow slide/cassette packages and learn of other audiovisual and written aids available.

Centers for Disease Control

Laboratory Program Office
Centers for Disease Control
Atlanta, GA 30333
(404) 329-3232

Request a copy of their catalog of laboratory safety training aids, "Laboratory Training and Consultation Division Training Aids."

American Society for Microbiology

Audiovisual Programs
1913 I Street, N.W.
Washington, DC 20006
(202) 833-9680

Request a copy of the catalog, "Multimedia Programs in Microbiology." The catalog lists a number of audiovisual aids available from ASM and other sources.

National Audiovisual Center

General Services Administration
Order Section CF
Washington, DC 20409
(301) 763-1896

Request copies of their catalog "Fundamentals for Safe Microbiological Research" as well as individual "Safety" and "Health" catalogs. The Center is the official repository for all U.S. Government-issued audiovisual aids.

Fisher Scientific

Contact your local sales representative for information on two films, "28 Ounces of Prevention" and "Safety—Isn't It Worth It?," or contact:

Fisher Scientific Co.
Educational Materials Division
4901 West LeMoyne Avenue
Chicago, IL 60651

and finally, whether the information learned has been effectively translated into safe actions or practices at the worksite.

To measure whether the material was initially accepted, one needs to develop a technique to measure the reactions of the participants to the training program. The distribution of an evaluation or feedback questionnaire on the last day of a program is the most common technique in obtaining paticipants' reactions not only to the content of the program, but also to the various elements of the learning environment (e.g., speakers, facilities, visual aids, exercises, etc.). The key features of such a questionnaire are as follows. (i) The questions should require minimal time to answer, and it should be easy to tabulate and quantify the results. (ii) The form should not require the participant to provide his or her name, to encourage honest and frank opinions, but should include the collection of certain demographic data to allow group responses to be analyzed against these characteristics. In some cases the analysis may indicate that the program was more suitable for a specific segment of the training population (e.g., technicians versus principal investigators). (iii) Space should be provided for additional comments not covered by questions that were designed to be tabulated and quantified. Specifically, a question asking participants to indicate what they will do differently as a result of the program forces them to review the presented information critically and quickly so as to make at least some initial practical decisions on their actions back in the laboratory. With such a question, the program organizers have an opportunity to initially assess whether the practices that have been emphasized throughout the program have in fact been "heard" by the audience.

The reaction of participants is often a critical factor in the decision to continue the program. By questioning the entire group, one will guard against making decisions based on a few very satisfied or disgruntled participants. The questionnaire should be used immediately after the training to assess initial reactions, and a slightly modified questionnaire should be distributed one or two months later to note any changes, such as whether the training was applicable to the person's specific job.

While it is important to receive favorable reactions to the program, it is equally important to recognize that such favorable reactions do not assure that the principles, facts, and techniques have been retained. Therefore, program organizers need also to determine objectively the amount of learning that has taken place. To do this, one needs to establish criteria or measures of success that are based on the objectives originally developed as a product of the front-end planning or assessment phase. When the objectives involve the acquisition of principles and facts, the most common evaluation technique to use is a set of multiple-choice questions. In constructing such questions it is a good idea to develop two questions that test, equivalently, the acquisition and retention of each objective. One of these questions may be used in a pretest that can be administered before the training to determine the participants' base-line knowledge, and the second question may be included in a set of questions that form a posttest to be administered after the training to assess whether any change has taken place.

If possible, the pretest and posttest should be given to a group of individuals who are comparable to the group who will be receiving the training in terms of their educational background and level of experience. This second group could be drawn from individuals in an organization who will receive the training later on. This second group will provide the organizer with a "control" group and enable him or her to determine whether the training program was in fact the reason for seeing a positive change.

A word of caution needs to be mentioned. Unless the set of questions accurately covers the material presented, it will not be a valid measure of the effectiveness of the learning. Frequently, when many presenters are involved and one person has developed the questions, only part of the material in the test will be covered in the program, largely due to individual presenters running out of time. To avoid this problem, the organizer should ask each presenter to develop his or her own questions based on the objectives to be fulfilled or, at a minimum, to review the questions before presenting the material. In addition, in designing pre- and posttests, it is often advisable to obtain the assistance of professional test constructors or evaluators of training programs.

In certain situations, other techniques besides a multiple-choice test can be used to assess objectively the amount of learning that has taken place. Case studies can be used not only as part of the training method but also as an evaluation technique to determine the group's ability to apply the principles of hazard recognition and control. For example, in lieu of a pretest, a case study involving a laboratory-associated infection could be presented to the group as an introduction to the program. Each person's responses to the series of questions mentioned earlier could be collected without discussion. The organizer could assess the strengths and weaknesses of the responses and thus be in a better position to emphasize critical areas that were not considered. The same or an equivalent case study could be presented at the end of the program, this time with group discussion on each of the questions. The use of case studies offers the advantage of presenting a relevant problem or issue that the group has encountered or may well encounter after attending the program.

If the initial assessment of the target population's educational background indicates that a formal written test or exam may be viewed as too threatening, or may be inappropriate for some other reason for evaluating the retention of material, the program organizer may opt to use a set of slides and refer to them as a "preview" and "review" of the program. For example, in a program for the housekeeping staff of a biomedical research institution where it is critical for such persons to understand the meaning behind certain warning signs and labels, slides of these could be projected and the audience would be required to identify the correct message to be conveyed, at both the beginning and the end of the program. A series of slides could also be used to illustrate hazards in the working environment. The audience would then be asked to identify the hazard and the recommended method of preventing, eliminating, or controlling it.

The evaluation process is incomplete without the measurement of safe performance. Just as a favorable reaction does not necessarily indicate that learning has occurred in the training program, knowledge of the principles and techniques, as demonstrated by superior performance on a posttest, does not always result in these techniques' being used in the work environment.

The evaluation of performance is an area in which behavioral psychologists have contributed a great deal in the last 10 years. One means of assessing a positive transfer of such safe performance back to the laboratory is through observational techniques. A checklist can be developed itemizing "behaviors" or specific practices that operationally define safe job performance. An index of safe performance can be calculated by dividing the number of safe behaviors or practices observed by the total number of observations and multiplying by 100. Observable behaviors may be grouped into categories such as the use of safety (e.g., personal protective) equipment, housekeeping, and general safety procedures. Examples of observable behaviors include using a mechanical pipetting aid instead of mouth pipetting, washing one's hands before and after a procedure, and disinfecting the work bench before beginning work. This list of observable behaviors can easily be assembled from the data collected during the initial planning or assessment phase.

Observable data should be collected before the training to develop a base-line reading of how frequently procedures are being performed safely. The same observations should be made during and after the training to assess whether or not the training program is responsible for any change.

The recording of observable behaviors can also reinforce safe performance in the laboratory long after the training has been completed. Intermittent collection of these data by laboratory supervisors, followed up by discussion of the results, can serve as an effective management tool as well as a reminder of good laboratory practices to all laboratory personnel.

Another strategy that has been used by designers of human relations programs to evaluate changes in behavior on the job is to design a questionnaire to be completed by the supervisors and subordinates of those participating in the training. The questionnaire is administered before the training and again approximately three months after the training to give the participants an opportunity to put into practice what they have learned. Providing this feedback to the participant may serve as a powerful eye opener as to the effect his or her behavior has on others and may aid in further changing the participants' actions in the laboratory.

FOLLOW-UP MEASURES

Training in and of itself is not likely to sustain performance of safe practices in the laboratory. The increased level of safety awareness that often is evident immediately after a training program must be consistently reinforced. Komaki et al. (2) looked at the effect of training and feedback in improving the safe performance of a city's vehicle maintenance division. Whereas employees showed only slight improvements during the training-only phase, their performance increased substantially during the combined training and feedback phase. Feedback consisted of section supervisors making randomly timed daily safety observations, setting a goal for the section, discussing the results of the observations with individuals receiving the training, and publicly posting a graph of the results.

While more research is needed regarding the content, frequency, and mode of the feedback provided as well as its application to a laboratory environment, the study of Komaki et al. indicates the need for designing follow-up measures to reinforce the principles and practices attained as a result of training.

The maintenance of a high level of safety awareness and the following of safe practices in the laboratory must not be neglected and can best be achieved through continual communication between safety professionals and laboratory personnel. The Division of Safety, NIH, issues one- to two-page bulletins called "Spot Hazards" on a quick-turnaround basis to alert the research community to accidents or incidents that have occurred. These bulletins describe succinctly the circumstances surrounding each incident and specific procedures to prevent future occurrences. The Division of Safety also issues periodic "Safety Data Announcements" in response to questions that may affect a wide segment of the NIH population. The intermittent time schedule on which these messages are distributed serves to maintain a high level of safety awareness.

Another mechanism for maintaining a certain degree of safety visibility is the distribution of a periodic newsletter or bulletin containing input from both the safety professionals and members of the biomedical community. Before such a project is undertaken, the potential readers should be surveyed to pinpoint the areas of greatest interest, to determine the format, length, and frequency of the publication, and to assess initial interest in contributing articles to the bulletin. The Division of Safety conducted such a survey before distributing its bulletin entitled "Lab Safety Views." The purpose of the bulletin is to provide an exchange of safety and health information for the biomedical community. Included in its distribution are NIH intramural researchers and safety and health managers or directors of institutions receiving NIH grants or contracts. The survey indicated that the top three areas of interest to both the safety professionals and biomedical researchers were: (i) descriptions of hazardous incidents, (ii) descriptions of new technological developments in the safety field that would have an impact on biomedical research, and (iii) guest articles contributed by experts in the field. Depending upon the extensiveness of the bulletin, article review by an editorial or advisory board consisting of members from both the safety and research community may enhance the publication's usefulness and acceptance.

The key point in selecting a follow-up approach to reinforce safe practices is to obtain input from the community that will benefit most from the information. It is critical that such information be conveyed in a manner that considers the perspective of this audience. By involving representatives from the community in the decision process and development of follow-up measures, one is more likely to succeed in improving safe performance in the laboratory.

SUMMARY

In meeting the challenge of improving the safe performance of laboratory personnel through safety training programs, the person(s) responsible for or organizing and developing the material needs to be able to answer all of the following questions. (i) What will the participants be able to do after completing the laboratory safety program? (ii) What instructional methods and training aids will maximize the participants' retention of the material not only during the training but back on the job as well? (iii) Did the training program accomplish what it was supposed to? (iv) What additional strategies can be employed to maintain the gains that were made in safety awareness as a result of the training efforts?

It is critical always to keep in mind the perspective of the training participants and to recognize that they have many demands on their time. They are primarily interested in acquiring new information and skills that can be immediately applied to their situation in the laboratory. The best way of maintaining this perspective is to involve representatives of the trainees in the planning, design, and implementation phases of the program.

LITERATURE CITED

1. **Anderson, R. H.** 1976. Selecting and developing media for instruction. Van Nostrand Reinhold Co. and American Society for Training and Development, New York.
2. **Komaki, J., A. Heinzmann, and L. Lawson.** 1980. Effect of training and feedback: component analysis of a behavioral safety program. J. Appl. Psychol. **65**:261–270.

tests selected. This variation reflects differences in job hazards at different institutes, but probably depends just as much on differences in the philosophy of the employee health physician and the capability of the clinical laboratory involved.

In general, analytical blood chemistry tests are designed to establish medical base-line data for individual program participants and to reassess such data periodicially and identify changes relating to possible exposure to hazardous materials within the work environment. Although changes may indicate systemic malfunction or disease, they generally do not specifically implicate the causative factor involved and thus represent only the first link in a chain of medical investigations needed to identify and correct the situation.

Information concerning analytical blood chemistry procedures required for laboratory medical examinations can be obtained from the National Institute of Occupational Safety and Health (3). Specific information on the methods and interpretation of various tests available can be found in texts on clinical laboratory methods (2, 5, 6).

Suggestions concerning the tests needed may be obtained from the institutional physician or the medical consultant employed by the agency. Pathologists or clinical laboratory scientists are excellent sources of information about test performance and interpretation.

A summary of the types of test batteries which may be performed as part of laboratory medical programs is presented in Table 1.

TABLE 1. Organ panels (Laboratory Diagnosis Related Groups)[a]

Anemia panels	**Hepatic panel**
Complete blood count with indices	Glutamic oxaloacetic transaminase (AST)
Reticolocyte count	Glutamic pyruvic transaminase (ALT)
Hypochromic microcytic	Alkaline phosphatase
Iron/iron-binding capacity/% saturation	γ-Glutamyl transpeptidase
Ferritin	Total bilirubin
Macrocytic	Conjugated bilirubin
B_{12}/folate	Total protein
	Albumin
Bone/joint panel	Prothrombin time
Uric acid	
Calcium	**Acute hepatitis, immunopathology panel**
Phosphorus	Hepatitis B surface antigen
Alkaline phosphatase	Antibody to hepatitis B surface antigen
Total protein	Antibody to hepatitis B core antigen
Albumin	Antibody to hepatitis B e antigen
	Anti-hepatitis A immunoglobulin M
Cardiac injury panel	
Lactate dehydrogenase	**Chronic hepatitis carrier, immunopathology panel**
Creatine kinase	Hepatitis B surface antigen
Creatine kinase isoenzymes	Anti-hepatitis B surface antigen
	Hepatitis B e antigen
Cardiac risk evaluation panel ("Executive Panel";	Antibody to hepatitis B e antigen
optional)	
Cholesterol	**Neoplasm (malignancy) panel**
Triglycerides	Alpha fetoprotein
High-density lipoprotein cholesterol	Carcinoembryonic antigen
Glucose	Prostatic acid phosphatase
	B-human chorionic gonadotropin
General health panel	Lactate dehydrogenase
Glucose	Alkaline phosphatase
Blood urea nitrogen/creatinine	
Cholesterol	**Renal panel**
Triglycerides	Blood urea nitrogen/creatinine
Glutamic oxaloacetic transaminase (AST)	Urinary creatinine, 24 h
Alkaline phosphatase	Urinary protein, 24 h
Uric acid	Creatinine clearance
Total protein	Total protein
Albumin	Albumin
Total bilirubin	Sodium
Lactate dehydrogenase	Potassium
Calcium	Chloride
Sodium	CO_2 content
Potassium	Glucose

[a]Reproduced in part from reference 2, with permission of the author.

Physical examinations

Physical examinations by a physician are a third major component of laboratory medical programs. Where possible, it is recommended that the physician be board certified in occupational medicine or be a member of the American Occupational Health Association (4). Experience also suggests that the participating physician be retained on a continuing basis to facilitate his familiarity with the agency and with the employees involved, thus fostering an optimal physician-patient relationship.

Like many other components of the medical surveillance programs for laboratory employees, the coverage provided in the physical examination can vary widely. Here again, the institutional or consulting physician plays a major role in providing advice on the degree of coverage required for the potential hazards inherent in work under way.

An example of the type and level of coverage which could be considered is provided in the U.S. Government G.S.A. Standard Form 88 Report of Medical Examination. The Environmental Protection Agency has recommended that this form, presently utilized by many federal agencies, be adopted to provide agency-wide comparability. A sample of the form may be obtained from the Agency.

Supplementary medical procedures

Medical procedures which could be considered for program inclusion on an "as-needed" basis or because of their general desirability include the following.

Chest X ray. The respiratory system is a major portal of entry to the body for potentially harmful materials. It is uniquely susceptible to inert materials, chemical irritants, and biological agents commonly aerosolized within the laboratory. Thus chest X rays are often included in laboratory medical programs.

Chest X ray may be recommended for employees working with pathogenic microorganisms, such as tuberculin-positive employees working with mycobacteria, and for employees exposed to fungi. Support personnel whose jobs involve work in dusty or smoky environments such as bulk feed handling areas or incinerator charging areas should also be given X rays on an annual basis.

X rays may be deferred for personnel working with materials which do not constitute a known hazard to the respiratory system, or they may be taken at intervals of 3 to 5 years or as recommended by the examining physician on the basis of physical findings or medical history.

Electrocardiography (EKG). The influence of stress in the workplace on the incidence of coronary disease suggests that an electrocardiogram is desirable in a medical program. Individuals 40 years or older should be evaluated with respect to the need for such testing. All individuals with a family history or physical findings suggestive of heart disease should be tested as a part of the laboratory program or advised to discuss the need for such testing with their private physician.

Tuberculin and hepatitis B surface antigen tests. Tine test for tuberculosis is frequently utilized for large-scale screening programs. The intradermal Mantoux (PPD) test is recommended for preemployment and annual testing of primate handlers and other at-risk personnel or for use after known or suspected accidents which entail probable exposure to tubercle bacilli.

Also to be considered is the desirability of an annual hepatitis B surface antigen test for personnel assigned to hemodialysis units, blood banks, clinical laboratories, or other biomedical or microbiological facilities where there is a potential for exposure to hepatitis B virus. Immunization with hepatitis B vaccine is preferred to routine or postexposure testing and reduces the need for administration of hepatitis B immunoglobulin.

Pap smears, pelvic examination. Although there are many who contend that laboratory hazards are essentially an "above-the-waist" problem, there are certain situations, including continuous exposure to sex hormones in research or pharmaceutical laboratories, which warrant annual Pap smears or pelvic examination. Where warranted, management should arrange for the services of a gynecologist.

In instances where Pap tests and pelvic examinations are not a component part of the medical program, the physician should recommend that the employee make private arrangements for such tests when indicated.

Cancer of the colon. Although screening tests to identify cancer of the colon are not necessarily justified by work exposure, the significance of this disease as a cause of death suggests that procedures designed for its detection could be considered as a component of laboratory medical programs. Simplified screening procedures, such as the Hemoccult R slide test (SmithKline) for occult blood, are available inexpensively or are sponsored and financed by volunteer organizations such as the American Cancer Society.

Pulmonary function tests. Pulmonary function tests should be considered for all individuals whose work entails respiratory exposure to dust, fumes, irritant gases, or chemical vapors. Such tests are helpful in the initial identification and assessment of respiratory impairment and for gauging the progression or regression of such diseases once they have been identified and corrective measures have been begun.

Such tests have specifically been recommended for workers exposed to acrylamide, acrylonitrile, alkanes, allyl chloride, ammonia, asbestos, beryllium, boron trifluoride, cadmium, carbaryl, carbon dioxide, carbon monoxide, chlorine, chlor-

TABLE 2. Recommendations for immunoprophylaxis in at-risk laboratory personnel[a]

Disease	Description of product	Recommended for use in:	Source of product
Anthrax	Inactivated vaccine	Personnel working regularly with cultures, diagnostic materials, or infected animals	Michigan Department of Public Health[b]
Botulism	Pentavalent toxoid (ABCDE) IND[c]	Personnel working regularly with cultures or toxin	Centers for Disease Control[d]
Cholera	Inactivated vaccine	Personnel working regularly with large volumes or high concentrations of infectious materials	Commercially available
Diphtheria-tetanus	Combined toxoid (adult)	All laboratory and animal care personnel irrespective of agents handled	Commercially available
Eastern equine encephalomyelitis	Inactivated vaccine (IND)	Personnel who work directly and regularly with this virus in the laboratory	Centers for Disease Control[d]
Hepatitis A	Immune serum globulin (human)	Animal care personnel working directly with chimpanzees naturally or experimentally infected with hepatitis A virus	Commercially available
Hepatitis B	Inactivated vaccine	Personnel working regularly with blood and blood components known or potentially infected with hepatitis B virus	Commercially available
Influenza	Inactivated vaccine	Vaccines prepared from contemporary strains may be of little value in personnel working with recent isolates from humans or animals	Commercially available
Japanese encephalitis	Inactivated vaccine (IND)	Personnel who work directly and regularly with this virus in the laboratory	Centers for Disease Control[d]
Measles	Live attenuated virus vaccine	Measles-susceptible personnel working with the agent or potentially infectious clinical materials	Commercially available
Meningococcal meningitis	Inactivated vaccine	Personnel working regularly with large volumes or high concentration of infectious materials	Commercially available
Plague	Inactivated vaccine	Laboratory and field personnel working regularly with cultures of Yersinia pestis, infected rodents, or fleas	Commercially available
Polio	Inactivated (IPV) and	Polio-susceptible personnel working with the virus	Commercially available

	live attenuated (OPV) vaccines	or entering laboratories or animal rooms where the virus is in use	
Poxviruses (vaccinia, cowpox, or monkey pox viruses)	Live (lyophilized) vaccinia virus	Personnel working with orthopox viruses transmissible to humans or animals infected with these agents, and persons entering areas where these viruses are in use	Centers for Disease Control[d]
Q fever	Live attenuated (phase II) vaccine	Personnel who have no demonstrable sensitivity to Q fever antigen and who are at high risk of exposure to infectious materials or infected animals	USAMRIID[e]
Rabies	Human diploid cell line inactivated vaccine	Personnel working with all strains of rabies virus or with infected animals, or persons entering areas where these activities are conducted	Commercially available
Rubella	Live attenuated virus vaccine	Rubella-susceptible personnel, especially women, working with "wild" strains or in areas where this virus is in use	Commercially available
Tuberculosis	Live, attenuated (BCG) bacterial vaccine	BCG vaccine is not ordinarily recommended or used in laboratory personnel in the United States	Commercially available
Tularemia	Live, attenuated bacterial vaccine (IND)	Personnel working regularly with cultures or infected animals, or persons entering areas where the agent or infected animals are in use	Centers for Disease Control[d]
Typhoid	Inactivated vaccine	Personnel who have no demonstrated sensitivity to the vaccine and who work regularly with cultures	Commercially available
Venezuelan equine encephalitis	Live, attenuated (TC83) virus vaccine (IND)	Personnel working with this virus and the related Cabassou, Everglades, Mucambo, and Tonate viruses or who enter areas where these viruses are in use	Centers for Disease Control[d]
Western equine encephalomeyelitis	Inactivated vaccine (IND)	Personnel who work directly and regularly with this virus in the laboratory	Centers for Disease Control[d]
Yellow fever	Live, attenuated (17D) virus vaccine	Personnel working with virulent and attenuated strains of yellow fever virus	Commercially available

[a]Adapted from recommendations of the Public Health Service Advisory Committee on Immunization Practices and reference 7 (March 1984).
[b]Michigan Department of Public Health, P.O. Box 30035, Lansing, MI 48909 [(404) 373-1290].
[c]IND, Investigational new drug.
[d]Clinical Medicine Branch, Division of Host Factors, Center for Infectious Disease, Centers for Disease Control, Atlanta, GA 30333 [(404) 329-3356].
[e]USAMRIID, U.S. Army Medical Research Institute for Infectious Diseases, Ft. Detrick, MD [(301) 935-7101].

oform, chloroprene, chromic acid, chromium (VI), cotton dust, crystalline silica, cyanide, dibromochloropropane, dinitro-*ortho*-creole, dioxane, epichlorohydrin, ethylene dibromide, ethylene dichloride, hydrogen fluoride, hydrogen sulfide, inorganic arsenic, inorganic fluoride, isopropyl alcohol, malathion, methyl parathion, methylene chloride, nitric acid, phenol, phosgene, sodium hydroxide, sulfur dioxide, sulfuric acid, tetrachloroethylene, toluene diisocyanate, 1,1,1-trichloroethane, trichloroethylene, xylene, and zinc oxide.

Blood cholinesterase determinations. It is recommended that employees working with organophosphate or carbonate insecticides should have a routine blood cholinesterase determination before handling these insecticides, to establish a base line. Similar tests should then be made during or after the spraying season to detect changes.

Special requirements for research laboratories. Organizations working with microbial agents may require special preventive medical programs. Such programs are designed to increase the resistance of individuals to the hazardous agents or toxins being handled and to improve the physician's ability to confirm presumptive diagnosis or to identify subclinical infections.

Immunization programs. Recommendations for prophylactic immunization of laboratory personnel working with infectious agents are summarized in Table 2.

Serum bank—serological screening programs. Serum samples for a serum bank program should be obtained from all new employees as early as possible, certainly before their assignment to risk areas. The sera should be stored at −70°C.

Employees working with microbiological agents for which reliable serological tests are available should be screened at periodic intervals not to exceed 1 year. The optimal frequency of serological screening should be jointly determined by the employee health physician and the senior scientist in charge of the work under way, who is best informed on the pathogenicity of the microbial strains being investigated, the concentration worked with, the aerosol potential of the procedures utilized, and the availability of containment systems or equipment.

Screening frequencies of 3 months (brucellosis, chlamydiosis) or 6 months (leptospirosis, bovine leukemia) have been selected as optimal for at-risk employees at the National Animal Disease Center. Employees may be screened at any time when accidental exposure has occurred or is suspected. Even if a reliable serological test is not available, stored sera may be valuable later on, e.g., for testing with newly discovered microorganisms or toxins.

LITERATURE CITED

1. **Center for Disease Control.** 1974. Lab safety at the Center for Disease Control. DHEW Publication no. CDC 758118. Center for Disease Control, Atlanta, Ga.
2. **Henry, J. B.** 1984. Todd-Sanford-Davidsohn: clinical diagnosis and management by laboratory methods, 17th ed. W. B. Saunders Co. Philadelphia.
3. **National Institute for Occupational Safety and Health.** 1978. Summary of NIOSH recommendations for occupational health standards, October 1978. National Institute for Occupational Safety and Health, Robert A. Taft Laboratories, Cincinnati.
4. **National Safety Council.** 1978. Health monitoring for laboratory employees. Research and development fact sheet. National Safety Council, Chicago.
5. **Sonnenwirth, A. C., and L. Jarett.** 1980. Gradwohl's clinical laboratory methods and diagnosis, 8th ed. C. V. Mosby Co. St. Louis.
6. **Tietz, N.** 1984. Fundamentals of clinical chemistry, 3rd ed. W. B. Saunders, Philadelphia.
7. **U.S. Department of Health and Human Services.** 1984. Biosafety in microbiological and biomedical laboratories. HHS Publication no. (CDC) 84-8395. U.S. Government Printing Office, Washington, D.C.
8. **U.S. Department of Health, Education and Welfare.** 1974. National Institutes of Health biohazards safety guide. U.S. Government Printing Office, Washington, D.C.
9. **U.S. Environmental Protection Agency.** 1977. Occupational medical questionnaire. Agency memorandum 13 January 1977. Guidelines for establishing laboratory employee health monitoring programs. Assistant Administrator for Planning and Management, U.S. Environmental Protection Agency, Washington, D.C.

Chapter 4

Reporting Laboratory Accidents, Injuries, and Illnesses

JAMES F. SULLIVAN

The accumulation and analysis of pertinent data in a specific area of interest have long been utilized by laboratory scientists in defining and resolving problems in major areas of concern. In most instances, the overall success of such an approach depends on both the quality and quantity of the source data available, as well as the thoroughness and efficiency of the review and analytical techniques employed.

Before the Occupational Safety and Health Administration (OSHA) was established in 1970, there were few federal regulations that required the recording and reporting of laboratory accidents, including those which resulted in unplanned infection of laboratory personnel. Many scientists or scientific institutions were hesitant to publish or release information concerning any subject which reflected upon the ability of the personnel involved.

Pike has indicated that only about 60% of the recorded cases of laboratory-acquired infections have been reported in the literature; the remaining cases were identified by means of personal communications or in response to survey questionnaires (4). The limited nature of such surveys, as well as the exclusion of subclinical cases, which may well outnumber the number of clinical cases observed, strongly suggests that only a small fraction of the actual number of cases of laboratory-related illness are available for compilation and analysis.

Obvious reporting deficiencies, as well as the lack of analytical tools needed to identify the specific infecting event responsible for a laboratory-associated infection (close to 80% of all such infections have been untraceable), present major obstacles to a complete understanding of the mode of unplanned transmission of infectious diseases in the biomedical laboratory. Such deficiencies have not, however, prevented the development of a large number of subjective working hypotheses on the subject of uncontrolled transmission of disease within the laboratory and animal isolation facility, many of which have subsequently been confirmed by controlled studies within the laboratory.

The Occupational Safety and Health Act

Although many aspects of the Occupational Safety and Health Act have been subject to criticism because of the unfavorable cost/benefit ratio involved in the implementation of certain sections of the act, or overzealous enforcement of other segments which appear to have minor impact on the health or well-being of the worker, there are few who dispute the need for improved recording and reporting of job-related injuries and illnesses. Mandatory reporting of a defined type of job-associated injuries or illnesses provides a much broader data base, which is needed to identify, assess, and control or eliminate hazards found in the workplace.

Although the act is designed to meet the general needs of all segments of the U.S. work force, a good deal of it has unusual relevance to individuals employed in a laboratory environment. The following sections of this chapter are intended to provide a capsular summary of the OSHA reporting requirements for injuries or illnesses occurring within the biomedical or biological laboratory.

Who

OSHA requires that all employers "engaged in a business affecting commerce" must comply with the record-keeping and annual reporting requirements detailed in the federal act or found in approved state OSHA programs. Federal agencies, while not covered by the act, are subject to similar record-keeping and reporting requirements. States and their political subdivisions are generally subject to record-keeping and reporting requirements under approved state plans. Employers with 10 employees or fewer do not have to keep records. As a general rule, biological laboratories are required to comply with the accident reporting requirements of either federal or state OSHA programs.

What

Injuries and illnesses must be recorded if they are job related and meet a defined level of severity. Injuries or illnesses which cause an employee fatality or result in the hospitalization of five or more employees must be reported to OSHA or the governing state agency within 48 h of the accident. Definitions of recordable injuries or illnesses are as follows.

An occupational injury involves a single traumatic event which produces a degree of physical impairment, such as cuts, lacerations, amputations, sprains, etc.

An occupational illness is a disorder or abnormal condition which is not the result of an occupational injury. Occupational illnesses are caused by harmful factors which exist within the working environment. Such illnesses have been categorized by OSHA in the following manner:

1. Occupational skin diseases
2. Lung diseases caused by dusts
3. Respiratory conditions due to toxic agents
4. Poisoning
5. Disorders due to physical agents
6. Disorders associated with repeated trauma
7. Other occupational illnesses, including infectious diseases

All work-related fatalities and occupational illnesses must be recorded. Occupational injuries must be reported if they result in:

1. Medical treatment other than first aid
2. Lost workdays
3. Loss of consciousness
4. Restriction of work or motion
5. Job reassignment
6. Termination of employment

How

Mandatory reports. OSHA requires that all fatalities resulting from an occupational injury or illness be reported to the controlling federal or state OSHA agency involved within 48 h of the accident. Such reports may be made by telephone or telegram.

Information regarding fatal injuries or illnesses must also be documented on OSHA Form 200, Log and Summary of Occupational Injuries and Illnesses, and OSHA Form 101, Supplementary Record of Occupational Injuries and Illnesses, or equivalent state OSHA forms in the same manner as all other recordable accidents.

Log and summary of occupational injuries and illnesses. A log and summary record of all occupational injuries and illnesses must be maintained by all employers or employing agencies. In the case of large companies with multiple work locations, such forms should be maintained in each work establishment.

In addition to pertinent information concerning the employing agency, log and summary statements must provide information about the affected employee (name, occupation, department) and the injury or illness reported. Such information usually includes the date it occurred, whether an injury or illness, the illness category, a brief description of the disability, the part(s) of the body affected, the resultant inability to perform assigned duties, and the need for reassignment or workdays lost.

Supplementary record of occupational injuries and illnesses. Supplementary records of recordable injuries or illnesses must be maintained at each establishment along with the required log and summary records. OSHA Form 101 or equivalent forms which provide the same basic information may be utilized.

All supplementary records of recordable occupational injuries and illnesses must provide the following information: (i) case number (CN); (ii) name; (iii) employer's mailing address; (iv) location where injury occurred, if different from mailing address; (v) social security number; (vi) home address; (vii) age; (viii) sex; (ix) occupation; (x) department; (xi) place of accident or exposure; (xii) was place of accident or exposure on employer's premises?; (xiii) what was employee doing when injured?; (xiv) how did the accident occur?; (xv) description of injury or illness, part of body affected; (xvi) object or substance which directly injured employee; (xvii) date of injury or initial diagnosis of illness; (xviii) did employee die?; (xix) name and address of physician; (xx) if hospitalized, name and address of hospital.

Optional reports. The reporting of laboratory accidents which cause death, or otherwise meet the severity requirements for recordable accidents as detailed by OSHA, provides a body of data which is helpful in identifying and correcting job-associated hazards in a laboratory setting. Unfortunately, recordable accidents represent only the tip of the iceberg. If only those accidents which OSHA defines as recordable are documented within a laboratory establishment, this would represent only a small percentage of the total accidents or laboratory mishaps experienced.

Examples of the disadvantages of documenting only recordable illnesses rather than all biological accidents that have the potential for initiating human infection are to be found in the accident files of many laboratories. For example, at the National Animal Disease Center and the National Veterinary Services Laboratories in Ames, Iowa, opened in 1960 and 1977, respectively, personnel have experienced a total of 28 job-related illnesses through 1983, including 1 subclinical infection and 8 infections that were of comparatively minor consequence and did not involve either time loss or inability to perform assigned tasks. During the same period of time, personnel assigned to these units reported 66 incidents which involved exposure to hazardous biological agents. Early reporting of such incidents afforded adequate time needed for review, consultation, and, when deemed advisable, the implementation of appropriate treatment. As a result, these exposures did not result in either clinical or subclinical infection, as gauged by relevant serological or cultural procedures.

Optional in-house accident or incident reporting programs. The ultimate consequences of minor laboratory accidents or suspected exposure to harmful biological or chemical agents are not always immediately apparent. This fact, coupled with the reluctance of many laboratory supervisors to provide relevant data in borderline situations which may or may not eventually be classified as a recordable injury or illness, strongly suggests the need for an intermediate reporting plan to cover such situations.

One such plan for an abbreviated in-house reporting system involves the use of a small (3 by 5 in.) file card listing the employee involved and documenting the date, time, and location of the accident or incident. A brief description of the accident or incident and a statement identifying contributing factors and suggesting means for their correction complete the information usually required. Such forms are frequently preaddressed to the health or safety unit of the agency involved. They are generally well accepted by employees because they are convenient and provide timely documentation of minor incidents which might lead to future complications entailing medical expense, temporary partial disability, or time loss. From a management standpoint, these forms prove to be extremely beneficial in that they provide more comprehensive coverage of work-related hazards and afford an improved opportunity for needed modification or correction.

When

Fatal injuries or illnesses, as well as accidents which result in the hospitalization of five or more employees, must be reported to the governing federal or state OSHA agency within 48 h of occurrence.

Entries into the log and summary record for occupational injuries and illnesses must be made no later than 6 working days after the employer has been notified of the occurrence of an occupational injury or diagnosis of an industrial illness.

Cumulative information for each calendar year must be posted in a conspicuous place for the information of employees during February of the following year.

Supplementary records (OSHA Form 101 or equivalent) must be completed by the employing agency within 6 working days of receipt of notice of the injury or illness. Such records must be maintained for a period of not less than 5 years after the end of the calendar year during which the injury or illness occurred.

Why

Accident reporting is essential to the achievement of a number of clearly defined management responsibilities, including the achievement of laboratory program objectives. Although the reasons for reporting laboratory accidents, injuries, or illnesses may be detailed with endless variety, the three major goals of all accident reporting systems involve the satisfaction of management's legal, moral, and financial obligations.

Penalties. Failure to comply with OSHA reporting requirements may result in the issuance of citations and the imposition of penalties of up to $400. Intentional falsification of reports by either the employer or the employee may result in the imposition of fines of up to $10,000, imprisonment for up to 6 months, or both.

Moral obligations. The documentation and in-house analysis of laboratory-associated mishaps provide laboratory management with a critical analytical tool needed in the early identification and timely correction of hazardous situations or operations. Accident reports are not intended to establish individual culpability. They are designed to identify deficiencies within the work environment, including attitude problems of the employee or supervisor, and to permit the development and implementation of corrective measures which are needed to assure a safe and healthful work environment.

Financial considerations. During calendar year 1982 the total cost of accidents within the United States exceeded $88.4 billion. Work-related accidents, second in cost only to motor vehicle accidents, were estimated to cost at least $31.4 billion, or $318 per year for each member of the American work force (1).

Hidden costs, relating to laboratory accidents which result in unplanned infection of animals utilized in research or diagnostic programs, add to the costs of the laboratory program involved and compromise the validity of the data obtained. The further fact that laboratory-associated illnesses are generally a good deal more costly than the typical industrial injury in terms of time loss, medical expense involved, and potential for permanent disability indicates a paramount need for hazard identification and control in a microbiological laboratory operation.

Preparation of injury/illness reports

The first-line supervisor plays a key role in the managerial chain of any organization. The supervisor serves as the major communications link between management and the work force, is responsible for the orientation and training of individuals assigned to his or her unit, and is frequently held accountable for both the quantity and quality of the product or service provided. The supervisor must also see that work is performed in an approved manner and that unsafe acts, work procedures, or equipment is promptly identified and eliminated.

In view of the supervisor's key role in hazard identification and control, it is not surprising that he or she is usually delegated the responsibility of investigating and reporting all job-related accidents, injuries, or illnesses that occur within the organizational unit involved. Many agencies, in fact, use an accident reporting form which specifically designates the supervisor as the official responsible for its completion. Form AD-278, entitled "Supervisor's Report of Accident,'' used within many federal agencies, is an example of one such form.

In investigating accidents and preparing reports the supervisor must be objective, observant, and honest. Witnesses or individuals involved must be assured that accident reports are designed only to identify and document existing hazards for the purpose of correcting them, not to determine individual culpability or guilt. In all too many instances, data contained in accident or injury reports are nonsupportive and may in fact be contradictory. Although the abbreviated narrative description called for might indicate that horseplay, distraction, or inattention contributed to the accident, such factors are often left unmentioned in later segments of the report for fear of jeopardizing job security or compromising the potential benefits awarded. In instances involving minor, first-aid, or other accidents or injuries which are not subject to mandatory reporting requirements, the individual involved must assume the responsibility for informing his or her supervisor and filling out appropriate ''Minor Injury/Property Damage'' reports. Such reports not only help in identifying hazardous situations or procedures but also document the event as job related and facilitate management's acceptance of related medical expenses in the event of unanticipated complications.

Accident/injury record processing review and maintenance

The centralized processing, review, and maintenance of records involving laboratory-associated accidents or illnesses is essential to a successful safety program. In large organizations where an institutional nurse or other full-time safety specialist or staff is available, these responsibilities may be delegated in a direct and logical manner. Optimal results are achieved with a knowledgeable individual familiar with administrative details relating to request for and payment for medical services required. At the National Animal Disease Center, the staff nurse also compiles a summary of all reportable injuries and illnesses on a monthly and annual basis for review by the center's Accident Review Committee.

All medical records involving the diagnosis, treatment, prognosis, and consequence of laboratory-associated injuries or illnesses are also maintained within the National Animal Disease Center's medical unit. Relevant information concerning both job-associated and serious non-work-related medical problems is incorporated in each employee's medical file. These files are reviewed annually to ensure that this information is incorporated in the updated medical history questionnaire which is a required component of the center's medical surveillance program. Maintaining current and comprehensive medical histories aids the institutional physician or medical consultant in selecting serological procedures, blood chemistry tests, or other specialized medical procedures needed to determine the health status of the employee and to assess the consequence of past work-associated accidents or injuries.

Accident history review and analysis

Review and analysis of laboratory injury or infection data are best performed by a safety specialist or professional whose position description specifically includes these responsibilities, or by a committee appointed by the institute director.

A general review of the make-up, responsibility, and authority of such a committee can be found in Chapter 1. One of the major committee responsibilities is the review of laboratory accidents, injuries, and illnesses and the development and implementation of corrective action.

Information concerning the occurrence, frequency, and severity of laboratory-acquired infections, including details of the mode of transmission, is a valuable tool in the identification and assessment of hazards attending work in a biological laboratory setting (4). Data obtained by review and analysis of laboratory-acquired infections serve as a basis for many of the procedures or practices which are found in current biosafety guidelines as well as for the design and operation of equipment or facilities presently used to control or contain hazardous biological or chemical materials.

Laboratory-acquired infection data accumulated since the early part of this century identify over 4,000 human infections caused by 159 different etiological agents (2). Analysis of such data by numerous investigators has concluded that 10 etiological agents are responsible for over 50% of all laboratory-acquired infections identified (2, 6) and that 26% of these infections result in a degree of permanent disability while an additional 4% of these patients die (2, 4, 7). The actual mode of infection can be identified in less than 20% of the cases studied (7, 8). This suggests that when a specific infecting event cannot be identified, the great majority of laboratory infections are probably transmitted by infectious aerosols created by all activities carried on within the laboratory or animal isolation facility (2, 3, 5, 7, 9).

Although analysis of laboratory-associated accidents or infection experience within a single laboratory facility cannot be expected to provide data needed to establish sweeping basic concepts, it can be relied upon at least to identify flaws or deficiencies associated with in-house activities and to suggest appropriate means for their control.

LITERATURE CITED

1. **National Safety Council.** 1983. Accident facts, 1983 ed. National Safety Council, Chicago.
2. **Pike, R. M.** 1978. Past and present hazards of working with infectious agents. Arch. Pathol. Lab. Med. **102:**333–336.

3. **Pike, R. M.** 1978. Laboratory associated infections: summary and analysis of 3,921 cases. Health Lab. Sci. **13:**105–114.

4. **Pike, R. M.** 1979. Laboratory associated infections: incidence, fatalities, causes, and prevention. Annu. Rev. Microbiol. **33:**41–66.

5. **Songer, J. R., and J. F. Sullivan.** 1974. Safety in the biological laboratory. J. Chem. Ed. **51:**A481–A485.

6. **Sullivan, J. F., J. R. Songer, and I. E. Estrem.** 1978. Laboratory-acquired infections at the National Animal Disease Center, 1960–1976. Health Lab. Sci. **15:**58–64.

7. **U.S. Department of Health, Education and Welfare.** 1974. National Institutes of Health Biohazards Safety Guide. U.S. Government Printing Office, Washington, D.C.

8. **U.S. Department of Health, Education and Welfare.** 1978. NIH laboratory monograph. U.S. Government Printing Office, Washington, D.C.

9. **Wedum, A. G.** 1974. Biohazard control, p. 191–210. *In* E. C. Melby and N. H. Altman (ed.), Handbook of laboratory animal science, vol. 1. CRC Press Inc., Cleveland.

Chapter 5

Safety in Clinical Microbiology Laboratories

DIETER H. M. GRÖSCHEL

Clinical and public health microbiologists process materials containing unknown microorganisms of unknown pathogenicity. Therefore, special precautions are necessary for the protection of both the laboratory worker and the environment. In this chapter, a safety program is outlined for laboratories dealing with specimens from patients. It is based on material covered in detail in other parts of this book and includes recommendations of various clinical laboratory inspection and accreditation agencies such as the Centers for Disease Control, the College of American Pathologists, and the Joint Commission for Accreditation of Hospitals (1–6). The reader must be reminded that each laboratory director has to develop his or her own safety program and describe it in a specialized safety manual available to all laboratory workers.

Personnel

Laboratory personnel are the key element in a safety program. Since most laboratory accidents occur in the course of performing a procedure, it is mandatory that only qualified personnel perform such procedures. The laboratory director must review an employee's qualifications and knowledge before assigning him or her to a new task, and each employee must be aware of the potential hazards of the assignment. New employees must be introduced to the safety program of the laboratory during orientation. They must learn about location and use of eye wash stations, safety showers, and fire-fighting equipment and the location of emergency exits. To maintain the level of safety consciousness in a laboratory, training sessions, awareness programs, and the use of visual aids (posters and stickers) are recommended.

Responsibility of supervisors

The responsibility for ensuring compliance with the safety policy of a laboratory lies with the first-line supervisor. Unfortunately, the development of clinical laboratory supervisors usually is limited to technical aspects, entailing little management and even less safety training. Professional organizations as well as personnel managers now are aware of these deficiencies and offer training programs in both areas.

The safety policy must be part of the laboratory's general operating or procedure manual. In addition, each procedure or methodology should include a statement about specific safety aspects, e.g., the danger of certain chemical components of a reagent or the infectivity of a microorganism. The director or manager of the clinical laboratory should establish a file of information and literature on laboratory safety, including reference books as well as documentation of safety meetings, training sessions, etc.

Infection control and the laboratories

Diagnostic microbiology personnel, especially in hospitals and clinics, have direct contact with patients, as members of the phlebotomy team and in specimen collection areas. Therefore, it is important that all personnel be familiar with the infection control policies of the institution. In addition to the general infection control information given to new employees, infection control practitioners should hold training sessions (with demonstrations) for laboratory workers, explain isolation procedures, and discuss specimen collection from isolated patients. Laboratory personnel must enforce the hospital policy to identify specimens obtained from patients with communicable diseases or nosocomial infections. They must also understand that unmarked specimens may be just as dangerous.

In diagnostic laboratories not directly connected with a medical care institution, it is strongly recommended that an infection control consultant assist the laboratory director or manager in the instruction of employees about patient-related infections. Personnel of these laboratories often have close patient contact, either in the specimen procurement area of the laboratory, in a clinic or a doctor's office, or during house calls.

Clinical laboratory personnel must be aware that neither the supervisor nor the infection control and safety program can completely prevent laboratory- and hospital-acquired infections. Each laboratory worker is responsible for his or her own health by adhering to the safety and infection control policies of the employer and maintaining a personal health program.

Personal protection of laboratory employees

Policies regarding personal protection may vary in different laboratories and depend on the type of work performed, the containment facilities present, and institutional rulings. The laboratory director is responsible for instituting these policies and for providing the necessary safety equipment and supplies. Like other laboratory policies, the safety regulations need to be reviewed regularly in close cooperation with the first-line supervisors. Infractions should be brought to the attention of the director, especially eating, drinking, smoking, and application of cosmetics in the laboratory. The institution should provide cafeterias or lounges for coffee and lunch breaks. Special refrigerators must be designated for food storage to avoid the improper storage of food in laboratory refrigerators.

Mouth pipetting is dangerous and not permitted in a clinical laboratory. It is the responsibility of laboratory directors and supervisors to provide sufficient functional pipetting aids for all technologists and technicians. Safety goggles or face shields must be available for droplet-producing work such as the use of certain tissue grinders and high-speed mixers, and heat-resistant gloves are needed for handling hot or cold materials.

Hand washing is a most important component of a microbiological safety program. Hand-washing facilities, preferably foot or elbow operated, must be available in the laboratory area and should not be used for other laboratory purposes. An antiseptic liquid soap or foam should be used. Eye wash facilities must be located in the laboratory. Eye wash stations with water-filled plastic bottles are not recommended since the water often is heavily contaminated with microorganisms. If separate eye wash stations cannot be installed, it is recommended that attachments such as the Splash Gard Eye Face Wash (Lab Safety Supply, Janesville, Wis.) be used.

Personnel with long hair or long beards must fasten them to avoid contact with contaminated materials, to prevent spreading organisms into the work area, and to keep them out of moving equipment such as centrifuges. Most laboratories have dress codes requiring safe clothing (not loose-hanging or flammable garments) which can be laundered easily. Shoes should be slip proof and cover the entire foot. Open-toed shoes expose the toes to injury such as cuts from broken glassware.

Employees are required to wear protective clothing, either a laboratory coat over street clothing or a laboratory uniform. Upon leaving the laboratory during breaks, or for patient contact (blood drawing), a clean coat should be worn. For work requiring protection with gloves and gown, such as in the mycobacteriology laboratory (Biosafety Level 3), the laboratory should provide permanent or disposable operating-room style gowns which close in the back and fit tightly around the wrists. Protective garments

for work with highly pathogenic microorganisms or dangerous chemicals are described in other chapters of this book.

Employee health care

Personnel in clinical laboratories are responsible for good health practices. Each employee should have a family physician who is aware of the type of work the employee is engaged in. Laboratories associated with medical centers usually participate in the center's employee health program. The physician in charge of the health service, usually in consultation with the infection control committee, the biosafety committee, and the laboratory director, will establish a routine health surveillance program which ought to include at least tuberculosis screening and may require certain immunizations (see also Chapter 3). Clinical laboratories outside of medical institutions are advised to contract for these services.

All laboratory accidents, even minor cuts and punctures, must be reported to the supervisor and documented. First aid information, including the telephone number of the local rescue squad, must be posted in all laboratories, and appropriate first aid supplies must be kept. The injured employee should be sent to the employee health service or to his or her personal physician for definitive treatment. Each accident should be investigated by a senior supervisor and the director, and corrective action should be taken. In large institutions the safety officer is responsible for this investigation. Refer to Chapter 4 for further details.

Laboratory design

The laboratory design should promote environmental safety and safety consciousness of the personnel. Even old and inadequate space can be remodeled into a safe laboratory for a reasonable amount of money. Other chapters (especially Chapters 17 and 18) describe safety-oriented building specifications in detail. Clinical laboratories should be located away from patient care and main traffic areas and have controlled access, separate ventilation and air conditioning systems, and a good housekeeping program. Windows should be protected with screens to prevent insects from entering the laboratory. The location of services within the laboratory area should be determined according to risk assessment of the procedures performed, and biohazard signs should be posted. Airflow pressure differentials, biological safety cabinets, and chemical hoods are supplemental safety features, and their use will depend on the biosafety level specified for the laboratory. Storage facilities for flammables and explosives must be available and comply with national and local codes. Tanks of compressed gases must be securely fastened to walls or other solid structures, and safe transport carts must be provided. Materials for floors and walls as well as furnishings must comply with all safety and housekeeping requirements for laboratories. Fire alarm systems and primary fire-fighting equipment such as fire blankets are required in all laboratories. An evacuation plan must be posted (Occupational Safety and Health Administration, 29CFR1910).

Laboratory safety manual

In addition to general information, the individual laboratory's safety manual must include detailed instructions, guidelines, and rules about the following:
1. Locations and permitted maximum quantities of flammable compounds for daily use; disposal of solvents in special containers until removal by the safety office or contractor
2. Storage and handling of compressed-gas cylinders
3. Electrical safety, especially use of extension cords, grounding, and repairs; use of personal radios, coffee makers, etc.
4. Labeling and handling of flammable, caustic, toxic, carcinogenic, and radioactive materials; storage close to the floor level to avoid accidental spillage or breaking during handling
5. Control of aerosolization of potentially infectious materials, e.g., by use of vented centrifuges or safety centrifuge shields, correct use of pipetting devices, flaming of tubes and inoculation loops, pouring of contaminated liquids, etc.
6. Selection, preparation, and use of disinfectants for discard containers, surfaces, and spills; use and maintenance of UV lights
7. Use of steam autoclaves for the sterilization of microbiological media and contaminated waste; biological (spore strip control) and chemical monitoring and recording of sterilization times and temperatures
8. Cleanup of spills
9. Use and maintenance of chemical and biological safety cabinets
10. Control of insects and other pests
11. Handling of experimental animals
12. Disposal of chemical and biological wastes; use of incinerators

Other chapters of this book address these items in detail and should be consulted for the formulation of a safety plan.

Specimen handling and processing

The microbiology laboratory is responsible for the selection of safe specimen containers, such as plastic tubes with or without transport medium, transport swabs, and urine, sputum, or stool cups which are leak proof and will not break under normal use conditions. Although all microbiological specimens should be considered infectious, the knowledge of locally prevalent infectious diseases, the indication of the suspected clinical diagnosis on the requisition, and the use of special identification labels and bags to flag specimens from infected patients (especially with hepatitis or AIDS [acquired immune deficiency syndrome]) will reinforce the safety awareness of messengers and laboratory personnel. The laboratory must establish criteria for the nonacceptance of specimens such as leaking or inappropriate containers (e.g., permeable paper cups for sputum or stool) and assist in writing a similar policy for the messenger service.

During processing of specimens, safety precautions are especially important. Most laboratories use a biological safety cabinet. Any procedure with the potential to generate aerosols (centrifugation, tissue grinding by hand or in an electric blender, and heat sterilization of inoculation loops, scalpels, or pipettes) must be written out in detail and with attention to the protection of personnel and environment. The laboratory worker must receive clear instructions on how to discard, disinfect, and reprocess all instruments and equipment used in preparing clinical material.

Inoculated media should be stored in containers during incubation and work-up to avoid dropping and aerosolization of microorganisms. Subculturing and identification tests must be performed according to the required biosafety procedure. Work with fungal, mycobacterial, and viral cultures requires the use of a biosafety cabinet, whereas bacteriological diagnostic work is usually performed on the laboratory bench. All workplaces must be cleaned regularly with disinfectants. Biosafety cabinets can also be decontaminated with UV light.

Needles and syringes used for aspirating blood culture bottles must de discarded into a safety container without recapping the needle. For the heat sterilization of inoculation loops and needles, safety incineration devices such as the Bacti-Cinerator (American Scientific Products, McGaw Park, Ill.) should be used to avoid aerosols. Laboratory reagents should be stored in safe containers and marked clearly as to toxic properties. Highly toxic chemicals (such as potassium cyanide) must be stored in a locked cabinet with controlled access. Stock cultures on agar slants, in agar tubes, or in lyophilization vials should not be stored in hard-to-reach wall cabinets because of the increased danger of an accident. For the shipment of microorganisms to reference laboratories, the applicable packaging, labeling, and shipping requirements of the Interstate Quarantine Regulations (42 CFR Part 72) must be followed (Chapter 25).

Disposal of contaminated waste

Spent media and specimen materials must be disposed of safely, usually after treatment in a steam autoclave. In laboratories with a large waste volume, only selected materials from the mycobacteriology, mycology, and virology laboratories may be steam sterilized, while other discards are incinerated. If contaminated materials are transported outside the laboratory area, they must be contained in closed boxes with leakproof plastic liners or in durable, double plastic bags. They must be recognizable to all personnel, especially housekeepers, as containers of biohazardous materials. The transport of waste should be performed only by qualified personnel and in covered, leakproof containers. A policy for cleanup must exist in the case of a leak or tear and spill of contaminated box or bag contents.

Discard pans on laboratory benches must contain an approved disinfectant at a concentration which, even after the addition of liquids, will ensure disinfectant activity. The selection of a disinfectant must be based on the need of the laboratory, not its availability in the storeroom (see Chapter 22 for recommendations). In some laboratories, discard pans with disinfectant also are steam sterilized. With some types of disinfectant, however, this may cause toxic fumes within the autoclave chamber and inhalation injury to the autoclave operator.

Steam autoclaves used exclusively for decontamination of waste require the same care as other laboratory and hospital autoclaves, i.e., a routine cleaning and maintenance program and regular quality control of their performance by chemical (autoclave tape, time-temperature-steam indicators), physical (melting point, time-temperature-pressure recorder), and biological (spore strip) monitoring. For a detailed discussion of sterilization see Chapter 22. Sterilized waste may be discarded with other waste. One should remember, however, that microbiological substrates will still support the growth of microorganisms and may increase the odor problem of waste storage containers.

Summary

The handling of patients' specimens in medical, dental, and veterinary clinical laboratories requires a special safety consciousness of all personnel. Work with unknown pathogens requires the establishment of, and adherence to, a clear safety policy contained in a safety manual. Furthermore, each laboratory procedure must be written with safety considerations in mind. While the laboratory director is responsible for providing a safe environment and developing a safety policy based on local regulations and the recommendations and standards of inspection and accreditation agencies, the first-line supervisor has the day-to-day responsibility for enforcement of safety rules.

LITERATURE CITED

1. **Carlson, D. J.** 1976. Guidelines for laboratory safety for medical technologists. Policies and procedures. College of American Pathologists, Chicago.
2. **Center for Disease Control.** 1974. Lab safety at the Centers for Disease Control. DHEW publication no. (CDC) 79-8118. U.S. Department of Health, Education and Welfare, Center for Disease Control, Atlanta, Ga.
3. **Commission of Laboratory Inspection and Accreditation.** 1974. Standards for accreditation of medical laboratories. College of American Pathologists, Chicago.
4. **Joint Commission for the Accreditation of Hospitals.** 1983. AMH/84. Accreditation manual for hospitals. Joint Commission for the Accreditation of Hospitals, Chicago.
5. **Richardson, J. H., and W. E. Barkley.** 1985. Biological safety in the clinical laboratory, p. 138–142. In E. H. Lennette, A. Balows, W. J. Hausler, Jr., and H. J. Shadomy (ed.), Manual of clinical microbiology, 4th ed. American Society for Microbiology, Washington, D.C.
6. **U.S. Department of Health and Human Services.** 1984. Biosafety in microbiological and biomedical laboratories. HHS publication no. (CDC) 84-8395. U.S. Government Printing Office, Washington, D.C.

Chapter 6

Introduction to Hazard Assessment

BRINTON M. MILLER

This section focuses on assessment of those factors that may put us at risk in the microbiological and biomedical laboratory. Generally speaking, these factors may be listed simply as ourselves, some chemicals we commonly use in the laboratory, and (especially) the targets or tools of our work, namely, infectious and toxin-producing organisms. Without an understanding of these factors, all of the safety management programs, containment equipment, and safety techniques for use in the labortory will have little meaning. In the introduction to the containment equipment section, Vesley reminds us that personal behavior factors contribute 90% toward a sound, workable safety program. While management must provide leadership and support for that safety program, actual conduct of the program falls squarely on the shoulders of the individual in action. Thus, we try in the following chapters to provide the individual with what is currently known regarding the safe handling of biological risks in the laboratory.

In chapter 7, Don C. Mackel and John E. Forney provide a look at the epidemiology of laboratory-acquired infections in general. This chapter, originally drafted in 1977, was updated in late 1984. Reading this and the introduction to "Biosafety in Microbiological and Biomedical Laboratories" (Appendix 1), the reader can readily understand the increasing need for good biosafety practices in the laboratory. G. Briggs Phillips in chapter 8 provides an enlightened view of how important the individual is in his own defense against undue risk. Phillips does not restrict his observations to the operator, however, but also calls attention to the responsibilities of leadership. Perhaps most importantly, Phillips reminds us that we must constantly review changes in the laboratory, especially human factors. For example, although someone from another discipline may bring new expertise and ideas to a laboratory, that same person may be ignorant of safety needs for the new task.

Chapters 9 through 13 deal with individual specialty areas of risk with microorganisms. They were especially selected because of current trends in research with these infectious agents. Favero presents an excellent current review on hepatitis infection, both its transmission and control. Blenden and Adldinger describe in detail the problems with viral zoonoses, which in the present fast-moving world of international transportation appear to increase yearly. The chapter on laboratory-acquired parasitic infections is appropriate for much the same reason. In addition, this field has increasingly attracted the interest of biochemists, pharmacologists, and others new to the field of infectious disease. The need for remarks on bacterial and rickettsial zoonoses and slow viruses is obvious.

In chapter 14, Gailya Walter and David Weathers provide this book with a much-needed system for managing the use of toxic chemicals. Biologists use or are exposed to numerous hazardous chemicals, many of which are carcinogenic. The system both classifies and references these compounds, providing us with a guide for their use, storage, etc.

Finally, in chapter 15, Joseph Songer reviews briefly the codification of risks in the laboratory as presented by the CDC/NIH publication "Biosafety in Microbiological and Biomedical Laboratories." More importantly, he presents a reminder that while risks may be measured by a variety of factors, evaluating the acceptability of risk involves a value judgment. Safety is a practice, based on good planning and knowledge of the materials and tools with which one works. Many of these essential facts are provided the reader in this book. Suggestions are made regarding programs, hazards, and equipment, but it is the practice of the individual worker which will determine whether or not risks remain risks or become actual accidents.

Chapter 7

Overview of the Epidemiology of Laboratory-Acquired Infections

DON C. MACKEL[†] AND JOHN E. FORNEY

Many clinical and research microbiologists are exposed to multiple sources of infection in the course of their day-to-day activities in the laboratory. However, laboratory-associated infections appear to occur relatively rarely in the United States. This may be due to the fact that the true incidence of laboratory-acquired infections is unknown because of inadequate reporting to health officials, lack of a central surveillance system, and unrecognized job association.

The epidemiology of disease transmission in the laboratory is diffuse and at times complex because the vehicles and routes of infection may be different from the usual sources of transmission outside the laboratory setting. Laboratorians are frequently exposed to infective doses greater than those necessary to produce disease in the normal person, and the period of incubation can be shorter than usual.

Laboratory personnel can be exposed to pathogens in the laboratory through direct contact (including droplets), indirect contact, infected animals and insect vectors, and airborne sources. Person-to-person spread is sometimes recognized. Whether a person becomes ill after being exposed to an organism depends on the extent of the contamination (i.e., dose), pathogenicity for humans, route of infection, and individual host susceptibility. Certain elements in the environment are known to cause greater risks for infection, particularly equipment and procedures that produce aerosols or droplets.

It is beyond the scope of this chapter to describe in detail all factors related to preventing and controlling laboratory-acquired infections, but some applicable information is discussed. The normal precautions needed for various manipulative procedures in the laboratory are described in greater detail in later chapters and in the Appendix.

EPIDEMIOLOGY OF DISEASE TRANSMISSION

Gordon (9) has defined epidemiology as part of a broader science called ecology which has to do with the mutual relationships of various living organisms in the environment and their reaction to animate and inanimate surroundings. This broad science can be termed "human medical ecology" to signify the interaction of humans with their total environment and the ways in which this interaction contributes to or significantly affects the development of disease or physiologic conditions (8).

The epidemiology of laboratory-acquired infection fits precisely into the framework of this definition, but in many respects it encompasses all of the different associated ecologic factors that can contribute to acquisition of an infectious disease (Table 1). It also involves interpretation of possible causes, using general principles and methods of study to explain the unusual.

Contact-transmitted diseases are spread by infected persons, animals, or contaminated environments (e.g., direct, droplet). It should be noted that infectious humans or animals may be symptomatic, asymptomatic, or colonized with a communicable agent. Indirect transmission includes the inanimate environment, such as the air, biological products,

TABLE 1. Ecologic factors to be considered in the epidemiology of disease transmission in the laboratory

I. Method of Transmission
  A. Direct contact
  B. Indirect contact
  C. Airborne
  D. Common vehicle
  E. Vector
II. Route of Infection
  A. Ingestion
  B. Inhalation
  C. Inoculation
  D. Penetration of skin and mucous membranes
  E. Animal bites
  F. Insect bites
III. Factors Determining Infection
  A. Pathogenicity of organisms
  B. Extent or magnitude of contamination (dose)
  C. Route of infection
  D. Host susceptibility
  E. Allergic reaction
IV. Environmental Factors
  A. Room air (ventilation)
  B. Equipment
  C. Procedures (techniques)
  D. Sanitation
  E. Crowding

[†]Deceased.

instruments, surfaces, food and water, etc. Vector-borne transmission can occur through the biological route (bite) or mechanical route (external contamination), depending upon the nature of the agent.

The pathogenicity or virulence of an organism can vary greatly among types and strains of microorganisms. Some organisms are known to be highly pathogenic, while others are only opportunistic pathogens. Infection by opportunistic agents usually results from a deficiency of the host defense mechanisms or acquisition of an overwhelming infectious dose. Most infections are usually acquired through a specific route into the host. The epidemiologic significance of the portal of entry should not be overlooked in determining mode of spread. Establishing certain modes of spread of some infections is often difficult; for example, it would be an error to imply that respiratory tract infections signify the "nose" as the portal of entry and thus that they implicate airborne spread. It is necessary to determine the specific infected tissue site within the respiratory tract. Infections of the nasal mucosa, tonsils, or the respiratory mucosa of the upper respiratory tract with, e.g., diphtheria, streptococcal pharyngitis, pertussis, or influenza can result from contact of droplets or from inhalation of large viable particles (airborne). In contrast, infections in the alveoli of the lung, such as pulmonary mycoses, psittacosis, Q fever, and primary tuberculosis, are exclusively airborne infections.

OCCURRENCE AND RISK OF INFECTION

Sulkin and Pike (21) used a mail questionnaire survey in 1960 to obtain information on infections that had occurred in approximately 5,000 U.S. laboratories in the previous 20 years. Reports disclosed that data on at least 875 of 1,342 (65%) cases had not appeared in published reports.

Pike (18) published an analysis of the 3,921 laboratory-acquired infections for the period 1924 to 1974 (Table 2) and later updated the reports to include 4,079 cases (19). Pike did not know how many persons were at risk during any of the periods. The case-fatality rate calculated for the 3,921 cases was 4.18%. More infections occurred in research laboratories than in clinical laboratories, but this may reflect better record keeping and reporting by research laboratories or a lack of adequate biosafety training for this group of investigators.

Pike also noted that only about 20% of these infections followed a recognized accident. The cases associated with a known accident usually involved an overt act: cut, bite, or scratch (6.6% of total infections), splatter or spill (6.0%), syringe or needle stick (5.6%), mouth pipetting (2.9%), and other causes (1.2%). Accidental sticks with a needle and syringe and aspiration while mouth pipetting accounted for approximately 3% of the reported

TABLE 2. Summary distribution of 3,921 laboratory-acquired infections (1924–1974)[a]

Type of infection	United States	Foreign	No. (%) of deaths
Bacterial	1,166	503	69 (4.13)
Viral	599	490	54 (5.15)
Rickettsial	381	192	23 (4.01)
Fungal	177	176	5 (1.42)
Chlamydial	84	44	10 (7.81)
Parasitic	70	45	2 (1.74)
Not specified	28	6	1 (2.94)

[a]Abstracted from reference 18.

accidents but only for about 9% of all reported laboratory-associated infections.

Many of the reported infections were caused by microorganisms known or suspected to be infective at low doses, although laboratory-acquired infections caused by organisms requiring higher concentrations to produce infection have occurred.

SOURCES AND TYPES OF INFECTION

When laboratory-acquired infections of unknown cause were tabulated according to most probable source (18), the sources and percentages of cases were: working with the infecting agent, 26.2%; animal, egg, or arthropod handling, 20.9%; aerosol, 16.6%; handling clinical specimens, 9.1%; human autopsy, 2.4%; discarded glassware, 1.5%; all others, 1.1%. Of the reported infections, 767 could not be associated with a known accident or source. These data also suggest that infections are most frequently associated with aerosols produced during laboratory manipulations or with the ingestion of microorganisms. Thus, the nose and mouth were the most frequently reported portals of entry.

Kenny and Sabel (14) showed that substantial numbers of aerosols and droplets can be generated during routine laboratory procedures or when accidents occur. For example, large numbers of organisms were suspended in the air when the lid of a blender jar was removed after the contents were blended, and organisms were also disseminated, but to a lesser degree, during the blending procedure with the lid on. They also showed that high concentrations of viable particles were disseminated in air when liquid cultures, lyophilized cultures, or infected eggs were dropped. Spilled liquids on surfaces also disseminated microorganisms into the air. Most of these events transmitted viable airborne particles within size ranges that would allow the inhaled microorganisms to reach the alveoli. As previously mentioned, sometimes single-source aerosols are disseminated widely in the laboratory. The effect of age of the aerosol on the level of infectivity for humans is unknown.

Publications by Pike (19, 20) summarize laboratory-acquired infections before 1975. Of the reported bacterial infections, 64% were brucellosis, typhoid fever, tularemia, or tuberculosis. Of the viral infections, 36% were hepatitis or Venezuelan equine encephalitis. Coccidioidomycosis and histoplasmosis represented 46% of the fungal infections. Dermatomycoses were equally prevalent (45%), but did not result in employee absenteeism. Chlamydia infections were particularly severe, with a death rate of 8.6 per 100 cases. No comparable systematic collection of incidence has been reported recently, although laboratory-acquired infections still occur.

To emphasize the continuing nature of the problem, 17 recently reported infections (76 cases) involved only a few of the infectious agents that continue to infect laboratory personnel. When multiple cases were associated with a single organism, no associated accident could be identified; this suggests a trend similar to that observed by Pike. Only in individual cases of infection could a specific accident such as aerosols, needle puncture, or a needle popping off a syringe be identified or associated with illness. These sources still account for over half of recognized accidents (19). We therefore surmise that laboratory personnel become infected while performing routine procedures, probably involving aerosols that are inhaled or contaminated surfaces and equipment.

The risk of laboratory-acquired brucellosis continues to rank high among bacterial agents. Between 1965 and 1974, 2,302 cases were reported in the United States; 34 of these resulted from laboratory accidents (7).

Twenty-three of 25 recently reported cases of typhoid fever apparently resulted from bringing *Salmonella typhi* cultures into the laboratory as teaching or proficiency-testing specimens. Some of these infections affected persons other than laboratorians, suggesting secondary spread from asymptomatic laboratorians. Doubtless the same pattern holds true for other infections whose mode of transmission is not obvious. The other two infections were acquired from diagnostic specimens (3, 4, 12).

In three recent outbreaks involving infectious mononucleosis, hepatitis, or Rocky Mountain spotted fever, the precise exposures were not identified, which emphasizes the need for any early epidemiologic approach. During the last 50 years, there has been a marked decrease in the relative frequency of laboratory-associated bacterial infections but an increase in viral infections (Fig 1). Numbers of rickettsial and bacterial infections peaked during and immediately after World War II but have greatly declined during the last 2 decades, whereas numbers of viral and fungal infections have declined only slightly (18, 19). Hepatitis and many new virus isolates may account for this marked shift. Also, because of better isolation techniques, workers are exposed to higher concentrations of viral agents.

Hospital laboratory personnel are at high risk of

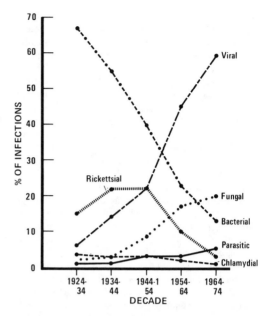

FIG. 1. Relative proportions of reported causes of laboratory-associated infections (1924 through 1974) (from reference 18).

acquiring viral hepatitis (16). The risks are highest for personnel in hematology, biochemistry, and serology laboratories. One survey of 51 U.S. clinical laboratories involving 731 laboratorians identified a 7.4% incidence of icteric and anicteric hepatitis and Australia antigen in 2.5% of the personnel (15). Hirschowitz et al., during a 26-month study of 310 clinical laboratory workers, found positive titers of antibody to hepatitis B surface antigen in about 25% and demonstrated the antigen in the blood of between 1 and 2% (11).

The Centers for Disease Control's Hospital Infections Program has established a nationwide voluntary needle stick surveillance system for AIDS (acquired immune deficiency syndrome) in hospital personnel. The Centers have published AIDS precautions for clinical and laboratory staffs (5).

INFECTIVE DOSE

Predisposing factors and the severity of laboratory infections are determined by host factors as well as the causative agent, including its virulence and dose. A heavy infective dose frequently shortens the usual incubation period, whereas a low dose can lengthen the time before clinical disease appears.

Although many causative agents have been studied to determine 50% infective doses in animal models, only a few studies with humans have been done with agents other than those causing infectious diarrhea. Investigators have conducted studies with normal, healthy volunteers. Hornick et al. (13) studied the infective dose of *S. typhi* for susceptible volunteers. An *S. typhi* isolate obtained from a carrier infected 40 (95%) of 42 volunteers who ingested 10^9 organisms. The median incubation period was 5 days. In another study, 116 subjects ingested 10^5 *S. typhi* organisms each, after which 28% became ill after a median incubation time of 9 days. None of 14 persons who ingested 10^3 organisms became ill.

DuPont and Hornick (6) used prison volunteers to study infective doses of ingested strains of *Shigella* sp., *Salmonella* sp., *Escherichia coli*, and *Vibrio cholerae*. The infective dose of *Shigella* sp. was $\leq 10^2$. *Salmonella* sp. at a dose of 10^5 organisms caused illness, and both *E. coli* and *V. cholerae* caused illness after a dose of at least 10^8 organisms.

Blacklow et al. (2) studied the infective dose of nonbacterial agents of gastroenteritis. Volunteers who ingested $\leq 10^2$ cysts of *Giardia lamblia* or *Entamoeba histolytica* became ill.

Wedum (23) stated that as few as 10 organisms of *Francisella tularensis* and *Coxiella burnetii* are sufficient to infect most susceptible persons. In comparison, one organism of *Mycobacterium tuberculosis* has been shown to produce disease in the lungs of nonhuman primates (1).

Although the exact number of virus particles necessary to produce hepatitis B infections is unknown, infectious serum diluted 10^{-8} has been shown to infect chimpanzees (see chapter 9).

CONTACT TRANSMISSION

In the absence of a known documented accident or other incident causing infectious organisms to be transmitted, we must presume that infectious material has been released into the environment (air, surfaces, or fomites) during routine laboratory operations in any of the following ways.

(i) Gross contamination. Contamination may be relatively gross, e.g., when a drop falls unnoticed from a pipette or a container leaks.

(ii) Inapparent contamination. Inapparent surface contamination may be caused by aerosols, e.g., droplets falling on such items as fingers, pipettes, or pencils, that can easily transfer the organism to the mouth or through inapparent or unnoticed breaks in the skin.

(iii) Person-to-person spread. Laboratory-acquired infections by person-to-person spread are believed to be rare.

(iv) Indirect spread: person to inanimate object to person. In one situation involving indirect spread, a laboratory technician developed tuberculosis after using pipettes that had been used by another technician with undiagnosed active tuberculosis (17).

(v) Accidental aerosols. Accidental release of aerosols into the environment can cause infection via the respiratory tract.

(vi) Specimens. Infections are sometimes caused by exposure to specimens (common-vehicle transmission), e.g., blood, urine, feces, exudate, or sputum. There have been undocumented reports of outbreaks caused by contaminated specimen-collecting equipment, antiseptics, and blood culture media.

(vii) Insect vector. Infection can occur in an environment that supports the presence of an insect vector which carries (biologically or mechanically) the causative agents of vector-borne diseases.

AIRBORNE TRANSMISSION

Authorities generally agree that contact is the most common method of spread of microbial infections. In most laboratories, persons are less likely to acquire infection directly from air than through direct contact. Nevertheless, air does play an important role in transmitting certain pathogens and in general contributing to the contamination of the environment. Airborne contamination can occur not only horizontally but vertically, i.e., above and below the source of contamination. Improperly maintained or operated air-conditioning systems can also be associated with the spread of contamination, whereas properly designed and maintained air-handling systems can help to remove contamination.

In high-risk areas, much can be done to reduce airborne spread of microbes, including minimizing nonessential traffic and employing biological safety cabinets. Air-handling systems have been designed or modified to substantially reduce airborne spread and contamination of laboratory surfaces by increasing the number of room air exchanges per hour and by adjusting positive and negative pressure between work and nonwork areas to prevent migration.

Airborne bacteria are seldom, if ever, found as exposed (naked) cells in the environment. Instead, such organisms are usually found as components of particles formed from such material as soil, lint fibers, organic material, or a combination of these materials. The carrier tends to protect the airborne or resting viable cell and is called a viable particle. The nature of airborne viable particles is frequently dictated by the source of contamination; i.e., they can appear as droplet nuclei, aerosols, or dust. Droplet nuclei are formed from splashes and aerosols.

A viable particle is usually considerably larger than the reported size of an individual bacterium, which is generally less than 2 μm in diameter. The sizes of naturally formed particles are important for several reasons. Particles 1 to 5 μm in diameter are most effective in penetrating lung tissue. Particles less than 0.5 μm in diameter are usually not retained in lung tissue effectively, and particles larger than 5 μm are usually filtered out by the upper respiratory tract before they reach the lung (10). Particles larger than 5 μm in diameter do not remain airborne very long and settle onto surfaces, inanimate objects, clothing, skin, etc., or may be retained in the upper respiratory tract.

Wedum (23) in reviewing the literature describes 94 cases of brucellosis caused by a tube that broke while being centrifuged in a basement laboratory; the organism ultimately spread to the third floor of the building. This review also describes the spread of other microbial contaminants (rickettsiae and arboviruses) throughout wide areas in buildings after laboratory accidents, causing multiple infections.

The nature of the agents that most commonly cause laboratory-acquired infections suggests that most pathogens are inhaled or otherwise ingested as aerosols during routine manipulations. However, only 17% of infected employees associated their infections with a specific aerosol-generating procedure (14).

Many viable particles contain more than one viable cell; e.g., the average number of staphylococci per particle is four and ranges from two to eight. The aggregation of microorganisms should be considered when attempting to determine infective doses.

The inanimate environment provides numerous reservoirs from which viable particles can become suspended in the air. All horizontal surfaces are readily contaminated, and floors are particularly important because they comprise a relatively large surface area. Such activities as inferior housekeeping procedures can resuspend contamination from the floor into the air, from which it subsequently deposits on clothing, on skin, into the nares, etc.

GENERAL CONTROL MEASURES

The most effective preventive measure is being certain that all approved procedures are consistently carried out (including methods for handling infectious material and operating equipment). Frequent and thorough hand washing is the second most important preventive measure that can be used. Transient skin organisms, often acquired by touching contaminated cultures, objects, or surfaces that may be contaminated with pathogens, are frequently implicated. The practice of good personal hygiene should also be emphasized. All appropriate vaccinations should be given and updated as needed.

Biological safety guidelines should call for an initial base-line serum specimen to be obtained from each employee and stored at at least − 20°C (− 70°C is preferable, to extend antibody activity over a long period); serum specimens should then be obtained every 1 to 3 years depending on the virulence of the microorganisms that the person is exposed to. It may be prudent to divide the serum samples into several or more aliquots to avoid refreezing when used. These sera are used to compare with acute- and convalescent-phase sera, which helps to determine suspected etiology by the presence or absence of specific antibody before infection. Additional control procedures and programs should include the following measures.

(i) Ongoing training sessions should be conducted by knowledgeable supervisors throughout the year in research, diagnostic, and academic laboratory settings. Untrained or inexperienced persons should not be permitted to handle potentially infectious microorganisms.

(ii) Proper methods for sterilization of contaminated equipment should be stressed. Principles and limitations of chemical disinfection of work surfaces should be clearly understood.

(iii) Ventilation systems should provide a negative pressure inside each laboratory and animal quarters relative to nonlaboratory areas.

(iv) Laboratory animals should be adequately housed and cared for so that they do not transmit disease to laboratory or other staff members or to other laboratory animals.

(v) Insectories should be properly maintained outside the laboratory area; wild species should be prevented from entering the laboratory.

(vi) All exposures to infectious material should be reported to an appointed laboratory safety officer. An infectious disease specialist should be consulted, and appropriate prophylactic therapy should be given when indicated. The status of persons exposed should be monitored at least throughout the known incubation period for the agent involved.

(vii) Ongoing routine surveillance for detection of laboratory-acquired infections should be maintained.

(viii) Optimally, all laboratorians should become familiar with the Centers for Disease Control/National Institutes of Health biosafety guidelines (22). Adapting these recommendations to particular laboratory operations, based on risk assessment of specific agents and activities, should minimize the risk of infections.

EPIDEMIOLOGIC INVESTIGATION OF LABORATORY-ACQUIRED INFECTIONS

The diagnosis of an infection should be confirmed through laboratory tests or clinical appraisal. When appropriate, serologic tests involving base-line (preinfection) serum from the patient should be done. Specific surveillance techniques should be used to determine whether the infection represents a sporadic case or one of an outbreak (i.e., more cases than would be expected in a population in a given time period). Information obtained about the case(s) should include date of onset, when exposed, procedures performed before onset, pathogens involved, age, race, sex, work location within the laboratory, known breaches of technique, and vaccination status. Information concerning activities and possible exposures outside the laboratory should be obtained. Case-control studies should be conducted (compare infected versus noninfected) for many of the parameters to define "population at risk" more precisely. This information is important in determining whether the case is a single event or part of an outbreak and what type of transmission it represents, e.g., common-source, person-to-person, etc. It may be appropriate to conduct additional case-control studies to test the importance of different types of exposures and to evaluate current control practices.

After the data collected are analyzed, a report of the episode should be written and referred to for taking action to prevent similar infections in the future. Laboratory safety training activities should be reviewed and redesigned when appropriate. A copy of the report should also be sent to the local health department.

Reporting the incident in the literature alerts others to take steps to prevent similar occurrences and provides a basis for better understanding the epidemiology of these infections.

SUMMARY

Control of communicable disease in the laboratory is governed by direct action against demonstrated causes, such as measures directed toward control of the environment and sometimes of both host and agents. The first principle is the recognition that causality is multifarious. Once the causes are recognized, procedures can be designed or modified to break the chain of transmission by intervening environmental controls, procedural practices, immunization, training, and infection surveillance.

LITERATURE CITED

1. **Barclay, W. R., R. L. Anacker, W. Brehmer, W. Leif, and E. Ribi.** 1970. Aerosol-induced tuberculosis in subhuman primates and the source of the disease after intravenous BCG vaccination. Infect. Immun. **2:**574–582.

2. **Blacklow, N. R., R. Dolin, D. S. Fedson, H. L. DuPont, R. B. Hornick, and R. M. Chanock.** 1972. Acute infectious non-bacterial gastroenteritis: etiology and pathogenesis. Ann. Intern. Med. **76:**993–1008.

3. **Blaser, M. J., and J. P. Lofgren.** 1981. Fatal salmonellosis originating in a clinical microbiology laboratory. J. Clin Microbiol. **13:**855–858.

4. **Center for Disease Control.** 1979. Follow-up on laboratory-associated typhoid fever. Morbid. Mortal. Weekly Rep. **28:**593–594.

5. **Centers for Disease Control.** 1982. Acquired immune deficiency syndrome (AIDS): precautions for clinical and laboratory staffs. Morbid. Mortal. Weekly Rep. **31:**577–580.

6. **DuPont, H. L., and R. B. Hornick.** 1973. Clinical approach to infectious diarrheas. Medicine **52:**265–270.

7. **Fox, M. D., and A. F. Kaufmann.** 1977. Brucellosis in the United States, 1965–1974. J. Infect. Dis. **136:**312–316.

8. **George, J., and M. Greenberg.** 1957. Medical officer's bookshelf on epidemiology and evaluation: part 1—epidemiology. Am. J. Public Health **47:**401–408.

9. **Gordon, J. E.** 1963. Preventive medicine and epidemiology: field epidemiology. Am. J. Med. Sci. **246:**354–376.

10. **Hatch, T. F.** 1961. Distribution and deposition of inhaled particles in respiratory tract. Bacteriol. Rev. **25:**237–240.

11. **Hirschowitz, B. I., C. A. Dasher, F. J. Whitt, and G. W. Cole.** 1980. Hepatitis B antigen and antibody and test of liver function. Am. J. Clin. Pathol. **73:**63–68.

12. **Holmes, M. B., D. L. Johnson, N. J. Furmara, and W. M. McCormack.** 1980. Acquisition of typhoid fever from proficency-testing specimens. N. Engl. J. Med. **303:**519–521.

13. **Hornick, R. B., S. E. Greisman, T. F. Woodward, H. L. DuPont, A. T. Dawkins, and M. T. Snyder.** 1970. Typhoid fever: pathogenesis and immunologic control. N. Engl. J. Med. **283:**686–692.

14. **Kenny, M. T., and F. L. Sabel.** 1968. Particle size distribution of *Serratia marcescens* aerosols created during common laboratory procedures and simulated laboratory accidents. Appl. Microbiol. **16:**1146–1150.

15. **Lo Grippo, G. A., and H. Hayoski.** 1973. Incidence of hepatitis and Australia antigenemia among laboratory workers. Health Lab. Sci. **10:**157–162.

16. **Moya, C. E., L. A. Guarda, and T. M. Sodemon.** 1980. Safety in the clinical laboratory, part 5: viral hepatitis. Lab. Med. **11:**589–590.

17. **Nottebart, H. C., Jr.** 1980. The law and infection control: nosocomial infections acquired by hospital employees. Infect. Control **1:**257–259.

18. **Pike, R. M.** 1976. Laboratory-associated infections: summary and analysis of 3,921 cases. Health Lab. Sci. **13:**105–114.

19. **Pike, R. M.** 1978. Past and present hazards of working with infectious agents. Arch. Pathol. Lab. Med. **102:**333–336.

20. **Pike, R. M.** 1979. Laboratory-associated infections: incidence, fatalities, cases, and prevention. Annu. Rev. Microbiol. **33:**41–66.

21. **Sulkin, S. E., and R. M. Pike.** 1951. Survey of laboratory-acquired infections. Am. J. Public Health **41:**769–781.

22. **U.S. Department of Health and Human Services.** 1984. Biosafety in microbiological and biomedical laboratories. HHS Publication no. (CDC) 84-8395. U.S. Government Printing Office, Washington, D.C.

23. **Wedum, A. G.** 1964. Laboratory safety in research with infectious aerosols. Public Health Rep. **79:**619–633.

Chapter 8

Human Factors in Microbiological Laboratory Accidents

G. BRIGGS PHILLIPS

Certain behavior patterns, attitudes, and perceptions of individuals are related to their involvement in accidents. Because of this, those interested in laboratory safety should be familiar with the role of human factors in the cause and prevention of accidents. This chapter will discuss various characteristics that, in at least one study, were more prevalent in accident-involved individuals than in those not having accidents.

It is useful for laboratory supervisors and workers to have some knowledge of the human factors that lead to, or avert, accidents. This understanding, in turn, can be helpful in accident prevention programs. Obviously, behavior patterns, attitudes, and perceptions are considered in the hiring of new employees. The information in this chapter may also be useful in organizing and presenting employee safety training programs. Finally, this information may be of help to laboratory managers who have responsibility for determining the causes of accidents and designing preventive measures.

BACKGROUND

The study reported here paired a group of accident-involved individuals with a matched accident-free group during a 2-year period. The individuals in both groups worked with infectious microorganisms in a large research laboratory. The study involved the examination of employee personnel and safety records and structured interviews with each individual; the similarities and differences between the two groups were then evaluated.

RESULTS

General

After appropriate pilot studies to establish and validate an interview outline and other study methods, 33 individuals involved in laboratory accidents and 33 not involved during the same 2-year period were selected. Selection was from a total laboratory population of several thousand people, but the two groups were matched by job classification. The number of individuals in each job category was in the same proportion as in the total laboratory population (Table 1). The results of the study are presented in four categories: (i) similarities between groups; (ii) differences between groups; (iii) causes of accidents in the accident-involved group; and (iv) comments by the accident-free group.

Similarities between groups

No differences between the accident-involved (AI) and accident-free (AF) individuals was detected for a number of characteristics, although these characteristics were not known or considered during the matching process. There were no significant differences in mean age (38 years), weight (170 lb), height (68.6 in.), length of employment (10.8 years), or amount of formal education (11.5 years). Likewise, each group had accumulated the same average amount of sick leave (60 days) and vacation time (30 days). There were 32 men and 1 woman in the AI group and 31 men and 2 women in the AF group.

There were no differences in the physical condition of the employees as judged from their use of prosthetic or corrective devices, from their statements as to the presence of constitutional disease or other symptoms and ailments, or from their statements as to the use of insulin, benzedrine, tranquilizers, and other drugs. Seventeen members of the AI group and 16 members of the AF group wore eyeglasses. The length of time since the last illness requiring a doctor's care and the last physical examination was the same in the two groups. The members of the two groups also did not differ with respect to living arrangements. The hobby and recreational activities of the two groups were the same.

No differences between the groups were detected with regard to off-the-job accident records and driving records. Six individuals (18%) in each group had had lost-time, off-the-job, non-motor vehicle accidents. (These were mostly falls resulting in broken limbs.) There were also no significant differences between the numbers who had had motor vehicle accidents and moving traffic violations. There was little difference in the total number of accidents and violations for the two groups. The seriousness of the accidents and violations was not different for the two groups. When points were assigned to the violations according to state traffic laws, the total points for

TABLE 1. Characteristics of the AI and AF groups

Job classification	No. of AI people	No. of AF people
Trained scientific personnel	5	5
Laboratory technical assistants	22	22
Animal caretakers	6	6

TABLE 2. Smoking and drinking habits

Item	AI group	AF group
Do you smoke?		
Yes	29	19
No	4	14
Chi square		11.196[a]
Do you drink alcoholic beverages?		
Yes	27	23
No	6	10
Chi square		4.288[a]

[a]At df = 1 and at the 0.05 level of significance, the hypothesis of equal frequencies is rejected.

each group were the same. One individual in each group had been the driver of an automobile involved in a fatal traffic accident.

There were no significant differences in the responses of the two groups to a number of "opinion" questions about safety. These dealt with the necessity of having safety programs, regulations, reporting systems, etc., and their influence in providing a more accident-free working environment. Both the AF and AI individuals agreed to this concept.

Each individual was asked to discuss immediate steps that could be taken in his laboratory to improve safety. He was also asked to discuss what additional things he would do if he possessed unlimited funds and authority. No differences were noted between the responses of the two groups.

From the above results, it is evident that the two groups shared many physical, cultural, and opinion characteristics, and that observing such characteristics would have been of little help in distinguishing between the two groups.

Differences between groups

The fact that subjects in the two groups possessed different characteristics or reacted differently to interview questions does not, of course, identify accident causes. However, differences may be suggestive of human elements that are associated with AI persons to a greater extent than with AF persons. Such characteristics vary in their significance, and subjective evaluation of their importance is necessary.

Although no group differences were detected in health status, there were significant differences in smoking and drinking habits. The AF group was composed of a significantly greater number of individuals who stated that they neither smoked nor drank alcohol (Table 2). There were four divorced individuals in the AI group, but none in the AF group. About the same number of people in each group were married and had children, but the mean number of children per parent was significantly higher in the AF group. On the average, each married AF individual who had children had one child more than the married parent in the AI group (Table 3).

Table 4 shows that differences in accident experience had existed within the two groups before the two-year test period. Because there were no significant differences in job classification and length of service, the differences were not due to different amounts of laboratory exposure. The dichotomy in accident experience is further illustrated by comparison of non-lost-time accidents reported by individuals in the two groups before the test period (Table 5). The AI group had had almost twice as many minor accidents as the AF group.

Compared to the AI group, AF individuals were more critical or conservative in rating supervisors. Employees were asked to estimate how frequently supervisors were fair to all persons, assumed their responsibilities, kept promises, kept employees informed, and gave credit where credit was due. The answers revealed that AI people as a group gave

TABLE 3. Marital status

Item	No. of AI people	No. of AF people
Married	27	29
Married, with children	24	25
Single	2	4
Divorced	4	0
Mean no. of children per parent	2.38	3.36
Standard deviation	1.06	1.68
Standard error of the difference between means	0.40	
t	2.450[a]	

[a]At df = 47 and at the 0.05 level of significance, the hypothesis of equal means is rejected.

TABLE 4. Lost-time accident records before the
2-year period studied

Previous lost-time accidents	AI group	AF group
Laboratory injuries		
No	28	33
Yes	5	0
Laboratory infections		
No	21	30
Yes	12	3
Chi square		29.700[a]

[a]At df = 1 and at the 0.05 level of significance the
hypothesis of equal frequencies is rejected.

TABLE 5. Non-lost-time accidents before the 2-year period
studied

Accident type	No. of accidents	
	AI group	AF group
Lacerations	40	24
Contusions	13	9
Exposures to infectious agents	11	1
Burns: hot liquids, steam, or UV	10	6
Strains and sprains	8	3
Eye injuries	7	5
Animal bites	4	4
Chemical splashes	3	1
Exposures to toxic fumes	2	1

higher ratings to supervisors than did the AF people
(Table 6). Because in almost all cases the paired
individuals were rating the same supervisor, the AF
individuals were more critical in their evaluation.

With regard to opinions about the safety consciousness of co-workers, the evaluations by the AF
group also were more conservative than those of the
AI group (Table 7). The AF group did not assume
that the safety habits of their co-workers were as high
as that assumed by the AI group. This could have a
relationship to the ability of a person to remain
accident free. In the same way that a good driver is a
defensive driver and does not assume that others will
always drive safely, a defensive attitude on the part
of a laboratory worker is a commendable quality.
Because he does not assume that his co-workers will
always perform safely, the good laboratory worker
will tend to avoid involvement in accident situations
created by others.

The AF subjects were more conservative in their
rating of working conditions on the job (Table 8).
Although none of the respondents graded conditions
as unsatisfactory, the AF people were more reluctant

than AI people to rate conditions as excellent or very
good. This indicates that their standards for a safe
working environment were higher.

Compared with the AI group, the AF group placed
greater importance on proper attitudes toward safety
(Table 9). A significantly larger proportion of the AF
group realized that an attitude reflecting a desire to
work without injury or infection was more important
than the provision of safety equipment and safety
techniques.

Respondents in each group were asked to rank the
importance of a number of safety techniques (Table
10). The AI group rated "personal experience from
previous accidents" as the second most important
factor in stimulating their thinking about safety. The
present and past accident experience of this group,
however, belies the value of this belief. Members of
the AF group, on the other hand, gave personal
experience a low rating relative to the other factors
listed.

Causes of accidents in the accident-involved group

Information was collected about the causes of
accidents occurring to the AI subjects. Each employee was asked to discuss what he felt caused his
accident. The primary cause was listed, and then
each subject was questioned regarding other possible

TABLE 6. Rating of supervisors

Does your supervisor fulfill his obligations?	% Responding	
	AI group	AF group
Always	55.2	38.2
Usually	38.8	49.1
Sometimes	3.6	7.9
Seldom	2.4	4.8
Chi square		21.781[a]

[a]At df = 3 and at the 0.05 level of significance, the
hypothesis of equal frequencies is rejected.

TABLE 7. Rating of co-workers

Do you feel your co-workers are safety conscious on their jobs?	AI group	AF group
Always	14	7
Usually	17	21
Sometimes or rarely	2	5
Chi square		9.562[a]

[a]At df = 2 and at the 0.05 level of significance, the
hypothesis of equal frequencies is rejected.

causes. Statements of principal cause were evaluated and grouped (Table 11).

Each subject was also questioned about the possible prevention of the accident. To the question, "Did you foresee that the accident was going to happen?," 2 answered, "Yes" and 31 answered, "No." The further question, "Would a different reaction on your part have prevented the accident or reduced its severity?," brought 11 "Yes" and 22 "No" responses. The discussions elicited the further information that AI individuals often lacked the ability to perceive accident situations in time to prevent them and that the inflexibility of techniques and working habits in the face of an impending accident may sometimes be a problem.

Situations that led to accidents which could have been prevented by different reactions were: (i) failed to hold animal securely for inoculation; continued to inoculate the animal even though it was struggling; (ii) failed to get help to lift heavy equipment; (iii) failed to wear available gloves to pick up broken glassware; (iv) placed a contaminated syringe in a hazardous location; noticed its hazardous location several times but failed to move it; (v) failed to move a long extension cord to a safe place; tripped on the cord several times before the accident occurred; (vi) removed safety glasses.

Discussions with the AI subjects also revealed that being in a hurry or working at an abnormal rate of

TABLE 8. Evaluation of working conditions

Rating of working conditions	AI group	AF group
Excellent or very good	23	16
Satisfactory	10	17
Chi square		5.126[a]

[a] At df = 1 and at the 0.05 level of significance, the hypothesis of equal frequencies is rejected.

TABLE 9. Opinions on attitudes and equipment

Item	AI group	AF group
Which is more important?		
Proper attitudes toward safety	20	27
Techniques and equipment for safety	13	6
Chi square		8.607[a]

[a] At df = 1 and at the 0.05 level of significance, the hypothesis of equal frequencies is rejected.

TABLE 10. Ranking of safety program techniques

Technique	Rank[a]	
	AI group	AF group
Safety meetings	1	1
Personal experience from previous accidents	2	6
Formal training programs	3	2
Safety bulletins	4	3
Safety inspections	5	4
Safety regulations	6	7
Safety posters	7	5

[a] Ranked order correlation coefficient, $r^1 = 0.57$, $t = 1.568$.

TABLE 11. Causes of accidents by the AI group

Principal cause	No. (%) of subjects	
Unsafe act or personal failure	20	(60.6)
Equipment failure	5	(15.2)
Combined equipment and personal failure	7	(21.2)
Unknown (laboratory infection)	1	(3.0)

speed contributed to a significant proportion of the accidents. Twelve of 33 accidents (36%) occurred while the employee was working at an abnormal speed. A frequent reason given was the desire to finish the task before lunch or before the end of the work day. Four individuals also stated that distractions, such as noise or the movement of other individuals in the room, contributed to their accidents. Two persons felt that poor illumination was a cause, and one individual blamed a high room temperature. A violation of safety regulations occurred in 10 accidents (30%). However, none of the violators felt that the regulations were unreasonable or needed revision. This contradicts the hypothesis that the AI people tend to be unfamiliar with the safety regulations.

The studies with AI individuals included a comparison of the information collected from each person with information from accident reports and investigations. The accident records tended to overstate the frequency of equipment failures and understate the personal failures and unsafe acts.

Table 12 shows the types of accidents sustained by the 33 AI individuals. Six resulted in lost work time; four of these were laboratory infections. Fifteen of the accidents were classified as biological, 15 were industrial, and 3 were combined biological and industrial. The tasks being performed at the time of the accidents, in order of significance, were (i) performing diluting and plating operations; (ii) washing,

TABLE 12. Accidents sustained by the AI group

Accident type	No.	(%)
Lacerations and contusions	9	(27.3)
Spills or exposures to infectious		
materials	7	(21.2)
Laboratory infections	4	(12.1)
Burns	4	(12.1)
Accidental self-inoculation with		
syringe and needle	4	(12.1)
Animal bites	3	(9.1)
Chemical exposures	2	(6.1)

handling, or sterilizing glassware; (iii) performing aerobiological experiments; (iv) exposing, injecting, or performing autopsy on animals; (v) handling quantities of infectious cultures.

Accidents often occurred when individuals failed to realize that a hazardous situation was building up to a point where an accident was probable. The failure in some cases may not have been the inability to recognize the situation so much as it was the failure to take appropriate action. This can be termed "excessive risk taking," but it is also appropriately related to a lack of accident perception. One reason for failing to act upon recognizing a potential accident was the excessive work speed being maintained by 30% of the individuals. There also was a tendency to "take the risk," illustrated by the fact that eight individuals were aware that a safety regulation was being violated and two others were undecided as to whether or not a violation was involved. None of these ten individuals, however, believed that the regulations involved were unreasonable.

Thus, it can be concluded that a lack of ability to foresee an accident, reluctance or inability to take precautions in the face of a recognized potential accident, and a willingness to take a chance by violating a safety regulation were important human characteristics distinguishing the AI group.

Comments by the AF group

Interviews with AF persons included discussions about factors each individual felt had been important in preventing involvement in accidents, and personal philosophies associated with freedom from accidents.

Thirty-two of the 33 individuals felt that an accident-free work record was "something to be proud of." One individual felt such a record to be a job responsibility.

In ranking significant items in accident prevention, the AF group rated safety equipment, such as ventilated cabinets, as the most important. Next was the training and guidelines from their supervisors, followed by protective equipment, safety training and orientation lectures, and safety regulations. Each person was asked to consider whether any of a list of items had been of value to him personally. The most frequently mentioned item was safety support from top management.

In an attempt to gather data on how individuals had maintained an accident-free record, each person was asked to imagine that he had been chosen for promotion and that he had been asked to instruct and train his replacement on how to work safely in the job he was leaving. To obtain the best response, role-playing was employed for this question, with the investigator playing the part of the person to be trained.

The most important observation from this part of the study was the awareness of the involvement of human factors. In the role-playing there was little emphasis on physical or mechanical items in accident prevention. For example, no mention was made of checking glassware before it was used, using proper personnel protective devices, checking for adequate ventilation of safety cabinets, or using safety containers, pipetters, needle-locking syringes, etc. Instead, the individuals spoke about matters relating to the person himself, his required personal actions, and his feelings and attitudes. It appeared that almost every employee assumed that the proper equipment and physical barriers would be present and that it was the human element that required emphasis in the instruction of the new person.

More specifically, the most frequent specific comments concerned the need to have respect for infectious agents and an awareness of the hazards. Some individuals were able to amplify their statements with specific details. Awareness of hazards meant (i) knowing that techniques or operations presented hazards, (ii) not forgetting these as time goes on, and (iii) letting this awareness guide the way techniques and procedures are performed. Respect for the infectious agent was a recommendation not to "live in fear" but rather to display confidence in laboratory work by purposeful planning approaches using recommended and safe techniques.

Eighteen individuals stressed the need to follow safety regulations or standard operating procedures. Even those who did not mention regulations specifically assumed that they were to be followed.

A summary of the role-playing comments by the AF individuals is shown in Table 13. Factors considered most important for an accident-free existence were related to attitudes toward safety and control of human error in accidents. It was clear that most individuals were not concerned with punitive action that might result from not following the regulations, but tended, rather, to think of the hazards to be avoided by following the rules. This reaction on the part of the AF subjects can be expressed as respect for safety regulations and an understanding of the hazards the regulations attempt to control. It is interesting to contrast this attitude with that of the AI people, 30% of whom knowingly violated a regula-

TABLE 13. Summary of comments by AF individuals during role-playing

Nature of comment	No. of individuals
Follow safety regulations and standard operating procedures...............	18
Have respect for infectious agents and hazards..........................	16
Be aware of the hazards	10
Be safety conscious	7
Think and use common sense	6
Plan the job carefully.................	5
Be cautious	4
Follow supervisor's instructions	3
Rely on previous accident experience....	2
Have no fear of biological agents.......	2
Have fear of biological agents..........	2
Develop good attitudes................	2
Use proper equipment	1
Develop efficient techniques	1
Insist on good safety management	1

tion and thereby caused an accident. From this it is reasonable to conclude that the AI employees, as a group, had less respect for and confidence in the regulations than did the AF group.

CONCLUSIONS

Many human factors characterizing AF and AI individuals may not be different. However, where differences exist, they may be significant tools in the planning and conduct of laboratory safety programs. On the basis of the study presented here, one may draw a number of conclusions.

(i) Accident-involved people may have a low opinion of laboratory safety programs and efforts.

(ii) People who smoke may have more accidents than those who do not.

(iii) People who drink alcohol may have more accidents than those who do not.

(iv) Accidents may be more prevalent among persons who are divorced than among single or married individuals.

(v) Persons with larger families and with close family ties may be safer workers.

(vi) People who have been accident free for several years may often remain so for longer periods.

(vii) A person's opinion on the adequacy of safety conditions may be related to safety performance. Those more critical in evaluating the safety efficiency of people and conditions may experience fewer accidents.

(viii) The development of "defensive" work habits may help an employee to avoid involvement in accidents.

(ix) Accident-free people may often acknowledge that a proper attitude toward safety helps to reduce accident involvement.

(x) People who place excessive reliance on experience gained from previous accidents in avoiding later accidents may not be entirely correct.

(xi) Accident-free people tend to develop an accident-perception ability. Inability to recognize building accident situations may be a significant cause of laboratory accidents.

(xii) Unsafe acts and personal failures are important causes of laboratory accidents.

(xiii) Inflexibility of work habits, precluding last-minute modifications when an accident situation is recognized, may contribute to laboratory accidents.

(xiv) Working too fast may contribute to the likelihood of accidents.

(xv) Physical factors such as noise, illumination, and temperature do not appear to have a major effect on laboratory accidents.

(xvi) Accidents sometimes occur when people are consciously taking excessive risks.

Chapter 9

Transmission and Control of Laboratory-Acquired Hepatitis Infection

MARTIN S. FAVERO, NORMAN J. PETERSEN, AND WALTER W. BOND

It is well recognized that viral hepatitis is an occupational risk for personnel working in health care fields, especially those who are employed in clinical laboratories. Viral hepatitis type B is the most frequently reported laboratory-acquired infection, and persons who frequently handle blood or blood-contaminated items are at higher risk than the general population for acquiring infection (8, 12, 14, 19). Viral hepatitis type A, a disease transmitted by the fecal-oral route, does not appear to be an occupational risk for laboratorians but has been shown to occur with high frequency among animal handlers working in areas housing chimpanzees, especially in laboratories conducting hepatitis A infectivity studies in these animals (19). Although non-A, non-B hepatitis has not been reported as a major laboratory-acquired infection, its epidemiology appears to be very similar to that of type B hepatitis in that it is a blood-borne disease (11). Specific information on high-risk groups is not available because of the lack of a serologic test. We assume, however, that like hepatitis B, non-A, non-B hepatitis constitutes a significant risk for persons in clinical laboratories.

The overall strategy for the prevention and control of viral hepatitis in laboratories necessarily must be considered from the standpoint of viral hepatitis type B and blood precautions. This is not to imply that one should not use care in handling fecal specimens or contaminated materials, but rather that standard precautions described elsewhere in this book are sufficient to prevent transmission of hepatitis A.

The basic reservoir of hepatitis B virus (HBV) is the human chronic or transient carrier of hepatitis B surface antigen (HBsAg), and consequently, the principal source of infectious virions is plasma or serum. There is general agreement that a fecal-oral route of hepatitis B transmission does not exist (10, 13). Many body fluids, secretions, and excretions, including urine, feces, saliva, nasopharyngeal washings, breast milk, bile, semen, synovial fluid, sweat, tears, and cerebrospinal fluid, have been hypothesized as being infective because they contain detectable HBsAg. However, only saliva and semen have been shown to be infective (1). Blood is acknowledged to be the most consistently effective vehicle of HBV transmission. The efficiency of the various mechanisms of hepatitis B transmission with-

in the hospital and clinical laboratory environments is related significantly, if not solely, to the extraordinary amounts of circulating HBV in blood of infected individuals who are either in the acute phase of infection or chronic HBsAg carriers and positive for hepatitis B e antigen. It has been shown, for example, that human serum positive for HBsAg and hepatitis B e antigen can be diluted to 10^{-8} and still produce hepatitis B infection when injected into susceptible chimpanzees (22).

As a result of this high viral titer, blood can be diluted to such an extent that even in the absence of visible or chemically detectable blood, HBV can be present in relatively small inocula or on laboratory environmental surfaces. Furthermore, HBV can be present in a variety of secretions, excretions, and other body fluids that contain small amounts of serum under normal conditions or under conditions of trauma. With the variables of exposure and host susceptibility being constant, the efficiency and probability of disease transmission would be expected to be directly related to the quantity of HBV in a particular body fluid or on fomites.

ROUTES AND MECHANISMS OF HEPATITIS TRANSMISSION IN LABORATORIES

The following are modes of hepatitis (primarily type B) transmission that can occur in a variety of epidemiologic settings, including laboratories, and they are listed in the probable order of efficiency of disease transmission.

Direct contact

(i) Direct percutaneous inoculation by needle of contaminated serum or plasma such as occurs by transfusion of contaminated blood or blood products and by accidental needle sticks

(ii) Percutaneous transfer of infective serum or plasma in the absence of overt needle puncture, such as contamination of minute cutaneous scratches, abrasions, burns, or any other lesions during laboratory manipulations

(iii) Contamination of mucosal surfaces by infective serum or plasma or infective secretions or excretions, such as occurs with accidents associated with mouth pipetting, label licking, splashes or spatterings, or other means of skin-to-mouth or skin-to-eye contact

Indirect contact

Hepatitis transmission, especially type B hepatitis, can occur by indirect means via common environmental surfaces in a laboratory, such as test tubes, laboratory benches, laboratory accessories, and other surfaces contaminated with infective blood, serum, secretions, or excretions which can be transferred to the skin or mucous membranes. The probability of disease transmission with a single exposure of this type may be remote, but the frequency of such exposures makes this mechanism of transmission potentially an efficient one over a long period of time. Activities in laboratories such as nail biting, smoking, eating, and a variety of hand-to-nose, -mouth, and -eye actions contribute to indirect transmission.

Fecal-oral transmission

The fecal-oral route does not appear to be an efficient mode of hepatitis B transmission. Therefore, handling feces does not constitute the same degree of hazard as handling blood specimens. HBV-contaminated materials can initiate infection when they enter the mouth, but the precise route of infection is associated with entering the host's vascular system via mucosal surfaces rather than via an intestinal route. This type of transmission is not enteric but rather peroral transmission.

Patients infected with hepatitis A virus (HAV) do not have a viremic stage that is extensive enough to pose a problem of disease transmission through direct contact with blood. Transmission of type A hepatitis can occur by direct or indirect contact via the fecal-oral route. The degree of conservatism practiced in strategies for handling feces or fecally contaminated specimens is controlled by a number of factors. First, the carrier state of HAV in stool has never been demonstrated (7). Second, mean levels of HAV in stool of infected patients occur during the incubation phase of the disease, and by the time a patient is hospitalized, the level of HAV in feces has been significantly reduced. Because HAV may be present in low numbers, stools being examined in laboratories, as well as contaminated specimens or objects, should be considered infective in the sense that if a sufficient amount of feces is introduced into susceptible human hosts, HAV infection would be expected to occur. However, the efficiency of HAV infection is considered to be positively correlated to high levels of HAV in stool. Mosley (15), for example, has pointed out that maximum (high-titer) HAV excretion in feces is equivalent, epidemiologically, to

optimum disease communicability. Third, patients' stools, except in highly specialized research laboratories, are not routinely examined for the presence of HAV since clinical diagnosis of the disease is based on serologic tests for antibodies to HAV. Consequently, a practical risk of disease transmission would be associated with those fecal specimens being examined microbiologically which are from patients hospitalized for other reasons, but who also may be in the incubation phase of type A hepatitis infection. This situation would be more likely for pediatric patients.

Nonetheless, the precautions for handling fecal specimens and contaminated items practiced in laboratories should be sufficient to prevent acquisition of type A hepatitis by laboratorians. Laboratory personnel conducting infectivity studies of type A hepatitis utilizing chimpanzees or marmosets, or caring for these animals, should be aware that handling of feces and cage-cleaning activities can create situations with the probability of fecal-oral transmission. These situations involve not only direct contact with infective feces and other excretions but also the creation of droplets which ultimately can enter the mouth and be ingested.

Airborne transmission

Although hepatitis B transmission by means of the airborne route has been hypothesized, it has not been documented. We have done two studies that show that true airborne transmission of hepatitis B from infective blood or saliva is not likely. In one study (18), a filter-rinse technique capable of detecting low levels of aerosolized airborne HBsAg was devised and evaluated. Laboratory tests showed the procedure had an efficiency of 22% and was capable of detecting as little as 5×10^{-5} ml of aerosolized HBsAg-positive serum in a single air sample. The technique was used in a hemodialysis center which served a patient population with a high prevalence of HBsAg seropositivity and in which conditions favored the production of aerosols. Samples were collected from the patient area during treatment, from the laboratory area while blood samples were being processed, and from the area where reusable dialyzers were disassembled and cleaned. HBsAg was not detected in any of 60 air samples collected. On the other hand, random swab-rinse samples of surfaces in and around hemodialysis centers showed that 15% of them, particularly ones frequently touched, were positive for HBsAg (4). In another study (17), air samples were collected from the dental operatory of an institution for the mentally retarded where residents had a high incidence of HBsAg seropositivity. Although gingival swab samples from nearly all of the patients showed the presence of HBsAg, air samples collected during procedures of scaling, extraction, high-speed drilling, and other procedures favoring aerosol production were uniformly negative for HBsAg. True aerosols, i.e., particles less than 100 μm in diameter, of blood, saliva, or mouth washings did not appear to be efficiently produced by these procedures.

It should be pointed out, however, that events such as splashing, centrifuge accidents, or removal of rubber stoppers from tubes can account for disease transmission by means of large-droplet transfer into the mouth, eyes, minor cuts, or scratches or onto abraded skin. This is not airborne transmission, but rather transmission by direct droplet contact.

SPECIFIC PRECAUTIONS FOR PREVENTING LABORATORY-ACQUIRED VIRAL HEPATITIS

The primary overall strategy that should be used in preventing laboratory-acquired viral hepatitis is the consistent practice of blood precautions. This is true not only in laboratories where blood, serum, and other specimens are processed from patients who are known to be infected with hepatitis, but in all clinical laboratories involved in the chemical, hematologic, and microbiologic assay of blood and blood products. The prevalence of HBsAg in serum of patients whose blood is being assayed varies with the population being served by the laboratory. However, one can assume that this prevalence can be as high as 1% of hospital admissions, and with specific high-risk groups such as hemodialysis patients, parenteral drug abusers, or male homosexuals the HBsAg carriage rate may be significantly higher. Laboratories that process hundreds to thousands of blood samples per day will no doubt handle a certain number of blood specimens that contain HBV but are not labeled as such. It is the practice of many hospitals to prominently identify specimens and specimen containers with a ''hepatitis'' label when patients are known to be hospitalized with hepatitis. Unfortunately, two methods of handling blood are sometimes practiced: very careful ones, with those tubes labeled ''hepatitis,'' and fairly lax ones with unlabeled specimens from other patients. It is emphasized that blood precautions should be employed at all times in clinical laboratories.

The following guidelines for the prevention of laboratory-acquired viral hepatitis are also applicable in routine laboratory practices to limit the acquisition of other infectious diseases.

Safety officer

The responsibility for laboratory safety resides ultimately with the director of the laboratory; however, from an operational standpoint a safety officer who is familiar with laboratory practices and biohazards should be appointed from among the laboratory staff. The safety officer should be responsible for giving advice and consultation to the laboratory staff in matters of biohazards; instructing new members in safety procedures; procuring protective equipment, disposal bags, and disinfectants; developing and maintaining a laboratory accident reporting system; periodically reviewing and updating safety procedures; and monitoring serologic surveillance data for the laboratory staff if such data are collected. In hospital laboratories, the laboratory safety officer should consult with and report new information to the hospital's infection control practitioner. Furthermore, when appropriate, reports should be made to the hospital's infection control committee, as well as to local and state health departments.

Reporting of accidents

Accidents such as cuts, needle sticks, and skin abrasions with instruments possibly contaminated with blood, soiling of broken skin, or contamination of the eyes or mouth must be reported promptly to the safety officer, who should maintain records and make sure that proper medical consultation and treatment, if necessary, are available. Prompt reporting is imperative because the efficacy of prophylactic post-exposure immune globulins is greatest when administered immediately after accidental exposure occurs. Spills of high-risk specimens such as documented or suspected HBsAg-positive blood, even if not associated with personnel contamination, should also be reported to the safety officer.

Hand washing

Frequent hand washing by laboratorians is an important safety precaution that should be practiced after contact with specimens and laboratory procedures, especially those associated with blood or blood products. Hands should always be washed before eating, drinking, or smoking and after completing analytical work. Frequent hand washing should be done even if gloves are used. Hand-washing facilities should be conveniently located for frequent use. A hand-washing product that is widely acceptable to personnel is desirable. Liquid soap, granule soap, or soap-impregnated tissues can be used. Ideally, sinks should be equipped with foot- or knee-operated faucets.

Gloves

All laboratorians who have direct or indirect contact with feces or articles contaminated with blood or feces should wear gloves. Wearing of gloves is obviously for the protection of the laboratorian. It should be realized that gloves themselves can become contaminated, so surfaces such as telephones, door knobs, laboratory equipment, etc., frequently touched by other laboratorians should not be touched with contaminated gloves. Disposable gloves are preferred because they can be changed frequently and it is not necessary for laboratorians to wear sterile gloves when performing laboratory activities.

Protective clothing

Every staff member should wear a gown with a closed front or a coat with an overlapping front when in any laboratory area. A disposable plastic apron also may be desirable. Disposable gloves should be used when opening or processing specimens. Gowns, aprons, and gloves must be removed and hands must be washed before staff members leave the laboratory for any purpose. Disposable gloves and aprons should be worn only once and then placed in an impervious bag for safe disposal. Gowns and coats should be put in a laundry bag at the end of each appropriate period of work. If a gown or coat is accidentally contaminated, it should be discarded in the laundry bag at once and a fresh one should be obtained. A face shield or safety goggles and mask should be worn when it is anticipated that there is a potential for blood and other types of specimens to be spattered.

Personal hygiene

Smoking, eating, and drinking beverages in the laboratory should be prohibited. Employees working in technical and analytical areas should be given an opportunity for regular breaks in accordance with laboratory policy, and there should be a lounge area where employees can have these breaks and eat lunch. Care should be taken not to put fingers, pencils, or other objects into the mouth, specimen tube labels must not be licked, and hands should be washed after every procedure in which they may have become contaminated.

Pipetting

Mouth pipetting should be prohibited; automatic pipettes with disposable plastic tips are recommended. If disposable pipette tips are used, they should be discarded after pipetting each specimen; e.g., they should not be rinsed several times in water between each specimen. Other pipettes should be used with rubber bulbs or an automatic suction device, and fluids should never be drawn up to the top of the pipette. The contents of the pipette should be expelled gently down the wall of the receptacle to avoid splashing, and pipettes should be held in a near vertical position while in use. Contaminated pipettes must not be placed on the laboratory bench; they should be placed gently in a flat discard pan and later completely submerged in disinfectant. Any rubber bulb that may have become contaminated internally during use should be placed in disinfectant and subsequently discarded.

Receipt of specimens

Incoming blood specimens should be received in a designated area of the laboratory and examined closely to be sure they have been properly closed and packed. Soiled or leaking containers should be brought to the attention of the safety officer to decide whether or not they should be autoclaved and discarded without being unpacked. When handling blood specimens, receiving technicians should wear disposable gloves and should open the specimen container slowly to reduce the possibility of spatter.

Labeling, processing, and storage of blood tubes

Blood, serum, and biologic specimens from patients who are known to be infected with hepatitis or for whom HBsAg or serum enzyme tests are being performed should be identified with special labels marked "hepatitis." It should be emphasized, however, that precautions employed by laboratorians when handling these types of high-risk specimens should not be ignored when handling specimens that are not labeled as such. Because blood tubes can be contaminated on the outside (6) as well as contain infective blood, they must be handled and stored with care. If blood tubes must be refrigerated, they should be capped and placed in a designated refrigerator or in a designated portion of a refrigerator. Blood tubes should never be kept in a refrigerator that contains food or beverages. If blood or serum is held in frozen storage, it should be placed in a container designed for low-temperature storage. Organized storage of these containers is desirable to prevent unnecessary handling of specimens.

Needles and syringes

Special precautions should be taken with needles and syringes contaminated with blood, especially when specimens are from hepatitis patients. Disposable needles and syringes should be used and should be discarded after a single use. Used needles should not be recapped; they should be placed in permanently labeled, leak- and puncture-resistant containers designed for this purpose. Needle nippers should not be used, and needles should not be purposely bent or broken by hand because accidental needle punctures can occur. Used syringes should be placed in a leak-resistant bag, and the containers of both needles and syringes should be either incinerated or autoclaved before being discarded. Reusable syringes, if used, should be rinsed thoroughly with cold water in a single designated sink by a person wearing gloves and should be wrapped using the double-bag technique. The outer bag should have a "hepatitis" label if the specimen is from a hepatitis patient, and the syringes should remain bagged until they have been decontaminated or sterilized.

Disposal of waste specimens and contaminated material

As mentioned previously, all blood and most biologic specimens from humans and from those primates involved in hepatitis research must be viewed as potentially infective. Accordingly, when these materials become waste, they must be disposed of in a safe manner. Each laboratory should have special receptacles for these wastes. Preferably, the wastes in these receptacles should be autoclaved in the laboratory or transported in impervious double bags to an autoclave or approved incinerator in another area. When it is not feasible to autoclave blood or other potentially infectious fluids, they may be poured down a single designated sink drain. Gloves should be worn during this procedure, and care should be taken to prevent splashing onto the walls of the sink. After a sink drain is used for fluid disposal, it should be thoroughly flushed with water (minimizing splashing) and the sink and drain should be flushed with an appropriate amount of a liquid disinfectant such as sodium hypochlorite.

Centrifuging

Specimens containing blood should be centrifuged with the tubes tightly capped. If a tube breaks in the centrifuge, the bucket containing the spilled blood and broken glass should be placed carefully in a pan of disinfectant. The surfaces of the centrifuge head, bowl, trunions, and remaining buckets should be swabbed with an appropriate disinfectant; alternatively, the trunions and buckets can be autoclaved. Microhematocrit centrifuges and bloodbank serofuges should be cleaned daily with a disinfectant. The top of the centrifuge should always be closed when the unit is in operation, and the top should not be opened until the unit has come to a complete standstill. Centrifuge tubes to be used in an anglehead centrifuge must never be filled to the point that liquid is in contact with the lip of the tube when it is placed in the rotor. When the tube lip becomes wet, liquid will be forced past the cap seal and over the outside of the tube.

Automated bioassay equipment

Automated equipment capable of performing a number of biochemical assays simultaneously is becoming commonplace in clinical laboratories. Although there are a variety of configurations of these types of equipment, we have not observed any which, because of their external and internal design, constitute significant risks of hepatitis transmission. Rather, the potential risks of hepatitis transmission are associated with procedures involving the handling, preparation, and delivery of specimens to the automated equipment. Tubes must not be packed too tightly, they must be capped up until the point of

insertion into the automated system, and blood-contaminated plastic tubing should be cleaned periodically during a day's work. Gloves should be worn by operators of this equipment at all times. If blood or serum ultimately is collected in a reservoir which is not piped into the sewer system, the contents of the reservoir should be autoclaved before disposal. If this is not feasible, the contents should be poured into a designated sink drain as described above for fluid wastes.

Care of laboratory bench tops

Each working area should be supplied with a wash bottle containing an appropriate disinfectant. The disinfectant solution should be mixed and renewed according to the directions on the manufacturer's label, and the bench surface must be cleaned and wiped with disinfectant at the beginning and end of each day or more frequently as spills or contamination occurs. Since accidents and errors are most likely to happen when the laboratory work area is crowded with equipment and materials, care should be taken to keep the laboratory work area tidy. Tubes and other containers should be placed only in the appropriate rack or tray, never directly on the bench. Disposable, absorbent, plastic-backed pads can be used to protect laboratory bench tops if spattering or spills are common or are anticipated.

Purification or concentration of infectious material

In some laboratories actively engaged in hepatitis research, biophysical and biochemical procedures are often used to concentrate or purify infective materials. These procedures increase the biohazard associated with these materials, and the safety practices and precautions that have been mentioned may not provide adequate protection. For example, although the hepatitis viruses collectively are designated as Biosafety Level 2 agents (23) for routine diagnostic procedures, they should be considered Biosafety Level 3 agents when purified, concentrated, or prepared in large volumes, and all laboratory procedures should be performed in a Class II biological safety cabinet (16). Also, it would be prudent to conduct potentially hazardous procedures, such as homogenization of infectious stool and liver preparations, in a biological safety cabinet even though purification and concentrations have not yet been performed. Ultracentrifuges used for procedures involving hepatitis viruses should be installed in a containment system that provides for the adequate single-pass-filtered exhaust of any material that might be released in the event of an accident.

Postmortem examinations

As elsewhere in the hospital environment, the number of carriers of HBV scheduled for autopsy will be larger than those identified by clinical serologic testing. Consequently, all autopsies should be conducted with the same precautions as would be observed with a known hepatitis patient. In general these precautions would include the wearing of gloves and protective clothing at all times. If procedures used can cause spattering of blood, feces, or other material, the use of face shields or eye goggles and masks should be considered. Appropriate disinfectants should be used routinely after each postmortem examination to disinfect those items or surfaces that become contaminated but cannot be autoclaved.

Animal areas

In laboratories where chimpanzees or marmosets are housed or are being used for hepatitis infectivity studies, both blood and enteric precautions should be practiced. Personnel, when performing ordinary types of animal manipulation such as feeding, injections, etc., should wear protective clothing, including gowns, gloves, caps, masks, and shoe covers. For cage cleaning, face shields or masks and eye protection are recommended. Areas should be cleaned daily with a detergent-disinfectant. An appropriate system should be designed for the disposal of feces, urine, and other wastes. In the event that soiling by animal waste occurs outside the immediate vicinity of the cages, the contaminated area should be cleaned immediately and then disinfected with an appropriate high-level disinfectant.

IMMUNE PROPHYLAXIS

The Advisory Committee on Immunization Practices (ACIP) of the Public Health Service periodically publishes recommendations and strategies for use of the hepatitis B vaccine as well as immune globulins for protection against several types of viral hepatitis (20, 21). These recommendations are updated periodically in the Centers for Disease Control's *Morbidity and Mortality Weekly Report*.

The HBV vaccine licensed for use in 1982 is currently available for both preexposure and postexposure prevention of hepatitis B. The vaccine is given in a three-dose series (initially and 1 and 6 months later) of 20 µg per dose for adults. The vaccine is over 90% effective in preventing HBV infection, has minimal side effects, and is recommended for preexposure prevention of hepatitis B. Laboratorians, especially those who frequently handle human or primate blood, are at high to moderate risk of hepatitis B infection and should consider receiving this vaccine.

Immune globulins are sterile solutions of antibodies (immunoglobulins) prepared from large pools of human plasma. Immune globulin (formerly referred to as "immune serum globulin" or "gamma globulin") produced in the United States contains antibodies against HAV and HBV. Tests of immune globulin since 1977 indicate that anti-HAV and anti-HBsAg have uniformly been present at stable titers. For example, all lots tested since that time have contained an anti-HBsAg titer of at least 1:100 by radioimmunoassay. Hepatitis B immune globulin (HBIG) is prepared from plasma preselected for high titer of anti-HBsAg and in the United States has an anti-HBsAg titer greater than 1:100,000 by radioimmunoassay. The price of a dose of HBIG is approximately 20 times that of a dose of immune globulin.

In the context of the hospital and laboratory environments, immune prophylaxis for protection against HAV infection usually is not warranted. In the event of a known exposure to HAV, such as accidental ingestion of fecal material known to contain HAV, a single intramuscular dose of 5 ml of immune globulin can be given. Preexposure prophylaxis is not recommended, but certain institutions that manage primates used in experimental HAV infection studies administer immune globulin periodically to animal handlers who are negative for anti-HAV. In this instance, 5 ml intramuscularly should be given every 5 to 6 months.

Postexposure prophylaxis to prevent HBV infection in the laboratory environment should be considered in the event of accidental percutaneous or permucosal exposure to blood known or thought to contain HBV. The decision to provide prophylaxis must take into account several factors: (i) whether the source of blood is known or unknown, (ii) whether the HBsAg status of the source is known or unknown, and (iii) the anti-HBsAg or HBV vaccination status of the exposed person.

Quite often exposures occur in laboratorians who have frequent contact with blood and who would be considered candidates for the HBV vaccine. If the vaccine has not been given previously, vaccination should be considered and given if warranted. HBIG and HBV vaccine can be administered simultaneously as long as they are administered at different injection sites.

The following summarizes prophylaxes for percutaneous or permucosal exposure according to the source of exposure. It is emphasized that for greatest effectiveness, prophylaxis should be given as soon as possible.

Source known, HBsAg status positive

For percutaneous (needle stick), ocular, or mucous membrane exposure to blood known to contain HBsAg, a single dose of HBIG (5.0 ml) should be given intramuscularly as soon as possible after exposure and within 24 h. HBV vaccine, 1 ml (20 µg), can be given intramuscularly at a second site as soon as possible but within 7 days of exposure, and the second and third doses should be given 1 and 6 months, respectively, after the first. If HBIG is unavailable, immune globulin may be given in an equivalent dosage. For persons who choose not to receive the HBV vaccine, a second dose of HBIG should be given one month after the first dose.

Source known, HBsAg status unknown

Two decisions are involved here: whether to test the source for HBsAg and what type of prophylaxis to administer. These decisions relate both to the relative probability that the source will be HBsAg positive and to the inherent delay in testing. The following guidelines are suggested.

(i) **High risk that the source is HBsAg positive.** Examples of high-risk situations include patients with acute, unconfirmed viral hepatitis; patients institutionalized with Down's Syndrome; hemodialysis patients; persons of Asian origin; homosexual men; and users of illicit intravenous drugs. If HBsAg test results can be known within 7 days of exposure, immune globulin (5.0 ml) should be given immediately (within 24 h). If test results are positive, HBIG (5 ml) should be given along with the first dose of HBV vaccine (at separate sites). The second and third doses of vaccine should be given 1 and 6 months later, respectively. If the result of HBsAg testing cannot be known within 7 days of exposure, the decision to provide prophylaxis must be based upon the epidemiologic characteristics of exposure, keeping in mind the importance of characterizing the source and providing prophylaxis as soon after exposure as possible. In this circumstance, treatment with the HBV vaccine will provide some protection and is preferable to treatment with HBIG alone.

(ii) **Low risk that the source is HBsAg positive.** The average hospitalized patient is unlikely to be HBsAg positive. In such cases prophylaxis is optional; HBsAg testing is not recommended. If an immune globulin is to be used, it should be given promptly (5 ml; within 24 h). Initiation of the HBV vaccine series should be considered.

Source unknown, HBsAg status unknown

Prophylaxis is optional. If it is to be used, immune globulin (5.0 ml) should be given as soon as possible, within 24 h. Initiation of the HBV vaccine series should be considered.

Currently, the value of phophylaxis with immune globulins against non-A, non-B hepatitis is not known. For accidental percutaneous or permucosal exposure to blood from a patient with non-A, non-B hepatitis, it may be reasonable to administer immune globulin (5.0 ml) as soon as possible after exposure.

DISINFECTION, STERILIZATION, AND DECONTAMINATION

The human hepatitis viruses have been difficult to study because some, such as HBV and the non-A, non-B viruses, have not yet been cultured in the laboratory. Comparative virucidal testing for the most part has not been performed as it has been for other types of viruses that can be conveniently cultured and tested in the laboratory. However, there is no evidence that any of these viruses are unusually resistant to physical or chemical agents, and it has been shown that HBV can be inactivated by several intermediate- to high-level disinfectants, including two glutaraldehyde-based formulations, 500 ppm free available chlorine, an iodophor hard-surface disinfectant, and 70% isopropanol (2). Although the human hepatitis viruses other than HBV may be comparatively more resistant to a variety of physical or chemical agents than most viruses, it is not reasonable to assume that they can approach the resistance levels of bacterial endospores. We therefore propose that the resistance of human hepatitis viruses that have not been studied in great detail (e.g., HAV and non-A, non-B viruses) be considered somewhat greater than that of the tubercle bacillus but much less than that of bacterial spores. The following recommendations for sterilization or disinfection are based on this proposed resistance level.

Conventional sterilization processes such as steam autoclaving or gaseous ethylene oxide can be relied upon effectively to inactivate hepatitis viruses. When one is faced with environmental and laboratory contamination, we recommend chemical formulations, concentrations, and contact times capable of producing at least an intermediate level of disinfectant action (Table 1); there are a number of chemical germicides which, if used correctly, can be considered effective for the inactivation of the hepatitis viruses. These disinfectants are used primarily for decontaminating spills of known HBsAg-positive blood. They are *not* recommended for routine housekeeping purposes.

In some high-risk areas, such as laboratories, hemodialysis units, and other health care environments, one is confronted with the problem of decontaminating large and small blood spills, patient care equipment that becomes contaminated with blood, and frequently touched instrument surfaces such as control knobs which may play a role in environmentally transmitted hepatitis. The strategies for applying the principles of HBV inactivation vary according to the item or surface being considered, its potential role in the risk of HBV transmission, and, to a certain extent, the thermal and chemical lability of the surface or instrument. For example, if a spill of known HBsAg-positive blood occurred on the floor or a counter top in a laboratory, the objective of the procedure to inactivate HBV would be one of decontamination or disinfection and not sterilization. Consequently, in such a situation we would recommend that gloves be worn and the blood spill be absorbed with disposable towels. The spill site should be cleaned of all visible blood, and then the area should be wiped down with clean towels soaked

TABLE 1. Some physical and chemical methods for inactivating hepatitis viruses[a]

Class	Concn or level	Activity
Sterilization		
Heat		
Moist heat (steam under pressure)	250°F (121°C), 15 min	
	Prevacuum cycle, 270°F (132°C), 5 min	
Dry heat	170°C, 1 h	
	160°C, 2 h	
	121°C, 16 h or longer	
Gas		
Ethylene oxide	450 to 500 mg/liter, 55 to 60°C	
Liquid[b]		
Glutaraldehyde, aqueous	2%	
Hydrogen peroxide, stabilized	6 to 10%	
Formaldehyde, aqueous	8 to 12%	
Disinfection		
Heat		
Moist heat	100°C	High
Liquid		
Glutaraldehyde, aqueous[c]	Variable	High
Hydrogen peroxide, stabilized	6 to 10%	High
Formaldehyde, aqueous[d]	3 to 8%	Intermediate to high
Iodophors[e]	30 to 50 mg/liter free iodine; 70 to 150 mg/liter available iodine	Intermediate
Chlorine compounds[f]	50 to 500 mg/liter free available chlorine	Intermediate

[a]*Comment.* Adequate precleaning of surfaces is the first prerequisite for any disinfection or sterilization procedure. The longer the exposure to a physical or chemical agent, the more likely it is that all pertinent microorganisms will be eliminated. Ten minutes of exposure may not be adequate to disinfect many objects, especially those that are difficult to clean because of narrow channels or other areas that can harbor organic material as well as microorganisms; thus, longer exposure times, i.e., 20 to 30 min, may be necessary. This is especially true when high-level disinfection is to be achieved. Although alcohols (e.g., isopropanol) have been shown to be effective in killing HBV, we do not recommend that they be used generally for this purpose because of their rapid evaporation and the consequent difficulty in maintaining proper contact times.

[b]This list of chemical germicides contains generic formulations. Other commercially available formulations can also be considered for use. Users should ensure that the formulations are registered with the Environmental Protection Agency. Information in the scientific literature or presented at symposia or scientific meetings can also be considered in determining the suitability of certain formulations.

[c]There are several glutaraldehyde-based proprietary formulations on the United States market, i.e., low-, neutral-, or high-pH formulations recommended for use at normal or elevated temperatures with or without ultrasonic energy and also a formulation containing 2% glutaraldehyde and 7% phenol. The manufacturer's instructions regarding use as a sterilant or disinfectant or anticipated dilution during use should be closely followed.

[d]Because of the continuing controversy over the role of formaldehyde as a potential occupational carcinogen, the use of formaldehyde is recommended only in limited circumstances under carefully controlled conditions, i.e., disinfection of certain hemodialysis equipment.

[e]Only those iodophors registered with the Environmental Protection Agency as hard-surface disinfectants should be used, and the manufacturer's instructions regarding proper use, dilution, and product stability should be closely followed. Iodophors formulated as antiseptics are not suitable for use as disinfectants.

[f]There currently is a chlorine dioxide formulation registered with the Environmental Protection Agency as a sterilant and disinfectant (depending on contact time). The manufacturer's instructions regarding its use as a sterilant or disinfectant or anticipated dilution during use should be closely followed. An inexpensive, broad-spectrum disinfectant for use on table tops and similar surfaces can be prepared by diluting common household bleach (5.25% sodium hypochlorite) to obtain at least 500 ppm free available chlorine (e.g., ¼ cup bleach per gallon of tap water). This dilution should be freshly made each day, and caution should be taken since chlorine compounds may corrode metals, especially aluminum.

in an appropriate intermediate- to high-level disinfectant such as a dilution of commercially available household bleach (sodium hypochlorite). All soiled towels should be put in a container which can either be placed with the infective waste of that particular department or, since it is sometimes easier, steam autoclaved. The concentration of disinfectant used depends primarily on the type of surface that is involved. For example, in the case of a direct spill on a porous surface that cannot be physically cleaned before disinfection, 0.5% sodium hypochlorite (5,000 mg of available chlorine per liter) should be used. On the other hand, if the surface is hard and smooth and has been cleaned appropriately, then 0.05% sodium hypochlorite (500 mg of available chlorine per liter) is sufficient.

If the item in question is a medical instrument that is classified as semicritical, such as a flexible fiberoptic endoscope (3), it is extremely important that meticulous physical cleaning precede the sterilization or high-level disinfection procedure. This is true whether or not ethylene oxide gas or a high-level disinfection procedure, using 2% alkaline glutaraldehyde with a contact time of 10 to 30 min, is used.

There are other types of environmental surfaces which could be classified as intermediate between noncritical and semicritical (9) and would include surfaces that are touched frequently, such as control knobs on hemodialysis systems, dental hand pieces in dental operatories, etc. Ideally, in these types of environments gloves should be worn to avoid ''finger painting'' of blood contamination and to avoid percutaneous exposure. Furthermore, these surfaces should be routinely cleaned and disinfected with cloths or swabs. In any case, the objective here would be to reduce the level of HBV contamination to such an extent that the chance of disease transmission is remote. In a practical sense, this could mean that a cloth soaked in either 0.05% sodium hypochlorite or a disinfectant-detergent could be used. In this context, the element of physical cleaning is as important as, if not more important than, the choice of the disinfectant. It obviously would not be cost effective, or in many cases even feasible, to attempt to achieve sterilization or high-level disinfection with some of these items or surfaces.

GENERAL HOUSEKEEPING PROCEDURES

As a rule, routine daily cleaning procedures that are used for general microbiologic laboratories can be used for laboratories in which blood specimens are processed. Obviously, special attention should be given to areas or items visibly contaminated with feces or blood. Furthermore, cleaning personnel must be alerted to the potential hazards associated with feces, blood, and serum contamination. Floors and other environmental surfaces contaminated in this manner should be thoroughly cleaned with a detergent-disinfectant. Cleaning personnel should wear gloves while performing these duties. However, in the case of large blood spills as mentioned above, this type of procedure may have to be augmented by specific site decontamination using a more effective chemical agent.

SUMMARY

Viral hepatitis is one of the most frequently reported laboratory-associated infections. Viral hepatitis type B and, presumably, non-A, non-B hepatitis constitute the major risks to laboratorians, in contrast to hepatitis A, which appears to be associated with personnel who handle experimental animals such as primates. Clinical laboratorians involved with handling blood and blood derivatives are at increased risk of acquiring hepatitis B. The primary mode of transmission is by direct contact with blood and serum specimens and environmental surfaces which are contaminated. The presence of blood or serum on hands, whether from direct or indirect sources, can result in the HBV gaining access to the vascular system subcutaneously by needle sticks or contamination of lesions or by nasal, oral, or ocular exposure (5). Although many biological and biochemical assays performed on blood are currently done by automated instrumentation, the element of automation in itself has not created additional hazardous environments. Rather, activities associated with sample reception, opening, transfer, and disposal of those specimens are most likely to be associated with hepatitis B exposure, especially when poor techniques are used. Infection control strategies should stress proper techniques for handling blood and containers, appropriate use of gloves, protective clothing and eyeware, frequent handwashing, good personal hygienic practices, and the effective use of disinfection and sterilization techniques. Infection control activities, including disease surveillance, should be performed by either a safety officer who is on the staff of the laboratory or a hospital infection control practitioner.

LITERATURE CITED

1. **Alter, H. J., R. H. Purcell, J. L. Gerin, W. T. London, P. M. Kaplan, V. J. McAuliffe, J. Wagner, and P. V. Holland.** 1977. Transmission of hepatitis B to chimpanzees by hepatitis B surface antigen-positive saliva and semen. Infect. Immun. **16:**928–933.
2. **Bond, W. W., M. S. Favero, N. J. Petersen, and J. W. Ebert.** 1983. Inactivation of hepatitis B virus by intermediate-to-high-level disinfectant chemicals. J. Clin. Microbiol. **18:**535–538.
3. **Bond, W. W., and R. E. Moncada.** 1978. Viral hepatitis B infection risk in flexible fiberoptic endoscopy. Gastrointest. Endosc. **24:**225–230.
4. **Bond, W. W., N. J. Peterson, and M. S. Favero.** 1977. Viral hepatitis B: aspects of environmental control. Health Lab. Sci. **14:**235–252.
5. **Bond, W. W., N. J. Peterson, M. S. Favero, J. W. Ebert, and J. E. Maynard.** 1982. Transmission of type B viral hepatitis via eye inoculation of a chimpanzee. J. Clin. Microbiol. **15:**533–534.
6. **Centers for Disease Control.** 1980. Hepatitis B contamination in a clinical laboratory—Colorado. Morbid. Mortal. Weekly Rep. **29:**459–460.

7. **Deinhardt, F.** 1980. Predictive value of markers of hepatitis virus infection. J. Infect. Dis. **141:**299–305.
8. **Dienstag, J. L., and D. M. Ryan.** 1982. Occupational exposure to hepatitis B virus in hospital personnel: infection or immunization? Am. J. Epidemiol. **115:**26–39.
9. **Favero, M. S.** 1983. Chemical disinfection of medical and surgical materials, p. 469–492. *In* S. S. Block (ed.), Disinfection, sterilization and preservation, 3rd ed. Lea & Febiger, Philadelphia.
10. **Favero, M. S., J. E. Maynard, R. T. Leger, D. R. Graham, and R. E. Dixon.** 1979. Guidelines for the care of patients hospitalized with viral hepatitis. Ann. Intern. Med. **91:**872–876.
11. **Francis, D. P., and J. E. Maynard.** 1979. The transmission and outcome of hepatitis A, B, and non-A, non-B: a review. Epidemiol. Rev. **1:**17–31.
12. **Levy, B. S., J. C. Harris, J. L. Smith, J. W. Washburn, J. Mature, A. Davis, J. T. Crosson, H. Polesky, and M. Hanson.** 1977. Hepatitis B in ward and clinical laboratory employees of a general hospital. Am. J. Epidemiol. **106:**330–335.
13. **Maynard, J. E.** 1978. Modes of hepatitis B virus transmission, p. 125–137. *In* Japan Medical Research Foundation (ed.), Hepatitis viruses. University of Tokyo Press, Tokyo.
14. **Maynard, J. E.** 1978. Viral hepatitis as an occupational hazard in the health care profession, p. 293–305. *In* Japan Medical Research Foundation (ed.), Hepatitis viruses. University of Tokyo Press, Tokyo.
15. **Mosley, J. W.** 1978. Epidemiology of HAV infection, p. 85–104. *In* G. N. Vyas, S. N. Cohen, and R. Schmid (ed.), Viral hepatitis: a contemporary assessment of epidemiology, pathogenesis and prevention. The Franklin Institute Press, Philadelphia.
16. **National Sanitation Foundation.** 1976. Standard 49. Class II (laminar flow) biohazard cabinetry. National Sanitation Foundation, Ann Arbor, Mich.
17. **Petersen, N. J., W. W. Bond, and M. S. Favero.** 1979. Air sampling for hepatitis B surface antigen in a dental operatory. J. Am. Dent. Assoc. **99:**465–467.
18. **Petersen, N. J., W. W. Bond, J. H. Marshall, M. S. Favero, and L. Raij.** 1976. An air sampling technique for hepatitis B surface antigen. Health Lab. Sci. **13:**233–237.
19. **Pike, R. M.** 1979. Laboratory-associated infections: incidence, fatalities, causes, and prevention. Annu. Rev. Microbiol. **33:**41–46.
20. **Public Health Service Advisory Committee on Immunization Practices.** 1972. Immune serum globulin for protection against viral hepatitis. Morbid. Mortal. Weekly Rep. **21**(Suppl.):6–9.
21. **Public Health Service Advisory Committee on Immunization Practices.** 1984. Immune serum globulin for protection against viral hepatitis. Morbid. Mortal. Weekly Rep. **33**(Suppl.):290.
22. **Shikata, T., T. Karasawa, K. Abe, T. Uzawa, H. Suzuki, T. Oda, M. Imai, M. Mayumi, and Y. Moritsugu.** 1977. Hepatitis B e antigen and infectivity of hepatitis B virus. J. Infect. Dis. **136:**571–576.
23. **U.S. Department of Health and Human Services.** 1984. Biosafety in microbiological and biomedical laboratories. HHS Publication no. (CDC) 84-8395. U.S. Government Printing Office, Washington, D.C.

Chapter 10

Slow Viral Infections: Safe Handling of the Agents of Subacute Spongiform Encephalopathies

DAVID M. ASHER, CLARENCE J. GIBBS, JR., AND D. CARLETON GAJDUSEK

The subacute spongiform encephalopathies (3, 34) are a group of at least five fatal degenerative diseases of the central nervous system (CNS): two are human diseases (kuru and Creutzfeldt-Jakob disease [CJD]) and three are diseases of animals (scrapie of sheep and goats, transmissible mink encephalopathy [TME], and wasting disease of mule deer and elk). All five diseases are caused by transmissible infectious agents that are subprotist in size and generally called unconventional viruses, although they have a spectrum of physical properties unlike those of known viruses and other names have been suggested for them (98). The spongiform encephalopathies have been described as "slow infections" (109) because of their very long asymptomatic incubation periods and protracted, unremitting clinical courses.

Kuru, the first human brain disease recognized to be a slow infection (35), occurs only in a small area of Papua New Guinea. Kuru formerly attacked both children and adults, but now is found only in adults born before 1959 (71). Transmission was clearly due to exposure to contaminated tissues during cannibalism, and incubation periods ranged from 4 to more than 25 years. The clinical picture of kuru is dominated by cerebellar ataxia, lack of coordination, and tremors, with signs of degeneration of the brainstem and other areas of the CNS as well. Since kuru was first described (40), the disease has become progressively rarer (34). Since kuru will never be encountered in most laboratories, it will not be discussed in greater detail here.

CJD occurs throughout the world. The reported incidence varies from 0.25 to more than 30 cases per million population per year (83), and about one case per million might be expected in most places. As many as 15% of cases in some series had a family history of presenile dementia consistent with CJD (4). The mean age at onset is in the sixth decade of life; the youngest reported patient was 17 and the oldest over 80. Incubation periods of iatrogenic cases were 15 months or more (10, 27, 55, 121). Illness typically begins with confusion or disturbed sensation followed by progressive dementia leading to stupor and coma, often within only a few months of onset. Many patients have ataxia early in disease. Most patients eventually develop typical myoclonic jerking movements (9). Focal signs suggestive of mass lesions are sometimes present. Fever occurs only with secondary infections. Hematology and blood chemistries are generally normal in CJD, except that liver function studies sometimes indicate mild hepatic parenchymal disease (R. Roos, D. C. Gajdusek, and C. J. Gibbs, Jr., Am. Acad. Neurol. Abstr. 23:34, 1971). Cerebrospinal fluid (CSF) protein levels are normal or slightly elevated. The electroencephalogram frequently shows typical periodic high-voltage complexes with slow background activity. Death usually occurs within less than 1 year after onset of illness (mean of 7 months in one series [83]), although about 10% of our cases survived more than 2 years (16).

Scrapie of sheep occurs in the United States, Canada, and many other countries throughout the world (57). It is characterized by progressive ataxia, lack of coordination, tremors, wasting, and sometimes itching as well (57). TME is probably a form of scrapie transmitted to mink by contaminated feed (82). A scrapie-like disease has also been observed among captive mule deer and elk in Colorado and Wyoming (122).

The histopathological lesions are generally similar in all five of the spongiform encephalopathies; changes are restricted to the CNS, although the infective agents are demonstrable in other tissues as well. Neurons become vacuolated, producing variable degrees of spongy change or "status spongiosus" of cerebral gray matter; this is followed by neuronal loss and cerebral atrophy. Neuronal loss frequently occurs in the cerebellum, where both Purkinje cells and granule cells may be strikingly affected. There are proliferation and hypertrophy of astrocytes in the areas of neuronal degeneration. White matter is not usually involved primarily, though there may be demyelination with reactive changes of microglia secondary to loss of neurons. An occasional case of CJD shows marked changes in white matter (113, 114). Amorphous "amyloid" plaques, as determined by periodic acid-Schiff staining, are present in the cerebellum of about 70% of patients with kuru, 15% of patients with CJD, and less often in the animal diseases. There are minimal signs of inflammation.

Diagnosis of the spongiform encephalopathies is made on the basis of clinical and histopathological findings, which may be confirmed by transmission of disease to susceptible animals. The demonstration in extracts of CJD-affected brain of helical fibrils by negative-stain electron microscopy and of low-molecular-weight protein bands by gel electrophoresis and immunoblotting has recently proven useful for confirming diagnosis as well (8, 11, 12, 85, 86; P. Brown, M. Coker-Vann, K. Pomeroy, D. M. Asher, C. J. Gibbs, Jr., and D. C. Gajdusek, submitted for publication).

Kuru and CJD were originally demonstrated to be infections by experimental transmission of disease to chimpanzees inoculated with extracts of brain tissue from patients. Animal assay is tedious and expensive, but remains the most reliable test for the presence of the infectious agent. Both infections are also transmissible to a variety of other animals including monkeys and rodents; in our experience, the squirrel monkey, *Saimiri sciureus*, is the most consistently susceptible animal readily available for assay. The agent of CJD has been demonstrated by animal assays to be present in the brains of more than 70% of unselected specimens referred to our laboratories with clinical diagnoses of CJD and in those of more than 90% when strict diagnostic criteria, including typical histopathological changes, were required; as many as one million lethal doses of the CJD agent per gram of human brain have been demonstrated. Other tissues such as spinal cord, liver, kidney, and spleen were sometimes but less regularly infectious, as were some spinal fluids, but no other body fluids, secretions, or excretions were found to be infectious in a small number of studies.

Titration studies by serial dilution demonstrate that the agents replicate. The size of the scrapie agent has been estimated by filtration to be between 27 and 50 nm in diameter (41, 45; Gibbs and Gajdusek, unpublished data). Characterization is made difficult by the association of infectivity with cellular membranes (64) and a tendency to aggregate into masses (106). The agents display a spectrum of extreme resistance to inactivation by heat, radiation, and a variety of chemicals (59), unlike the resistances of known viruses. Several treatments that degrade proteins inactivated the scrapie agent, implying that it must contain a protein (22, 98), as do all other known pathogens except viroids of plants. It was even suggested that the diseases may be transmitted by unique infectious "proteinaceous" pathogens, or "prions," that are subviral in size and do not contain nucleic acids (98). However, reinterpretation of heat and irradiation-inactivation studies found the behavior of the scrapie agent to be consistent with a structure containing a nucleic acid genome the size of a small virus (105–107). The failure to demonstrate convincingly a subviral size of the scrapie agent (26, 102) and the failure to retain infectivity in purified protein extracts purported to contain the agent (100) also call the infectious-protein hypothesis into question. The scrapie-associated fibril mentioned above

and a low-molecular-weight protein (PrP 27–30 [12, 13]) that seems to be a component of the scrapie-associated fibril (25, 104) both "copurify" with scrapie infectivity, but the recent demonstration that PrP 27–30 is encoded by a normal host gene and that a protein antigenically related to PrP 27–30 is present in normal tissue (92) casts doubt on their being the actual infectious agent and suggests that they are more likely to be pathological host proteins. So the structure of the spongiform encephalopathy agents, whether they are unusual viruses or unique subviral pathogens, remains unknown and the subject of controversy. We prefer to refer to these agents as viruses until they are proven otherwise. Viruslike particles have been described in neurons of several types of animals with spongiform encephalopathies (90) but have yet to be identified or associated with infectivity.

It has been hypothesized that other degenerative diseases of the human brain, especially the much more common Alzheimer's disease, which shares the histopathological finding of amyloid plaques, may also be caused by infectious agents similar to those of the spongiform encephalopathies (99). However, Alzheimer's disease has not been convincingly transmitted to animals (49), and no other human disease besides kuru and CJD and its variants has been attributed to the spongiform encephalopathy agents.

ASSESSMENT OF RISK

It is difficult to assess risk of infection for humans with the spongiform encephalopathy agents because the diseases are rare, their incubation periods are very long, and the sources of infection and portals of entry are generally not known. Barkley and Wedum (6) suggested several factors to be considered in assessing the risk of infectious agents in microbiological laboratories, and Chatigny and Prusiner (20, 21) addressed the spongiform encephalopathies in particular. In examining risk factors for the spongiform encephalopathies, we must first consider what is known about the natural and experimental transmission of the diseases.

Laboratory-acquired and iatrogenic infections. Most important to consider is the record of known accidents—the acquisition of infection in the laboratory or hospital. No accidental transmission of kuru under these circumstances has been recognized (34, 83). CJD was accidentally transmitted to one patient by corneal transplant from an infected donor (27) and to two patients by electrocorticography with contaminated electrodes (10); the intervals between surgery and the onset of clinical CJD in these patients were 16 to 20 months. Three other patients had craniotomies performed in the same surgical suite of an English hospital 15 to 17 months before the onset of CJD (91, 121), suggesting iatrogenic spread. Recently, four cases of CJD were recognized in young adults who had received repeated injections during childhood and adolescence of growth hormone prepared from human pituitary glands (18a, 46a, 55, 71a, 97a), indicating that the agent need not be introduced directly into the eye or CNS to infect humans. In most cases of CJD the mechanism of transmission is unknown; several epidemiological studies have investigated other potential iatrogenic mechanisms, including dentistry and head, eye, and general trauma surgery (23, 83) and ocular tonometry (23). Although several intriguing statistical associations were observed, none established a convincing etiological link with CJD.

No accidental infection of hospital or laboratory personnel with spongiform encephalopathy has been documented. During the more than 60 years since CJD was first described (69), it has not been reported to occur in any pathologist (83). CJD was diagnosed in one neurosurgeon (39) and in 17 other medical workers among 308 patients in one survey (83); however, it is not clear that health professionals were significantly more common among those patients than in the general population.

Potential sources and routes of infection. The only natural source of kuru infection seems to be tissues handled by relatives and friends of patients during ritual cannibalism; kuru has not occurred in people born since those rituals ended (34, 71), even in the children of mothers with kuru or among those who cared for terminally ill patients. Consistent with those observations, in the limited attempts possible to date, kuru agent has not been found in blood, urine, feces, milk, placenta, or CSF of patients, although small amounts of the agent were detected in a variety of organs besides the CNS (3). There is no evidence of spread of kuru by fomites or nonhuman vectors.

The sources of CJD are less certain. Brain, pituitary, and eye tissues are demonstrated sources of infection (10, 27, 55). The agent of CJD is found in other tissues throughout the human body, though not so consistently or in amounts as great as in brain (3, 43). CJD agent has not been found in the blood, serum, feces, saliva, or urine of patients, but it has been demonstrated in 25% of CSF specimens tested (14, 43). Studies of CJD in animals showed a similar distribution of agent (3, 43). Although viremia has not been found in primates with CJD, the agent has been demonstrated in the blood of experimentally infected rodents (73, 78). Only a limited number of specimens besides brains have been adequately tested for the presence of CJD agent, and human materials become available only late in the course of illness when the diagnosis of CJD becomes clear; virus could thus easily escape detection if excreted from the body intermittently, in only a small percentage of cases, or during the incubation period.

The natural sources of scrapie, TME, and mule deer wasting disease are also not completely understood. Scrapie appears to spread in sheep and goats by direct contact. Although early observations

in the United Kingdom (48) and Iceland (93) suggested that fomites or vectors might spread scrapie, all recorded outbreaks of scrapie in the United States have followed direct contact with clinically sick animals or with animals later found to have been incubating the disease (57). The nature of the contacts and fomites, if any, are still unknown. In the United States, scrapie virus has not been found in sheep or goat feces, urine, saliva, or other body fluids or in grass or manure from sheds of infected animals (51, 57); the agent has been found in sheep stomach worm (57) and the placentas of sheep (89) and goats (57). Tissues of infected animals are a known source of contagion in sheep (47; W. S. Gordon, A. Brownlee, and D. R. Wilson, Abstr. 3rd Int. Congr. Microbiol., p. 2–9, 1939). Scrapie agent can be detected in spleen, lymph nodes, bone marrow, salivary glands, intestines, and other tissues of clinically normal animals before appearing in the CNS (31, 50, 51). Scrapie agent has sometimes been found in the CSF of goats (50, 96) and sheep (51), and it has been reported occasionally to occur in blood (3), though most studies failed to find viremia (50, 51). Sources of infection with TME seem to be generally similar to those of scrapie; the agent is widely disseminated throughout many tissues of infected mink, though amounts are greatest in brain. Virus has occasionally been found in feces, but not in urine or serum (79). Little is known of mule deer wasting disease, but it generally seems to resemble the other two spongiform encephalopathies of animals.

The potential of animals with scrapie or similar diseases to serve as sources of human spongiform encephalopathy is not known. The absence of diagnosed CJD in scrapie research investigators, the similar rates of occurrence of CJD in places such as Western Europe, where scrapie is highly prevalent, and Australia and New Zealand, where scrapie has not occurred for decades, and the modest representation of veterinary and farm workers among CJD patients (14, 83), all militate against scrapie being an important source of human disease. Also working against that possibility is the fact that chimpanzees, highly sensitive to kuru and CJD viruses, have been resistant to intracerebral inoculation with high doses of scrapie and TME agents (43), though only a few strains have been adequately tested. However, at least two species of monkeys are susceptible to experimental infection with both scrapie and TME viruses (42, 79), and the diseases produced are clinically and histopathologically indistinguishable from experimental kuru and CJD in the same species. So, although it seems unlikely that the animal diseases are a reservoir for human kuru or CJD, that possibility has not been completely excluded.

The natural portals of entry for kuru virus are probably breaks in the skin and mucous membranes (34). All patients have had a history of direct contact with infected tissues (71); there is nothing to suggest that airborne spread has occurred. It is interesting that the careful introduction of large amounts of kuru agent directly into the stomachs of chimpanzees through nasogastric tubes did not transmit infection (3), whereas monkeys allowed to handle and chew infected tissues did acquire the disease (42). This suggests that portals in the skin and oral mucosa may be more important than those in the gastrointestinal tract.

The portals of entry for CJD agent are less well understood. Accidental direct introduction of the virus into brain (10, 91), eye (27), and muscle (18a, 46a, 55, 71a, 97a) of humans obviously creates a portal. CJD is readily transmitted to animals experimentally by intracerebral inoculation (44). Manuelidis et al. (77) transmitted CJD to guinea pigs by inoculation of infected cornea into the anterior chamber of the eye; attempts to infect monkeys by grafting infected corneas have been unsuccessful for more than 10 years (Gibbs et al., unpublished data). As discussed above, many patients with CJD have a history of surgery during an interval before onset of illness which is compatible with the incubation periods of iatrogenic cases and with experimental incubation periods in animals (83). However, most patients have no history of surgery, and there are probably other portals. The rare occurrence of conjugal cases of CJD suggests that contact spread may take place, though one reported couple became ill almost simultaneously, which seems more consistent with a common exposure (65); in the other family, not well documented, one spouse became ill 3 years after the other (84). CJD, like kuru, has been experimentally transmitted to animals by feeding them raw infected tissue (42), as well as by subcutaneous and intradermal inoculations (43). These observations suggest that CJD virus, like kuru virus, may enter the body through breaks in the integument.

Scrapie is clearly spread by contact of normal sheep and goats with infected animals (48, 57, 93). Animals can also be infected by subcutaneous inoculation (52), as well as by intramuscular and intraperitoneal injections and by direct introduction of virus into the CNS (3). The finding of scrapie virus in the intestines and mesenteric lymph nodes of otherwise healthy sheep in which no agent was detected in the CNS or other tissues (51) suggests that there may also be an enteric portal in natural scrapie. TME has also been transmitted by the oral route (19, 80), and its portals and those of mule deer disease are probably similar to those for scrapie.

The infectious dose for kuru and CJD is not known for any route of infection in humans. More scrapie virus is required to infect animals by peripheral routes than by direct intracerebral inoculation (50, 66). Limited evidence suggests that the same thing is true for experimental kuru and CJD in animals (Gibbs and Gajdusek, unpublished data).

Evidence of accidental infection of animals. Neither kuru nor CJD has been observed to spread from infected animals to normal animals caged with them. Scrapie spreads readily among sheep and goats; the accidental spread of scrapie from infected rodents to normal animals appears to be very uncommon (3).

Possible factors increasing susceptibility to infection. Different breeds of sheep differ in susceptibility to scrapie. Goats and mice are uniformly susceptible, though there are marked differences in incubation periods and patterns of histological changes in mice of different genotypes (24). Genetic differences in susceptibility of mink to TME have not been recognized (79). No genetic differences in susceptibility of humans to kuru have been proven. The occurrence of CJD in families with a pattern suggestive of an autosomal dominant might be due to hereditary susceptibility, but that remains conjectural (4).

Younger goats and sheep appear to be more susceptible than older animals to naturally acquired scrapie (57). A study of mice inoculated intraperitoneally with scrapie agent suggested the opposite relationship of age to susceptibility (56). Kuru has occurred among children as well as adults, and CJD has occurred in adolescents, so people of all ages are clearly susceptible. No other predisposing factors have been identified or studied.

Prophylaxis and therapy. There are no known effective prophylactic or therapeutic measures for the spongiform encephalopathies. Several substances interfered with experimental scrapie when administered before or shortly after peripheral inoculation, but these did not reverse the established infection of the CNS (29, 66, 68). A report of successful treatment of human CJD (108) has not been confirmed.

In short, the demonstrated risk to hospital and laboratory personnel from patients with spongiform encephalopathy is very small, even in laboratories with the closest potential exposure to the most highly infectious material. There is a demonstrated risk to other patients undergoing surgery or receiving products derived from human brains. However, the rare occurrence of possible contact infection with CJD, the demonstration that breaks in the integument are likely to be adequate portals of entry, and the knowledge that tissues and spinal fluids of patients may contain substantial amounts of infectious agent all dictate that reasonable precautions should be taken in handling materials potentially contaminated with the viruses of spongiform encephalopathies (5, 20, 21, 36, 37, 115, 116).

PHYSICAL INACTIVATION

Heat. Scrapie agent is highly thermostable (111). There is very little loss of infectivity after prolonged exposure to temperatures up to 80°C, though higher temperatures reduce infectivity markedly (30, 46, 63). A small amount of infectious virus sometimes remains after heating crude suspensions of scrapie-infected tissue suspensions at 100°C for 1 h (46, 58, 63) or at 118°C for 10 min (63); most of the infectivity remaining was apparently in the coagulated sediment. In several studies no infectious virus was detected in preparations treated for 30 min or more at 121°C (15 lb/in^2 [120]) in the steam autoclave (37, 111; C. J. Gibbs, Jr., and D. C. Gajdusek, Abstr. 3rd Int. Congr. Virol., p. 47, 1975), in a solvent bath (105), or at 141°C for 10 min in a solvent bath (63). However, it must be remembered that the temperatures actually achieved by materials inside an autoclave are often not the same as those indicated on the thermometer of the device, for several reasons (32), and detectable amounts of infectious scrapie virus have survived exposure for 1 h in the autoclave (15).

Radiation. The scrapie agent is also uniquely resistant to inactivation by ionizing and UV irradiations (2). It has an action spectrum, the profile of inactivation produced by various wavelengths of UV light, different from those for other known viruses and nucleic acids (74). (These observations prompted the inference that the scrapie agent might be a novel pathogen not containing nucleic acids [2].) A 50-kGy dose of ionizing radiation produced less than a 10-fold reduction of scrapie virus infectivity, and 200 kGy produced less than a 1,000-fold reduction, giving a D_{37} (i.e., the dose leaving 37% of original infectious activity) of 45 kGy (45). The D_{37} of UV light at 254 nm, the wavelength of germicidal lamps, was 21.5 to 30 kJ/m^2 (74). The viruses of TME (80), kuru, and CJD (74) have been less thoroughly studied but seem to be similar in their resistance to irradiation. Conditions adequate to sterilize the viruses of spongiform encephalopathies with radiation cannot be practically achieved in the microbiology laboratory.

CHEMICAL INACTIVATION

The scrapie agent is highly resistant to many substances commonly used to inactivate viruses. The usual problems inherent in estimating the virucidal activity of disinfectants (28, 70, 72) are compounded for the spongiform encephalopathy agents because their assays require large numbers of animals and long periods of observation. Therefore, only limited data on chemical inactivation of scrapie agent, and even less for other members of the group, are available. Some substances reported to have significant inactivating activity against the spongiform encephalopathy viruses (20, 37, 60, 87), when examined more carefully under conditions closer to those available in practice, have proven to be much less effective than had been claimed. Repeated studies have often yielded inconsistent results. The effects of various classes of chemicals on infectivity of the scrapie agent have been reviewed (87).

Aldehydes. The earliest indication of the unusual

properties of the scrapie agent came from its accidental introduction into a batch of Formalin-inactivated louping-ill vaccine (Gordon et al., Abstr. 3rd Int. Congr. Microbiol. 1939); the resulting outbreak of scrapie among sheep inoculated with the vaccine demonstrated that the casual agent had survived exposure for more than 3 months to 0.3% Formalin, a concentration ordinarily considered to be virucidal (47, 112). Substantial amounts of infectious scrapie virus survived following exposure for as long as 28 months to concentrations of Formalin as high as 20% (8% formaldehyde) (94). Similar resistance to Formalin was later demonstrated for the viruses of TME (81) and CJD (10, 17, 38). Dialdehydes, which are more effective in disinfecting conventional viruses than formaldehyde (28), have not been adequately studied, but limited data indicate that glutaraldehyde is not effective in inactivating the scrapie agent (17).

Alkylating agents. β-Propiolactone at concentrations of 0.2 to 1.0% produced variable decreases in the titers of scrapie agent, but never more than 2 log$_{10}$ LD$_{50}$ (50% lethal dose) (53). Acetyl ethyleneimine is reported to have been ineffective in inactivating scrapie virus (112). Ethylene oxide, both at ambient and high temperatures and pressures, was also found to be ineffective (18; D. M. Asher, C. J. Gibbs, Jr., A. Diwan, D. T. Kingsbury, M. P. Sulima, and D. C. Gajdusek, Abstr. 12th World Congr. Neurol., p. 255, 1981).

Halogen compounds. The effects of various solutions on the infectivity of spongiform encephalopathy agents are listed in Table 1. Sodium hypochlorite (commercial household liquid chlorine bleach) at concentrations of 0.5 to 2.5% (54; Asher et al., Abstr. 12th World Congr. Neurol. 1981) inactivated large amounts of scrapie virus, in some cases reducing titers to undetectable levels; however, lower concentrations of hypochlorite were far less effective (Gibbs et al., unpublished data). Similar results were obtained with 0.5 and 1.3% sodium hypochlorite and the CJD agent (18). Chlorine dioxide, another chlorine generator, gave less inactivation of the scrapie agent (14, 15). Sodium perchlorate (0.01 M) produced only negligible reduction in infectivity of scrapie agent (1). A solution of about 0.3% iodine in 70% ethanol produced a marked reduction in infectivity of scrapie agent (more than 5 log$_{10}$) after 3 h at room temperature, but inactivation was not complete in one study (7). A more recent study found little inactivation of the scrapie agent by 2% iodine (15). Sodium periodate reportedly reduced titers of scrapie markedly (more than 3.5 log$_{10}$ LD$_{50}$ in the study of Hunter et al. [60]); that finding was only partially confirmed in one study (15) and not confirmed in another (1). Sodium iodate (0.01 M) was not effective in inactivating scrapie virus (60). Iodophore disinfectants have been recommended for inactivating CJD on theoretical grounds (37), but they proved to be ineffective against scrapie virus (Asher et al.,

TABLE 1. Effects of several salt and other solutions on the infectivity of spongiform encephalopathy agents[a]

Solution	Concn (M)	Reduction in infectivity ($\log_{10}$ LD_{50})	Detectable residual infectivity	References
Cesium chloride	2.5	1.1–1.8	Yes	41, 89
Guanidium hydrogen bromide	6	5	?	G. C. Millson, unpublished data[b]
Lithium chloride	6	>1.0–1.5	?	64
	12	4.0	Yes	G. D. Hunter, unpublished data[b]
Lithium diiodosalicylate	0.3	1.0	?	G. C. Millson, unpublished data[b]
Potassium permanganate	0.024	1.1–1.3	Yes	18*
	0.012	<0.05	Yes	18*
	0.006	0.5–1.2	Yes	15, 18*
	0.002	≥2.4	?	1
Potassium tartrate	1.1–1.4	<1	Yes	67
Sodium chloride	1.0	<1	Yes	64
Sodium hypochlorite	0.15	≥2.0	Yes	D. M. Asher et al.[c]
	0.08	≥3.8	Yes	18*
	0.03	≥3.5	Yes	15, 18*
	0.015	<1	Yes	D. M. Asher et al.[c]; C. J. Gibbs, Jr., et al., unpublished data
Sodium iodate	0.01	1	Yes	60
Sodium periodate	0.1	1.5	Yes/no	1, 60
	0.01	3.6	Yes	15
Sodium thiocyanate	1.0	5.9	No	103
Sodium trichloroacetate	1.0	6	Yes	101
Urea	3.0	1.7	Yes	98
	6.0	2.0	Yes	60
	8.0	>5	No/yes	D. L. Mould, unpublished data[d]; 17*

[a]Adapted from Millson et al. (87). All results are for scrapie except where indicated (*) for CJD. Times and conditions of exposure varied. In some reports it is not clear whether concentration of solute was expressed as initial or final.
[b]Cited in reference 87.
[c]Asher et al., Abstr. 12th World Congr. Neurol. 1981.
[d]Cited in reference 60.

Abstr. 12th World Congr. Neurol. 1981; unpublished data).

Other salts. Potassium permanganate (0.002 M) reduced infectivity of scrapie virus by 2.4 $\log_{10}$ LD_{50} to undetectable levels in one study (1); this finding was not confirmed (15; Asher et al., Abstr. 12th World Congr. Neurol. 1981). Higher concentrations of potassium permanganate were found in one study to inactivate substantial amounts of scrapie agent (Asher et al., Abstr. 12th World Congr. Neurol. 1981), but other studies showed little or no inactivation and highly inconsistent results (15, 18; Asher et al., unpublished data). Guanidium hydrogen bromide (6 M) (87), 12 M lithium chloride (87), and the chaotropic ions thiocyanate and trichloroacetate (101, 103) have also been reported to reduce titers of scrapie virus markedly, but their practical application as disinfectants has not been investigated. Urea (8 M) was reported to reduce infectivity of scrapie markedly (60), but a recent study failed to confirm this for the CJD agent (16, 17).

Organic solvents. The effects of several solvents and other organic compounds on the infectivity of spongiform encephalopathy agents are presented in

Table 2. Extraction of scrapie-infected tissue suspensions with 90% aqueous phenol removed all detectable infectivity in several studies (60, 87, 101). Similar observations were made with TME virus (81). These observations led to the suggestion that phenolic disinfectants be used to clean surfaces contaminated with the CJD agent (37). However, substantial amounts of scrapie virus survived exposure to 2% (60) and 5% (Asher et al., unpublished data) solutions of phenol. The recommended concentration of a commercial preparation of o-phenylphenol was also ineffective against the scrapie agent (15; Asher et al., unpublished data). Ethanol (50 and 70%) did not effectively inactivate scrapie virus (87), nor did 70% ethanol disinfect electrodes contaminated with CJD virus in clinical use (10; Gibbs et al., unpublished data). An 80% solution of 2-chloroethanol produced a marked reduction (5 $\log_{10}$ LD_{50}) in scrapie virus infectivity (61); however, 50% 2-chloroethanol and a variety of other solvents and solvent mixtures, including 10% n-butanol, n-pentanol, chloroform (61), diethyl ether (30, 46, 76), chloroform-methanol (2:1), and aqueous methoxyethanol, produced much less inactivation (61, 76).

TABLE 2. Effects of several solvents and other organic compounds on the infectivity of spongiform encephalopathy agents[a]

Solvent	Reduction in infectivity ($\log_{10}$ LD_{50})	Detectable residual infectivity	References
Acetone	2.0	Yes	89
Acetone-ethyl ether	1.8	Yes	62, 96
n-Butanol (10%)	<1	Yes	61, 89
Butanol-chloroform (2:1)	1.5	Yes	64
2-Chloroethanol			
80%	5.0	Yes	61
50%	1.9	Yes	61
Chloroform	1.0	Yes	76
Chloroform-methanol (2:1)	1.0	Yes	61, 89
Ethanol			
70%	?	Yes	10*
50%	?	Yes	54
Ethyl ether	2.0	Yes	30, 46, 76, 89
Fluorocarbons (trichlorotrifluoroethane)	<1	Yes	52, 64, 76, 87
2-Mercaptoethanol (0.1%)	<1	Yes	110
Methoxyethanol (80%)	1.4	Yes	61
n-Pentanol	<1	Yes	61
Phenolics, substituted (Lysol 10% = 0.5% phenolics)	1.5–2.3	Yes	15
Phenol			
90%	>4.5	No	60, 81,** 87
5%	ND	Yes	D. M. Asher et al.[b], unpublished data
2%		Yes	60

[a]Adapted from Millson et al. (87). All results are for scrapie except where indicated for CJD (*) and TME (**). Times and conditions of exposure varied. ND, Not determined.
[b]Asher et al., Abstr. 12th World Congr. Neurol. 1981.

Fluorocarbons (trichlorofluoroethane) produced little if any decrease in infectivity of scrapie virus (64, 76, 87).

Detergents. The effects of several detergents on the infectivity of spongiform encephalopathy agents are listed in Table 3. Most detergents, including lysolecithin, sodium cholate, sodium deoxycholate, cetyltrimethylammonium bromide, and Tween 80 had no significant effect on the infectivity of scrapie virus, while 0.5% sodium dodecyl sarcosinate and Triton X-100 produced small decreases (less than 2 $\log_{10}$ LD_{50}). Sodium dodecyl sulfate, however, produced marked inactivation of scrapie virus (102), and at concentrations of 5% or greater reduced infectivity by more than 5 $\log_{10}$ LD_{50} to undetectable levels (61, 88). However, a recent attempt to remove scrapie virus infectivity from surfaces with 5% sodium dodecyl sulfate was not successful (Asher et al., unpublished data).

Acids and bases. Scrapie virus was stable when incubated for 24 h at a pH varying from 2.1 to 10.5 (52, 89). However, 5 N NaOH (Clarke, cited in reference 87) through 0.3 N (103) and 0.25 N (60) HCl inactivated substantial amounts of scrapie virus.

Other substances. Hydrogen peroxide (1.5 and 3%) was not effective for inactivation of scrapie virus (15; Asher et al., Abstr. 12th World Congr. Neurol. 1981).

TABLE 3. Effects of several detergents on the infectivity of spongiform encephalopathy agents[a]

Detergent	Concn (%)	Reduction in infectivity ($\log_{10}$ LD$_{50}$)	Detectable residual infectivity	References
Anionic:				
Lithium dodecyl sulfate	3	4.0	Yes	101
Lysolecithin	0.8	<1	Yes	61, 88
Sodium cholate	0.5	<1	Yes	S. C. Collis, unpublished data[b]
Sodium deoxycholate	6	<1	Yes	88
	5	<1	Yes	18*
	0.5	<1	Yes	64
Sodium dodecyl sarcosinate	0.5	1.5	Yes	87, 101
Sodium dodecyl sulfate	10	>5	No	88
	5	>5	No	61, 88
	2.5	4.5	Yes	88
	2.0	≤1.5	Yes	18*
	0.5	1–2	Yes	61
Cationic: cetyl-trimethyl-ammonium bromide	0.5	<1	Yes	87, 101
Nonionic:				
Triton X-100	1.0	≤1.5	Yes	18,* 110
	0.5	<1	Yes	110
	0.05	<1–1.5	Yes	64, 110
Tween 80	0.01	<1	Yes	63

[a]Adapted from Millson et al. (87). All results are for scrapie except where indicated (*) for CJD. Times and conditions of exposure varied.
[b]Cited in reference 86.

APPROPRIATE HANDLING OF SPONGIFORM ENCEPHALOPATHY AGENTS IN THE LABORATORY

We have modified slightly the recommendations made previously (5, 20, 21, 36, 37, 115, 116). The procedures recommended here are based on the information reviewed above, on generally accepted practices in medical virology laboratories, and in a few cases on our own preferences.

General. Good basic microbiological laboratory techniques are probably the best protection against infection with the spongiform encephalopathy agents. If "hepatitis" precautions are observed at all times in diagnostic laboratories, accidental infections with spongiform encephalopathy agents will in all likelihood be prevented. All workers must be given adequate training in biosafety. Laboratory practices should require the appropriate wearing of gloves, stress the avoidance of needle and glass cuts, and forbid eating, drinking, smoking, and mouth pipetting in bench areas. There must be especially careful handling of all materials contaminated with tissues and spinal fluid. Laboratory benches must be cleaned and disinfected frequently.

Whenever possible, objects potentially contaminated with spongiform encephalopathy viruses should be disinfected with heat. We now recommend the steam autoclave at the elevated temperature of 132°C instead of the conventional temperature of 121°C. Although a 30-min exposure may be adequate, we prefer to autoclave for at least 1 h. If 121°C is used it is wise to increase the time beyond 1 h. The general rules for the effective use of all steam autoclaves must be observed (118): objects must be properly loaded to admit steam to all potentially contaminated surfaces. When there is doubt about the temperatures actually achieved in contaminated materials, it is especially important to augment conventional autoclaving by raising the temperature or increasing the time of exposure (105), or reducing the amount of infectious agent by preliminary chemical treatment (15). Disposable equipment, contaminated organs, animal carcasses, and the like should be carefully incinerated, observing the usual precautions of capping all needles, covering all blades and broken glass, and placing everything in several plastic bags within a rigid leakproof container so that those who transport trash will not be injured.

When chemical disinfectants must be used for

decontaminating surfaces or objects that cannot be incinerated or autoclaved, or when it is desirable to reduce the levels of infectious agent present before autoclaving, only two treatments are currently recommended: sodium hydroxide (1 N or more) and 5.25% sodium hypochlorite (undiluted fresh liquid household chlorine bleach). We suggest that surfaces or objects be exposed for at least 1 h. It must be recognized that no chemical disinfectant has been as effective as heat in inactivating the spongiform encephalopathy agents, and that some residual infectivity may remain where original amounts are large. It must be stressed again that ethylene oxide gas sterilization and most cold liquid sterilizing solutions appear to be ineffective and must not be used.

Each laboratory has the responsibility for assessing the potential contamination of its own materials, selecting appropriate decontamination procedures, and monitoring the efficacy of its own reagents, equipment, and operating procedures (118). The laboratory itself must also determine whether the corrosive properties of sodium hydroxide or of bleach and its vapors, or the damage to objects caused by heating, are acceptable for a specific task.

Containment. It has been proposed by an interagency working group of the Centers for Disease Control and the National Institutes of Health (117) that the agents of kuru and CJD and materials potentially contaminated with them should be manipulated using "Biosafety Level 2 practices, containment equipment, and facilities." Chatigny and Prusiner (20, 21) suggested multiplying the pathogen's "class number" (2 for scrapie and TME and 3 for kuru and CJD) by a "composite risk factor" appropriate for the laboratory procedure to be performed. We prefer simply to perform each operation under the highest level of containment practically feasible. In our laboratories almost all bench-top procedures involving the spongiform encephalopathy agents are currently performed in rooms with access limited to necessary personnel, with flow of air negative to the corridors, and inside Class II laminar-flow biological safety cabinets whenever possible (117). (However, for 15 years, procedures were performed on open bench tops in isolation cubicles or still-air "Pasteur" hoods, and no recognized infections of investigators resulted.) Inoculations and necropsies of animals are still performed on open bench tops by operators wearing closed-front long-sleeved gowns, caps, glasses, and fresh disposable surgical face masks and rubber gloves. For centrifugations of infectious materials, it is desirable to use commercially available safety shields over the centrifuge tubes.

When laboratory procedures are complete, benches and other potentially contaminated surfaces are wiped with sodium hydroxide (1 N stock solution) or full-strength household chlorine bleach. All materials used, including protective clothing, towels, and disposable items, are autoclaved as described above. Waste tissues and animal carcasses are packaged either in heavy plastic-lined double paper bags or in three heavy plastic bags in a strong cardboard carton, sealed with nylon filament tape, clearly marked as infectious, and transported to the incinerator. Some surfaces that are especially difficult to clean, such as pipetting machines, microtomes, and cryostats, we consider to be contaminated at all times, and no attempt is made to disinfect them until they need adjustment or repair.

Shipment of specimens. Specimens for diagnosis of spongiform encephalopathies and materials known to contain the agents of kuru or CJD must be shipped interstate as specified by federal regulation (33, 117).

LITERATURE CITED

1. **Adams, D. H., E. J. Field, and G. Joyce.** 1972. Periodate—an inhibitor of the scrapie agent? Res. Vet. Sci. **13:**195–198.
2. **Alper, T., D. A. Haig, and M. C. Clarke.** 1966. The exceptionally small size of the scrapie agent. Biochem. Biophys. Res. Commun. **22:**278–284.
3. **Asher, D. M., C. J. Gibbs, Jr., and D. C. Gajdusek.** 1976. Pathogenesis of subacute spongiform encephalopathies. Ann. Clin. Lab. Sci. **6:**84–103.
4. **Asher, D. M., C. L. Masters, D. C. Gajdusek, and C. J. Gibbs, Jr.** 1983. Familial spongiform encephalopathies, p. 273–291. *In* S. S. Kety, L. P. Rowland, R. L. Sidman, and S. W. Matthysse (ed.), Genetics of neurological and psychiatric disorders. Association for Research in Nervous and Mental Diseases Research Publications, vol. 60. Raven Press, New York.
5. **Baringer, J. R., D. C. Gajdusek, C. J. Gibbs, Jr., C. L. Masters, N. E. Stern, and R. D. Terry.** 1980. Transmissible dementias: current problems in tissue handling. Neurology **30:**302–303.
6. **Barkley, W. E., and A. G. Wedum.** 1977. The hazard of infectious agents in microbiological laboratories, p. 740–753. *In* S. S. Block (ed.), Disinfection, sterilization, and preservation, 2nd ed. Lea & Febiger, Philadelphia.
7. **Bell, T. M., E. J. Field, and G. Joyce.** 1972. Action of an alcoholic solution of iodine on the scrapie agent. Res. Vet. Sci. **13:**198–199.
8. **Bendheim, P. E., J. M. Bockman, M. P. McKinley, D. T. Kingsbury, and S. B. Prusiner.** 1985. Scrapie and Creutzfeldt-Jakob prion proteins share physical and antigenic determinants. Proc. Natl. Acad. Sci. USA **82:**997–1001.
9. **Bernoulli, C. C., C. L. Masters, D. C. Gajdusek, C. J. Gibbs, Jr., and J. O. Harris.** 1979. Early clinical features of Creutzfeldt-Jakob disease (subacute spongiform encephalopathy), p. 229–251. *In* S. B. Prusiner and W. J. Hadlow (ed.), Slow transmissible diseases of the nervous system, vol. 1. Academic Press, Inc., New York.
10. **Bernoulli, C. C., J. Siegfried, G. Baumgartner, F. Regli, T. Rabinowicz, D. C. Gajdusek, and C. J. Gibbs, Jr.** 1977. Danger of accidental person-to-person transmission of Creutzfeldt-Jakob disease by surgery. Lancet **i:**478–479.
11. **Bockman, J. M., D. T. Kingsbury, M. P. McKinley, P. E. Bendheim, and S. B. Prusiner.** 1985.

Creutzfeldt-Jakob disease prion proteins in human brains. N. Engl. J. Med. **312:**73–78.

12. **Bolton, D. C., M. P. McKinley, and S. B. Prusiner.** 1982. Identification of a protein that purifies with the scrapie prion. Science **218:**1309–1311.

13. **Bolton, D. C., R. K. Meyer, and S. B. Prusiner.** 1985. Scrapie PrP 27–30 is a sialoprotein. J. Virol. **53:**596–606.

14. **Brown, P.** 1980. An epidemiologic critique of Creutzfeldt-Jakob disease. Epidemiol. Rev. **2:**113–135.

15. **Brown, P., C. J. Gibbs, Jr., H. L. Amyx, D. T. Kingsbury, R. G. Rohwer, M. P. Sulima, and D. C. Gajdusek.** 1982. Chemical disinfection of Creutzfeldt-Jakob disease virus. N. Engl. J. Med. **306:**1279–1282.

16. **Brown, P., P. Rodgers-Johnson, F. Cathala, C. J. Gibbs, Jr., and D. C. Gajdusek.** 1984. Creutzfeldt-Jakob disease of long duration: clinicopathological characteristics, transmissibility, and differential diagnosis. Ann. Neurol. **16:**295–304.

17. **Brown, P., R. G. Rohwer, and D. C. Gajdusek.** 1984. Sodium hydroxide decontamination of Creutzfeldt-Jakob disease virus. N. Engl. J. Med. **310:**727.

18. **Brown, P., R. G. Rohwer, E. Green, and D. C. Gajdusek.** 1982. Effect of chemicals, heat and histopathologic processing on high-infectivity hamster-adapted scrapie virus. J. Infect. Dis. **145:**683–687.

18a. **Brown, P. W., D. C. Gajdusek, C. J. Gibbs, Jr., and D. M. Asher.** 1985. Potential epidemic of Creutzfeldt-Jakob disease from human growth hormone therapy. N. Engl. J. Med. **313:**728–730.

19. **Burger, D., and G. R. Hartsough.** 1965. Encephalopathy of mink. II. Experimental and natural transmission. J. Infect. Dis. **115:**393–399.

20. **Chatigny, M. A., and S. B. Prusiner.** 1979. Biohazards and risk assessment of laboratory studies of the agents causing the spongiform encephalopathies, p. 491–514. In S. B. Prusiner and W. J. Hadlow (ed.), Slow transmissible diseases of the nervous system, vol. 2. Academic Press, Inc., New York.

21. **Chatigny, M. A., and S. B. Prusiner.** 1980. Biohazards of investigations on the transmissible spongiform encephalopathies. Rev. Infect. Dis. **2:**712–724.

22. **Cho, H. J.** 1980. Requirement of a protein component for scrapie infectivity. Intervirology. **14:**213–216.

23. **Davanipour, Z., L. Goodman, M. Alter, E. Sobel, D. M. Asher, and D. C. Gajdusek.** 1984. Possible modes of transmission of Creutzfeldt-Jakob disease. N. Engl. J. Med. **311:**1582–1583.

24. **Dickinson, A. G., and V. M. H. Meikle.** 1969. A comparison of some biological characteristics of the mouse-passaged scrapie agents 22-A and ME-7. Genet. Res. **13:**213–225.

25. **Diringer, H., H. Gelderblom, H. Hilmert, M. Özel, and C. Edelbluth.** 1983. Scrapie infectivity, fibrils and low molecular weight protein. Nature (London) **306:**476–478.

26. **Diringer, H., and R. H. Kimberlin.** 1983. Infectious scrapie agent is apparently not as small as recent claims suggest. Biosci. Rep. **3:**563–568.

27. **Duffy, P., G. Collins, A. G. DeVoe, B. Streeten, and D. Cohen.** 1974. Possible person-to-person transmission of Creutzfeld-Jakob disease. N. Engl. J. Med. **290:**693.

28. **Dunham, W. B.** 1977. Virucidal agents, p. 426–441. In S. S. Block (ed.), Disinfection, sterilization, and preservation, 2nd ed. Lea & Febiger, Philadelphia.

29. **Ehlers, B., and H. Diringer.** 1984. Dextran sulphate 500 delays and prevents mouse scrapie by impairment of agent replication in spleen. J. Gen. Virol. **65:**1325–1330.

30. **Eklund, C. M., W. J. Hadlow, and R. C. Kennedy.** 1963. Some properties of the scrapie agent and its behavior in mice. Proc. Soc. Exp. Biol. Med. **112:**974–979.

31. **Eklund, C. M., R. C. Kennedy, and W. J. Hadlow.** 1967. Pathogenesis of scrapie virus infection in the mouse. J. Infect. Dis. **117:**15–22.

32. **Ernst, R. R.** 1977. Sterilization by heat, p. 481–521. In S. S. Block (ed.) Disinfection, sterilization, and preservation, 2nd ed. Lea & Febiger, Philadelphia.

33. **Federal Register.** 1980. Interstate shipment of etiologic agents. Fed. Regist. **45:**48626–48627.

34. **Gajdusek, D. C.** 1977. Unconventional viruses and the origin and disappearance of kuru. Science **197:**943–960.

35. **Gajdusek, D. C., C. J. Gibbs, Jr., and M. Alpers.** 1966. Experimental transmission of a kuru-like syndrome to chimpanzees. Nature (London) **209:**794–796.

36. **Gajdusek, D. C., C. J. Gibbs, Jr., and D. M. Asher.** 1978. Precautions with Creutzfeldt-Jakob disease. N. Engl. J. Med. **298:**975–976.

37. **Gajdusek, D. C., C. J. Gibbs, Jr., D. M. Asher, P. Brown, A. Diwan, P. Hoffman, G. Nemo, R. Rohwer, and L. White.** 1977. Precautions in medical care of, and in handling tissues from, patients with transmissible virus dementia (Creutzfeldt-Jakob disease). N. Engl. J. Med. **297:**1253–1258.

38. **Gajdusek, D. C., C. J. Gibbs, Jr., G. Collins, and R. D. Traub.** 1976. Survival of Creutzfeldt-Jakob disease virus in formol-fixed brain tissue. N. Engl. J. Med. **294:**553.

39. **Gajdusek, D. C., C. J. Gibbs, Jr., K. Earle, G. J. Dammin, W. C. Schoene, and H. R. Tyler.** 1974. Transmission of subacute spongiform encephalopathy to the chimpanzee and squirrel monkey from a patient with papulosis maligna of Köhlmeier-Degos. Excerpta Med. Int. Congr. Ser. **319:**390–392.

40. **Gajdusek, D. C., and V. Zigas.** 1957. Degenerative disease of the central nervous system in New Guinea. The endemic occurrence of "kuru" in the native population. N. Engl. J. Med. **257:**954–978.

41. **Gibbs, C. J., Jr.** 1967. Search for infectious etiology in chronic and subacute degenerative diseases of the central nervous system. Curr. Top. Microbiol. Immunol. **40:**44–58.

42. **Gibbs, C. J., Jr., H. L. Amyx, A. Bacote, C. L. Masters, and D. C. Gajdusek.** 1980. Oral transmission of kuru, Creutzfeldt-Jakob disease, and scrapie to nonhuman primates. J. Infect. Dis. **142:**205–207.

43. **Gibbs, C. J., Jr., and D. C. Gajdusek.** 1978.

Atypical viruses as the cause of sporadic, epidemic, and familial chronic disease in man: slow viruses and human diseases, p. 161–198. *In* Perspectives in virology, vol. 10. Raven Press, New York.

44. **Gibbs, C. J., Jr., D. C. Gajdusek, D. M. Asher, M. P. Alpers, E. Beck, P. M. Daniels, and W. B. Matthews.** 1968. Creutzfeldt-Jakob disease (spongiform encephalopathy): transmission to the chimpanzee. Science **61:**388–389.

45. **Gibbs, C. J., Jr., D. C. Gajdusek, and R. Latarjet.** 1978. Unusual resistance to ionizing radiation of the viruses of kuru, Creutzfeldt-Jakob disease and scrapie. Proc. Natl. Acad. Sci. USA **75:**6268–6270.

46. **Gibbs, C. J., Jr., D. C. Gajdusek, and J. A. Morris.** 1965. Viral characteristics of the scrapie agent in mice, p. 195–202. *In* D. C. Gajdusek, C. J. Gibbs, Jr., and M. P. Alpers (ed.), Slow, latent, and temperate virus infections. NINDB monograph no. 2. U.S. Government Printing Office, Washington, D.C.

46a. **Gibbs, C. J., Jr., A. Joy, R. Heffner, M. Franko, M. Miyazaki, D. M. Asher, J. E. Parisi, P. W. Brown, and D. C. Gajdusek.** 1985. Clinical and pathological features and laboratory confirmation of Creutzfeldt-Jakob disease in a recipient of pituitary-derived human growth hormone. N. Engl. J. Med. **313:**734–738.

47. **Gordon, W. S.** 1946. Advances in veterinary research. Vet. Rec. **58:**516–520.

48. **Gordon, W. S.** 1966. Transmission of scrapie and evidence of spread of infection in sheep at pasture, p. 8–18. *In* Report of scrapie seminar. U.S. Government Printing Office, Washington, D.C.

49. **Goudsmit, J., C. H. Morrow, D. M. Asher, R. T. Yanagihara, C. L. Masters, C. J. Gibbs, Jr., and D. C. Gajdusek.** 1980. Evidence for and against the transmissibility of Alzheimer disease. Neurology **30:**945–950.

50. **Hadlow, W. J., R. C. Kennedy, T. A. Jackson, H. W. Whitford, and C. C. Boyle.** 1974. Course of experimental scrapie virus infection in the goat. J. Infect. Dis. **129:**559–567.

51. **Hadlow, W. J., R. E. Race, R. C. Kennedy, and C. M. Eklund.** 1979. Natural infection of sheep with scrapie virus, p. 3–12. *In* S. B. Prusiner and W. J. Hadlow (ed.), Slow transmissible diseases of the nervous system, vol. 2. Academic Press, Inc., New York.

52. **Haig, D. A., and M. C. Clarke.** 1965. Observations on the agent of scrapie, p. 215–219. *In* D. C. Gajdusek, C. J. Gibbs, Jr., and M. P. Alpers (ed.), Slow, latent, and temperate virus infections. NINDB monograph no. 2. U.S. Government Printing Office, Washington, D.C.

53. **Haig, D. A., and M. C. Clarke.** 1968. The effect of β-propiolactone on the scrapie agent. J. Gen. Virol. **3:**281–283.

54. **Hartley, E. G.** 1967. Action of disinfectants on experimental mouse scrapie. Nature (London) **213:**1135.

55. **Hintz, R., M. MacGillivray, A. Joy, and R. Tintner.** 1985. Fatal degenerative neurological disease in patients who received pituitary-derived human growth hormone. Morbid. Mortal. Weekly Rep. **34:**359–366.

56. **Hotchin, J.** 1979. Scrapie as a slow virus, p. 55–69.

In S. B. Prusiner and W. J. Hadlow (ed.), Slow transmissible diseases of the nervous system, vol. 2. Academic Press, Inc., New York.

57. **Hourrigan, J., A. Klingsporn, W. W. Clark, and M. DeCamp.** 1979. Epidemiology of scrapie in the United States, p. 331–356. *In* S. B. Prusiner and W. J. Hadlow (ed.), Slow transmissible diseases of the nervous system, vol. 1. Academic Press, Inc., New York.

58. **Hunter, G. D.** 1965. Progress toward the isolation and characterization of the scrapie agent, p. 259–262. *In* D. C. Gajdusek, C. J. Gibbs, Jr., and M. P. Alpers (ed.), Slow, latent, and temperate virus infections. NINDB monograph no. 2. U.S. Government Printing Office, Washington, D.C.

59. **Hunter, G. D.** 1979. The enigma of the scrapie agent: biochemical properties and the involvement of membranes and nucleic acids, p. 365–385. *In* S. B. Prusiner and W. J. Hadlow (ed.), Slow transmissible diseases of the nervous system, vol. 2. Academic Press, Inc., New York.

60. **Hunter, G. D., R. A. Gibbons, R. H. Kimberlin, and G. C. Millson.** 1969. Further studies of the infectivity and stability of extracts and homogenates derived from scrapie affected mouse brains. J. Comp. Pathol. **79:**101–108.

61. **Hunter, G. D., R. M. Kimberlin, G. C. Millson, and R. A. Gibbons.** 1971. An experimental examination of the scrapie agent in cell membrane mixtures. I. Stability and physicochemical properties of the scrapie agent. J. Comp. Pathol. **81:**23–32.

62. **Hunter, G. D., and G. C. Millson.** 1964. Further experiments on the comparative potency of tissue extracts from mice infected with scrapie. Res. Vet. Sci. **5:**149–153.

63. **Hunter, G. D., and G. C. Millson.** 1964. Studies on the heat stability and chromatographic behavior of the scrapie agent. J. Gen. Microbiol. **37:**251–258.

64. **Hunter, G. D., and G. C. Millson.** 1967. Attempts to release the scrapie agent from tissue debris. J. Comp. Pathol. **77:**301–307.

65. **Jellinger, K., F. Seitelberger, W. D. Heiss, and W. Holczabek.** 1972. Konjugale Form der subakuten spongiosen Enzephalopathie (Jakob-Creutzfeldt Erkrankung). Wein. Klin. Wochenschr. **84:**245–249.

66. **Kimberlin, R. H.** 1979. Early events in the pathogenesis of scrapie in mice: biological and biochemical events, p. 33–54. *In* S. B. Prusiner and W. J. Hadlow (ed.), Slow transmissible diseases of the nervous system, vol. 2. Academic Press, Inc., New York.

67. **Kimberlin, R. H., C. F. Millson, and G. D. Hunter.** 1971. Experimental examination of the scrapie agent in cell membrane mixtures. J. Comp. Pathol. **81:**383–391.

68. **Kimberlin, R. H., and C. A. Walker.** 1983. The antiviral compound HPA-23 can prevent scrapie when administered at the time of infection. Arch. Virol. **78:**9–18.

69. **Kirschbaum, W. R.** 1968. Jakob-Creutzfeldt disease. Elsevier, New York.

70. **Klein, M., and A. DeForest.** 1963. Antiviral action of germicides. Soap Chem. Spec. **39:**70–72; 95–97.

71. **Klitzman, R. L., M. P. Alpers, and D. C. Gajdusek.** 1984. The natural incubation period of kuru

and the episodes of transmission in three clusters of patients. Neuroepidemiology **3:**3–20.

71a. **Koch, T. K., B. O. Berg, S. J. DeArmond, and R. F. Gravina.** 1985. Creutzfeldt-Jakob disease in a young adult with idiopathic hypopituitarism: possible relationship to administration of cadaveric human growth hormone. N. Engl. J. Med. **313:**731–733.

72. **Koski, T. A., and J. H. S. Chen.** 1977. Methods of testing virucides, p. 116–134. *In* S. S. Block (ed.), Disinfection, sterilization, and preservation, 2nd ed. Lea & Febiger, Philadelphia.

73. **Kuroda, Y., C. J. Gibbs, Jr., H. L. Amyx, and D. C. Gajdusek.** 1983. Creutzfeldt-Jakob disease in mice: persistent viremia and preferential replication of virus in low-density lymphocytes. Infect. Immun. **41:**154–161.

74. **Latarjet, R.** 1979. Inactivation of the agents of scrapie, Creutzfeldt-Jakob disease, and kuru by radiations, p. 387–407. *In* S. B. Prusiner and W. J. Hadlow (ed.), Slow transmissible diseases of the nervous system, vol. 2. Academic Press, Inc., New York.

75. **Latarjet, R., B. Muel, D. A. Haig, M. C. Clarke, and T. Alper.** 1970. Inactivation of the scrapie agent by near monochromatic ultraviolet light. Nature (London) **227:**1341–1343.

76. **Lavelle, G. C.** 1972. Large fraction of scrapie virus resistant to lipid solvents. Proc. Soc. Exp. Biol. Med. **141:**460–462.

77. **Manuelidis, E. E., J. N. Angelo, E. J. Gorgacz, J. H. Kim, and L. Manuelidis.** 1977. Experimental Creutzfeldt-Jakob disease transmitted via the eye with infected cornea. N. Engl. J. Med. **296:**1334–1336.

78. **Manuelidis, E. E., E. J. Gorgacz, and L. Manuelidis.** 1978. Viremia in experimental Creutzfeldt-Jakob disease. Science **200:**1069–1071.

79. **Marsh, R. F., D. Burger, R. Eckroade, G. M. Zu Rhein, and R. P. Hanson.** 1969. A preliminary report on the experimental host range of the transmissible mink encephalopathy agent. J. Infect. Dis. **120:**713–719.

80. **Marsh, R. F., D. Burger, and R. P. Hanson.** 1969. Transmissible mink encephalopathy: behavior of the disease agent in mink. Am. J. Vet. Res. **30:**1637–1642.

81. **Marsh, R. F., and R. P. Hanson.** 1969. Physical and chemical properties of the transmissible mink encephalopathy agent. J. Virol. **3:**176–180.

82. **Marsh, R. F., and R. P. Hanson.** 1979. On the origin of transmissible mink encephalopathy, p. 451–460. *In* S. B. Prusiner and W. J. Hadlow (ed.), Slow transmissible diseases of the nervous system, vol. 1. Academic Press, Inc., New York.

83. **Masters, C. L., J. O. Harris, D. C. Gajdusek, C. J. Gibbs, Jr., C. Bernoulli, and D. M. Asher.** 1979. Creutzfeldt-Jakob disease: patterns of worldwide occurrence and the significance of familial and sporadic clustering. Ann. Neurol. **5:**177–188.

84. **Matthews, W. B.** 1975. Epidemiology of Creutzfeldt-Jakob disease in England and Wales. J. Neurol. Neurosurg. Psychiatry **38:**210–213.

85. **Merz, P. A., R. G. Rohwer, H. M. Kascsak, H. M. Wisniewski, R. A. Somerville, C. J. Gibbs, and D. C. Gajdusek.** 1984. Infection-specific particle from the unconventional slow virus diseases. Science **225:**437–440.

86. **Merz, P. A., R. A. Somerville, H. M. Wisniewski, and K. Iqbal.** 1981. Abnormal fibrils from scrapie-infected brain. Acta Neuropathol. **54:**63–74.

87. **Millson, G. C., G. D. Hunter, and R. H. Kimberlin.** 1976. The physico-chemical nature of the scrapie agent, p. 243–266. *In* R. H. Kimberlin (ed.), Slow virus diseases of animals and man. North-Holland Publishing Co., Amsterdam.

88. **Millson, G. C., and E. J. Manning.** 1979. The effect of selected detergents on scrapie infectivity, p. 409–424. *In* S. B. Prusiner and W. J. Hadlow (ed.), Slow transmissible diseases of the nervous system, vol. 2. Academic Press, Inc., New York.

89. **Mould, D. L., A. M. Dawson, and W. Smith.** 1965. Scrapie in mice: the stability of the agent to various suspending media, pH and solvent extraction. Res. Vet. Sci. **6:**151–154.

90. **Narang, H. K.** 1974. An electron microscopic study of natural scrapie sheep brain: further observations on virus-like particles and paramyxovirus-like tubules. Acta Neuropathol. **28:**317–329.

91. **Nevin, S., W. H. McMenemy, D. Behrman, and D. P. Jones.** 1960. Subacute spongiform encephalopathy: a subacute form of encephalopathy attributable to vascular dysfunction (spongiform cerebral atrophy). Brain **83:**519–564.

92. **Oesch, B., D. Westaway, M. Wälchli, M. P. McKinley, S. H. B. Kent, R. Aebersold, R. A. Barry, P. Tempst, D. B. Teplow, L. E. Hood, S. B. Prusiner, and C. Weissmann.** 1985. A cellular gene encodes scrapie PrP 27–30 protein. Cell **40:**735–746.

93. **Palsson, P. A.** 1979. Rida (scrapie) in Iceland and its epidemiology, p. 357–366. *In* S. B. Prusiner and W. J. Hadlow (ed.), Slow transmissible diseases of the nervous system, vol. 1. Academic Press, Inc., New York.

94. **Pattison, I. H.** 1965. Experiments with scrapie with special reference to the nature of the agent and the pathology of the disease, p. 249–257. *In* D. C. Gajdusek, C. J. Gibbs, Jr., and M. Alpers (ed.), Slow, latent, and temperate virus infections. NINDB monograph no. 2. U.S. Government Printing Office, Washington, D.C.

95. **Pattison, I. H., M. N. Hoare, J. Jebbett, and W. A. Watson.** 1972. Spread of scrapie to sheep and goats by oral dosing with foetal membranes from scrapie-affected sheep. Vet. Rec. **90:**465–468.

96. **Pattison, I. H., and G. C. Millson.** 1960. Further observations on the experimental production of scrapie in goats and sheep. J. Comp. Pathol. **70:**182–196.

97. **Pattison, I. H., and G. C. Millson.** 1961. Scrapie produced experimentally in goats with special reference to the clinical syndrome. J. Comp. Pathol. **71:**101–108.

97a. **Powell-Jackson, J., R. O. Weller, P. Kennedy, M. A. Preece, E. M. Whitcombe, and J. Newsome-Davis.** 1985. Creutzfeldt-Jakob disease after administration of human growth hormone. Lancet **ii:**244–246.

98. **Prusiner, S. B.** 1982. Novel proteinaceous infectious particles cause scrapie. Science **216:**136–144.

99. **Prusiner, S. B.** 1984. Some speculations about prions, amyloid and Alzheimer's disease. N. Engl. J. Med. **310:**661–663.

100. **Prusiner, S. B., D. F. Groth, D. C. Bolton, S. B. Kent, and L. E. Hood.** 1984. Purification and structural studies of a major scrapie prion protein. Cell **38**:127–134.

101. **Prusiner, S. B., D. F. Groth, S. P. Cochran, F. R. Masiarz, M. P. McKinley, and H. M. Martinez.** 1980. Molecular properties, partial purification, and assay by incubation period measurements of the hamster scrapie agent. Biochemistry **19**:4883–4891.

102. **Prusiner, S. B., D. F. Groth, S. P. Cochran, M. P. McKinley, and F. R. Masiarz.** 1980. Gel electrophoresis and glass permeation chromatography of the hamster scrapie agent after enzymatic digestion and detergent extraction. Biochemistry **19**:4892–4898.

103. **Prusiner, S. B., D. F. Groth, M. P. McKinley, S. P. Cochran, K. A. Bowman, K. C. Kasper, and F. R. Masiarz.** 1981. Thiocyanate and hydroxyl ions inactivate the scrapie agent. Proc. Natl. Acad. Sci. USA **78**:6675–6679.

104. **Prusiner, S. B., M. P. McKinley, K. A. Bowman, D. C. Bolton, P. E. Bendheim, D. F. Groth, and G. G. Glenner.** 1983. Scrapie proteins aggregate to form amyloid-like birefringent rods. Cell **35**:349–358.

105. **Rohwer, R. G.** 1984. Virus-like sensitivity of scrapie agent to heat inactivation. Science **223**:600–602.

106. **Rohwer, R. G.** 1984. Scrapie infectious agent is virus-like in size and susceptibility to inactivation. Nature (London) **308**:658–662.

107. **Rohwer, R. G., and D. G. Gajdusek.** 1980. Scrapie—virus or viroid. The case for a virus, p. 333–355. *In* A. Boese (ed.), Search for the cause of multiple sclerosis and other chronic diseases of the central nervous system. Verlag Chemie, Weinheim, Federal Republic of Germany.

108. **Sanders, W. L., and T. L. Dunn.** 1973. Creutzfeldt-Jakob disease treated with amantadine. J. Neurol. Neurosurg. Psychiatry **36**:581–584.

109. **Sigurdsson, B.** 1954. Observations on three slow infections of sheep. Br. Vet. J. **110**:255–270; 307–322; 341–354.

110. **Somerville, R. A., G. C. Millson, and R. H. Kimberlin.** 1980. Sensitivity of scrapie infectivity to detergents and 2-mercaptoethanol. Intervirology **13**:126–129.

111. **Stamp, J. T.** 1959. Symposium on scrapie, p. 295–299. *In* 63rd Annual Meeting of U.S. Livestock Sanitary Association. MacCrellish and Quigley, Trenton, N.J.

112. **Stamp, J. T., J. G. Brotherston, I. Zlotnik, J. M. K. Mackay, and W. Smith.** 1959. Further studies on scrapie. J. Comp. Pathol. **69**:268–280.

113. **Tateishi, J., M. Ohata, M. Koga, Y. Sato, and Y. Kuroiwa.** 1979. Transmission of chronic spongiform encephalopathy with kuru plaques from humans to small rodents. Ann. Neurol. **5**:581–584.

114. **Tateishi, J., Y. Sato, H. Koga, M. Koga, H. Doi, and M. Ohta.** 1980. Experimental transmission of human spongiform encephalopathy to small rodents. I. Clinical and histological observations. Acta Neuropathol. **51**:127–134.

115. **Traub, R. D., D. C. Gajdusek, and C. J. Gibbs, Jr.** 1974. Precautions in conducting biopsies and autopsies on patients with presenile dementia. J. Neurosurg. **41**:394–395.

116. **Traub, R. D., D. C. Gajdusek, and C. J. Gibbs, Jr.** 1975. Precautions in autopsies on Creutzfeldt-Jakob disease. Am. J. Clin. Pathol. **64**:287.

117. **U.S. Department of Health and Human Services.** 1984. Biosafety in microbiological and biomedical laboratories. HHS Publication no. (CDC)84–8395. U.S. Government Printing Office, Washington, D.C.

118. **U.S. Environmental Protection Agency.** 1982. Draft manual for infectious waste management. EPA report SW-957. U.S. Government Printing Office, Washington, D.C.

119. **Valenti, W. M., J. F. Hruska, M. A. Mengus, and M. J. Freburn.** 1981. Nosocomial viral infections. III. Guidelines for prevention and control of exanthematous viruses, gastroenteritis viruses, and uncommonly seen viruses. Infect. Control. **2**:38–49.

120. **Walter, C. W.** 1978. Precautions with Creutzfeldt-Jakob disease. N. Engl. J. Med. **298**:975.

121. **Will, R. G., and W. B. Matthews.** 1982. Evidence for case-to-case transmission of Creutzfeldt-Jakob disease. J. Neurol. Neurosurg. Psychiatry **45**:235–238.

122. **Williams, E. S., and S. Young.** 1980. Chronic wasting disease of mule deer: a spongiform encephalopathy. J. Wild. Dis. **16**:89–98.

Chapter 11

Transmission and Control of Viral Zoonoses in the Laboratory

DONALD C. BLENDEN AND HANS K. ADLDINGER

Accidentally acquired infections are an inherent risk in any laboratory which purposefully or unknowingly handles microbes. Even though laboratory-acquired infections are likely underreported, there is ample evidence of their importance, and some have the potential of being very serious or fatal. It is the objective of this chapter to focus on the risks involved in the handling of various viruses causing zoonotic disease, with the ultimate goal of minimizing or preventing the risk of exposure and thus preventing the disease from occurring in laboratory workers. Studies have shown that exposures to zoonotic infections are in fact a type of accident, and that accident-prone persons are therefore at greater risk of acquiring zoonotic disease along with other types of accidents (50). Transferring this information to the laboratory setting, the supervisor may wish to assess the accident-prone tendency of potential employees during recruitment. The World Health Organization has published detailed guidelines for the classification of risks and the prevention and management of laboratory accidents involving infectious agents. The reader seeking detailed information in general laboratory management or regarding specific risks is encouraged to seek out this discussion (63).

Of the approximately 150 diseases transmitted between animals and the human being (the zoonoses), a significant proportion are induced by viruses. The laboratory-associated risk of human exposure to and subsequent infection from these viruses falls into two categories. First, laboratory personnel working with known agents, such as in a research, teaching, or vaccine production laboratory, can and should be ready for containment of certain agents under routine and accidental release conditions. Second, a different type of risk exists in the diagnostic laboratory, where an agent of high virulence can suddenly be identified in clinical specimens after considerable exposure to personnel has already occurred. Prearranged infection containment plans of the diagnostic laboratory should be sufficiently comprehensive to contain the most extreme possible risk to be so encountered. Even a specialized diagnostic laboratory (e.g., a rabies laboratory, wherein specimens submitted have been preselected because of tentative diagnoses or the need to determine the presence of a particular agent) may discover that an agent is being handled for which preparation is inadequate. Laboratory personnel may thus be unexpectedly exposed to a new risk, under compromised circumstances. It is under such circumstances that human infections may result from an unsuspected or previously unrecognized agent and a new "emerging zoonosis" be the result (35, 44).

It has also been recommended that preemployment and yearly physical examinations be used as a means of employee protection and surveillance (with special provision and selected duty for pregnant employees) and that employees receive annual or biannual formal training in laboratory safety (53).

Microbiological agents and the conditions under which they should be handled have been classified according to the risk posed by the agent (57, 60). These containment levels and animal biosafety levels are described elsewhere in this volume. This chapter will deal with individual families of zoonotic viruses, based on the most recent taxonomic classification, by presenting each family within the format of general characteristics and transmission, laboratory-associated infections, and prevention and control measures in the laboratory.

ARENAVIRIDAE

General characteristics

The small family *Arenaviridae* contains the viruses of lymphocytic choriomeningitis (LCM), Lassa fever, and the Tacaribe complex. Arenaviruses are restricted in their natural occurrence to South American countries and Florida (New World viruses: Tacaribe complex) and several African states (Old World viruses: Lassa fever and Mozambique virus). An exception is the possibly worldwide LCM virus, the type species of the family. The serologic relationships among arenaviruses correspond to their geographic distribution. Old World viruses, for example, are distinguishable by immunofluorescence analysis from New World viruses, which are closely related to each other, and all are related to LCM virus (25, 34).

Most arenaviruses are restricted in nature to a single rodent host (exception: Tacaribe virus maintained in bats) constituting the reservoir. They cause persistent, largely inapparent infections and probably in all cases involve insufficient or delayed immune reactivity, producing lifelong virus carriers and shedders. Infection of humans and other animals appears to be accidental and infrequent. LCM, Lassa fever, Junin, and Machupo viruses are the presently known human pathogens among the arenaviruses. LCM causes influenzalike, meningeal or meningo-encephalomyelitic illnesses that rarely take a severe or fatal course; fetal infections seem more likely to be serious. In contrast, Lassa fever, Junin (Argentine hemorrhagic fever), and Machupo (Bolivian hemorrhagic fever) viruses may cause severe hemorrhagic infections. Outbreaks of the hemorrhagic fever infections in Argentina and Bolivia, mostly among farm workers during the harvest season, have resulted in case-fatality rates of over 20%. Occasional secondary human contact infections have been reported. Secondary and even tertiary cases of infection with Lassa fever virus occur frequently, especially among hospitalized patients and hospital personnel, with case-fatality rates approaching 50%. Lassa fever has a lesser hemorrhagic component than its South American counterparts but involves more organ systems and, in fatal cases, leads to circulatory collapse. While clinically inapparent infections with Junin and Machupo viruses are uncommon, they also may occur with Lassa fever, posing a serious risk of contact exposure (1, 10, 25).

Muridae (mice and rats) provide a natural reservoir for LCM and for Lassa virus, while the Cricetidae (voles, lemmings, gerbils, etc.) are the natural reservoirs for the other arenaviruses. Transuterine, transovarian, and postpartum infections (virus is contained in milk, urine, and feces) maintain the viruses in their natural hosts. Spread to humans occurs horizontally via contaminated urine, feces, aerosols, grain, water, and dust. Characteristically, Argentine hemorrhagic fever infections occur predominantly during the major harvest season of April to June when agricultural workers move into the endemic areas (1, 25).

It is also noteworthy that in recent years 181 human LCM infections in the United States were associated with pet hamsters (23). Although arenavirus isolations have been made from arthropods, they are not considered an essential component in the cycle of natural transmission (51).

Laboratory-associated infections

The virus of LCM is easily transferred to humans by aerosol or direct contact when infected laboratory animals are handled. Susceptible laboratory species include albino and gray mice, rats, dogs, hamsters, guinea pigs, monkeys, chimpanzees, and chicken embryos. The most susceptible laboratory animals are guinea pigs and white mice. Rabbits, pigs, and birds are refractory to infection and present minimal risk. The excreta of laboratory animals, whether dry or moist, can be dust-borne or aerosolized, resulting in transmission by inhalation. Vertical transmission of the viruses easily occurs, and animal colonies can be infected for many generations but show no signs of infection. This possibility can only be detected by prospective serological surveillance, commonly carried out for the detection of subclinical LCM, or retrospectively upon the occurrence of a human case. Presymptomatic shedding of virus does not seem likely, with the exception of LCM (1, 10).

The other arenaviruses are transmitted by routes comparable to that of LCM, although there is presumably no concern of the persistent, subclinical infection in laboratory animals which characterizes LCM. All tissues, fluids, or excreta from animals and any possible in vitro sources of virus must be considered as infectious and highly dangerous.

Lassa fever virus from in vitro sources is highly infectious for laboratory animals and human beings, and therefore extreme measures of protection must be employed. The virus has occurred in laboratory workers, and risk potential is always high. Medical personnel having contact with cases of both Lassa and Machupo fever have been secondarily infected. The long incubation period of some infections may tend to mask the true source of infection. The agent of Korean hemorrhagic fever has not been well described, and little is known about the laboratory risk involved in its handling, although a few laboratory-acquired cases are reported (48).

Prevention and control

The highest standards of hygiene and containment in laboratory and animal quarters are imperative to prevent exposure to arenaviruses. Careful and stringent steps must be taken to prevent aerosol generation and inhalation of aerosols, remembering that common-source exposure can infect several persons at the same time. Direct contact with skin or mucous membranes and puncture wounds is an excellent means of exposure. Special care must be exercised with the bedding and excreta of potentially infective laboratory animals. Rodent control in the laboratory environment is mandatory. Vaccines are not available to immunize personnel against arenaviruses.

The arenaviruses are high in risk potential for the laboratory worker. LCM virus is classified in Biosafety Level 3 (57) for risk management and facility design and operation. Lassa fever virus and the remaining arenaviruses are placed in Biosafety Level 4, a maximum containment facility (57), which is available at few locations in the world. Details of the design and utilization of these biosafety level facilities are easily available and must be consulted and followed (57).

BUNYAVIRIDAE

General characteristics

At least 145 of over 200 arboviruses assigned to the family *Bunyaviridae* are related antigenically and form the genus *Bunyavirus*. They are subdivided serologically into 16 groups and contain the type species Bunyamwera virus as well as Germiston, Oropouche, California encephalitis, and La Crosse viruses, among others. Other genera of bunyaviruses are *Phlebovirus* (sandfly fever group), *Nairovirus* (Nairobi sheep disease and related viruses), and *Uukuvirus* (Uukuniemi and related viruses), all formed on the basis of serologic relatedness (5, 34, 51).

Members of the *Bunyaviridae* infect various warm- and cold-blooded vertebrates and arthropods, humans being accidental and terminal hosts (Table 1). While inapparent infection in humans is common in geographic areas of *Bunyavirus* occurrence, as detected by the presence of humoral antibodies, clinical disease is rare. Dengue-like fevers, however, sometimes complicated by hemorrhages, jaundice, and encephalitis and resulting in death, have been caused by Rift Valley fever virus. After disease outbreaks in sheep and cattle, several severe epidemics have occurred in South Africa, Zimbabwe, and Egypt, involving tens of thousands of human cases and the death of hundreds of thousands of animals. Several other members of the *Bunyaviridae* have caused sporadic human cases of dengue-like fevers (sandfly fever group, Bwamba, Bunyamwera, and Oropouche viruses and others) and encephalitis (California, Tensaw, and La Crosse viruses) (1, 5, 51).

Most of the bunyaviruses are transmitted by mosquitoes, and a few (the Tete subgroup) also by ticks. For some viruses assigned to the other genera, transmission by mosquitoes, ticks, or sandflies has been determined. The vertebrate reservoir hosts, i.e., those capable of producing a viremia of a level and duration sufficient to ensure infection of the arthropod host, have not been determined for all viruses. Wild rodents, rabbits, and squirrels and in some instances larger wild and domestic mammals were found to be involved in the biological cycle of many of the viruses. Reptiles and amphibians may play a role as over-wintering hosts, and birds may serve in distributing the virus over long distances. Transova-

TABLE 1. Bunyaviruses of zoonotic importance

Genus	Subgroup	Examples of viruses	Vector	Geographical distribution	Human disease syndromes
Bunyavirus	Bunyamwera	Bunyamwera	Mosquito	Africa	Fever
		Germiston	Mosquito	Africa	Fever
		Tensaw	Mosquito	United States	Encephalitis
	C-Group	Apeu	Mosquito	South America	Fever
		Caraparu	Mosquito	South America	Fever
		Marituba	Mosquito	South America	Fever
	California	California	Mosquito	North America	Encephalitis
		La Crosse	Mosquito	North America	Encephalitis
		Keystone	Mosquito	North America	Encephalitis
		Tahyna	Mosquito	Europe	Fever
	Guama	Catu	Mosquito	South America	Fever
	Simbu	Oropouche	Mosquito	South America	Fever
	Bwamba	Bwamba	Mosquito	Africa	Fever
		Pongola	Mosquito	Africa	Fever
	Tete	Tete	Tick	Africa	Fever
Phlebovirus	None	Sicilian SF	Phlebotomine	Europe, Africa	Fever with rash and myalgia
		Candiru	Phlebotomine	South America	Fever with rash and myalgia
		Rift Valley fever	Mosquito	Africa	Fever, hemorrhagic fever, encephalitis
Nairovirus	Crimean-Congo hemorrhagic fever	Congo strain	Tick	Africa, Asia, Europe	Hemorrhagic fever
	Nairobi sheep disease	Nairobi sheep disease	Tick	Africa	Fever
		Dugbe	Tick	Africa	Fever

rial and venereal transmission for some mosquito-borne viruses (e.g., La Crosse) was also demonstrated. In addition to arthropod transmission, infection by aerosol and direct contact with infected animals, fresh meat, and tissues or clinical specimens may be of epidemiological significance, as has been shown during Rift Valley fever outbreaks and laboratory infections. This fact establishes an occupational risk for veterinarians, farmers, butchers, and laboratory personnel (1, 22, 48, 51).

Laboratory-associated infections

The most important agent of the family *Bunyaviridae* is the virus of Rift Valley fever (48). Numerous laboratory infections have occurred with this agent, presumably because of contact with infected tissue, although the specific source of many cases has been unknown. Germiston and Apeu viruses are also known to have resulted in laboratory-acquired infections. It must be presumed that all members of the *Bunyaviridae* are potentially infectious for humans, and that the lack of recognition of the agent in the literature should not be confused with its absence or lack of virulence.

Prevention and control

An unlicensed vaccine is available for those who must handle Rift Valley fever virus for any purpose (48). Vaccines for protection against other bunyaviruses are not available. General preventive measures should include special attention to outer clothing, personal hygiene, careful handling of animals and their excreta, and prevention of aerosol or inoculation exposure. As of 31 December 1979 the American Committee on Arthropod-Borne Viruses had registered 424 arboviruses. The Committee's Subcommittee on Arbovirus Laboratory Safety has, on the basis of risks, assigned these 424 arboviruses to Biosafety Levels 2 through 4, as shown in the Subcommittee's publication *Laboratory Safety for Arboviruses and Certain Other Viruses of Vertebrates* (48). Ninety-four are assigned to Levels 3 and 4; seven, which are indigenous or commonly used in laboratories, are assigned to Level 2 at least under some circumstances. The handling conditions for these 101 most dangerous of the arboviruses are included elsewhere in this book (Appendix 1; 57).

HERPESVIRIDAE

General characteristics

The herpesviruses form a large group of morphologically similar viruses and probably can be isolated from all vertebrates. Based on genome characteristics and properties such as host range, reproductive cycle, and antigenic relationship, three subfamilies have been established: *Alphaherpesvirinae* (herpes simplex viruses 1 and 2, varicella-zoster virus, bovine mammillitis virus, pseudorabies virus of swine, equine abortion virus, simian herpes [B] virus, and others as probable members of the group); *Betaherpesvirinae* (cytomegaloviruses of humans and animals); and *Gammaherpesvirinae* (Epstein-Barr virus of humans and several other probable and possible members, including herpesvirus saimiri, herpesvirus ateles, and Marek's disease virus of chickens [34]).

The simian B virus has long been recognized as one of the most dangerous zoonotic agents because of the severe consequences of infection in humans. Pseudorabies virus of swine (Aujeszky's disease) can cause mild illness in humans, while herpes simplex and varicella-zoster viruses may cause more severe disease in nonhuman primates which then could become potential sources for human infection (22, 27, 62).

Herpesviruses generally establish persistent, often latent infections in the young host, resulting in mild disease or clinically inapparent infection. Infection with herpes simplex viruses may cause the well-known local vesicular lesions on the skin and mucous membranes of the mouth, gums, and genitals (cold sores, fever blisters). Infection with varicella-zoster virus causes the rashes associated with chickenpox and shingles, while cytomegalovirus infections usually remain asymptomatic. The Epstein-Barr virus causes infectious mononucleosis and has some still undefined role in human neoplasms (e.g., Burkitt's lymphoma, nasopharyngeal carcinoma). Frequently, latent herpesviruses are provoked into recurrent episodes of activity with or without clinical manifestations. Severe or life-threatening disease (primary or recurrent) is the exception but can be precipitated by circumstances such as virus exposure of the fetus or newborn, immune deficiencies, eczema and other concurrent disease or secondary infection, and severe stresses such as high fever, traumas, massive blood transfusions, immunosuppressive drug treatment, and others. Among the serious complications are encephalitis, meningitis, eczema herpeticum, hepatitis, cytomegalic inclusion disease syndromes following congenital infection, and pneumonitis. The capacity of herpesviruses to cause severe or fatal disease is most strikingly expressed in some cases of infection of the atypical (unadapted) host. The accidental infection of humans with B virus has in nearly all reported cases resulted in either fatal encephalomyelitis or severe neurologic sequelae in most of the few survivors. In monkeys of the *Macaca* genus (reservoir host) this virus causes a persistent, usually mild or asymptomatic infection much like herpes simplex virus infections in humans (36, 46, 64).

Herpesviruses of humans and animals are relatively labile in the extracellular environment; spread between infected and susceptible hosts usually requires close contact. Vesicular skin lesions are a rich source of infectious virus, and direct contact with abraded skin of a recipient will transfer the infection. Equally effective transmission occurs also in the absence of apparent lesions, and virus shedding by asymptomatic carriers is probably the most important mechanism for the maintenance of herpesviruses in nature. Infectious virus has been demonstrated in secretions of the oral-respiratory and urogenital tracts, in the conjunctival sacs, and in milk and urine. Spread via aerosols, dust, contaminated fomites, bites, and scratches, as well as venereally, is well documented. Transmission from mother to fetus in utero or at birth occurs frequently with beta herpesviruses and several members of the alpha subfamily (22, 46).

Laboratory-associated infections

Humans are the only known reservoir host of herpes simplex viruses, but several nonhuman primate species are susceptible to experimental as well as natural infection and disease. Apparently, animal caretakers were the sources of herpesvirus that caused generalized and fatal infections in gibbons, owl monkeys, marmosets, and tree shrews. The possibility of accidental transmission (beginning during the incubation period) from monkeys back to laboratory personnel obviously exists also. However, the actual risk of developing disease from these animal sources is probably negligible unless aggravating circumstances in the human host such as mentioned above coincide (1, 46, 62).

The overt or latent presence of B virus in captive monkeys and diagnostic or experimental specimens presents a serious hazard to animal handlers and laboratory personnel. At least 25 human cases of B virus infection appear to have occurred, including the first reported case in 1934 (32). Clinical signs developing after infection may depend in part on the portal of virus entry. Local herpetic skin lesions with regional lymphadenopathy, pharyngitis, pneumonia, abdominal pain, diarrhea, and others have been described. These are accompanied or followed by ascending myelitis, paralysis, and various other neurological signs. Only five patients have survived the final encephalopathy, three with severe neurological sequelae. All patients were involved in research or vaccine production utilizing monkeys (mostly rhesus) or monkey tissues. The modes of transmission of the infections in the majority of cases were monkey bite and scratch wounds, contamination of existing skin laceration, and skin perforations with injection needles or broken glass from a vessel containing primary monkey kidney cell culture. For several cases the source of the infection was not determined, indicating that virus transmission could have been effected by aerosol, dust, or other contaminated fomites. In addition, direct contact of a less dramatic nature than bites could have gone unnoticed and also resulted in infection via the conjuctiva, the oral route, or the respiratory tract. Contributing to the uncertainties related to the transmission of the virus is the occasional occurrence of incubation periods of several weeks instead of days as documented in some cases, as well as the report of one case which apparently was caused by reactivation of a latent infection (27, 46, 58, 62).

Prevention and control

In the absence of a licensed vaccine for human use, the emphasis of safety measures must be on prevention of exposure. B virus is extremely hazardous to laboratory personnel and requires Biosafety Level 4 containment when working with the virus or diagnostic materials with the potential of containing B virus (57).

Several additional points deserve special consideration. On the basis of observations of prolonged incubation periods and of at least one probable case of reactivation of a latent infection, the possibility that latent B virus infection in humans can indeed occur must be taken into account. Provocation of recurrent B virus infection in monkeys by stress factors such as shipping, crowding, and experimentation will increase the chance and dose of exposure to humans or other animals. Although not observed so far, person-to-person spread cannot be ruled out. In case of infection by bite or other accident, immediate cleaning of the wound with soap and water and local application of antiseptics have been recommended. The wound should be scrubbed vigorously to make it bleed. Expert medical follow-up must be sought immediately for passive immunization, the collection of samples from patient and animal, and further treatment as needed (27, 46). Other viruses of the family *Herpesviridae* fall into Biosafety Levels 2 and 3 for handling.

ORTHOMYXOVIRIDAE

General characteristics

The *Orthomyxoviridae* family is formed by human and animal influenza viruses. They are morphologically and antigenically complex viruses, subdivided into types A, B, and C according to the specificity of complement-fixing internal nucleoprotein and matrix antigens. Further subdivision of types into subtypes and strains presently is based on differences between multiple determinants of viral surface antigens (hemagglutinin and neuraminidase). Numerous different antigen subtypes, as well as antigenic variation within subtypes, have been demonstrated among influenza A virus strains isolated from humans, pigs, horses, and birds. Less antigenic variation was found in influenza B strains; few C strains have been studied so far (8, 15, 34).

Influenza viruses of types A and B produce respiratory and systemic manifestations in over half of persons infected, whereas illness produced by type C virus is mild, common cold-like, or subclinical. Recovery is usually rapid, but complications in the form of bacterial superinfection pose a serious threat to the aged, those with concurrent lung and heart disease, and children (14). Certain influenza A virus strains infect a variety of mammals and birds, both domestic and wild, and can cause enzootic and epizootic respiratory or generalized disease. Influenza B and C viruses have so far not been shown to occur naturally in animal populations. Several cases of human infection and disease with animal influenza A viruses have been reported (15).

Large numbers of virus particles are carried from infected persons and animals by aerosols produced by sneezing and coughing. Infection of new hosts occurs via inhalation of these aerosols. Transmission by direct contact, contaminated fomites, feces, and water is also possible, but the virus loses infectivity rapidly when exposed to heat, sunlight, and acid conditions. Human influenza occurs sporadically as a subclinical infection or as mild (and often unrecognized) respiratory illness interspersed between outbreaks in populations with acquired immunity to the prevalent virus subtype. Minor antigenic variations of the virus (antigenic drift) occur, resulting in fast-spreading outbreaks every 2 to 3 years. Residual immunity to the previous virus subtype provides partial protection against the cross-reacting new virus, resulting in low mortality although morbidity may be high. A major antigenic change (antigenic shift with appearance of a new hemagglutinin or neuraminidase subtype, or both), which is thought to be caused by reassortment of whole segments of the viral genome in mixed infections, can lead to devastating pandemics with high mortality (8, 14, 52).

Similar mechanisms of transmission, the immune status of the population, and antigenic variation of the virus appear to influence the ecology and distribution of animal influenza. Waterfowl and pigs have been incriminated as important reservoirs for human influenza A viruses, with animals constituting long-term hosts for previously epidemic subtypes as well as being carriers of current ones. Seroepidemiological studies, experimental subtype recombination in vitro and in vivo, and the epidemic experiences with swine influenza (or closely related) virus in human cases have provided evidence supporting the above concepts (1, 15, 22, 35, 52).

Laboratory-associated infections

Type A virus is the only known zoonotic risk to humans among the influenza viruses. The innocuous nature of these viruses may, in fact, result in careless handling. Thus, with the increased risk of an accident, the risk of exposure to high concentrations of influenza virus also increases. Pike lists a total of 15 laboratory-associated infections in the world literature; 1 of these cases was fatal (43). In addition to laboratory sources of high concentrations of virus generated into aerosols, the possible risk of contact with infected nonhuman primates and ferrets (57) must be considered.

Prevention and control

Influenza virus A/PR/8/34 is categorized for Biosafety Level 1 risk with no restrictions placed on its use and handling. All other strains of influenza virus are placed in Biosafety Level 2 for handling and containment of aerosols (57).

The nature and risk of laboratory-acquired infection via exposure to an artificially large dose of the virus are not well described. Immunization with the antigenic strains being handled may be advisable if high concentrations of virus are to be involved, especially if any personnel have any of the known high-risk characteristics, e.g., predisposing illness.

PARAMYXOVIRIDAE

General characteristics

Three genera are within the family *Paramyxoviridae*: *Paramyxovirus* (parainfluenza, mumps, Newcastle disease viruses), *Morbillivirus* (measles, canine distemper, rinderpest viruses), and *Pneumovirus* (respiratory syncytial virus [RSV]). All family members have a common general morphology, but antigenic relationships exist only within genera (34). Diseases caused in humans by parainfluenza virus and RSV involve the respiratory tract and are mild, common cold-like, or inapparent. However, in infants and young children these agents can be important causes of severe croup, bronchiolitis, and pneumonia, frequently as nosocomial infections. Mumps, a rather trivial and often inapparent infection of children and young adults, may be complicated by orchitis, by inflammation of salivary glands other than the parotids, and by aseptic meningitis. Successful vaccination against measles has reduced the occurrence and awareness of this potentially dangerous disease. Complications such as otitis media, pneumonia, and encephalomyelitis occur commonly and are sometimes severe. An increase in the number of nonimmune individuals raises morbidity and mortality dramatically, as is known from epidemics in isolated or undernourished populations. Accidental infection with the avian pathogen Newcastle disease virus (NDV) may result in conjunctivitis which is often unilateral. Although headache is frequently associated with NDV conjunctivitis, other systemic reactions are unusual (1, 11, 21, 24, 40, 41).

Human NDV infections have a high occupational risk and occur in poultry workers, meat processors in slaughterhouses, and research personnel handling infected birds or virus preparations. A possible source of human measles virus infection may be nonhuman primates that can develop a mild form of measles (may be severe in marmosets) in captivity as a result of exposure to infected humans. Similarly, RSV was isolated from chimpanzees with coryza and a laboratory worker who had been in contact with the animals, even though RSV is unusual in that species. There have been reports of mumps virus infections in pet cats and dogs during outbreaks of mumps in the owners' households, but as with primates, these animals probably do not constitute significant natural reservoir hosts (1, 11, 21, 22, 24, 40, 41, 62).

Laboratory-associated infection

NDV infection of humans was first discovered by accidental splashing of virus into the eye of a laboratory worker (7). There have been numerous reports of laboratory-associated infections with NDV since that time. The virus may be disseminated in a widespread fashion by aerosolizing suspensions of virus-containing media, by transferring the virus to the face or eyes by the hands, or by direct contact with infected birds.

The other viruses in this family, measles, mumps, RSV, and parainfluenza, while obviously able to produce an infection in human beings, presently are not recognized as zoonotic infections and are hence not further discussed here. There are likely no laboratory animals which might serve as natural reservoirs, with the possible exception of the primates. Successfully inoculated animals which develop active infection may, however, serve as a source of virus for transfer to the laboratory worker.

The combination of (i) high concentrations of virus and close contact which can be present in the laboratory and (ii) a laboratory worker with compromised resistance (e.g., immunosuppressed) could possibly present circumstances leading to an active and highly significant infection.

Prevention and control

Prevention of infection with these agents is primarily by limiting or preventing exposure. Laboratory precautions should be taken to avoid the formation of aerosols and contact of the mucous membranes of laboratory personnel with contaminated hands.

Parainfluenza virus 3, strain SF4, has not been associated with disease in healthy human adults and may be handled at Biosafety Level 1. The same applies to the strains of NDV which are now licensed for use as vaccines. All the other members of the *Paramyxoviridae* are placed in Biosafety Level 2 or 3 for laboratory handling (57).

The foreign strains of exotic NDV, which cause devastating epizootics in birds and domestic poultry when introduced into the United States, seem to present no increased risk for the human being.

PICORNAVIRIDAE

General characteristics

Picronaviruses are subdivided into the genera *Enterovirus* (polio, coxsackie, swine vesicular disease, ECHO, animal enteroviruses), *Cardiovirus* (encephalomyocarditis viruses), *Rhinovirus* (common cold, animal rhinoviruses), and *Aphthovirus* (foot-and-mouth disease [FMD] virus) on the basis of sensitivity to acid, virion buoyant density, and antigenic characteristics as well as clinical disease syndromes in the host (12, 34).

The acid-stable enteroviruses (over 70 serotypes) primarily infect the gastrointestinal tract, frequently without causing clinical manifestations. However, viremia and subsequent invasion of other tissues (e.g., spinal cord, brain, meninges, myocardium, respiratory tract, liver, conjunctivae, or skin) may occur, resulting in a variety of clinical syndromes. These may range from paralytic poliomyelitis, aseptic meningitis, and myocarditis to conjunctivitis, skin rashes, herpangina, or common cold-like illness. Enteroviruses are usually host species specific; however, certain primates are susceptible to natural infection with human viruses (e.g., chimpanzees; paralytic poliomyelitis in captive gorillas and orangutans), and the swine vesicular disease virus, which is closely related (if not a variant of the human coxsackievirus B 5), can infect humans, causing aseptic meningitis (1, 38, 62).

The acid-labile rhinoviruses cause mild upper respiratory tract diseases such as the common cold and perhaps also bronchitis and bronchopneumonia. Cardioviruses (various strains serologically indistinguishable: Columbia-SK, encephalomyocarditis, murine encephalomyelitis, mengo-encephalomyelitis, MM, and 3-day fever viruses) usually cause inapparent infections but can be pathogenic for a variety of small and large mammals including humans and primates. In humans a "3-day fever" with severe headache and pharyngitis, rarely with encephalitis and paralysis, has been related to cardiovirus infection. Myocarditis does not occur in humans but is the most prominent lesion in cattle, monkeys, and especially swine, which in outbreaks suffer the heaviest losses as a species. Both myocarditis and encephalitis are seen in rodents. Members of the seven serotypes of aphthoviruses cause FMD in cloven-hoofed animals; types O, C, and A may also infect humans, producing a benign febrile illness with similar symptomatology and vesicular oral and skin lesions (1, 38, 62).

Several important factors contribute to the high degree of communicability of picornaviruses. All (with the exception of rhinovirus) are relatively resistant to inactivation by chemical and physical agents and survive for long periods in the environment, even in treated sewage. Most picornaviruses are shed in abundance in secretions and excretions and from skin lesions of the infected host. In addition, they frequently cause subclinical infections or mild illness and produce asymptomatic carriers after recovery. With few exceptions, picornaviruses are species specific and are maintained in the host population mainly through inapparent infections of the young protected by passive immunity.

Transmission of the enteroviruses is mostly by fecal contamination of hands and fomites, food or water supplies, and insects serving as mechanical carriers. However, because the viruses initially replicate in the pharynx (polioviruses) and upper respiratory tract (coxsackie, ECHO viruses), saliva and respiratory droplet transmission also occur. Little is known about the natural transmission of cardioviruses among rodents, which are the probable natural hosts, serving as a reservoir for other species including humans. Virus is shed in feces and urine, but because of virus isolations from blood and mosquitoes, an arthropod vector cannot be ruled out. Respiratory droplet transmission is the accepted route for the spread of rhinoviruses (12, 22, 38).

Because of the great economic impact of FMD, the spread of this disease has been studied in great detail. It can be said that FMD virus is one of the most versatile viruses in terms of modes of transmission. The virus is present in large quantities in vesicular fluid and saliva and also in milk, semen, meat, and hides. Fomites, arthropods, large and small animals, birds, and humans can transmit the infection as mechanical or biological, sometimes asymptomatic, reservoirs. As with the other zoonotic picornaviruses, humans are an accidental host, and the resulting disease, if any, is usually mild and self-limiting (1, 22).

Laboratory-associated infections

The cardioviruses can infect some laboratory animal species as well as humans and therefore have the potential of being transmitted from infected laboratory swine, cattle, primates, and suckling mice. The virus is widespread throughout the infected animal's body and is shed in large quantities in feces and urine. It must be assumed, then, that direct contact and handling of these species presents a risk of this infection. The portal of entry for the virus is likely oral or mucous membranes. Necropsy procedures could well present an increased risk of infection to the persons involved.

The enteroviruses, with the possible exception of poliomyelitis, seem to present little or no risk of animal-to-human transmission. Poliomyelitis virus is readily infective for certain primates (38), but natural transmission to a human by a primate host perhaps has not occurred or at least has not been reported. The rhinoviruses are not recognized as having important animal hosts.

FMD virus also presents little risk of human infection, at least of producing an infection of consequence. Humans are highly resistant to the clinical effects of the infection. Persons working with the virus or infected animals will likely have virus in the saliva, nasopharyngeal passages, respiratory secretions, and expired air with no clinical evidence of infection, with or without the development of specific antibody.

Prevention and control

The risk of human disease produced by infection with the picornaviruses, mediated by an animal host system or laboratory procedure, is limited. The fact that they also produce mild, even nuisance, rather than consequential infections in humans makes them marginally important for discussion as zoonotic diseases. They are recommended to be handled using Biosafety Level 2 precautions in the laboratory.

POXVIRIDAE

General characteristics

Poxviruses are presently subdivided into six genera of mammalian and avian viruses (and a subfamily of insect poxviruses) on the basis of morphology, antigenic relationship, and host range. The members with zoonotic significance so far have been found only in the genera *Orthopoxvirus, Parapoxvirus,* and *Capripoxvirus* and among still unclassified poxviruses (Table 2) (18, 22, 34).

Diseases caused by orthopoxviruses in humans and animals are characterized by pock lesions on the skin, often also on mucous membranes of the mouth and throat, that are either localized or generalized as part of a rash. Typically, the lesions progress through stages of macules, papules, vesicles, and pustules which eventually dry up and form scabs, sometimes leaving scars after healing. The severity and clinical manifestations of the disease range from generalized overt and often fatal cases to localized, mild, and even clinically inapparent infections. Severe systemic disease may occur in the susceptible natural host (e.g., smallpox in humans, cowpox in cattle), whereas localized and milder forms are usually observed in accidental and experimental hosts (smallpox in monkeys, cowpox in humans, etc.), although severe and sometimes fatal infections of humans with monkeypox virus have occurred in central and western Africa (6). The course of the disease and the ex-

TABLE 2. Poxviruses with zoonotic potential

Genus	Virus species	Host range	Potential reservoir hosts
Orthopoxvirus	Variola, alastrim	Humans, monkeys,	Humans, monkeys, rodents
	Whitepox	mice	Monkeys, rodents
	Monkeypox	Monkeys, rodents	Monkeys, rodents
	Cowpox	Monkeys, humans,	Unknown
	Cowpoxlike viruses	rodents	Rodents
	Poxviruses of buffalo, camel,	Cattle, humans, other	Species of association
	mouse, other mammals	mammals	Unknown
	Vaccinia	Rodents, carnivores,	
		elephants, humans	
		Species of association,	
		possibly humans	
		Humans, other mammals,	
		birds	
Parapoxvirus	Pseudocowpox (milker's node,	Cattle, humans	Cattle
	paravaccinia)	Sheep, goats, cattle humans	Sheep, goats
	Orf (contagious ecthyma)	Cattle, humans	Cattle
	Bovine papular stomatitis		
Capripoxvirus	Goatpox	Goats, humans	Goats
Unclassified	Molluscum contagiosum	Humans, chimpanzees	Humans
	Tanapox	Monkeys, humans	Monkeys
	Yaba monkey tumor pox	Monkeys, humans	Monkeys

anthema may be modified with different routes of infection or by preexisting immunity; lesions may not progress through all stages or may be absent altogether. Diseases caused by parapoxviruses and goatpox virus are usually localized and mild in humans and animals, with the possible exception of orf complicated by secondary infection in sheep and goats. Rare cases of generalized but uneventful orf in humans have been reported (1, 22, 39).

The unclassified poxviruses listed here cause skin lesions in which the proliferative phase, temporarily present in all pox lesions (papule), is exaggerated and prolonged. Pustulation is absent. Lesions are always few in number and constitute benign proliferations that later regress, although this process may take periods of months, even years (molluscum contagiosum). Tanapox virus infection in humans only initially appears similar to smallpox with regard to both clinical symptoms and early lesion development (1, 18, 39).

Vesicular fluid, pus, scrapings, and crusts from lesions are the most important sources of infectious virus. When associated with these materials, especially in a desiccated state, the virus is resistant to chemical and physical agents and can remain infectious for months. Skin, hair, and clothing can therefore be contaminated long after the healing of lesions and shedding of scabs. Infections affecting the eyelids and mucous membranes of the upper respiratory tract result in infectious secretions from these sites. During viremic phases of systemic infections, virus can also be isolated from the blood. Virus transmission among natural hosts commonly is by direct and indirect contact with skin lesions and secretions, but occurs also via mechanical means by arthropod and fomites vectors. The transmission of poxviruses from animals to humans is usually restricted to direct contact, as is well documented by occupationally related cases of poxvirus infections (e.g., cowpox, monkeypox, milker's node, and orf in milkers, ranchers, veterinarians, and animal handlers). The virus enters through abrasions, scratches, and bite wounds of the skin, generally causing localized infections. Generalization can occur, however, in the immune-deficient host or in the presence of concurrent disease. As poxviruses easily become airborne (aerosol, dust) upon shedding, infection may occur via the respiratory route. Also, recent virus isolations and serologic test results have incriminated rodents as an important reservoir host for certain poxviruses (4, 22, 39).

Laboratory-associated infections

Smallpox vaccination has prevented orthopoxviruses from posing a major human health hazard in laboratory settings in the past 30 years of systematic worldwide recording. Of all 1,179 reported laboratory-associated viral infections, only 29 were caused by vaccinia and variola viruses, mostly in unvaccinated persons, and none were fatal. Of the other poxviruses, only Yaba and tanapox viruses are known to have caused a total of 24 laboratory-acquired infections. Accidents with syringe and needle or skin laceration and scratches during the handling of infected monkeys allowed the entrance of the viruses (1, 6, 44).

Prevention and control

Because of the apparent success of smallpox eradication programs, the only remaining sources of variola virus may well be stocks retained in laboratories; virtually all countries have already terminated compulsory smallpox vaccination as a condition of immigration and entry. This measure, which on the one hand will avoid the complications associated with vaccination and reduce the spread of vaccinia virus to domestic animals, will on the other hand also eliminate the protection against disease caused by animal orthopoxviruses. Although there is no evidence for the existence of a natural reservoir for variola virus at present, whitepox viruses harbored by apparently healthy monkeys are as yet indistinguishable from variola virus. Although these viruses may not cause smallpox in humans, they could cause confusion in the diagnosis of other poxvirus infections. A somewhat similar situation exists with regard to monkeypox virus infection in humans. The disease is clinically indistinguishable from smallpox; however, secondary contact infections and deaths are rare. Because of the above facts, taken together with the possible existence of still undefined animal reservoirs for orthopoxviruses, immunization should be available for all laboratory personnel in diagnostic laboratories and institutes where the potential of exposure to the orthopoxviruses exists, especially those workers using freshly imported nonhuman primates. Strict rodent and arthropod control is also essential, particularly where animal colonies are maintained. Routine precautions in the laboratory and in animal holding facilities and avoiding aerosols and skin penetrations will have to suffice in protecting against disease caused by poxviruses of the other genera (4, 6, 35, 39, 44).

The possession and use of orthopoxviruses (variola and alastrim viruses) is restricted to the designated World Health Organization Collaborating Centers for Smallpox and Other Poxvirus Infections in Atlanta, Ga. and in Moscow, USSR. The remaining poxviruses are to be handled by immunized personnel using Biosafety Level 2 practices (57).

REOVIRIDAE

General characteristics

The animal viruses included in the family *Reoviridae* belong to the genera *Reovirus, Orbivirus,* and *Rotavirus.* Orbiviruses are arboviruses infecting humans (e.g., Colorado tick fever virus), horses (e.g., African horse sickness virus), and sheep and other ruminants (e.g., bluetongue virus), among other mammals. Presently, 17 serological subgroups are recognized among the orbiviruses. Rotaviruses were assigned to the *Reoviridae* on the basis of their similar morphology and genome structure. Although viruses within each of the three genera share antigenic determinants, members of the different genera do not cross-react (34, 51).

Reoviruses occur in three mammalian and several avian serotypes and have been recovered from humans, many mammalian species, and domestic fowl. Recent surveys have demonstrated the simultaneous presence of anti-hemagglutinins against all three mammalian serotypes in a high percentage (30 to 90% depending upon the serotype and species) of human and domestic animal populations. Mild respiratory disease was apparently reproduced with reoviruses in cats, dogs, and calves and more recently in a horse. Mild respiratory disease accompanied by headache and nasal hemorrhage also occurred in two veterinarians and two animal keepers during experimentation with reovirus type 3 in horses. All patients had seroconversions, and the virus was isolated successfully in one case. Stress factors (such as concomitant infections) affecting host resistance may be required to produce clinical disease after infection. These findings, coupled with the widespread occurrence of these viruses in humans and animals, suggest that reoviruses should be understood as facultative zoonotic agents (22, 35, 55, 57).

Orbiviruses of the Colorado tick fever, Changuinola, and Kemerovo subgroups produce acute, febrile, dengue-like disease usually without rash. Although Colorado tick fever is characteristically mild, the disease may be severe in children with occasional encephalitis or a tendency to bleed; deaths are uncommon. Because of the viremia of up to 110 days post onset and the erythrocyte association of the virus, there is a considerable risk of transmitting the infection via blood transfusion (1, 22, 51).

Rotaviruses are ubiquitous throughout the world and a major cause of acute gastroenteritis in infants and young children and the newborn of numerous domestic and wild mammals, chickens, and turkeys. Subclinical infections commonly occur in adults. Mortality in animals can reach 80% and is mainly due to severe dehydration. Deaths of human infants have been reported, but hospitalization usually leads to complete recovery. Human and animal rotaviruses are antigenically related, and several were shown to cross species barriers; they therefore are considered potential zoonotic viruses (20, 28, 35).

Reoviruses of the three mammalian serotypes are widespread in human and animal populations, as well as wild and laboratory mice, monkeys, and chimpanzees. These viruses are relatively resistant to physical and chemical agents and are transmitted mostly by the fecal-oral route but also via the respiratory tract (55). Orbiviruses are strictly arthropod borne, with reservoir hosts among rodents, squirrels, and large mammals and with biological vectors such as mosquitoes, midges, ticks, and phlebotomines (51).

Rotaviruses appear even more widespread than reoviruses due to several factors promoting their spread and maintenance in nature. They are quite resistant to heat, acid, lipid solvents, and most common disinfectants. They are produced in and excreted in large quantities from the intestinal tract. Infection in the newborn protected by colostral antibody is asymptomatic. Rotaviruses may also cross the placenta and show little species specificity. Obviously the fecal-oral route is the principal pathway of transmission, whether directly or indirectly via environmental contamination (e.g., water, dust, aerosols, fomites) (20, 28, 35, 37).

Laboratory-associated infections

Transfer of reo- or rotaviruses to any person handling them seems quite likely to occur. The risk of a serious illness resulting from the tranfer is considered to be quite low. The risk is likely, however, to be dose dependent; the laboratory is exactly the environment which could place high doses of virus in a small space for the human to acquire. The principal route of exposure is via the mouth and secondarily via the respiratory tract. Accidental skin puncture with virus-contaminated laboratory ware seems a possible route of entry for infectious doses of members of the *Reoviridae*.

Control and prevention

Any steps taken to prevent fecal contamination of food or water is a control for the reoviruses. While person-to-person transfer of the virus is difficult to prevent, and the casual laboratory worker handling the virus will receive exposure, the consequences may be insignificant. There are no individualized control measures that can or need to be taken for the reoviruses. No vaccine is available. All routine laboratory procedures should have the goal of minimizing the risk of aerosols, oral entry, or skin puncture, especially when handling large concentrations of virus. These precautions, used as a general practice to prevent infection with any agent, will apply to this group. In the case of Colorado tick fever, avoidance of mammalian or insect bites is a logical preventive step. Members of the *Reoviridae* are placed in Biosafety Level 2 for containment facilities and precaution (57).

RHABDOVIRIDAE

General characteristics

Rhabdovirus has become recognized as a generic term for RNA-containing viruses which morphologically resemble a bullet or bacterium. The rhabdoviruses are a large and growing group derived from vertebrate, invertebrate, and plant sources. In many cases, morphologic criteria are the sole taxonomic factors placing viruses in the group, while detailed physicochemical properties are as yet inadequately studied. Considerable knowledge has accumulated, however, on the major viruses in the family, including the type species of the genus *Vesiculovirus*, vesicular stomatitis virus (VSV). The important animal and human pathogens in the group are several serotypes of VSV, Chandipura virus, the members of the genus *Lyssavirus* (rabies, Lagos bat, Mokola, and Duvenhage viruses), and the tentatively assigned virus of bovine ephemeral fever (30).

The diseases associated with this group vary considerably. Vesicular stomatitis occurs naturally among horses, cattle, and swine. Feral swine and other wildlife species have been shown to have antibodies against VSV, suggesting the possibility of wildlife reservoirs. As the disease in animals is not very serious in terms of morbidity and mortality, the major concern with the disease in nature is its differentiation from vesicular diseases of great importance, e.g., FMD. VSV will infect humans, and in fact the best-documented cases are in laboratory workers. The infection is characterized by a mild influenzalike illness and is rarely reported outside of the laboratory or experimental setting. However, because of the nondescript nature of the illness, other cases may occur that are not accurately diagnosed. The incubation period of the infection is short, with experimentally inoculated cases showing early signs after 24 h. The course of illness is also short, with a duration of 3 or 4 days (22, 26).

The knowledge about rabies is now in a state of rapid change, with certain historically derived conceptions slowly being eliminated. As an example, there is considerable documentation that recoveries from rabies occur with surprising frequency.

Other lyssaviruses (rabieslike viruses) have been discovered; it seems possible that a concerted search might divulge even more. Those described to date are Lagos bat and Mokola viruses, which have not clearly been found pathogenic for humans but do produce disease in inoculated rhesus monkeys. Mokola virus has been accidentally inoculated into the conjunctiva of a laboratory worker, with no effect other than the development of antibody. Mokola virus has been isolated from a patient with fatal central nervous system disease (17). Duvenhage virus has been isolated from a man in nature, but little further is known about its significance. Obodhiang virus, one of the rabies-related viruses, is unusual in that it seems to have the potential to be insect borne. It and Kotonkan virus have not been implicated in human illness. All of these rabies-related viruses have serological or morphological similarities to rabies virus. Their importance in human infection is unclear, and therefore their risk to the laboratory worker is unknown. Immunization against rabies does not seem to protect against these agents. Although they have been found only in Africa to date, a concerted effort may well disclose their presence elsewhere, and perhaps new and additional types as well (30).

Bovine ephemeral fever is caused by a cone-shaped virus tentatively assigned to the rhabdoviruses pending further knowledge. The disease is widespread in Africa and other areas of the eastern hemisphere. It does not occur in the United States and is not known to be infective for humans (30).

Laboratory-associated infections

VSV is present in the saliva and vesicular fluid of infected animals. The chief mode of transmission in nature is by direct contact. The portal of entry for the virus into a new host is probably multiple, including inhalation and direct inoculation into mucous membranes or abraded skin. Because many routine laboratory procedures produce aerosols, the mere act of handling the virus constitutes a significant risk of exposure. The virus has average susceptibility to disinfectants but can withstand a wide range of pH values and resists normal pasteurization temperatures. While it behooves the prudent laboratory worker to take adequate precautions against any aerosol or droplet contact from virus cultivation vessels or any contact with the body fluids of infected animals (saliva and vesicular fluid in particular), the clinical course of VSV infection is likely to be quite mild. No vaccine is available for use in humans, and the use of vaccine in animals is not well studied (22, 26).

Rabies is usually produced by bite exposure, involving the direct inoculation of virus into tissue surrounding peripheral nerves. Inhalation exposure also occurs and can produce infection in individuals in specialized environments, such as a cave or laboratory, where virus may be aerosolized and remain airborne in a viable state. Saliva protects the infectivity of rabies virus, whereas it is rapidly inactivated by any properly used disinfectant, an acid pH, drying, or sunlight.

There are reasonably clear guidelines for evaluating the potential for rabies infection resulting from bites, but these guidelines apply only to the laboratory worker in direct contact with potential carriers in laboratory animal colonies. A period of observation of a dog or cat after it has bitten can be adequate, but the same is not advisable if any other species is involved. The wildlife species (skunk, fox, and bat, among others) may have virus in their saliva for long periods of time before symptoms develop, if indeed they do develop symptoms. (In case of question, immediate laboratory examination of the biting animal is the only alternative in any species with a rabies risk.) The caged species (mice, guinea pigs, hamsters, nonhuman primates, etc.) are susceptible but usually lack any potential for exposure and thus are only rarely considered a risk. Laboratory-inoculated rodents can obviously present a risk.

Puncture exposures can occur by accidental inoculation with a scalpel or syringe and needle, or by cuts from broken laboratory ware. The risk of inhalation exposure is less well understood. As propagation of rabies virus in cell culture or other replication systems may provide high concentrations of virus, exceedingly dangerous human exposures may result from inhalation of aerosols from these sources. The proven occurrence of aerosol-induced rabies in several humans (all cave or laboratory acquired) and in animals provides ample stimulus for laboratory workers to evaluate procedures thoroughly before their use for their potential for puncture exposures or the generation and containment of aerosols (1, 3, 26).

Prevention and control

Preexposure immunization, routinely advised for high-risk persons including laboratory personnel, provides excellent protection against puncture exposure, but may not protect well against an exposure by inhalation. Once a worker is immunized, his or her immune status should be essentially ignored (except for periodic review and maintenance), so that all other primary and secondary containment precautions are placed into full effect. The possibility that little immunity has been induced against aerosol exposure to rabies virus should remain paramount. Should an accidental exposure occur despite all precautions, preventive steps should immediately be applied. In the case of a puncture exposure, the wound or puncture site should be encouraged to bleed and be scrubbed vigorously with plenty of soap and water. The immune status of the person should then be ascertained and considered in making the decisions related to further immunization with antirabies vaccine and rabies immune globulin.

The limited knowledge of the rabies-related viruses makes it impossible to define their laboratory risk with certainty. It would seem however, that the laboratory dealing with concentrations of these agents might provide the ideal ecological setting by which the infectivity might be proven. It must therefore be assumed that they are infective for humans, may have multiple portals of entry into the human body, and while serorelated to rabies virus, will not be inhibited by rabies immunization. These agents must be handled with extreme caution, and in specialty laboratories prepared for the purpose, until better knowledge is accumulated.

Specific containment precautions are essential for most of the rhabdoviruses. Rabies virus is recommended to be handled using Biosafety Level 2 procedures, except during activities having potential for generating droplets, aerosols, or large concentrations of virus. Under these conditions, precautions approximating Biosafety Level 3 should be used. The remainder of the family, except for VSV and bovine ephemeral fever virus, should be handled under Biosafety Level 3 precautions (57).

TOGAVIRIDAE

General characteristics

Currently, over 80 viruses have been assigned to the family *Togaviridae*, a somewhat artificial family of viruses. They are subdivided into four genera, *Alphavirus* (arbovirus group A), *Flavivirus* (arbovirus group B), *Rubivirus* (rubella virus), and *Pestivirus* (hog cholera and other animal viruses), based on antigenic relationships and similarities of structure, composition, and replication. However, members of a genus are related serologically only to each other, but not to other members of the togavirus family. Alphaviruses and most flaviviruses are arthropod borne and of zoonotic importance (Table 3); viruses of the other two genera are not. In nature, arthropod-borne togaviruses have established alternating infectious cycles in arthropod vectors and vertebrate reservoir hosts. Apparently, disease is rarely produced in the arthropod vector, which may serve as reservoir in certain cases, but often occurs in humans and animals inserting themselves inadvertently into the natural cycle. Once the virus is introduced into accidental host populations, the spread of infection and disease may assume epidemic proportions (9, 34, 49, 51).

Most infections of humans and animals are asymptomatic, but several togaviruses can cause severe and potentially lethal disease. The clinical syndromes are varied and complex (Table 4). Practically all alpha- and flaviviruses cause fevers of variable severity, but some of them are complicated by rashes, arthralgia, and hemorrhage or may progress to encephalitis, hepatitis, nephritis, and hypotensive shock. The best known of the togaviruses capable of causing encephalitis are Eastern, Western, and Venezuelan equine encephalitis (EEE, WEE, and VEE) viruses. EEE is the most serious infection, producing low morbidity but high mortality, whereas the reverse is true for WEE. VEE, on the other hand, is usually a mild infection in humans. St. Louis encephalitis virus may also cause encephalitis in humans, especially in older persons. Animal reservoirs are wild and domestic birds and bats (1, 22, 51).

Most of the important togaviruses are transmitted by mosquitoes. Although there is evidence that direct-contact and droplet or indirect-contact transmission does occur, the majority of disease in nature is related to the risk of exposure to infected arthropods. Both EEE and WEE viruses are transmitted by mosquitoes from a reservoir of wild birds and perhaps other wild animals. The equine is not an efficient reservoir for these viruses because of a limited viremic phase, reducing accessibility to biting mosquitoes. However, the virus of VEE is present in high titer in both the blood and saliva of infected horses, and the infection can be acquired either from mosquitoes or by direct contact with an infected animal (22, 47–49, 51).

Laboratory-associated infections

The history of laboratory-acquired viral infections is enriched by encounters with the togaviruses, especially yellow fever virus. It was learned quite early in this century that these agents were dangerous to work with and could cause fatal infection in laboratory workers. Semliki Forest virus has recently resulted in infections of laboratory workers. While

TABLE 3. Some togaviruses of zoonotic importance

Genus	Subgroup	Vector	Examples of viruses	Host Reservoir	Host Diseased
Alphavirus	VEE	Mosquito	VEE	Rodents, equines	Humans, equines
			Semliki Forest	Chimpanzees, rodents	Humans
			Chikungunya	Humans, monkeys	Humans
	WEE	Mosquito	WEE	Birds	Humans, equines
			Sindbis	Mammals, birds	Humans
	EEE	Mosquito	EEE	Birds, turtles	Humans, equines, pheasants
Flavivirus	Tick-borne	Tick	Kyasanur Forest	Monkeys, rodents, ticks	Humans, monkeys
			Louping ill	Sheep, rodents	Humans, sheep
			Hypr	Rodents, birds	Humans
			Russian spring-summer encephalitis	Rodents, birds	Humans
	Mosquito-borne	Mosquito	Yellow fever	Humans, monkeys	Humans
	West Nile		West Nile	Birds	Humans
			Japanese (B) encephalitis	Birds, pigs	Humans
			Wesselsbron	Humans, sheep	Humans, sheep
	Dengue		Dengue, types 1 through 4	Humans, monkeys	Humans
	Spondweni		Spondweni	Cattle, sheep, goats	Humans
			Zika	Humans, monkeys	Humans

TABLE 4. Typical human disease syndromes caused by togaviruses

Clinical syndrome	Genus	Prototype viruses	Geographic distribution
Fever with rash, arthralgia, myalgia	*Flavivirus* *Alphavirus*	Dengue types 1 through 4 Chikungunya	Southeast Asia, Pacific, Caribbean, Africa Asia, Africa
Encephalitis or meningo-encephalitis	*Alphavirus* *Flavivirus*	VEE, WEE, EEE Japanese encephalitis West Nile Russian spring-summer encephalitis	Americas East Asia Africa, Europe USSR, Europe
Hemorrhagic fever	*Flavivirus*	Yellow fever Kyasanur Forest Dengue types 1 through 4 Omsk Chikungunya	Africa, South America India Southeast Asia, Pacific, Caribbean, Africa USSR Asia, Africa

most of the togaviruses are implicated in laboratory-acquired infections and all must be suspected for risk, the most severe are Semliki Forest, WEE, EEE, VEE, yellow fever, and St. Louis encephalitis viruses and the Far Eastern form of tick-borne encephalitis virus (49, 51).

The route of transmission of laboratory infections with togaviruses is often not totally known. Needle puncture exposure is easily identified, and aerosol exposure should be recognized in advance for its risk by the nature of the laboratory manipulations being carried out. One person has been infected in a laboratory who was not even working directly with the agent. Laboratory animals and their excreta may serve as a source of high concentrations of togaviruses.

Prevention and control

The togaviruses deserve a high level of respect in the laboratory. To work otherwise is inviting infection of "unknown origin." The use of carefully maintained biological safety cabinets is mandatory for safe handling of these viruses. Needle punctures or other contact with sharp laboratory ware should be prevented, bearing in mind that accident-proneness and exposure to zoonotic infection are known to be related. Protective immunization should be secured by persons working in risk areas when possible; e.g., yellow fever immunization is considered a highly effective preventive measure.

Laboratory protection is usually at Biosafety Level 2 or 3, but several viruses require Biosafety Level 4 protection. Laboratory function and procedure should be at the highest level needed for the virus of highest risk which could be handled. Precautions for a particular virus should be carefully researched before its handling (49, 57, 63). Those arboviruses which require Level 3 and 4 handling and those which are assigned to Level 2 are tabulated elsewhere in this book (Appendix 1).

UNCLASSIFIED, POTENTIAL, AND EMERGING ZOONOTIC VIRUSES

It can be expected that the continuing improvement of diagnostic and research methodologies, comparative epidemiologic and pathogenetic studies, and the availability of computerized data banks and retrieval systems will reveal new or potential zoonotic agents and the human-animal interactions facilitating their transmission. Such agents, and the diseases they cause, may always have been present in susceptible hosts and have gone unnoticed or misdiagnosed. It is equally possible that changes in socioeconomic conditions, population density, migrations of people and animals, rapid world travel, new food sources and habits, and changing patterns of human and animal interrelationships may allow certain agents to spread to new hosts and acquire status as being zoonotic. To qualify for inclusion in a category of potential and emerging zoonotic agents, a virus might be expected to be widespread in human or animal populations, not to be strictly host specific, generally to cause mild or asymptomatic infections at least in the species of origin, to be antigenically related to group members infecting other host species, and to share the human habitat or to occur in foodstuffs for human consumption. Several viruses or viruslike agents meeting one or several of these criteria are known and briefly discussed below.

Examples are the yet unclassified Marburg and Ebola viruses. Because of some morphological and chemical similarities these agents share with the rhabdoviruses, they have been grouped occasionally with this family. There is, however, no antigenic relationship between Marburg and Ebola viruses, and so far no relationship of either virus with other animal viruses has been found. Recently, separate family status has been proposed for these viruses (29). Morphologically they are indistinguishable and also cause identical clinical symptoms in humans (42, 61). The Marburg or Ebola virus diseases (synonyms: African hemorrhagic fever, "green monkey disease") are characterized by abrupt onset with severe headache followed by high fever, diarrhea, nausea, vomiting, myalgia, and a maculopapular rash. Severe hemorrhagic manifestations in multiple sites often develop (as in the hemorrhagic fevers of South America) in the gastrointestinal tract and the lungs and always precede fatal outcome. The mortality in primary cases is high (29 to over 90%), less so in secondary cases. Marburg disease was first recognized in 1967 during outbreaks among laboratory personnel handling organs and tissues of African green monkeys (*Cercopithecus aethiops* from Uganda) in Marburg and Frankfurt, Germany, and in Belgrade, Yugoslavia. Subsequent cases of Marburg disease in South Africa and Ebola virus disease epidemics in the Sudan and Zaire (the latter with 395 deaths of 561 total cases) could not be linked to monkeys as the source of infection; the role of monkeys as reservoir host for Marburg virus therefore remains doubtful, although they are susceptible to experimental infection and develop the disease. Although the natural transmission cycle, including the hosts, is still unknown, the large number of secondary human cases (frequently among attending medical and nursing personnel) was found to result from the high virus content in the patients' blood, excretions and secretions, and tissues handled on autopsy, and the extended presence of infectious virus in recovered patients. Close contact with patients appears to be necessary, but fomites, aerosol, and even arthropod transmission are also likely. It is clear that these dangerous agents (Biosafety Level 4) require maximum containment facilities for diagnosis and high-security patient isolation. No vaccine is available (1, 42, 61, 62).

The so-called "slow viruses," unclassified agents of scrapie of sheep and transmissible mink encephalopathy, causing subacute spongiform encephalopathies in their respective hosts and in nonhuman primates, are virtually indistinguishable from the agents of kuru and Creutzfeldt-Jakob diseases in humans. They are potentially zoonotic agents and are discussed elsewhere in this volume. Also covered separately are the still unclassified viruses of infectious (type A) and serum (type B) hepatitis. The transmission of infectious hepatitis from primates to humans has been well documented (59).

The agent of cat scratch disease, causing a self-limiting, subacute granulomatous lymphadenitis in humans, has been suspected of being viral or chlamydial in nature but so far has eluded isolation and characterization. Likewise, the true nature of the disease and its status as a zoonosis have not been clarified (16, 61).

Among the emerging zoonotic viruses, the coronaviruses are widespread in nature and infect humans and a variety of domestic and wild mammals and birds. Presently, their host range appears to be restricted to the species of origin. However, the experimental host range of some coronaviruses includes other species. For example, transmissible gastroenteritis virus of pigs extends to cats, dogs, humans, muskrats, opossums, skunks, and starlings. Antigenic similarities between coronaviruses of human and animal origin have also been demonstrated. Diseases caused in humans by coronaviruses are upper respiratory, common cold-like illnesses with diarrhea. Respiratory and gastrointestinal diseases are also the most common manifestations of infection by the respective coronavirus of animals. Transmission of coronaviruses occurs via aerosols, fomites, and the fecal-oral route. So far, laboratory-acquired human infections with animal coronavirus have not been reported. Coronaviruses have been placed in Biosafety Level 2 for the purpose of laboratory handling (22, 35, 56, 57).

In recent years the application of direct and immune electron microscopy to fecal preparations from cases of nonbacterial gastroenteritis, a major cause of morbidity and mortality in infants and young animals, has revealed that new or previously unsuspected viruses are associated with these conditions (28, 33). Several viruses from different groups are potential or emerging zoonotic agents, e.g., the rotaviruses (discussed under *Reoviridae*). Others are astro-, parvo-, and adeno-associated viruses, small unenveloped DNA-containing viruses that are widespread in domestic and wild mammals and birds and show various species- and interspecies-specific antigenic relationships between some members of the respective virus groups (2, 28). Adeno- and caliciviruses have been similarly implicated in the causation of gastroenteritic syndromes (19, 28, 54). Papovaviruses are considered generally species specific although widespread among natural mammalian hosts. However, recent seroepidemiological surveys have discovered evidence of interspecies infections (45). It is also known that the incidence of papillomavirus infection as evidenced by warts is higher in butchers, veterinarians, and farmhands. The human risk resulting from infection with animal gastroenteric viruses and papovaviruses appears minimal at the present time, and no specific preventive measures are recommended (13, 60).

The National Cancer Institute has published safety standards for research involving oncogenic viruses of presumed low, moderate, and high risk (60). Presently no high-risk viruses, by definition those proved to induce cancer in humans, are known. Criteria for moderate-risk viruses include the capacity of the virus to transform human cells in vitro or to produce cancer without experimental host modification in nonhuman primates or across a species barrier in other mammals. Also classified as moderate risks are any suspected oncogenic isolates from humans and any concentrated oncogenic virus or infectious, transforming viral nucleic acid and genetic recombinants between an animal oncogenic virus and a microorganism infectious for humans. Accordingly, several oncogenic animal viruses are moderate-risk agents: simian virus 40 of the *Papovaviridae,* herpesvirus saimiri and herpesvirus ateles of nonhuman primates, simian sarcoma virus type 1, gibbon ape leukemia virus, Yaba monkey tumor poxvirus, adenovirus type 2-simian virus 40 hybrid viruses, and the feline leukemia/sarcoma viruses. The human Epstein-Barr virus is also considered a moderate-risk agent because it can induce malignant lymphomas in certain New World primates. Several Epstein-Barr-like viruses recently isolated from chimpanzees, orangutans, and baboons are also worthy of inclusion in this category (64). Other animal tumor viruses are in the low-risk category. Human exposures to several low- and moderate-risk animal viruses are known to have occurred, most of them among laboratory personnel and animal caretakers, some of them in the form of accidents. Although seroconversion has been observed in several instances, no illnesses have resulted (60).

GLOSSARY

Control. Limitation of the occurrence of a disease to a predetermined, low, but acceptable level.

Natural transmission. Any transmission other than by artificial inoculation; for this purpose, transmission within the confines of the laboratory is considered as natural.

Prevention. Measures resulting in the complete and absolute absence of a disease from occurrence within a circumscribed area.

Zoonosis. Those diseases and infections which are naturally transmitted between animals and humans. For the purpose of this chapter, all are infectious in nature, and "natural" transmission includes transmission within the laboratory environment.

Zoonosis, emerging. As a result of new observations, certain previously known viruses may suddenly be found to have an animal reservoir, or animals newly introduced for any purpose may suddenly be found to harbor a new undescribed agent. The emerging zoonoses usually are manifest by humans making a previously untried use of a particular species or invading the ecological niche of the species in a new relationship.

Zoonosis, laboratory-associated. Those zoonotic diseases which are acquired during manipulation of the agent or the reservoir in the laboratory environment. The laboratory may provide opportunity for unusually large infective doses of an agent previously thought to be of marginal virulence.

LITERATURE CITED

1. **Acha, P. N., and B. Szyfres.** 1980. Zoonoses and communicable diseases common to man and animals. Scientific Publication no. 354. Pan American Health Organization, Washington, D.C.

2. **Bachman, P. A., M. D. Hoggan, E. Kurstak, J. L. Melnick, H. G. Pereira, P. Tattersall, and C. Vago.** 1979. Parvoviridae: second report. Intervirology **11:**248–254.

3. **Baer, G. M.** 1975. The natural history of rabies, vol. I and II. Academic Press, Inc., New York.

4. **Baxby, D.** 1977. Poxvirus hosts and reservoirs. Arch. Virol. **55:**169–179.

5. **Bishop, D. H. L., C. H. Calisher, J. Casals, M. P. Chumakov, S. Y. Gaidanovich, C. Hannoun, D. K. Lvov, I. D. Marshal, N. Oker-Blom, R. F. Petterson, J. S. Porterfield, P. K. Russell, R. E. Shope, and E. G. Westaway.** 1980. Bunyaviridae. Intervirology **14:**125–143.

6. **Breman, J. G.** 1980. Human monkeypox 1970–1979. Bull. W.H.O. **58:**165–182.

7. **Burnet, F. M.** 1943. Human infection with the virus of Newcastle disease of fowls. Med. J. Austral. **2:**313–314.

8. **Burnet, F. M.** 1979. Portraits of viruses: influenza virus A. Intervirology **11:**201–214.

9. **Calisher, C. H., R. E. Shope, W. Brandt, J. Casals, N. Karabatsos, F. A. Murphy, R. B. Tesh, and M. E. Wiebe.** 1980. Proposed antigenic classification of registered arboviruses. I. Togaviridae, alphavirus. Intervirology **14:**229–232.

10. **Casals, J.** 1979. Arenaviruses, p. 815–841. *In* E. H. Lennette and H. J. Schmidt (ed.), Diagnostic procedures for viral, rickettsial and chlamydial infections, 5th ed. American Public Health Association, Washington, D.C.

11. **Chanock, R. M.** 1979. Parainfluenzaviruses, p. 611–632. *In* E. H. Lennette and N. J. Schmidt (ed.), Diagnostic procedures for viral, rickettsial and chlamydial infections, 5th ed. American Public Health Association, Washington, D.C.

12. **Cooper, P. D., V. I. Agol, H. L. Bachrach, F. Brown, Y. Ghendon, A. J. Gibbs, J. H. Gillespie, K. Lonberg-Holm, B. Mandel, J. L. Melnick, S. B. Mohanty, R. C. Povey, R. R. Rueckert, F. L. Schaffer, and D. A. J. Tyrrell.** 1978. Picornaviridae: second report. Intervirology **10:**165–180.

13. **DePeuter, M., B. DeClerco, A. Minette, and J. M. Lachapelle.** 1977. An epidemiological survey of virus warts of the hands among butchers. Br. J. Dermatol. **96:**427–431.

14. **Dowdle, W. A., A. P. Kendal, and G. R. Noble.** 1979. Influenza viruses, p. 585–609. *In* E. H. Lennette and N. J. Schmidt (ed.), Diagnostic procedures for viral, rickettsial and chlamydial infections, 5th ed. American Public Health Association, Washington, D.C.

15. **Easterday, B. C.** 1975. Animal influenza, p. 449–481. *In* E. D. Kilbourne (ed.), The influenza viruses and influenza. Academic Press, Inc., New York.

16. **Emmons, R. W., J. L. Riggs, and P. Schachter.** 1976. Continuing search for the etiology of cat scratch disease. J. Clin. Microbiol. **4:**112–114.

17. **Familusi, J. B., B. O. Osunkoya, D. L. Moore, F. Odelola, and A. Fabijc.** 1972. A fatal human infection with Mokola virus. Am. J. Trop. Med. Hyg. **21:**959–963.

18. **Fenner, F.** 1979. Portraits of viruses: the poxviruses. Intervirology **11:**137–157.

19. **Flewett, T. H., and H. Daries.** 1976. Caliciviruses in man. Lancet **i**:311.

20. **Flewett, T. H., and G. N. Woode.** 1978. The rotaviruses. Arch. Virol. **57**:1–23.

21. **Gershon, A. A., and S. Krugman.** 1979. Measles virus, p. 665–693. *In* E. H. Lennette and N. J. Schmidt (ed.), Diagnostic procedures for viral, rickettsial and chlamydial infections, 5th ed. American Public Health Association, Washington, D.C.

22. **Gillespie, J. H., and J. F. Timoney.** 1981. Hagan and Bruner's infectious diseases of domestic animals, 7th ed. Cornell University Press, Ithaca, N.Y.

23. **Gregg, M. B.** 1975. Recent outbreaks of lymphocytic choriomeningitis in the USA. Bull. W.H.O. **52**:549–554.

24. **Hopps, H. E., and P. D. Parkman.** 1979. Mumps virus, p. 633–653. *In* E. H. Lennette and N. J. Schmidt (ed.), Diagnostic procedures for viral, rickettsial and chlamydial infections, 5th ed. American Public Health Association, Washington, D.C.

25. **Howard, C. R., and D. I. H. Simpson.** 1980. The biology of the arenaviruses. J. Gen. Virol. **51**:1–14.

26. **Hubbert, W. T., W. F. McCulloch, and P. R. Schnurrenberger (ed.).** 1975. Diseases transmitted from animals to man, 6th ed. Charles C Thomas, Publisher, Springfield, Ill.

27. **Hull, R. N.** 1973. The simian herpesviruses, p. 390–426. *In* A. S. Kaplan (ed.), The herpesviruses. Academic Press, Inc., New York.

28. **Kapikian, A. Z., R. H. Yolken, H. B. Greenberg, R. G. Wyatt, A. R. Kalica, R. M. Chanock, and H. W. Kim.** 1979. Gastroenteritis viruses, p. 927–995. *In* E. H. Lennette and N. J. Schmidt (ed.), Diagnostic procedures for viral, rickettsial and chlamydial infections, 5th ed. American Public Health Association, Washington, D.C.

29. **Kiley, M. P., E. T. W. Bowen, G. A. Eddy, M. Isaacson, K. M. Johnson, J. B. McCormick, F. A. Murphy, S. R. Pattyn, D. Peters, O. W. Prozesky, R. L. Regnery, D. I. H. Simpson, W. Slenczka, P. Sureau, G. van der Groen, P. A. Webb, and H. Wulff.** 1982. Filoviridae: a taxonomic home for Marburg and Ebola viruses? Intervirology **18**:24–32.

30. **Knudson, D. L.** 1973. Rhabdoviruses. J. Gen. Virol. **20**:(Suppl.):105–130.

31. **Lennette, E. H., and N. J. Schmidt (ed.).** 1979. Diagnostic procedures for viral, rickettsial and chlamydial infections, 5th ed. American Public Health Association, Washington, D.C.

32. **Ludwig, H., G. Pauli, H. Gelderblom, G. Darai, H. G. Koch, R. M. Flugel, B. Norrild, and M. D. Daniel.** 1983. B virus (Herpesvirus simiae), p. 385–428. *In* B. Roizman (ed.), The herpesviruses, vol. 2. Plenum Publishing Corp., New York.

33. **Madeley, C. R., and B. P. Cosgrove.** 1975. 28 nm particles in feces in infantile gastroenteritis. Lancet **ii**:451–452.

34. **Matthews, R. E. F.** 1982. Classification and nomenclature of viruses (report of ICTV). S. Karger, New York.

35. **Mayr, A.** 1980. New emerging viral zoonoses. Vet. Rec. **106**:503–506.

36. **McCarthy, K., and F. A. Tosolini.** 1975. A review of primate herpesviruses. Proc. R. Soc. Med. **68**:145–150.

37. **McNulty, M. S.** 1978. Rotaviruses. J. Gen. Virol. **40**:1–18.

38. **Melnick, J. L., H. A. Wenner, and C. A. Phillips.** 1979. Enteroviruses, p. 471–534. *In* E. H. Lennette and N. J. Schmidt (ed.), Diagnostic procedures for viral, rickettsial and chlamydial infections, 5th ed. American Public Health Association, Washington, D.C.

39. **Nakano, J. H.** 1979. Poxviruses, p. 257–308. *In* E. H. Lennette and N. J. Schmidt (ed.), Diagnostic procedures for viral, rickettsial and chlamydial infections, 5th ed. American Public Health Association, Washington, D.C.

40. **Parkman, P. D., and H. E. Hopps.** 1979. Newcastle disease virus, p. 655–663. *In* E. H. Lennette and N. J. Schmidt (ed.), Diagnostic procedures for viral, rickettsial and chlamydial infections, 5th ed. American Public Health Association, Washington, D.C.

41. **Parrott, R. H., H. W. Kim, C. D. Brandt, M. O. Beem, L. Richardson, J. L. Gerin, and R. M. Chanock.** 1979. Respiratory syncytial virus, p. 695–708. *In* E. H. Lennette and N. J. Schmidt (ed.), Diagnostic procedures for viral, rickettsial and chlamydial infections, 5th ed. American Public Health Association, Washington, D.C.

42. **Pattyn, S. E. (ed.).** 1978. Ebola virus hemorrhagic fever. Elsevier/North-Holland Biomedical Press, Amsterdam.

43. **Pike, R. M.** 1976. Laboratory-associated infections: summary and analysis of 3921 cases. Health Lab. Sci. **13**:105–114.

44. **Pike, R. M., and J. H. Richardson.** 1979. Prevention of laboratory infections, p. 49–63. *In* E. H. Lennette and N. J. Schmidt (ed.), Diagnostic procedures for viral, rickettsial and chlamydial infections, 5th ed. American Public Health Association, Washington, D.C.

45. **Pyrhonen, S., and E. Neuvonen.** 1978. The occurrence of human wart-virus antibodies in dogs, pigs and cattle. Arch. Virol. **57**:297–305.

46. **Rawls, W. E.** 1979. Herpes simplex virus types 1 and 2 and herpesvirus simiae, p. 309–373. *In* E. H. Lennette and N. J. Schmidt (ed.), Diagnostic procedures for viral, rickettsial and chlamydial infections, 5th ed. American Public Health Association, Washington, D.C.

47. **Reeves, W. C.** 1974. Overwintering of arboviruses. Progr. Med. Virol. **17**:193–220.

48. **Scherer, W. F.** 1980. Laboratory safety for arboviruses and certain other viruses of vertebrates (The Subcommittee on Arbovirus Laboratory Safety of the American Committee on Arthropod-Borne Viruses). Am. J. Trop. Med. Hyg. **29**:1359–1381.

49. **Schlesinger, R. W.** 1980. The togaviruses—biology, structure, replication. Academic Press, Inc., New York.

50. **Schnurrenberger, P. R., J. K. Grigor, J. F. Walker, and R. J. Martin.** 1978. The zoonosis-prone veterinarian. J. Am. Vet. Med. Assoc. **173**:373–376.

51. **Shope, R. E., and G. E. Sather.** 1979. Arboviruses, p. 767–814. *In* E. H. Lennette and N. J. Schmidt (ed.), Diagnostic procedures for viral, rickettsial and chlamydial infections, 5th ed. American Public Health Association, Washington, D.C.

52. **Shortridge, K. F., and R. G. Webster.** 1979. Geographical distribution of swine (Hsw 1 N 1) and Hong

Kong (H3 H2) influenza virus variants in pigs in Southeast Asia. Intervirology **11**:9–15.

53. **Stark, A.** 1975. Policy and procedural guidelines for the health and safety of workers in virus laboratories. Am. Ind. Hyg. Assoc. J. **36**:234–240.

54. **Studdert, M. J.** 1978. Caliciviruses. Arch. Virol. **58**:157–191.

55. **Thein, P., and R. Scheid.** 1981. Reoviral infections, p. 191–216. *In* J. H. Steele (ed.), Zoonoses, section B: Viral zoonoses, vol. II. CRC Press Inc., Boca Raton, Fla.

56. **Tyrrell, D. A. J., D. J. Alexander, J. D. Almeida, C. H. Cunningham, B. C. Easterday, D. J. Garwes, J. C. Hierholzer, A. Kapikian, M. R. Macnaughton, and K. McIntosh.** 1978. Coronaviridae: second report. Intervirology **10**:321–328.

57. **U.S. Department of Health and Human Services.** 1984. Biosafety in microbiological and biomedical laboratories. HHS Publication no. (CDC) 84-8395. U.S. Government Printing Office, Washington, D.C.

58. **U.S. Department of Health, Education and Welfare.** 1973. Herpes B encephalitis—California. Morbid. Mortal. Weekly Rep. **22**:333.

59. **U.S. Department of Health, Education and Welfare.** 1973. Hepatitis A in humans associated with nonhuman primates—Ohio. Morbid. Mortal. Weekly Rep. **22**:407.

60. **U.S. Department of Health, Education and Welfare.** 1977. Biological safety manual for research involving oncogenic viruses. DHEW Publication no. (NIH) 76-1165. Office of Research Safety, National Cancer Institute, Bethesda, Md.

61. **Warren, J.** 1979. Miscellaneous viruses, p. 997–1019. *In* E. H. Lennette and N. J. Schmidt (ed.), Diagnostic procedures for viral, rickettsial and chlamydial infections, 5th ed. American Public Health Association, Washington, D.C.

62. **Whitney, R. A., Jr.** 1976. Important primate diseases (biohazards and zoonoses), p. 23–50. *In* M. L. Simmons (ed.), Biohazards and zoonotic problems of primate procurement, quarantine and research. DHEW Publication no. 76-890. U.S. Dept. of Health, Education and Welfare, Washington, D.C.

63. **World Health Organization.** 1980. Guidelines for the management of accidents involving microorganisms: a memorandum. Bull. W.H.O. **58**:245–256.

64. **zur Hausen, H.** 1980. Oncogenic herpesviruses, p. 747–795. *In* J. Tooze (ed.), Molecular biology of tumor viruses, 2nd ed., part 2: DNA tumor viruses. Cold Spring Harbor Laboratory, Cold Spring Harbor, N.Y.

Chapter 12

Transmission of Bacterial and Rickettsial Zoonoses in the Laboratory

JOHN M. BOYCE AND ARNOLD F. KAUFMANN

Over the years, diseases that affect both humans and animals (zoonoses) have been among the most commonly reported occupational illnesses of laboratory workers (18, 19, 25, 26). Agents that are responsible for a variety of bacterial and rickettsial zoonoses accounted for 35 to 40% of all laboratory-acquired infections reported by Pike in 1976 (18). Fully 50% of the fatal bacterial infections listed by Pike were due to agents normally transmitted from animals to humans. Published information on human infective dose is included in the Agent Summary Statements in Appendix 1.

Although some agents such as *Brucella* spp., *Francisella tularensis, Coxiella burnetii,* and *Chlamydia psittaci,* which formerly caused many laboratory-associated infections, are not handled as frequently as they once were, infections caused by such zoonotic agents continue to affect personnel working in research and reference laboratories. This chapter discusses the more common bacterial and rickettsial zoonoses which have been transmitted to laboratory personnel, the modes of transmission for these diseases, and how this transmission can be controlled.

ANTHRAX

Human anthrax has three major clinical forms, cutaneous, inhalation, and gastrointestinal. Cutaneous anthrax is associated with a characteristic skin lesion developing at the site where *Bacillus anthracis* is introduced beneath the skin, e.g., by needle puncture or through a cut. After a 2- to 7-day incubation period, the lesion begins as a papule, which gradually enlarges and develops a central vesicle. The vesicle ruptures shortly after formation, revealing an underlying ulcer. A depressed scab or eschar rapidly forms over the ulcer's surface. The surrounding skin is commonly edematous, and secondary vesicles may develop in the edematous zone. Death from overwhelming septicemia or other complications occurs in about 5 to 20% of untreated patients but is uncommon if effective antibiotic therapy is administered.

Inhalation anthrax results from breathing aerosols of *B. anthracis* spores generated, e.g., during specimen centrifugation. The spores are transported from the lungs to the mediastinal lymph nodes, where germination occurs. The disease is characterized by a primary hemorrhagic mediastinitis with secondary septicemia. The clinical disease begins as a mild febrile illness resembling the common cold. After 2 to 4 days, the second stage of acute toxicity begins with sudden onset of dyspnea, cyanosis, and profuse sweating. Death usually ensues 24 h after symptoms appear, even with therapy.

Gastrointestinal anthrax, a rare form of the disease characterized by severe abdominal distress, fever, and septicemia, results almost exclusively from eating raw or undercooked meat from infected animals. Gastrointestinal anthrax has not occurred in laboratory personnel.

B. anthracis normally resides in soil. Animal anthrax results from grazing on infective pastures or eating contaminated feeds, and human anthrax usually results from exposure to anthrax-infected animals or their by-products.

Laboratory-acquired anthrax has been primarily a problem at facilities conducting anthrax research. Most cases have been cutaneous, with inhalation anthrax being rarely reported. In one series, cutaneous infection occurred at the sites of known preceding trauma in only 5 of 25 cases (6). Since the introduction of the human anthrax vaccine in the late 1950s, no laboratory-acquired anthrax cases have been reported in the United States. Occasional cases have occurred in persons performing necropsies in nonlaboratory settings on animals that died of anthrax.

Persons who are frequently exposed to anthrax-infected animals or pure cultures of *B. anthracis* should be immunized with anthrax vaccine (prepared by the Michigan Department of Public Health, Lansing). Biosafety Level 3 precautions should be followed for procedures involving large volumes, high concentrations, or potential aerosols of *B. anthracis;* otherwise, Biosafety Level 2 precautions are adequate.

Antibiotic treatment with penicillin or, alternatively, tetracyclines or other broad-spectrum antibiotics is usually effective in curing cutaneous anthrax.

Treatment is less successful in inhalation and gastrointestinal anthrax.

BRUCELLOSIS

Human brucellosis is characterized by fever that may be irregular, intermittent, or continuous, chills, profuse sweating particularly at night, headache, weakness, anorexia, weight loss, arthralgia, generalized aches, and depression. After an incubation period ranging from 5 days to a month or more, the illness begins abruptly or insidiously and lasts for days to months or, occasionally, several years. A variety of complications may occur, including septic arthritis, spondylitis, orchitis, endocarditis, meningitis, and, rarely, death.

Hypersensitivity to *Brucella* antigens is also a biohazard in *Brucella* research laboratories. The clinical presentation varies from an intensely burning or itching rash on exposed skin areas to an illness resembling acute brucellosis.

Brucellosis has been the most commonly reported bacterial infection acquired in laboratories. Most cases have occurred in research laboratories and have involved exposure to *Brucella* organisms being grown in large quantities. All four *Brucella* spp. with known human pathogenicity (*B. abortus*, *B. canis*, *B. melitensis*, and *B. suis*) have caused illness in laboratory personnel (15, 23).

Human infections can result from exposure to *Brucella* organisms by contact, inhalation, or ingestion. Direct contact with cultures of *Brucella* organisms or with infectious materials such as uterine discharges or blood of infected animals is an important mode of transmission. The intact skin is an effective barrier against invasion by *Brucella* organisms, but even minor scratches and abrasions may act as portals of entry. *Brucella* organisms can also readily invade the body through the conjunctiva if infectious material is rubbed or sprayed into the eyes.

Aerosol transmission of *Brucella* organisms is a well-documented hazard in laboratories. The largest single laboratory-associated epidemic (45 cases, 1 death) and many smaller outbreaks have been due to this mode of transmission (14). Ingestion is probably the least frequent mode of transmission to laboratory workers but has resulted from mouth pipetting.

Brucella infections frequently result in long-lasting immunity. A safe and effective human brucellosis vaccine, however, is not available. Protection of laboratory personnel is dependent on adherence to Biosafety Level 3 precautions. Because *Brucella* infection or antibiotics administered to patients with brucellosis may cause fetal injury, pregnant women should not work with *Brucella* organisms.

Tetracycline and streptomycin are the most commonly used drugs in treatment of brucellosis. Relapses occur in about 5 to 10% of patients, particularly if treatment is delayed or administered for an inadequate period of time. Prophylactic antibiotic therapy is commonly administered to personnel exposed in high-risk accidents, but the optimal treatment regimen for this purpose has not been determined.

GLANDERS AND MELIOIDOSIS

Glanders and melioidosis have clinical and pathologic similarities but differ significantly in their epidemiology. Both diseases are characterized by localized to disseminated suppurative lesions involving a variety of organ systems. In glanders, abscess formation tends to be more extensive than in melioidosis, and involvement of the upper respiratory tract is very common. Lung abscesses occur more frequently in melioidosis than in glanders. The clinical course of both diseases ranges from acute to extremely chronic, with illnesses that lasted more than 10 years having been reported. Untreated glanders and melioidosis cases commonly result in death. Specific antimicrobial therapy results in a significant reduction of mortality.

The incubation period of both diseases after traumatic inoculation ranges from 1 to 5 days. In glanders, the incubation period after inhalation of infectious aerosols is usually 10 to 14 days. The initial infection in melioidosis and, less commonly, glanders may result in latent disease that becomes clinically apparent months to years later.

Glanders is caused by *Pseudomonas mallei* and has its reservoir in horses and other solipeds. Glanders has been eradicated from most areas of the world, and its distribution is currently restricted to Asia. Melioidosis is caused by *Pseudomonas pseudomallei*, a soil saprophyte. Although *P. pseudomallei* is widely distributed in tropical and subtropical areas of Asia, Africa, Australia, and the Americas, melioidosis occurs most commonly in Southeast Asia.

Numerous laboratory-acquired glanders infections have occurred in research and veterinary diagnostic laboratories (9, 20). Cases have resulted from contact exposure of broken skin and nasal mucous membranes, accidental inoculations, and inhalation of aerosols. Sources of infection included cultures as well as infected animal tissues. In contrast to glanders, few cases of laboratory-acquired melioidosis have been reported (8). Infections have resulted from accidental inoculation of cultures and inhalation of infectious aerosols.

Vaccines are not available for either glanders or melioidosis. Biosafety Level 3 precautions should be followed for all work with *P. mallei* and *P. pseudomallei*.

Optimal antimicrobial therapy regimens have not been determined for human glanders due to its rarity. Sulfadiazine, streptomycin, and tetracyclines have been used with apparent success. In melioidosis, antimicrobial therapy should be based on susceptibility studies and administered for a minimum of 30 days. Successful therapy has been reported with various combinations of tetracyclines, chloramphenicol, kanamycin, sulfonamides, and trimethoprim-sulfamethoxazole.

LEPTOSPIROSIS

Leptospires are motile, finely coiled spirochetes that are cultivable in artificial media. Over 180 serovars (serotypes) of pathogenic leptospires (*Leptospira interrogans*) have been recovered from feral and domestic animals. Seven to 14 days (rarely up to 26 days) after exposure, affected individuals experience sudden onset of fever, headache, nausea, conjunctival suffusion, and myalgia. Pharyngitis, evanescent rashes, and cough with pulmonary infiltrates occur in some patients. Jaundice occurs in only 5 to 10% of patients. After 4 to 8 days, the fever usually disappears. One to 3 days after becoming afebrile, many patients have recurrence of fever that may be accompanied by meningitis or, rarely, uveitis. Several laboratory workers who became infected with serovar *ballum* after exposure to Swiss albino mice developed unilateral or bilateral orchitis during the second phase of the illness. Subclinical leptospirosis has also been reported in workers who frequently handled Swiss albino mice (24).

About 70 cases of leptospirosis among laboratory personnel have been reported, making it the sixth most common bacterial cause of laboratory-acquired infections (18). The most common sources responsible for such infections are cultures of leptospires and naturally or experimentally infected animals. Mice, rats, rabbits, and guinea pigs are the animals most frequently implicated.

Cases have been acquired by needle stick or by accidental spraying of the conjunctiva or other mucous membranes with a leptospiral culture. The latter type of accident usually occurs when a needle suddenly becomes separated from a syringe being used to inoculate animals with leptospires. Mouth pipetting has also resulted in cases of laboratory-acquired leptosprosis. A few laboratory personnel appear to have become infected when bitten by an infected animal. Serovars implicated in cases of laboratory-acquired leptospirosis include *icterohaemorrhagiae, georgia, ballum,* and members of the Javanica serogroup.

No leptospiral vaccine is approved for human use in the United States. Biosafety Level 3 precautions should be followed for procedures involving large volumes, high concentrations, or potential aerosols of leptospires; otherwise, Biosafety Level 2 precautions are adequate. Preventive measures should include appropriate training of personnel involved in handling experimental animals and maintaining their

awareness that laboratory animals may be asymptomatically infected with leptospires. Use of disposable gloves by persons with abrasions on their hands and careful handwashing after contact with known or potentially infected animals should minimize the risks of acquiring leptospirosis from experimental animals.

Although the role of antimicrobial therapy in treating leptospirosis has not been clearly resolved, several studies suggest that doxycycline, penicillin, or tetracycline reduce the severity and duration of symptoms, but only if initiated within the first 4 to 5 days of the illness (14a). Doxycycline has also been shown to be effective for chemoprophylaxis against leptospirosis (26a). In at least one reported case, prompt administration of prophylactic oral penicillin failed to prevent symptomatic leptospirosis.

PASTEURELLOSIS

Pasteurella spp. are small, nonmotile gram-negative coccobacilli that cause disease in a wide variety of domestic and laboratory animals. The genus includes *P. multocida, P. pneumotropica, P. hemolytica, P. ureae, P. aerogenes,* and *Pasteurella* sp. new species 1. *Pasteurella* infections in laboratory workers are almost always caused by *P. multocida* and are associated with animal bites. Local swelling, tenderness, and erythema develop in less than 72 h (often within 6 to 8 h) after a bite. The rapidity with which symptoms and signs develop and the severity of the pain, which seems out of proportion to the size of the injury, are characteristic of *P. multocida* bite wound infections. Regional lymphadenopathy and fever occur in some patients. Local suppurative complications include tenosynovitis, arthritis, and osteomyelitis. A few asymptomatic infections (oropharyngeal colonization) have been reported.

Very few human cases of laboratory-acquired pasteurellosis have been reported. Animals (mainly dogs and cats) used in research work represent the major source from which laboratory personnel become infected. About 10 to 55% of dogs and 50 to 90% of healthy cats carry *P. multocida* in their oropharynx. Personnel who sustain a dog or cat bite (or cat scratch) have about a 5 to 15% chance of developing a *P. multocida* infection. One case caused by a rat bite and another by a rabbit bite have also been reported.

Avoiding animal bites is the only effective way to minimize the risk of acquiring *P. multocida* infections. There is no vaccine available for animal caretakers or other personnel who must handle laboratory animals. Although Biosafety Level 2 precautions are generally adequate for work with *Pasteurella* spp., Biosafety Level 3 precautions should be followed when working with large-volume cultures of these organisms.

Penicillin is the drug of choice for *P. multocida* infections. Tetracycline, ampicillin, or cephalothin are reasonable alternative drugs. Although there are no adequate studies of the efficacy of penicillin in preventing animal (particularly cat) bite-associated *P. multocida* infections, postexposure prophylaxis with penicillin is reasonable (27).

PLAGUE AND OTHER YERSINIOSES

Yersinia pestis is the etiologic agent of plague. The most common clinical syndromes caused by *Y. pestis* are bubonic, primary septicemic, and pneumonic plague. The incubation period is usually 2 to 6 days in bubonic and septicemic plague and 2 to 5 days in pneumonic plague. When direct contact with infected animal tissue or a flea bite results in percutaneous inoculation of *Y. pestis*, a primary lesion may develop at the site of entry. Although such lesions are often clinically insignificant, bullous or ulcerative lesions may develop. The infection spreads within a few days via the lymphatics to the regional lymph nodes, which become swollen and painful (a bubo). Shortly before or at the time the bubo becomes apparent, fever and nonspecific signs and symptoms of toxemia develop. Bacteremia is common and may result in various secondary complications such as pneumonia and meningitis. Death occurs in 50% or more of untreated patients but is substantially reduced by early, appropriate antibiotic therapy.

Primary septicemic plague is similar to bubonic plague except that bubo formation does not occur or is clinically inapparent. The prognosis is worse for primary septicemic than for bubonic plague patients, but cure is possible with early appropriate therapy.

Persons with pneumonic plague experience fever, chills, and malaise that may be accompanied by vague pain in the chest. Eighteen to 24 h after onset, patients develop dyspnea and a cough that later becomes productive of blood-streaked sputum. Roentgenographic evidence of pneumonia is often present on day 1 or 2 of the illness. Unvaccinated, untreated persons often succumb on day 2 or 3 of the illness.

Surprisingly few cases of laboratory-acquired plague have been reported. Before 1936, only 4 bubonic and 11 pneumonic plague cases were reported as being laboratory associated (30, 31). Since 1936, five cases of laboratory-acquired pneumonic plague, including three in the United States, have been documented (4). A variety of personnel have been affected, including physicians, bacteriologists, students, a chemist, an animal caretaker, and other unspecified laboratory workers.

The sources from which personnel are most likely to acquire plague include cultures of virulent *Y. pestis*, infected animals, and tissues from plague patients. In the laboratory, the plague bacillus can be transmitted to personnel by infectious aerosols created during centrifugation of cultures, inoculation of solid media, and postmortem examination of infected animals. One case has resulted from mouth pipetting. Research projects and field studies dealing with *Y. pestis* or animals with naturally acquired or experimentally induced plague should be conducted in laboratories that meet the containment requirements specified for Biosafety Level 3 agents.

Plague vaccines have been used since the late 19th century, but their effectiveness has never been measured precisely. Field experience indicates that plague vaccine (Cutter Laboratories, Inc., Berkeley, Calif.) reduces the incidence and severity of disease resulting from the bite of infected fleas. Whether comparable protection is afforded against an accidental injection of a larger inoculum than that resulting from a flea bite (>10,000 CFU) is unknown. The degree of protection against primary pneumonic infection is also not known. Because of this uncertainty, persons who have been exposed to known *Y. pestis* aerosols or accidental inoculations should be given a 7- to 10-day course of antibiotic therapy regardless of immunization history.

Plague vaccination should be a routine requirement for: laboratory personnel who are working with *Y. pestis* organisms resistant to antibiotics that are normally used in plague therapy; laboratory personnel frequently working with *Y. pestis* or *Y. pestis*-infected animals; and persons engaged in aerosol experiments with *Y. pestis*.

Routine bacteriologic precautions, including the use of a biological safety cabinet to isolate procedures that may produce aerosols, are sufficient to prevent accidental infection with plague in clinical laboratory workers at hospitals. No cases acquired during the course of activities normally conducted in hospital-based clinical laboratories have been reported. Immunization might further reduce the already minimal risk but is not clearly indicated.

The primary vaccination series consists of three doses of vaccine, with an interval of 4 weeks between the first two doses and 6 months between the first and third doses. Whenever possible, the first two doses of the primary series should be administered before any anticipated exposures.

Serum antibody to Fraction 1 capsular antigen, as measured by the passive hemagglutination (PHA) test, is correlated with resistance to *Y. pestis* infection in experimental animals. Although direct evidence does not exist, a comparable correlation between PHA titer and immunity probably occurs in humans. After the primary immunization series, about 7% of individuals may not produce PHA antibody. Other persons occasionally fail to achieve a PHA titer of 1:128, the level correlated with immunity in experimental animals. PHA titers should be determined for individuals who have unusually high risk of infection or who have a history of serious reactions to the vaccine.

Laboratory workers who develop signs or symptoms compatible with pneumonic or bubonic plague should be hospitalized promptly and placed in strict isolation. As soon as appropriate cultures have been obtained, streptomycin or tetracycline therapy should be instituted without waiting for bacteriologic confirmation of the diagnosis. Chemoprophylaxis is recommended for persons who have had close contact with a patient with suspected or confirmed primary pneumonic plague or plague pneumonia. If a researcher develops pneumonic plague after a recognized accident, other similarly exposed individuals should be given chemoprophylaxis, usually tetracycline, for 7 days. Persons allergic to tetracycline may be given an appropriate oral sulfonamide. However, sulfonamides are not recommended for patients with symptomatic pneumonic plague.

Although *Yersinia pseudotuberculosis* and to a lesser extent *Yersinia enterocolitica* are capable of causing disease in feral and domestic animals and in humans, these agents have apparently caused very few laboratory-acquired infections.

PSITTACOSIS

Psittacosis is an acute systemic disease caused by *Chlamydia psittaci*. After an incubation period of 4 to 15 days, the illness usually begins abruptly with fever, chills, headache, backache, and photophobia. Although respiratory signs and symptoms tend to be limited to a nonproductive cough, X rays commonly reveal a patchy, lobular pneumonia resembling the interstitial pneumonias due to various viral agents. A variety of other clinical manifestations such as splenomegaly, hepatitis, myocarditis, thrombophlebitis, and meningoencephalitis may occur. The case-fatality rate is currently about 1%.

Psittacosis has its reservoir in numerous species of wild and domestic birds. Human infections are acquired primarily from pet birds and domestic poultry. Infected birds shed *C. psittaci* in their respiratory secretions and feces, with the latter having particular epidemiologic significance. Infection is primarily spread by aerosolized bird feces (21). In the laboratory, infectious aerosols may be produced while dissecting infected animals or processing diagnostic specimens. Direct contact with or traumatic inoculation of infectious material may also result in disease. Person-to-person transmission has occurred but is rare.

Prevention of laboratory-acquired psittacosis is based upon control of potentially infectious aerosols. Biosafety Level 3 precautions are usually appropriate, but Biosafety Level 2 precautions may suffice when working with small specimens or materials that contain low concentrations of *C. psittaci*. No vaccine is available. Recovery from a psittacosis infection does not confer immunity, and multiple infections have occurred in persons with ongoing exposure.

Tetracycline is the drug of choice for treatment of psittacosis, with erythromycin and chloramphenicol being alternatives.

RAT-BITE FEVER

Rat-bite fever (RBF) is actually two different diseases with clinical and epidemiologic similarities. Streptobacillary RBF is caused by *Streptobacillus moniliformis*, and spirillary RBF (soduku) is caused by *Spirillum minus*.

The clinical spectra of streptobacillary and spirillary RBF overlap, preventing reliable clinical differentiation. Streptobacillary RBF has a short incubation period (3 to 10 days, rarely longer). If a rat bite has occurred, the wound usually heals uneventfully, although swelling and tenderness may be present initially. The illness begins abruptly with fever, chills, vomiting, headache, and myalgia. Shortly after onset, a maculopapular rash, most pronounced on the extremities, commonly develops. Arthritis, often involving multiple joints, occurs in over 70% of cases. Endocarditis and focal abscesses may be late complications of untreated cases.

Spirillary RBF has a longer incubation period (1 to 3 weeks, usually more than 10 days). The bite wound often heals initially, but an inflammatory recrudescence occurs at the site coincidental with onset of systemic illness. The systemic illness is characterized by fever, headache, myalgia, nausea, vomiting, regional lymphadenopathy, and rash. The rash may be maculopapular or consist of red to purple plaques. Arthritis is unusual in spirillary RBF.

Untreated, both forms of RBF tend to run a chronic course, frequently taking months to resolve. About 7 to 10% of untreated cases result in death. Both diseases respond promptly to penicillin therapy, with tetracycline and streptomycin being effective alternative drugs.

Most cases of streptobacillary and spirillary RBF, including those acquired in the laboratory, result from rat bites (2, 5). Cases have infrequently resulted from the bite of other animals such as mice, weasels, cats, and dogs. Occasional cases have occurred in individuals with no history of animal bites or scratches. Experimental treatment of syphilis with spirillary RBF induced by inoculation of naturally infected rodent blood has demonstrated that accidental inoculation of rodent blood could also potentially lead to disease. Streptobacillary fever has resulted from ingestion of contaminated milk. Aerosol transmission has not been documented, but available evidence suggests that this mode of transmission is a potential hazard.

Prevention is based on elimination of the infection from laboratory rodent colonies and on training personnel in appropriate animal handling techniques. A vaccine is not available. Prophylactic antimicrobial therapy of personnel who are bitten by rats from known infected colonies has been recommended, but the efficacy of prophylactic therapy has not been evaluated.

RELAPSING FEVER

Epidemic (louse-borne) and endemic (tick-borne) relapsing fevers are both caused by spirochetes that belong to the genus *Borrelia*. After an incubation period of 5 to 15 days, affected individuals experience sudden onset of fever, chills, myalgia, arthralgias, photophobia, and headache. Physical examination often reveals hepatosplenomegaly, injection of the sclerae, petechiae, epistaxis, and occasionally jaundice, lymphadenopathy, or nuchal rigidity. The initial febrile illness lasts for 3 to 7 days and ends suddenly. The patient remains afebrile for 7 to 10 days, and then the fever and other symptoms return abruptly and persist for several days. Patients with endemic relapsing fever usually have more relapses than those with the epidemic form.

At least 45 cases of laboratory-acquired relapsing fever have been reported, making it the seventh most common bacterial cause of laboratory-associated infections (18).

Laboratory-acquired relapsing fever is most likely to occur when infectious materials from humans or animals (including hemolymph from lice and ticks) are accidentally injected or come in direct contact with broken skin or oral, nasal, or conjunctival mucous membranes of laboratory personnel (7). At least one case of probable relapsing fever occurred when a syringe containing spirochetes suddenly became separated from the attached needle and a worker sprayed infectious material into his eye. Bites of *Borrelia*-positive ectoparasites have also probably accounted for some cases of relapsing fever among persons employed in research units.

Biosafety Level 2 precautions are adequate for almost all work with *Borrelia* spp. Specific preventive measures include the use of syringes with locking hubs when handling *Borrelia* cultures or potentially infected blood and the use of disposable gloves when personnel responsible for handling *Borrelia* have obvious abrasions or open lesions on their hands. Research units that house infected ectoparasites should follow procedures that minimize the chances of employees being bitten by tick vectors.

Tetracyclines are the drugs of choice for both endemic and epidemic relapsing fever. Recent studies have shown that a single 500-mg dose of tetracycline or 100 mg of doxycycline will cure most patients with epidemic relapsing fever. Patients should be watched carefully for signs and symptoms of a Jarisch-Herxheimer reaction, which often occurs within several hours of the institution of antibiotic therapy. Patients that are allergic to tetracycline may be treated with chloramphenicol.

RICKETTSIAL INFECTIONS

Rocky Mountain spotted fever

Rickettsia rickettsii is the etiologic agent of Rocky Mountain spotted fever (RMSF). The incubation period in laboratory-acquired RMSF varies from 1 to 8 days. Affected personnel usually have sudden onset of fever, chills, photophobia, myalgia, and moderate to severe headaches. Although patients often develop a maculopapular rash that later becomes petechial, some individuals with laboratory-acquired RMSF never develop a rash (11). Rhinorrhea, sneezing, and cough may occur in personnel who acquire RMSF via the respiratory route.

About 75 cases of laboratory-acquired RMSF have been reported (18). Eleven of the cases, including two fatal cases, occurred in the 1970s. Most of the affected personnel worked directly with *R. rickettsii* or infected ticks, while others merely entered laboratories where the organism was being handled. Cultures of *R. rickettsii* grown in yolk sacs or chicken embryo fibroblast cultures are the sources implicated in most cases. Modes of transmission that have been responsible for laboratory-acquired cases include spread via droplets or the airborne route, direct inoculation during accidents involving needles and syringes, and bites inflicted by infected ticks. Direct person-to-person spread via droplet nuclei produced by the cough of a patient with respiratory involvement has been suggested as a possible means of transmission, but this theory has never been proven.

Guidelines for containment of Biosafety Level 3 agents should be observed. Inoculation of cell cultures or yolk sacs of embryonated eggs, harvesting and centrifugation of infectious suspensions, and animal inoculation studies should be carried out in biosafety cabinets. Mouth pipetting should be strictly forbidden.

For many years exposed personnel were immunized with a killed-cell yolk sac vaccine; however, direct-challenge studies have shown that this vaccine is relatively ineffective. An improved Formalin-inactivated vaccine prepared in cell cultures is currently being evaluated; if this vaccine is shown to be efficacious, then laboratory personnel at high risk of acquiring RMSF should be immunized.

Anecdotal reports suggest that prophylactic administration of tetracycline may protect persons involved in laboratory accidents with *R. rickettsii*. Exposed individuals who develop signs and symptoms compatible with RMSF should be treated with tetracycline or chloramphenicol for 5 to 7 days.

Q fever

Coxiella burnetii is the rickettsia that causes Q fever. The incubation period in laboratory-associated disease varies from 13 to 28 days (usually 14 to 18) in cases acquired via the respiratory route, but may be as short as 1 day in persons who have been accidentally inoculated with the organism. Affected individuals often have sudden onset of fever, chills, headache, myalgia, cough, and pleuritic chest pain. About 50% of cases have roentgenographic evidence of pneumonitis. Granulomatous hepatitis occurs in some patients. Relatively mild cases are often mistaken for influenza since affected personnel do not develop a rash. Asymptomatic infections have been documented during serosurveys of exposed laboratory personnel. Weil-Felix agglutinin titers, which become elevated in other rickettsial diseases, remain negative in Q fever.

C. burnetii is one of the most common causes of laboratory-acquired infections, with more than 400 cases having been reported (1, 18). Individuals handling the organism, as well as support personnel (e.g., maintenance workers, clerical personnel) employed in the same building, are often affected in institutional outbreaks (10, 22). Sources that been implicated in cases of laboratory-associated Q fever include yolk sac suspensions of C. burnetii, naturally infected sheep and goats and possibly experimentally infected guinea pigs (or their excreta), contaminated laundry and clothing, and infected human tissues handled during postmortem examination.

Q fever is usually spread to laboratory workers via infectious aerosols produced when yolk sac suspensions are emulsified, centrifuged, pipetted, or inoculated into animals. If yolk sac antigens are prepared in rooms under positive air pressure, persons in adjacent laboratories or in remote areas of the same building may acquire the disease via airborne transmission. In the last decade, numerous research personnel and animal caretakers exposed to sheep (and their excreta) have acquired Q fever, presumably by inhaling the organism. Occasionally, recognized laboratory accidents such as breaking glass vials containing C. burnetii or sudden leaks in aerosol chambers or animal exposure bags have caused Q fever in exposed personnel. At least one laboratory worker has become infected by accidentally inoculating himself with viable organisms.

Adherence to recommended safety precautions is of paramount importance in preventing laboratory-associated Q fever. Inoculation and harvesting of yolk sacs of chicken embryos and further preparation of yolk sac antigens should be carried out in laboratories that meet Biosafety Level 3 containment standards. Personnel working with suspensions of viable organisms must be familiar with established procedures for handling hazardous agents. Carcasses of infected animals should be placed in impervious containers until disposed of, and animal caretakers should use measures that prevent aerosolization of dried animal excreta. Autoclaving contaminated media, glassware, and laundry before removal from the laboratory is desirable.

No Q fever vaccine is approved for use in humans in the United States. Prophylactic administration of tetracycline to persons involved in a known laboratory accident is of questionable value. Exposed employees who develop signs and symptoms suggestive of Q fever should be treated with tetracycline or chloramphenicol.

Epidemic and endemic typhus

Rickettsia prowazekii and R. typhi are the etiologic agents of epidemic and endemic typhus, respectively. The incubation period in cases of laboratory-acquired typhus varies from 4 to 14 days. Affected personnel often have sudden onset of fever, chills, headache, and myalgia. A macular rash may appear on the trunk on day 4 to 7 of the illness in unimmunized workers, but is frequently absent in immunized individuals. Occasionally, patients will complain of right upper quadrant tenderness and may have elevated serum aspartate aminotransferase values.

Approximately 110 cases of laboratory-acquired typhus have been reported (18). Infections occur primarily in persons working with the organisms, and to a lesser extent among employees with other responsibilities in laboratories where the agents are being handled. Suspensions of rickettsia prepared in yolk sacs or passaged in guinea pigs are usually the source of laboratory-acquired typhus. Dried feces from infected lice has also been suggested as a possible source. Exposure to infectious droplets and airborne transmission account for most cases (29). Intranasal inoculation of mice and emulsifying suspensions of rickettsia in blenders are procedures that are likely to create infectious aerosols. A few laboratory personnel have acquired typhus by accidentally splashing or spraying suspensions of virulent organisms into their eyes.

To prevent laboratory-acquired typhus, research projects and vaccine production should be carried out in facilities that meet Biosafety Level 3 containment standards. For many years, persons at risk were immunized with typhus vaccines prepared from formaldehyde-inactivated R. prowazekii. However, American and Canadian manufacturers have discontinued production of these vaccines, whose efficacy had not been established in appropriate controlled trials. Laboratory personnel exposed to R. prowazekii or R. typhi should be treated for 4 to 5 days with tetracycline or chloramphenicol if they develop signs and symptoms suggestive of typhus. Recent studies have shown that a single dose of doxycycline may be effective therapy for epidemic typhus.

Scrub typhus

Rickettsia tsutsugamushi is the etiologic agent of scrub typhus. After an incubation period of 1 to 3 weeks, affected individuals experience sudden onset of fever, chills, headache, and myalgia and sometimes have a nonproductive cough. An eschar may develop at the site where the rickettsiae have entered the skin, but is absent in patients who acquire the disease via the respiratory route. A macular or maculopapular rash appears 5 to 8 days after onset of fever in some, but not all, patients.

Thirty-five cases of laboratory-acquired scrub typhus have been reported (18). Most cases have occurred in personnel who worked directly with the organism. Suspensions of virulent *R. tsutsugamushi* are almost always the source of laboratory-acquired infections, although tissues obtained during postmortem examination of affected patients or experimentally infected animals have also served as sources. The disease has been transmitted to research workers via exposure to droplet nuclei with subsequent inhalation of the organism, direct contact (contamination of conjunctival mucosa or abraded skin), accidental inoculation during accidents involving needles or glassware, rat bite, and possibly mite bites (17).

Biosafety Level 3 precautions should be followed for work with *R. tsutsugamushi*. A satisfactory scrub typhus vaccine is not available.

Although prophylactic administration of chloramphenicol to exposed personnel will prevent the disease, chemoprophylaxis is not indicated in most situations. Personnel who develop an illness compatible with scrub typhus should be treated with tetracycline or chloramphenicol for 7 to 14 days. Relapses are reasonably common if either drug is given for only 3 or 4 days, but can be avoided by giving a second short course of therapy 4 to 5 days after stopping the initial treatment. A single 200-mg dose of doxycycline also appears to be effective.

TULAREMIA

Francisella tularensis may cause several clinical syndromes, including ulceroglandular, glandular, typhoidal, oculoglandar, primary pulmonary, and pharyngeal tularemia. In the ulceroglandular form, a papular lesion develops at the site where organisms enter the skin. The incubation period varies from 1 to 5 days. The papule frequently ulcerates and is accompanied by painful regional lymphadenopathy. Immunized laboratory personnel may or may not develop concomitant fever, myalgia, headache, and diarrhea. In primary pulmonary and typhoidal tularemia, fever, sore throat, substernal discomfort, and cough develop after an incubation period of 1 to 10 days. Occasionally, affected individuals also complain of cervical lymphadenopathy. Although 30 to 45% of laboratory workers with typhoidal tularemia have evidence of bronchopneumonia by chest roentgenogram, physical examination of the chest is often unimpressive. Oculoglandular tularemia is characterized by painful, purulent conjunctivitis with preauricular or cervical lymphadenopathy. Pharyngeal tularemia produces an acute exudative pharyngotonsillitis with cervical lymphadenopathy.

Multiple episodes of laboratory-acquired tularemia (reinfections) have been documented in persons who received killed-cell vaccines and in unimmunized individuals. In some cases of reinfection the severity and duration of symptoms may approach that seen in initial infections.

In the last 50 years, tularemia has been the third most common bacterial cause of laboratory-acquired infections, with at least 225 cases having been reported (18). Almost all cases have been associated with facilities involved in tularemia research, with very few cases among personnel in clinical microbiology laboratories. Bacteriologists, physicians, and laboratory technologists that work with the organism are at greatest risk, but cases have also occurred in animal caretakers, maintenance personnel, and employees responsible for washing glassware used in research laboratories.

Cultures of *F. tularensis* are the source most commonly implicated in cases of laboratory-acquired tularemia (16). As few as 10 to 50 organisms can result in disease after inhalation as an aerosol or inoculation beneath the skin (20a, 20b). On a few occasions, naturally or experimentally infected animals have been the source. Laboratory workers may acquire tularemia in several ways. Infection may occur when organisms come into direct contact with cuts, cracks, or abrasions on the employee's skin. Several cases have occurred when individuals cut themselves with contaminated autopsy instruments, glassware, or ampoules containing *F. tularensis*. Accidental needle punctures sustained during animal inoculation studies have caused ulceroglandular tularemia, and at least one immunized worker acquired oculoglandular disease when a needle came off a syringe and virulent organisms were sprayed into his eye. Accidents associated with mouth pipetting may cause pharyngeal tularemia.

Numerous laboratory workers have acquired typhoidal tularemia via infectious droplets and by the airborne route. The respiratory tract is usually the portal of entry when these modes of transmission are implicated. Personnel who perform high-risk procedures such as centrifugation and lyophilization of cultures are usually affected, but a number of persons who merely entered laboratories have acquired the disease via airborne transmission.

An investigational live, attenuated tularemia vaccine is currently available (Immunobiologics Activity, Centers for Disease Control, Atlanta, Ga.). This vaccine does not prevent overt clinical illness, but does reduce severity of illness (3). Prevention of infection is based on adherence to Biosafety Level 3 standards.

The role of chemoprophylaxis in immunized persons exposed to a laboratory accident involving infectious material is unknown. Studies conducted before the availability of live attenuated vaccine demonstrated that a 5-day course of streptomycin usually prevented laboratory-acquired tularemia. Tetracycline has also been demonstrated to be an effective chemoprophylactid drug (20c). Since typhoidal tularemia is uncommon among immunized personnel, postexposure chemoprophylaxis is probably not indicated.

Persons who develop laboratory-acquired tularemia should be treated with streptomycin given intramuscularly for 7 to 10 days. Tetracycline and chloramphenicol are acceptable alternatives, but clinical relapses occur more frequently when these drugs are used.

NATURALLY INFECTED LABORATORY ANIMALS

Naturally infected laboratory animals are another potential biohazard for laboratory personnel. However, the reported spectrum of bacterial infections acquired from the commonly used laboratory animals is limited.

Rodents and rabbits

Rodents and rabbits comprise more than 99% of all animals used in laboratories. Although many thousands of people are directly exposed each year, few bacterial infections have been reported as being acquired from these animals in the laboratory. RBF is the most commonly reported bacterial disease associated with laboratory rodents, but only 24 cases had been reported as being laboratory associated through 1984. Leptospirosis has been reported slightly less frequently. Most leptospirosis infections resulted from mouse or rat bites. The infecting *Leptospira* serovar has been *ballum* in cases associated with mice and *icterohaemorrhagiae* in cases associated with rats.

Other bacterial infections are so rare as to hardly merit discussion. For example, two cases of possible *Bordetella bronchicanis* (*bronchiseptica*) infection have been reported in animal caretakers with, respectively, a chronic rhinorrhea and an influenzalike illness, and a single case of *Pasteurella multocida* (*lepiseptica*) wound infection due to a rabbit bite has been reported (1a, 28).

Nonhuman primates

Tuberculosis and bacterial gastroenteritis are the primary bacterial diseases associated with nonhuman primates (12). Tuberculosis has been reported as a spontaneous disease in most primate species commonly used in the laboratory. The incidence in recently imported animals varies with their geographic origin, being most common (1 to 2% incidence) in primates from Asia, less frequent in African species, and rare in American species. Exposure to tuberculous primates in holding facilities after im-

portation, however, may result in a high infection rate regardless of the group's original geographic origin.

Simian tuberculosis may be caused by either *Mycobacterium tuberculosis* or *Mycobacterium bovis*. Infections by other mycobacteria such as *M. kansasii* have been reported, but secondary transmission of these infections to laboratory personnel has not been documented.

Simian tuberculosis is primarily spread by aerosols of infectious droplet nuclei. Shedding of *M. tuberculosis* or *M. bovis* via the respiratory tract usually precedes the development of tuberculin hypersensitivity, a factor that complicates control programs. Transmission via bites, ingestion, and fomites such as tattoo needles and thermometers has been reported, but these mechanisms of spread are of lesser importance. Primates are extremely susceptible to tuberculosis, and the disease spreads rapidly once introduced into a colony.

Persons working with primates are at high risk of tuberculosis infection. In the United States, the annual tuberculin conversion rate in persons occupationally exposed to primates is about 70/10,000 compared with less than 3/10,000 in the general population (13). Most infections appear to result from inhaling infectious aerosols while working in areas where tuberculous animals are housed. At least one case of cutaneous tuberculosis due to a scalpel puncture during necropsy of a monkey has been reported.

Simian tuberculosis control is based on routine quarantine and tuberculin testing of all new additions to the colony, as well as periodic tuberculin testing of all primates in the maintenance colony (13). All tuberculin-positive animals should be immediately killed or placed in strict isolation if further evaluation is indicated. Treatment is not recommended except in unusual circumstances.

Because tuberculosis can be transmitted from nonhuman primates to their handlers as well as the reverse, all persons who work around these animals should be included in a tuberculosis screening program. Face masks should be routinely worn while working in primate-holding areas, particularly in quarantine areas.

Bacterial gastroenteritis, primarily due to *Shigella, Salmonella,* and *Campylobacter jejuni,* is the leading cause of morbidity and mortality in laboratory primates. Anecdotal information suggests that persons working with primates are also at high risk of acquiring diarrheal disease; at least 71 *Shigella,* 2 *Salmonella,* and 1 *C. jejuni* infections have been reported as being acquired from these animals. The fecal-oral route of transmission is of primary importance. Control is based on adherence to scrupulous personal hygiene practices while working with primates. Water-impervious gloves should be worn while handling primates and their excretions, and no eating or smoking should be allowed in primate-holding areas.

LITERATURE CITED

1. **Bernard, K. W., G. L. Parham, W. G. Winkler, and C. G Helmick.** 1982. Q fever control measures: recommendations for research facilities using sheep. Infect. Control **3:**461–465.

1a. **Boisvert, P. L., and M. D. Fousek.** 1941. Human infection with *Pasteurella lepiseptica* following a rabbit bite. J. Am. Med. Assoc. **116:**1902–1903.

2. **Brown, T. M., and J. C. Nunemaker.** 1942. Rat-bite fever: a review of the American cases with reevaluation of etiology; report of cases. Bull. Johns Hopkins Hosp. **70:**201–328.

3. **Burke, D. S.** 1977. Immunization against tularemia: analysis of the effectiveness of live *Francisella tularensis* vaccine in prevention of laboratory-acquired tularemia. J. Infect. Dis. **135:**55–60.

4. **Burmeister, R. W., W. D. Tigertt, and E. L. Overholt.** 1962. Laboratory-acquired pneumonic plague: report of a case and review of previous cases. Ann. Intern. Med. **56:**789–800.

5. **Cole, J. S., R. W. Stoll, and R. J. Bulger.** 1969. Rat-bite fever: report of three cases. Ann. Intern. Med. **71:**979–981.

6. **Ellingson, H. V., P. J. Kadull, H. L. Bookwalter, and C. Howe.** 1946. Cutaneous anthrax: report of twenty-five cases. J. Am. Med. Assoc. **131:**1105–1108.

7. **Felsenfeld, O.** 1971. Borrelia: strains, vectors, human and animal boreliosis, p. 71. Warren H. Green, Inc., St. Louis.

8. **Green, R. N., and P. G. Tuffnell.** 1968. Laboratory-acquired melioidosis. Am. J. Med. **44:**599–605.

9. **Howe, C., and W. R. Miller.** 1947. Human glanders: report of six cases. Ann. Intern. Med. **26:**93–115.

10. **Johnson, J. E., III, and P. J. Kadull.** 1966. Laboratory-acquired Q fever: a report of fifty cases. Am. J. Med. **41:**391–403.

11. **Johnson, J. E., III, and P. J. Kadull.** 1967. Rocky Mountain spotted fever acquired in a laboratory. N. Engl. J. Med. **277:**842–847.

12. **Kaufmann, A.F.** 1972. Nonhuman primate zoonoses surveillance in the United States, p. 58–67. *In* E. I. Goldsmith and J. Moor-Jankowski (ed.), Medical primatology 1972, part III. S. Karger, Basel.

13. **Kaufmann, A. F., and D. C. Anderson.** 1978. Tuberculosis control in nonhuman primate colonies, p. 227–234. *In* R. J. Montali (ed.), Mycobacterial infections of zoo animals. Smithsonian Institution Press, Washington, D.C.

14. **Kaufmann, A. F., M. D. Fox, J. M. Boyce, D. C. Anderson, M. E. Potter, W. J. Martone, and C. M. Patton.** 1980. Airborne spread of brucellosis. Ann. N. Y. Acad. Sci. **353:**105–114.

14a. **McClain, B. L., W. R. Ballou, S. M. Harrison, and D. L. Steinweg.** 1984. Doxycycline therapy for leptospirosis. Ann. Intern. Med. **100:**696–698.

15. **Morisset, R., and W. W. Spink.** 1969. Epidemic canine brucellosis due to a new species, *Brucella canis*. Lancet **ii:**1000–1002.

16. **Overholt, E. L., W. L. Tigertt, P. J. Kadull, M. K. Ward, N. D. Charkes, R. M. Rene, T. E. Salzman, and M. Stephens.** 1961. An analysis of forty-two cases of laboratory-acquired tularemia. Am. J. Med. **30:**785–806.

17. **Philip, C. B.** 1948. Tsutsugamushi disease (scrub typhus) in World War II. J. Parasitol. **34:**169–191.

18. **Pike, R. M.** 1976. Laboratory-associated infections: summary and analysis of 3921 cases. Health Lab. Sci. **13:**105–114.

19. **Pike, R. M., S. E. Sulkin, and M. L. Schulze.** 1965. Continuing importance of laboratory-acquired infections. Am. J. Public Health **55:**190–199.

20. **Redfearn, M. S., and N. J. Palleroni.** 1975. Glanders and melioidosis, p. 110–128. *In* W. T. Hubbert, W. F. McCulloch, and P. R. Schnurrenberger (ed.), Diseases transmitted from animals to man, 8th ed. Charles C Thomas, Publisher, Springfield.

20a. **Saslaw, S., H. T. Eigelsbach, H. R. Wilson, J. A. Prior, and S. R. Carhart.** 1961. Tularemia vaccine study. I. Intracutaneous study. A.M.A. Arch. Intern. Med. **107:**689–701.

20b. **Saslaw, S., H. T. Eigelsbach, H. R. Wilson, J. A. Prior, and S. R. Carhart.** 1961. Tularemia vaccine study. II. Respiratory challenge. A.M.A. Arch. Intern Med. **107:**702–714.

20c. **Sawyer, W. D., H. G. Dangerfield, A. L. Hogge, and D. Crozier.** 1966. Antibiotic prophylaxis and therapy of airborne tularemia. Bacteriol. Rev. **30:**542–548.

21. **Schacter, J., and C. R. Dawson.** 1977. Human chlamydial infections, p. 9–43. PSG Publishing Co., Littleton, Mass.

22. **Schacter, J., M. Sung, and K. F. Meyer.** 1971. Potential danger of Q fever in a university hospital environment. J. Infect. Dis. **123:**301–304.

23. **Spink, W. W.** 1956. The nature of brucellosis, p. 106–108. University of Minnesota Press, Minneapolis.

24. **Stoenner, H. G., and D. MacLean.** 1958. Leptospirosis (ballum) contracted from Swiss albino mice. Arch. Intern. Med. **101:**606–610.

25. **Sulkin, S. E., and R. M. Pike.** 1951. Survey of laboratory-acquired infections. Am. J. Public Health **41:**769–781.

26. **Sullivan, J. F., J. R. Songer, and I. R. Estrem.** 1978. Laboratory-acquired infections at the National Animal Disease Center 1960–1976. Health Lab. Sci. **15:**58–64.

26a. **Takafugi, E. T., J. W. Kirkpatrick, R. N. Miller, J. J. Karwacki, P. W. Kelley, M. R. Gray, K. M. McNeil, H. L. Timboe, R. E. Kane, and J. L. Sanchez.** 1984. An efficacy trial of doxycycline chemoprophylaxis against leptospirosis. N. Engl. J. Med. **310:**497–500.

27. **Tindall, J. P., and C. M. Harrison.** 1972. *Pasteurella multocida* infections following animal injuries, expecially cat bites. Arch. Dermatol. **105:**412–416.

28. **Winsser, J.** 1960. A study of *Bordetella bronchiseptica*. Proc. Anim. Care Panel **10:**87–104.

29. **Wright, L. J., L. F. Barker, I. D. Mickenberg, and S. M. Wolff.** 1968. Laboratory-acquired typhus fevers. Ann. Intern. Med. **69:**731–738.

30. **Wu, L. T.** 1926. A treatise on pneumonic plague, p. 100–105. League of Nations, Geneva.

31. **Wu, L. T., J. W. H. Chun, R. Pollitzer, and C. Y. Wu.** 1936. Plague: a manual for medical and public health workers, p. 516–521. The Mercury Press, Shanghai.

Chapter 13

Laboratory-Acquired Parasitic Infections

G. ALEXANDER CARDEN, DENNIS D. JURANEK, AND DOROTHY M. MELVIN

The life cycles and modes of infection associated with parasitic organisms responsible for human disease cover a wide spectrum. For this reason, the precautions necessary for handling different types of clinical specimens, parasites in culture, or infected laboratory animals vary over a wide range, depending on the type of infective organism. This chapter will present the risks associated with the handling of all parasites in the laboratory and will further elaborate those measures which are indicated for the management of specific types of organisms. It must be kept in mind that the handling of specimens or laboratory animals harboring parasitic infections may expose the worker to the risk of other nonparasitic hazards such as bacterial or viral infections or chemical toxins. The basic principles of personal hygienic practices, as well as the Centers for Disease Control (CDC) classification of laboratory risks, environmental exposure precautions, and specimen disposal techniques within microbiology laboratories, are outlined elsewhere in this volume. This chapter will focus, therefore, on those problems and hazards peculiar to the handling of specimens associated with parasitic infections, as well as the diseases they produce. The diagnostic evaluation and clinical management of certain infections requires an individualized approach. In such instances it is suggested that the individual or his or her physician contact the Parasitic Diseases Division at CDC (404) 329-3843 (day) or (404) 329-3644 (nights and weekends, emergency only) or the Parasitic Disease Drug Service at (404) 329-3670 (8:00 a.m.–4:30 p.m., Monday through Friday) for consultation.

PROTOZOAN INFECTIONS

Tissue Protozoa

Toxoplasma gondii

Toxoplasma gondii infection may be acquired from clinical specimens, either tissue (human, animal) or fecal (feline), as well as experimental laboratory preparations (cell culture). Both the cysts in tissue and the sporulated oocysts in cat feces are infective for many animal species, including humans, by the oral route. Autopsy workers as well as bench laboratory technicians (8) are at risk. Pike (14) has reported 16 cases of laboratory-acquired *Toxoplasma gondii* infection between 1949 and 1974, with one fatality in 1951 (15). The exact mode of transmission (aerosol versus fomites) was not identified in every instance, although the oral cavity was the apparent portal of entry in all cases (8). Percutaneous infection is theoretically possible. Use of a laboratory biological safety cabinet, gloves, mask, and manual pipetting equipment is indicated. Contaminated instruments and laboratory glassware should be boiled or cleaned with an iodine-containing compound if oocysts are being handled. Centrifugation should be carried out in closed containers.

Female laboratory personnel of child-bearing age should be serologically screened for *Toxoplasma* antibodies before working with *Toxoplasma* spp. in either diagnostic or research laboratories. Seronegative women in this age group who are or may become pregnant should be excluded from working with these organisms. Women whose sera are positive for *Toxoplasma* antibodies are protected from congenital infection of the fetus should they become pregnant and are, therefore, not at any greater risk than seropositive men who work with this organism in the laboratory. The most widely used serological screening test is the total globulin or immunoglobulin G (IgG)-based immunofluorescence assay (IFA).

If acute infection is suspected in a laboratory worker, the most useful diagnostic test is by IgM-IFA serology, the results of which should rise sharply 14 to 21 days after the infection. This titer will usually fall to negative within 3 to 6 months of infection. IgM-IFA is available at most major medical centers. If the test is not available locally, specimens may be submitted to the state laboratory for IgM-IFA analysis. A fourfold rise in the IgG titer determined by either IFA, indirect hemagglutination assay, or the Sabin-Feldman dye test technique also strongly supports the diagnosis of acute infection. A single elevated level of IgG, however, may reflect a remote prior exposure to infection and is of little value in diagnosing acute infection with *T. gondii*. The clinical course may vary from the complete absence of symptoms to a syndrome of fever and lymphadenopathy and, in rare cases, to diffuse involvement of multiple organs. The severity of the clinical course is probably related to the strain of *T.*

gondii with which the laboratory is working and possibly to the developmental stage of the parasite. Oocysts have been shown to be more pathogenic in mice than tissue cysts from the same strain of *T. gondii*.

Treatment is reserved for those persons in whom organ damage is apparent by clinical criteria (abnormal liver function tests, retinal damage, central nervous system manifestations) or symptoms remain severe for prolonged periods. Treatment of nonocular disease involves a 3- to 4-week course of pyrimethamine (25 mg per day) and trisulfapyrimidines (2 to 6 g per day) in conjunction with folinic acid (3 to 10 mg once a week). Recurrent or persistent retinochoroiditis is treated with clindamycin (300 mg) and sulfadiazine (1 g, orally, four times a day for 4 weeks) (18). Concomitant administration of corticosteroids (prednisone, 60 mg per day) is recommended by some consultants. However, steroid therapy without antimicrobial coverage is not recommended.

Sarcocystis

Three species of *Sarcocystis* are recognized as infective for humans. Humans are the intermediate host for the parasite *S. lindemanni,* for which the definitive host is unknown. Humans may acquire intestinal infection with *S. hominis* by ingesting undercooked infective beef and aquire *S. suihominis* from undercooked infective pork. Clinical symptoms of anorexia, nausea, vomiting, abdominal pain, and diarrhea have only been associated with *S. suihominis* infections. Accidental laboratory infections have not been documented. Sporocysts shed in the stool of infected humans and animals have not been shown to be infective for humans (5). There are no available serological tests, and treatment is not required. Humans experimentally infected with tissue cysts of *S. hominis* shed sporocysts for periods of 9 to 179 days with no reported symptoms.

Leishmania

Transmission of *Leishmania* is generally thought to require competent vectors (phlebotomines), so that the hazard of handling laboratory specimens in the absence of these is considerably less. In research laboratories where vector colonies are maintained, strict containment and security measures must be taken. Nonetheless, the handling of infected animals (mice, hamsters) or in vitro-cultured parasites requires precautions against contamination of cutaneous abrasions or injuries, especially of the fingers or nails. Gloves and manual pipettes should therefore be used. The incubation period for leishmanial infection is variable depending on the species, but infection is generally apparent within 3 to 4 weeks, frequently less. Cutaneous ulceration or erythema at the site of inoculation, at an adjacent

site, or on a mucous membrane of the nasopharynx would raise sufficient suspicion to warrant clinical evaluation in the form of: (i) IFA serology; (ii) biopsy of the lesion and submission of the specimens for pathology, in vitro cultivation, and hamster inoculation; and (iii) saline aspiration at the margin of the lesion and in vitro cell-free cultivation of the aspirate in Schneider medium (10). Cultivation of aspirated fluid has recently been demonstrated to be an efficient and sensitive alternative to incisional or punch biopsy.

Treatment of active disease, except for that with *Leishmania tropica*, which is usually self-limited, is a 10-day course of pentavalent antimonial therapy (sodium stibogluconate; Pentostam), available through the Parasitic Disease Drug Service at CDC (see above).

Laboratory-acquired cases of kala azar (visceral leishmaniasis) have not been reported. The hypothetical risk of percutaneous transmission remains a possibility if a needle stick or skin break contaminated with laboratory-cultured parasites were to occur.

Any worker subjected to such exposure with a viscerotropic *Leishmania* strain (e.g., *L. donovani, L. chagasi, L. infantum*) should be followed serologically at 1- to 2-month intervals for 6 to 12 months. If the worker develops clinical illness (unexplained fever, hepatosplenomegaly) or seroconversion, a bone marrow aspiration and further work-up (e.g., liver biopsy, etc.) should be performed.

Trypanosoma

Eighteen cases of trypanosomal infection in laboratory workers have been reported in the world literature; many more cases have probably gone unreported (14). In at least two of these cases, *T. cruzi* infection was acquired through small or inapparent cutaneous injuries of the finger which provided the parasite with a portal of entry to the bloodstream (3). This mechanism closely mimics the natural mode of infection of *T. cruzi*. The ability of these parasites to enter the bloodstream across an apparently intact mucous membrane provides a further mandate for the use of eye protection (in case of a spill or splash) as well as a mask and gloves when handling infected animals, reduviid bugs, or in vitro culture specimens (9). It must be remembered that the infective metacyclic trypanosomes are contained in the reduviid feces. The bugs and their excreta must, therefore, be handled with equal precautions. Blood specimens from infected patients or laboratory animals should be considered infective and handled with gloves as well (9).

Acute infection with *T. cruzi* usually presents with a trypanosomal chancre at the site of inoculation within one week of infection. This lesion is characterized by erythema and induration and, in case of infections not acquired via a reduviid bite, would be expected to occur at the site of a previous cutaneous injury. Clinical symptoms may or may not ensue over subsequent weeks, but if present, they will include fever, malaise, myalgias, headache, and occasionally chest or abdominal pain. The presence of a morbilliform rash and regional adenopathy are also common findings, as are tachycardia, signs of cardiac dilatation, irregular pulse, and encephalopathic findings.

Diagnosis is based upon identification of trypanosomes in a peripheral blood smear or cultivation in vitro (21). When available, xenodiagnosis using *T. cruzi*-free reduviid bugs is thought by some investigators to be the most sensitive diagnostic method. Others feel that cultivation of the parasite in biphasic medium (liver infusion-tryptose medium over blood agar) at 28°C is a more practical and equally sensitive method. The sensitivity of blood smear examination can be enhanced by centrifugation of fresh blood and examination of the fraction immediately above the buffy coat in a Giemsa-stained preparation. Low-power microscopic examination of fresh blood preparations in microhematocrit tubes may reveal motile organisms in the buffy coat. Serology is useful in following clinical response to treatment. In a laboratory worker with no antecedent exposure to infection (i.e., no residence in endemic areas), serology may also serve as a useful supporting diagnostic test. In cases of suspected primary infection where parasitemia is difficult to demonstrate, an initial titer taken at the time of or shortly after exposure should be negative, whereas a second titer taken 2 to 3 weeks later should demonstrate a significant rise if infection has occurred. Treatment of acute Chagas' disease (*T. cruzi* infection) consists of 12 to 14 weeks of gradually increased doses of nifurtimox, available through the Parasitic Disease Drug Service at CDC (see above).

Laboratory-acquired African trypanosomiasis has been documented (16). Workers in both diagnostic and research laboratories dealing with African trypanosomiasis in infective or possibly infective clinical specimens, animals, or trypanosomal cultures should observe glove and Class I or IIA biological safety cabinet precautions as well as added care in the handling and disposal of contaminated instruments (especially needles), glassware, and other equipment. These should be decontaminated with laboratory detergent or autoclaved before disposal. Laboratory-acquired infection with *T. rhodesiense* or *T. gambiense* is likely to take place by a route similar to that of *T. cruzi*, i.e., inoculation through a skin break by contact with infective blood or tissue from animal or human specimens. The trypanosomal chancre is more likely to appear with the former infection, but will usually be apparent within 2 to 3 days with either organism. Incubation of *T. rhodesiense* infection would be expected to progress to clinical disease within 5 to 10 days, whereas a *T. gambiense* infec-

tion is likely to remain subclinical for periods varying from a few weeks to many months.

The presenting symptoms of the more acute infection are frequently headache, malaise, nausea, vomiting, and diarrhea lasting for a few days and usually accompanied by episodes of fever and chills once or twice per day. Focal neurological deficits, diminished level of consciousness, or even unexplained behavioral changes may herald the onset of clinical *T. gambiense* infection. The diagnosis is based upon identification of trypanosomes in peripheral blood early in the course of infection and in cerebrospinal fluid (CSF) specimens during the later stages. Identification of the Morula cell of Mott (a plasma cell with abundant amphophilic cytoplasmic globules) in the CSF is pathognomonic of central nervous system (CNS) invasion by trypanosomes. Blood concentration and staining techniques similar to those used for the diagnosis of *T. cruzi* are appropriate. Animal inoculations and in vitro cultivation are also useful, but the species of laboratory animal and the type of culture media that should be used vary depending on the species of trypanosome that is suspected. Modifications of liver infusion-tryptose medium are generally used for this purpose. Serology may be helpful in acute disease and is usually accompanied by a marked rise in serum total IgM level. Early CNS involvement is frequently accompanied by a concomitant CSF IgM elevation in conjunction with increased total CSF protein and pleocytosis. A falling titer after treatment correlates well with a clinical cure or remission. Late relapses may occur in both South American and African trypanosomiasis.

Treatment of *T. rhodesiense* infection involves the use of suramin for hemolymphatic infection, melarsoprol when CNS involvement is apparent, or both. *T. gambiense* infections are usually best treated with pentamidine for hemolymphatic disease or melarsoprol for CNS involvement. Suramin and melarsoprol are available only through the Parasitic Disease Drug Service at the CDC (see above).

Free-living amebae

Free-living species of amebae known to be pathogenic for humans generally belong to two genera, *Naegleria* and *Acanthamoeba*. *Naegleria* spp., particularly *N. fowleri*, are associated with acute fulminant disease and a known portal of entry (the nasopharynx) to the brain. *Acanthamoeba* spp. are associated with a more insidious onset of symptoms and an unknown mechanism of infection.

Because of the nearly uniformly fatal outcome of meningoencephalitic infections with these organisms (2, 4, 7) and the lack of more definitive information regarding the mechanism of its transmission in humans (1, 6), strict precautionary measures including the use of gowns, gloves, masks, negative-pressure hoods, or any or all of these are recommended for the handling of CNS tissue or cultured material containing these organisms. Pathologic observations of human infection with *N. fowleri*, as well as the production of fatal encephalitic infections by intranasal inoculation of bovine and murine species in the laboratory, provide a strong body of evidence which should warn laboratory workers to protect their nasopharyngeal mucosa while handling specimens known or thought to contain free-living amebae. It should also be kept in mind that airborne spread has not been excluded as a possible mode of inoculation, particularly with the *Acanthamoeba* spp. Culturing operations should, therefore, be kept away from areas of patient care. Individuals with immunologic impairment of any kind should not be permitted in areas where free-living amebae are being cultured. To date, no case reports of primary amebic meningoencephalitis have been reported in laboratory personnel working with these organisms. The incubation period for *Naegleria* infection is thought to be 3 to 6 days after exposure, based on histories of patients swimming in fresh water heavily infested with these organisms. The incubation period for *Acanthamoeba* sp. infections is unknown but thought to be considerably longer than that for *Naegleria* sp., perhaps weeks or even months. *Naegleria* infections are characterized by a rapid evolution of symptoms, usually starting with headache, drowsiness, and malaise and progressing rapidly over 24 to 48 h to a picture of fulminant meningoencephalitis, including CSF pleocytosis with a polymorphonuclear predominance and, in many cases, motile amebae visualized in the fresh CSF. Confirmation of a suspected case should be carried out by the protocol outlined by Krogstad et al. (12).

Acanthamoeba infections, by contrast, have a much more gradual onset with slow, progressive development of neurological impairment, frequently of a focal or localized nature. The evolution of disease may last as long as several months. CSF lymphocytosis is more prominent in this form of primary amebic meningoencephalitis, and in contrast to the *Naegleria* infections, amebae are not found in the CSF. When clinical suspicion of this infection is high, brain biopsy is the diagnostic measure of choice.

Treatment of primary amebic meningoencephalitis involves the use of multiple drugs, including amphotericin B and miconazole, by intravenous and intrathecal routes. Rifampin and polymyxin B as well as other antimicrobial agents may have a role in therapy, depending on the species of infecting organism. If such infection is suspected, the physician in charge of the patient's case should contact the Parasitic Diseases Division at CDC, where consultation is available on a 24-h basis (see above).

Pneumocystis carinii

Pneumocystis carinii pneumonia has not been reported in laboratory workers handling either infected animals or human specimens. The question of whether human disease with this pathogen represents hypercolonization resulting from a breakdown of ecological equilibrium between an immunologically normal host and his endogenous respiratory flora, or whether clinical disease results from inoculation of the immunologically compromised respiratory tract from the outside, remains unresolved (17). It seems possible that some component of both mechanisms is at work. For this reason it is safer to handle infected specimens with gloves and mask or gloves and hood protection, even though no laboratory-acquired cases of clinical illness have been reported to date. As with the other opportunistic organisms, immunocompromised persons should not be permitted in a laboratory where this organism is being handled. Although the incubation period is unknown in humans, patients undergoing immunosuppressive chemotherapy may develop this infection within a week after treatment is started. The disease may present as a diffuse bilateral pneumonia, sometimes accompanied by a nonproductive cough and fever. Tachypnea and hypoxia, of a severity disproportionate to the initially observed radiographic changes, most commonly accompany the initial presentation of this infection. Diagnosis can usually only be made by bronchoalveolar lavage, transthoracic needle biopsy, or open-lung biopsy. Impression smears or sections of lung tissue should be stained to demonstrate the cysts (methenamine silver nitrate stain) and intracystic bodies or trophozoites (Giemsa stain). The drug combination of first choice in treatment of these infections is trimethoprim (20 mg/kg) and sulfamethoxazole (100 mg/kg), in four divided doses daily for 14 days. Pentamidine isethionate available from the Parasitic Disease Drug Service at CDC (see above) is an alternative in the case of treatment failure or a drug reaction to the first-line regimen.

Blood Protozoa

Malaria

Malaria infection may be acquired in the laboratory by three mechanisms. First, and most common, is inadvertent contact with an infective vector in laboratories where mosquito colonies are maintained. Pike (14, 15) has stressed this as an important mechanism of laboratory-acquired malaria even with *Plasmodium* species generally associated with animal infections (e.g., *P. cynomolgi*), and has warned that mosquitoes infected with these species deserve the same degree of respect accorded those species more commonly associated with human infection: *P. vivax*, *P. falciparum*, *P. malariae*, and *P. ovale*.

The second mechanism of transmission is in the handling of infected primate hosts, especially when these are being bled. Inadequate protection of animal handlers, inadequate sedation of animals being bled, and poor control of bleeding sites can result in increased exposure to infective material through cuts or scratches in the handler's skin.

The third mechanism of transmission within the laboratory is accidental inoculation of a skin wound with a needle, syringe, or glass slide contaminated with infective blood. The use of appropriate precautions and protective gloves is essential for the safe handling of blood containing infective plasmodia.

The spectrum of symptomatology and the range of incubation periods associated with various forms of malaria infection are too broad to discuss in detail in this text, but it should be noted that any worker exposed to a laboratory where malaria is being investigated who develops an unexplained febrile illness should have serial thick and thin blood smears examined at frequent intervals for a period of several days. Serology may prove diagnostically useful if a negative base-line titer has been obtained previously and a subsequent fourfold elevation is noted. However, in acute infections, the serologic changes will follow the appearance of acute infection, and therefore, blood smear examination is the primary diagnostic tool.

Treatment will, of course, depend on the species of malaria isolated and its susceptibilities, if known. When doubt exists as to the identity of the species in a laboratory-acquired malaria infection, the appropriate procedure is to obtain blood for culture and primate inoculation, if these are available or can be arranged through CDC (see above); to make additional thick and thin smears for subsequent review and quantitation of parasitemia; and to treat the patient for *P. falciparum* malaria if this is one of the species to which he or she may have been exposed. Knowledge of the drug susceptibilities of the parasites within a given laboratory may be helpful in this respect. When chloroquine or other drug resistance is a possibility, the treatment regimen should be designed to cover the resistant strain until the exact sensitivity and species can be determined. Therapeutic regimens recommended by the Parasitic Disease Division at CDC are outlined in Table 1.

Babesia

Babesia infections present only a slight risk for laboratory workers since the vector ticks are more easily controlled in the laboratory than are mosquitoes, and the infection in immunologically normal individuals with intact spleens appears to be self-limited and, in many cases, subclinical. Since there appears to be high correlation between severe *Babesia* infection and the absence of an intact, functional spleen, it is recommended that splenectomized or immunocompromised persons avoid work in laboratories where infective ticks or *Babesia*-infected animals or specimens are being handled. Although no cases of laboratory-acquired babesiosis have been reported at the time of this writing, reports of two

cases of severe clinical infection associated with the transfusion of human blood from infected donors demonstrate the potential for direct transmission within the laboratory via accidental needle stick or infective blood spilled on an open cut or wound. While this risk is small, it should be kept in mind when handling infective material.

The incubation period for babesiosis may vary from a few days to several weeks, and initial symptoms usually involve periodic fever, chills, malaise, and headache. Treatment is controversial. Drugs used to date have failed to eradicate parasites from patients' blood. Some drugs, however, do provide symptomatic relief. Inquiries regarding treatment should be referred to the Parasitic Disease Division, CDC (see above).

Since *Babesia* organisms on blood smears may closely resemble those of malaria, it is suggested that thick and thin blood films be made and referred through a state laboratory to the CDC for diagnosis. Acute- and convalescent-phase serology specimens drawn 2 to 3 weeks apart will also be useful in supporting the diagnosis of acute *Babesia* infection and should be submitted by the same route. In cases of low-grade infections, inapparent on thick smears of peripheral blood, hamster inoculations may be of diagnostic value, although an incubation interval of 2 to 4 weeks may be necessary for infection to become patent in the hamster.

Intestinal Protozoa

Entamoeba histolytica and Giardia lamblia

The major intestinal protozoa of concern to the laboratory worker are *Entamoeba histolytica* and *Giardia lamblia*. Both species are infective for humans in the cyst form, which may be contained in proctoscopic or stool specimens (both) or small-intestinal aspirates (*Giardia*). Normal personal hygienic measures in the laboratory as discussed earlier in the text are sufficient protection. Disposal of stool samples, if obtained in large numbers as during an outbreak investigation, may present a problem. Incineration is the method of choice but may not be feasible if specimens are collected in plastic containers or if local air pollution regulations prohibit this practice. Alternatively, viable cysts may be inactivated by heat or chemical means.

Since these infections may be frequently subclinical, the incubation period may vary from a few days to several months, with some infected individuals never recognizing any symptoms. Diarrhea, abdominal cramps associated with constipation, excessive flatulence, bulky foul-smelling stools, blood or mucus in the stool, or simply malaise with loss of appetite may herald such an infection. Diagnosis based on stool examination should include at least three fresh stool specimens obtained over a 6- to

TABLE 1. Treatment of active malaria infection

Organism	Illness	Drug of choice	Treatment[a]	Alternative drug	Treatment[a]
All plasmodia except chloroquine-resistant P. falciparum	Uncomplicated attack	Chloroquine phosphate	1 g (600 mg base), then 500 mg (300 mg base) in 6 h, then 500 mg (300 mg base) per day for 2 days	Amodiaquine hydrochloride	780 mg (600 mg base) day 1, then 520 mg (400 mg base) per day for 2 days
	Severe illness, or if oral dose cannot be administered	Chloroquine hydrochloride	250 mg (200 mg base) IM every 6 h	Quinine dihydrochloride	600 mg in 300 ml of normal saline IV over at least 1 h; repeat in 6 or 9 h (maximum, 1,800 mg per day)
	Prevention of relapse[b]	Primaquine phosphate[c]	26.3 mg (15 mg base) per day for 14 days or 79 mg (45 mg base) per week for 8 weeks		
Chloroquine-resistant P. falciparum	Uncomplicated attack	Quinine sulfate + Pyrimethamine + Sulfadiazine	650 mg tid for 3 days 25 mg bid for 3 days 500 mg qid for 5 days	[d]Quinine sulfate + Tetracycline or [e]Quinine sulfate + Trimethoprim + Sulfamethoxazole	650 mg tid for 3 days 500 mg qid for 7 days 650 mg tid for 3 days 320 mg 1.6 g bid for 5 days
	Severe illness	Quinine dihydrochloride	600 mg in 300 ml of normal saline IV over at least 1 h; repeat in 6 to 8 h (maximum, 1,800 mg per day)		

[a] Abbreviations: IM, intramuscularly; IV, intravenously; tid, three time a day; bid, twice a day; qid, four times a day.
[b] "Radical" cure after "clinical" cure; P. vivax and P. ovale only.
[c] Primaquine should only be given after a negative guanosine-6-phosphate dehydrogenase screen is obtained since this drug may cause severe hemolysis, methemoglobinemia, or both in deficient patients.
[d] This is the regimen of choice for treatment of P. falciparum infections resistant to both chloroquine and pyrimethamine/sulfadoxine (Fansidar).
[e] Although it is of less proven efficacy than the other regimens listed, this combination is recommended as an additional alternative because of its availability.

9-day period, preserved separately in 10% Formalin and in polyvinyl alcohol fixative, and examined by Formalin-ether or a comparable concentration technique and by permanent stain, if necessary. If clinical suspicion of a *Giardia* infection remains high despite multiple negative stool examinations, duodenal sampling by an "Enterotest" (string test) may be in order (19).

Treatment of *G. lamblia* infection is best accomplished with quinacrine hydrochloride (100 mg three times a day for 5 days). Metronidazole (250 mg three times a day for 5 days) is a second-choice treatment for giardiasis. Symptomatic amebiasis may be treated with metronidazole (750 mg three times a day for 5 to 10 days), whereas asymptomatic individuals who are passing only cysts may be treated with diloxanide furoate (500 mg three times a day for 10 days), available through the CDC Parasitic Disease Drug Service (see above).

Other intestinal protozoa

Other intestinal protozoa which may occasionally cause disease in humans include *Dientamoeba fragilis, Blastocystis hominis, Isospora belli,* and *Balantidium coli.* These organisms may be present in small numbers in the intestine without symptoms and are, therefore, considered very low-risk material to handle in the laboratory. Treatment of such low-grade infections is generally not required.

Trichomonas

Trichomonas infections, both *T. vaginalis* and *T. hominis,* are of comparatively low risk in the laboratory. These infections are generally spread only by person-to-person contact; orally ingested organisms will be destroyed in the stomach.

Cryptosporidium

Recognized for many years as the cause of infections in numerous animal species, *Cryptosporidium* sp. has only recently been recognized as a human pathogen. This coccidian protozoan organism has been found to infect the intestinal tract of immunocompromised human hosts, most notably those with AIDS (acquired immunodeficiency syndrome) (11). Occasional infections in immunocompetent hosts have also been observed (20).

Since it is only recognized by acid-fast stain or electron microscopic examination of stool or small intestinal biopsy specimens, the infection is likely to go unrecognized unless specifically looked for. In immunocompetent human hosts, infections are usually either asymptomatic or mild and self-limited to a course of 10 days or less. For this reason, the risk to the healthy laboratory worker handling these specimens is minimal. Routine precautions for handling fecally contaminated material should be observed,

including careful hand washing after specimen handling and appropriate disposal and handling of the specimen and any contaminated laboratory materials. The infection in immunologically normal hosts requires no specific treatment. Unlike other coccidia such as *Sarcocystis* and *Isospora,* antifolate treatment has not been shown to be effective. A limited number of cases responding to spiramycin (3 g per day for 3 days) have been reported.

HELMINTH INFECTIONS

Nematodes (Intestinal)

Strongyloides

Simple primary human infection with *Strongyloides stercoralis* usually results in the shedding of noninfective rhabditiform larvae in the stool. An interval of 12 to 24 h is required for these larvae to reach the infective (filariform) stage. In the immunosuppressed patient, however, multiple organ involvement may occur with an overwhelming autoinfective cycle. In this instance, filariform (infective) larvae may be found in fresh stools and respiratory secretions, posing an added risk to laboratory workers handling these specimens. However, this appears to be a rare occurrence (13). The risk of acquiring infection from contact with such specimens is slight since an infective larva requires several minutes of contact with the skin in a moist environment to effect penetration (8). Nonetheless, this risk should be recognized, and clinical laboratory specimens, as well as unfixed autopsy material, should be regarded as potentially infective and be handled accordingly. Pike's series of 3,921 (viral, bacterial, fungal, and parasitic) laboratory-acquired infections include two cases of strongyloidiasis (14). Skin penetration by larvae is accompanied by intense pruritus and petechiae. Subsequent symptoms caused by migration and maturation of large numbers of the worms may result in chest pain and cough followed by abdominal cramping, nausea, vomiting, and occasionally diarrhea over a period of 3 to 4 weeks. Eosinophilia is frequently present. Light infections are usually asymptomatic after the initial cutaneous reaction at the site of larval penetration.

The diagnosis of primary infection is made after 3 to 4 weeks of incubation, when viable larvae or even adult worms may be found in the stool specimen.

Thiabendazole (25 mg/kg twice a day for 2 days) is the treatment of choice for primary infection. In light infections, the larvae will migrate asymptomatically to various tissues and encyst. These infections may remain subclinical for the life of the individual if no immunosuppressive condition supervenes, and for this reason some physicians choose not to treat such light primary infections in immunologically normal hosts. The encysted larvae may remain viable, although inactive, for many years. Persons with a history of a documented and untreated prior *Strongyloides* infection, who are given steroids or other immunosuppressive therapy or who develop a debilitating illness or other form of immunocompromise, should be treated promptly with a 5-day course of thiabendazole. Stool and respiratory secretions of such persons should be followed closely for the presence of *Strongyloides* larvae even after treatment has been given. Serologic diagnosis by immunoabsorbent hemagglutination assay is only applicable for identification of antecedent infection, epidemiologic surveillance, or follow-up of late disease to rule out reactivation.

Ascaris, hookworm, *Trichuris,* and *Enterobius*

Ascaris, hookworm, and *Trichuris* infections may be acquired from stool specimens in the laboratory if these are allowed to incubate for a period of time sufficient to allow the eggs or larvae to develop to the infective stage. This requires a period of 8 to 14 days under optimal conditions. Specimens discarded within a week of being collected from the patient should not present an infectious risk. Suspected infections in laboratory personnel should be evaluated by stool examination using direct and concentration techniques. Heavy infections are treated with mebendazole (100 mg twice a day for 3 days). Hookworm and *Ascaris* may also be treated with pyrantel pamoate (11 mg/kg in a single oral dose).

Some animal hookworm species such as *Ancylostoma brazilensis* or *Ancylostoma caninum* can produce a cutaneous larva migrans syndrome if third-stage larvae are allowed to come in contact with human skin. This condition requires no specific therapy. An additional problem sometimes encountered with *Ascaris* infections is the development of immediate hypersensitivity after directly handling adult worms. This problem is best managed by avoiding direct skin contact with adult worms. Reactions, which after sensitization may be provoked simply by the near presence of adult ascarids, are best managed with antihistamines and anti-inflammatory agents. *Enterobius vermicularis* may be transmitted within the laboratory if strict hand-washing precautions are not observed. Aerosol dispersion of *Enterobius* eggs, a route of transmission frequently mentioned in texts, is considered a much less likely route of infection than hand-to-mouth or fomites transmission (8). Pyrantel pamoate (11 mg/kg) and mebendazole (100 mg orally in a single dose) are equally effective treatments for this infection. The treatment should be repeated after 2 weeks to remove worms which may have been embryonating at the time of the initial therapy. When adult worms are not readily apparent, the "Scotch tape test" will confirm the diagnosis by microscopic identification of ova on a piece of transparent tape applied briefly to the perianal skin.

Nematodes (Tissue)

Trichinella

Trichinella spiralis is the only tissue nematode presenting a significant risk to the laboratory worker. It should be remembered that fresh tissue preparations and even those digested with pepsin hydrochloride may contain *Trichinella* cysts, which are infective if ingested. Use of gloves and careful cleaning of fingernails are, therefore, important.

Symptoms of acute infection may include abdominal and muscle pain accompanied by a high eosinophil count 1 to 2 weeks after ingestion of infective cysts. Light infections may remain asymptomatic. A positive muscle biopsy confirms the diagnosis. After 3 to 4 weeks, titers obtained by the bentonite flocculation test begin to rise. In laboratory workers, who, if infected, are likely to receive only a small inoculum, serological conversion is the most sensitive means by which to confirm the diagnosis. A positive muscle biopsy may only be found 6 weeks after infection and then only if the number of ingested larvae is sufficiently large. Eosinophilia frequently accompanies and provides supporting evidence of infection. Treatment of this infection is usually supportive, although there is some clinical evidence that thiabendazole (25 mg/kg twice a day for 5 days) may be of benefit in heavy infections (which would be unlikely in a laboratory worker) diagnosed within 24 h of the ingestion of infective cysts.

Other tissue nematodes

The various filarial infections (*Wuchereria, Onchocerca, Mansonella, Dipetalonema,* etc.) and such rare organisms as *Angiostrongylus* and *Gnathostoma* do not pose a risk for the laboratory worker since they require intermediate hosts to develop to the infective stage. None has been associated with reported laboratory-acquired infections.

Cestodes

Taenia solium

Taenia solium constitutes an important infective risk for the laboratory worker handling cestode infections. The eggs of *T. solium* are capable of producing cysticercosis in humans and thus should be handled and disposed of with due caution. Treatment of discarded specimens with heat (autoclave or incinerator) is recommended. Cysticercosis in humans may involve larval migration to a number of organ systems. Subcutaneous tissues are the most common site of migration of these larvae, but the brain and other organs are frequently involved as well. The disease may thus present with a number of abnormal

TABLE 2. Common parasitic infections to which laboratory workers are exposed

Organism	Route of infection	Biosafety level	Protective measures	Diagnostic tests	Prepatent period	Signs and symptoms	Treatment
Toxoplasma	Oral	2	Hand washing	IgM-IFA titer; liver function tests, CBC	Variable, usually 5 to 14 days	Adenopathy, fever, malaise	Usually self-limited; if severe: pyramethamine + trisulfapyrimidines
Leishmania	Percutaneous	2	Gloves, wound protection	Biopsy; in vitro cultivation, hamster inoculation serology	Variable; 1 to 4 weeks	Skin ulceration, cutaneous	Pentostam; consult CDC
Acanthamoeba	Unknown	3	Mask, gloves, gown; negative-pressure hood	Brain biopsy serology	Unknown, probably several weeks	Headache, neurological impairment	Consult CDC
Naegleria	Transmucosal (nasopharynx)	3	Same	CSF exam and culture; E stains (7)	2 to 3 days	Headache, obtundation, neurological impairment	Amphotericin IV + IT[a]; miconazole IV + IT; oral rifampin
Entamoeba histolyica	Oral	2	Hand washing	Three stool exams with concentration; serology (for invasive disease); IHA	Ofen asymptomatic; variable, 4 to 8 days avg	Diarrhea (may be bloody), abdominal pain	Metronidazole (750 tid for 10 days); asymptomatic, use diloxanide furoate and call CDC
Giardia lamblia	Oral	2	Hand washing	Three stool exams with concentration	4 to 8 days	Diarrhea, abdominal cramps, flatulence, bloating	Quinacrine HCl (100 mg tid for 5 days) or metronidazole (250 mg tid for 5 days)
Strongyloides	Percutaneous	2	Gloves and caution when handling specimens	Stool exam; motile larvae may be seen in wet prep	18 to 25 days	[b]Cough, chest pain, followed by mid-abdominal pain, crampy	Thiabendazole (25 mg/kg bid for 2 days)

	Route		Precautions	Diagnosis	Incubation	Symptoms	Treatment
Trichinella	Oral	2	Careful specimen handling; hand washing	Muscle biopsy	6 to 8 days	[b]Abdominal and muscle pain	Symptomatic treatment only
Ascaris	Oral		Hand washing	Stool exam	5 to 10 days	[b]Cough, fever, pneumonia, abdominal cramps, diarrhea or constipation	Mebendazole (100 mg tid for 2 days), or pyrantel pamoate (11 mg/kg, one dose)
	Skin contact	2	Gloves	Exposure	Immediate	Hypersensitivity reactions, especially pruritus	Anti-inflammatory; avoid future exposure
Hookworm	Percutaneous	2	Gloves, hand washing	Stool exam	20 to 30 days	[b]Animal species: creeping eruption or "ground itch" (skin); human species: diarrhea, abdominal pain, anemia	Mebendazole (100 mg tid for 2 days) or pyrantel pamoate (11 mg/kg, one dose)
Trichuris	Oral	2	Gloves, hand washing	Stool exam	20 to 30 days	[b]Abdominal pain, tenesmus	Mebendazole (100 mg tid for 2 days)
Enterobius	Oral	2	Gloves, hand washing, nail cleaning	Scotch tape test	14 days	Perianal pruritus	Pyrantel pamoate (11 mg/kg, one oral dose)
Taenia solium	Oral	2	Gloves, hood, careful hand washing and specimen disposal	Stool exam, serology (cysticercosis), brain scan + X ray	6 to 12 weeks	Abdominal pain, neurological symptoms, cysticercosis	Niclosamide for adult tapeworm; surgery for cysticercosis
Hymenolepis nana	Oral	2	Gloves, hand washing	Stool exam	2 to 6 weeks	Abdominal pain, diarrhea	Niclosamide; call CDC

[a]IV, Intravenous; IT, intrathecal.
[b]Symptoms are unusual unless infecting inoculum is heavy, which would be unlikely in most laboratory-acquired infections.

neurological findings which may partially or completely resolve as the cysts wall off and calcify. These can be detected by nuclear brain scans during acute disease and, if calcified, by skull film later in the course. Indirect hemagglutination serology is helpful in distinguishing cysticercosis from other focal neurological disease. There are no drugs available for treatment of cysticercosis; surgical excision of the cysts, if their presence is life-threatening, may be employed. Acute neurological manifestations are sometimes ameliorated by the administration of short courses of corticosteroids. The current treatment of choice for cysticercosis is praziquantel (50 mg/kg per day in three divided doses for 14 days).

Hymenolepis nana

Hymenolepis nana is also noteworthy for the laboratory worker because of the infectivity of eggs when passed in stool. No intermediate host is required, and this should be kept in mind, as with *T. solium* (see above), when handling or disposing of stools containing this parasite. The treatment of choice for adults is niclosamide (2 g per day for 5 days), available through the Parasitic Disease Drug Service at CDC (see above).

Echinococcus

Echinococcus cysts in animal or human tissues are not infective for humans. Only the tapeworm's eggs in the feces of a definitive host (dog, fox, wolf, cat) are infectious for humans. Treatment of the larval (cystic) stage in humans is primarily surgical, although prolonged courses of high-dose mebendazole have been tried with limited success in preliminary studies. Therapeutic consultations should be referred to the Parasitic Disease Division at CDC (see above).

Other cestodes

Other cestode infections require an intermediate host to reach an infective stage and are, therefore, of little risk to laboratory workers unless they are involved in research with these organisms in their natural animal hosts. If acquired in this way, these infections (Taenia saginata, Taenia solium, Diphyllobothrium latum, and Dipylidium caninum) may be diagnosed by stool examination and treated with niclosamide (2 g in a single oral dose).

Trematodes

The flukes in general do not pose a risk to the diagnostic laboratory worker since their infectivity is dependent upon the intermediate host. In investigational laboratories where competent intermediate hosts are cultivated, workers should be aware of exposure risks, whether they be percutaneous cercarial invasion (schistosomes) or oral ingestion (*Fasciola, Fasciolopsis, Clonorchis,* etc.).

CONCLUSION

In summary, most transmissible parasitic infections can be easily avoided in the laboratory by following simple hygienic principles. Persons working consistently with stool specimens would probably be wise to keep their polio immunization status up to date, and the annual collection and storage of a serum sample for reference use is also a recommended precaution for all laboratory workers in both diagnostic and research laboratories who are working with parasitic organisms. A summary of the common parasitic infections to which laboratory workers are exposed is given in Table 2.

LITERATURE CITED

1. **Centers for Disease Control.** 1978. Primary amebic meningoencephalitis, California, Florida, New York. Morbid. Mortal. Weekly Rep. **27:**343–347.
2. **Centers for Disease Control.** 1980. Primary amebic meningoencephalitis, Fort Bliss, Texas. Morbid. Mortal. Weekly Rep. **29:**117–119.
3. **Centers for Disease Control.** 1980. Chagas disease, Kalamazoo, Michigan. Morbid. Mortal. Weekly Rep. **29:**147–148.
4. **Centers for Disease Control.** 1980. Primary amebic meningoencephalitis, United States. Morbid. Mortal. Weekly Rep. **29:**405–407.
5. **Dubey, J. P.** 1977. Toxoplasma, Hammondia, Besnoitia, Sarcocystis, and other tissue cyst-forming coccidia of man and animals, p. 101–237. *In* J. P. Krier (ed.), Parasitic protozoa, vol. 3. Academic Press, Inc., New York.
6. **Duma, R. J., and R. Finley.** 1976. In vitro susceptibility of pathogenic *Naegleria* and *Acanthamoeba* to a variety of therapeutic agents. Antimicrob. Agents Chemother. **10:**370–376.
7. **Duma, R. J., W. I. Rosenblum, R. F. McGehee, M. J. Jones, and E. C. Nelson.** Primary amoebic meningoencephalitis caused by *Naegleria.* Ann. Intern. Med. **74:**923–931.
8. **Faust, E. C., and P. F. Russell (ed.).** 1970. Clinical parasitology, 8th ed. Lea & Febiger, Philadelphia.
9. **Gutteridge, W. E., B. Cover, and A. J. D. Cooke.** 1974. Safety precautions for working with *Trypanosoma cruzi.* Trans. R. Soc. Trop. Med. Hyg. **68:**161.
10. **Hendricks, L., and N. Wright.** 1979. Diagnosis of cutaneous leishmaniasis by *in vitro* cultivation of saline aspirates in Schneider's Drosophila Medium. Am. J. Trop. Med. Hyg. **28:**962.
11. **Jones, T. C.** 1985. Other protozoa including *Cryptosporidium, Sarcocystis, Isospora,* and *Balantidium coli,* p. 1560–1651. *In* G. L. Mandell, R. G. Douglas, and I. E. Bennett (ed.), Principles and practice of infectious disease, 2nd ed. John Wiley & Sons, New York.
12. **Krogstad, D. J., G.S. Visvesvara, K. W. Walls, and J. W. Smith.** 1985. Blood and tissue protozoa, p.

612–630. *In* E. H. Lennette, A. Balows, W. J. Hausler, Jr., and H. J. Shadomy (ed.), Manual of clinical microbiology, 4th ed. American Society for Microbiology, Washington, D.C.

13. **Kyle, L. H., D. G. McKay, and H. J. Sparling.** 1948. Strongyloidiasis. Ann. Intern. Med. **29:**1014–1042.

14. **Pike, R. M.** 1976. Laboratory associated infections—summary and analysis of 3,921 cases. Health Lab. Sci. **13:**105–114.

15. **Pike, R. M.** 1979. Laboratory associated infections—incidence, fatalities, causes, and prevention. Annu. Rev. Microbiol. **33:**41–66.

16. **Robertson, D. H. H., S. Pickens, J. H. Lawson, and B. Lennex.** 1980. An accidental laboratory infection with African trypanosomes of a defined stock. I and II. J. Infect. Dis. **2:**105–112, 113–124.

17. **Ruebush, T. K., R. A. Weinstein, R. L. Baehner, D. Wolff, M. Bartlett, F. Gonzales-Crussi, A. J. Sulzer, and M. G. Schultz.** 1978. An outbreak of *Pneumocystis* pneumonia in children with acute lymphocytic leukemia. Am. J. Dis. Child. **132:**143–148.

18. **Tabbura, K. F., and G. R. O'Conner.** 1980. Treatment of ocular toxoplasmosis with clindamycin and sulfadiazine. Ophthalmology **87:**129–134.

19. **Wolfe, M. S.** 1979. Giardiasis. Ped. Clin. North Am. **26:**295–303.

20. **Wolfson, J. S., J. M. Richter, M. A. Waldron, D. J. Weber, D. M. McCarthy, and C. C. Hopkins.** 1985. Cryptosporidiosis in immunocompetent patients. N. Engl. J. Med. **312:**1278–1282.

21. **Yeager, R. G.** 1960. A method of isolating trypanosomes from blood. J. Parasitol. **56:**228.

Chapter 14

System for Classifying and Managing Toxic and Carcinogenic Chemicals

GAILYA P. WALTER AND DAVID WEATHERS

Managing hazardous chemicals in chemical and microbiological laboratories is becoming increasingly impotant for a number of reasons. Federal regulations, such as those promulgated by the Occupational Safety and Health Administration (OSHA) (4, 5) to control use of certain chemical carcinogens and those of the Environmental Protection Agency (2, 6) to control manufacture and disposal of toxic and otherwise hazardous chemicals of all kinds, have made it mandatory to control the procurement, use, and disposal of such chemicals. In addition, employees and their organizations are becoming increasingly aware of hazards associated with chemicals and are demanding that employers provide training, equipment, facilities, and procedures which will minimize exposure. Furthermore, the potential economic loss associated with employee exposure to hazards is of increasing concern to employers.

To manage hazardous chemicals properly, one must first define categories or levels of hazards and then group chemicals according to certain criteria. Then, on the basis of the severity of the potential hazard, appropriate procedures for managing these categories can be established.

A convenient system, used by the National Institute for Occupational Safety and Health (NIOSH), categorizes a chemical as hazardous if it meets one of the following criteria.

(i) Toxicity. A toxic substance is one that has demonstrated the potential to endanger human life by exposure via any route found in the workplace (for example, inhalation, ingestion, skin contact), produce short- or long-term disease or bodily injury, affect health adversely, induce cancer or other neoplastic effects in humans or experimental animals, induce a transmissible change in the characteristics of an offspring from those of its human or experimental animal parents, or cause the production of physical defects in the developing human or experimental animal embryo.

(ii) Flammability. A flammable substance is one that will burn in air when exposed to a temperature of 1,500°F (815°C) for 5 min or less.

(iii) Reactivity. A reactive substance is one that will release hazardous amounts of energy when subjected to shock; will spark or light during uncontrolled polymerization or when contacted by common substances, for example, water, air, or steel; or is a strong oxidizing or reducing material.

As a laboratory user of a wide variety of chemicals, the Centers for Disease Control (CDC) has developed a system which enables CDC to manage the procurement, storage, use, and disposal of hazardous chemicals of all kinds. The first phase in implementing this system, which includes only carcinogenic or extremely toxic chemicals, is described in this chapter.

THE CDC CLASSIFICATION SYSTEM

For their classification system, CDC developed three categories of relative toxicity as shown in Table 1. In the first phase of implementation, we chose to deal only with Category A chemicals, which are the most hazardous. The list of 120 chemicals (Table 2) was compiled from those regulated by OSHA, those listed by the Department of Health and Human Services (3) as being carcinogenic, and a group of chemicals selected by CDC as "ultratoxic" on the basis of published toxicity data. This list may be expanded as additional information is available.

MANAGEMENT OF TOXIC CHEMICALS

The CDC system for managing toxic chemicals involves five basic elements:
1. Compilation of information
2. Inventory of carcinogens/ultratoxins
3. Control of procurement
4. Control of work practices, storage, and spills
5. Disposal of carcinogens/ultratoxins

Compilation of information

RTECS. To provide the laboratorian with basic toxicity information, CDC has made the data file for the NIOSH Registry of Toxic Effects of Chemical Substances (RTECS) accessible through computer terminals located throughout CDC. The RTECS file also contains chemical identifiers (Chemical Abstracts Services [CAS] registry number) and information necessary for the preparation of safety directives

TABLE 1. Relative acute toxicity criteria

CDC category	Key words	50% Lethal dose (LD$_{50}$): single oral dose, rats (mg/kg)	50% Lethal concn (LC$_{50}$): inhalation vapor, exposure, rats (ppm)	50% Lethal dose (LD$_{50}$): skin, rabbits (mg/kg)
A	Extremely to highly hazardous	≤ 50	≤ 100	≤ 42
B	Moderately to slightly hazardous	$<5,000$	$<10,000$	$<2,800$
C	No significant hazard	$\geq 5,000$	$\geq 10,000$	$\geq 2,800$

and for hazard evaluation of chemical substances. Toxicity data are available for an estimated 100,000 or more substances. The RTECS file will be updated monthly. The user can, by entering the prime name, synonym, or CAS number of a given compound, retrieve information needed to assess its toxicity. The CAS number is the key identifier used in the inventory of chemicals. The manufacturer's Safety Data Sheet (SDS) (example, Fig. 1) will provide specific technical and safety information pertaining to use of the chemical. The SDS should include:

1. Material identification: CAS number is given here
2. Ingredients and hazards: toxic hazard data are stated in terms of concentration, citing specific standards, for example, OSHA or the American Conference of Governmental Industrial Hygienists (ACGIH).
3. Physical data
4. Fire and explosion data: flash points, recommended fire extinguisher media, and unusual fire and explosive hazard information
5. Reactivity data: information on chemical properties
6. Health hazard information: toxicity data, first aid provisions
7. Spill, leak, disposal procedures
8. Special protection: information on engineering considerations and on personal protection
9. Special precautions and comments

The SDS is filed in a location readily accessible to workers potentially exposed to the hazardous or toxic material. The SDS can be used as a training aid and as a basis for discussion during training sessions for new employees and at safety meetings. It should assist management by directing attention to the specific engineering controls, work practices, and protective measures needed to ensure the safe handling and use of the material. It aids in planning a safe and healthful work environment and suggests appropriate emergency procedures and sources of help in the event employees are exposed.

Inventory of carcinogens/ultratoxins

On the basis of the classification criteria described earlier, all Category A chemicals (carcinogens and ultratoxins) have been inventoried to establish (i) the chemical agents used at CDC, (ii) how they are used, (iii) where they are used, (iv) where they are stored, and (v) the personnel authorized to use them (see Fig. 2). This information is entered into the system by using ADABAS (data base software package), and the programs are written in the programming language NATURAL. The system processes include the following functions:

1. Interactive input and editing of data for possible errors and omissions
2. Inventory control of carcinogens and ultratoxins used at CDC
3. Employee exposure surveillance and control
4. Interactive updating of the controlled chemical inventory file and the employee exposure file
5. Batch updating of the RTECS file
6. Interactive inquiry capability into each of the following files:
   (a) RTECS file
   (b) Controlled chemical inventory file Category A
   (c) Employee exposure file
7. Scheduled reports
   (a) Controlled chemical inventory verification
   (b) Authorized employee exposure verification
   (c) Employee exposure control notices

Although this inventory system was designed for the control of carcinogens and ultratoxins, CDC has attempted to make the design flexible so that it can be modified. Other categories of hazardous chemicals, including flammables and radioactive substances, used at CDC may be entered and controlled with little or no change in the system. Biological and radiological hazard information can be added to the employee exposure file. As changes occur in personnel and stocks of chemicals, the laboratorians will submit an updated transaction form so that the information can be entered into the system. Adjustments in the inventory due to laboratory usage will be picked up in the annual verification of inventory which the investigator completes.

This computerized inventory system has a number of additional benefits. For example, the file can be queried, generating a printout that will indicate the current availability of a particular chemical. In many cases, this could eliminate unnecessary procurement.

TABLE 2. Controlled chemical carcinogens and ultratoxins

Primary name	CAS no.	Primary name	CAS no.
Abrin	001393620	Hyoscyamine	000101315
N-Acetoxy-2-acetylaminofluorene	006093448	Inorganic arsenic	007440382
2-Acetylaminofluorene	000053963	Isobenzan	000297789
Aflatoxin B$_1$	001162658	K-strophanthin	008001512
Aflatoxin B$_2$	007220817	Lanatoside	MX8006960
Aflatoxin G$_1$	001165395	Lysergic acid diethylamide (LSD)	000050373
Aflatoxin G$_2$	007241987	3-Methylcholanthrene	000056495
Aldicarb	000116063	Methyl chloromethyl ether	000107302
alpha-Naphthylamine	000134327	4,4'-Methylene bis-(2-chloroaniline)	000101144
o-Aminoazobenzene	000060093	Methylhydrazine	000060344
4-Aminodiphenyl	000092671	Methyl methanesulfonate	000066273
2-Aminofluorene	000153786	1-Methyl-3-nitro-1-nitrosoguanidine	000070257
Anabasine	000494520	Nicotine salicylate	000321459
Apholate	000052460	4-Nitrobiphenyl	000092933
Arsenic trioxide	001327533	N-[4-(5-Nitro-2-furyl)-2-thiazoly]-formamide	024554264
Atropine	000051558	Nitroglycerin	000055630
Asbestos	001332214	4-Nitroquinoline-1-oxide	000056575
Benz [a] anthracene	000056553	N-Nitrosodiethylamine	000055185
Benzene	000071432	N-Nitroso-N-ethylurethane	000614959
benzidine	000092875	N-Nitrosodi-n-butylamine	000924163
Benzo [a] pyrene	000050328	N-Nitrosodimethylamine	000062759
beta-Naphthylamine	000091598	N-Nitrosod-N-methylurea	000684935
beta-Propriolactone	000057578	N-Nitrosodi-n-propylamine	000621647
bis-Chloromethyl ether	000542881	N-Nitrosopiperidine	000100754
Bromoethyl methanesulfonate	004239105	N-Nitroso-N-ethylurea	000759739
Cantharidin	000056257	N-Nitroso-N-methylurethane	000615532
Carbon tetrachloride	000056235	Pantopon	MX8002764
Chlorambucil	000305033	Parathion	000056382
2-chloro-4-Dimenthl-amino-6-methyl-pyrimidine	000535897	Paroxon	000070702
Chloroform	000067663	Phosphine	007803512
Cycasin	014901087	Phosphorodithioic acid	000298044
Diazomethane	000334883	Phosphorous (yellow)	007723140
Dibenz[a,h] anthrancene	000053703	Polychlorinated biphenyls	001336363
1,2-Dibromo-3-chloropropane	000096128	Procarbazine	003660701
3,3'-Dichlorobenzidine	000091941	Propylenimine	000075558
Diepoxygutane	001464535	1,3-Propane sultone	001120714
Digalen	MX8031694	2-propyl-peperidine	000458888
Digifolin	MX8001330	Ricin	009009863
Digoxin	020830755	Scopolamine	000051343
7,12-Dimethylbenz [a] anthracene	000057976	Sarin	000107488
4-Dimethylaminoazobenzene	000060117	Sodium Azide	026628228
3,3'-Dimethoxybenzidine	000119904	Sodium selenate	013410010
3,3'-Dimethylbenzidine	000119937	Sulfotepp	003689245
Dimethylethylenimine	002658244	Tabun	000077816
1,1-Dimethylhydrazine	000057147	Tepp	000107493
1,2-Dimethylhydrazine	000054073	2,3,7,8 tetrachlorodibenzodioxin	001746016
1,4-Dinitrosopiperazine	000140794	Thimet	000298022
p-Dioxane	000123911	m-Toluenediamine	000095807
Duboisine	000101315	Uracil mustard	000066751
Ethionine	013073353	Urethane	000051796
Ethylene Dibromide	000106934	Urginin	MX8002355
Ethylenimine	000151564	Vinyl chloride	000075104
Ethylene glycol dinitrate	000628966		
Ethyl methanesulfonate	000062500		
Fluoroacetic acid	000144490	Coke oven emissions	
Gitalin	001391759	Arsenites	
Heroin	000561273	Nornicotine	
Hexaethyl tetraphosphate (HETP)	000075784	2,3,7,8, tetrachlorodibenzofuran	
Hydrazine	000302012	Penta- and Hex-chloro- or bromo-dibenzodioxins	
Hydrazoic acid	007782798	or furans	
Hydrocyanic acid	000074908	Amino methyl carbamate	
N-Hydroxy-2-acetylaminofluorene	000053952	fiber glass	

The file can also be used to select appropriate training for employees using hazardous chemicals. Furthermore, the immediate availability of the stored information will be useful in emergency situations.

Control of procurement

If the procurement of toxic chemicals is to be monitored and controlled, the investigator must de-termine the category of hazard (toxicity) of chemicals and obtain the CAS number before initiating a purchase request. If the chemical has not been categorized, the investigator, with appropriate assistance from the Office of Biosafety, must evaluate toxicity and other properties of the chemical to determine the appropriate hazard category. The laboratorian will provide the specific category and CAS number of the chemical to the procurement office. Procurement

EASTMAN KODAK COMPANY
APPROVED BY U.S. DEPARTMENT OF LABOR
"ESSENTIALLY SIMILAR" TO FORM OSHA-20

MATERIAL SAFETY DATA SHEET

919

SECTION I

PRODUCT NAME	Cyanogen Bromide	SIZE 25 g, 100 g
CHEMICAL NAME	Cyanogen Bromide	
FORMULA	BrCN	
MANUFACTURER	EASTMAN KODAK COMPANY	
ADDRESS	343 STATE STREET, ROCHESTER, NEW YORK 14650	

FOR INFORMATION ON HEALTH HAZARDS CALL   (716) 722-5151

INFORMATION EFFECTIVE AS OF 8/20/79

SECTION II HAZARDOUS INGREDIENTS OF MIXTURES

PRINCIPAL HAZARDOUS COMPONENT(S)	%	TLV (Units)

SECTION III PHYSICAL DATA

BOILING POINT (°F) 144°F (62°C)	SPECIFIC GRAVITY (H_2O=1) 2.015 (20/4)	
VAPOR PRESSURE (mm Hg) 100 mm at 22.6°C	PERCENT VOLATILE BY VOLUME (%) 100%	
VAPOR DENSITY (AIR=1) 3.65	EVAPORATION RATE (=1)	

SOLUBILITY IN WATER   Appreciable

APPEARANCE AND ODOR   Colorless needles; penetrating odor

SECTION IV FIRE AND EXPLOSION HAZARD DATA

FLASH POINT (Method Used)　*	FLAMMABLE LIMITS Not Available	Lel	Uel
EXTINGUISHING MEDIA	CO_2, dry chemical, water		
SPECIAL FIRE-FIGHTING PROCEDURES	Air mask should be worn		
UNUSUAL FIRE AND EXPLOSION HAZARDS	None		

*Nonflammable solid by DOT Regulations, Section 173.150

FIG 1. Sample of manufacturer's SDS.

919

SECTION V   HEALTH HAZARD DATA

THRESHOLD LIMIT VALUE

None assigned

EFFECTS OF OVEREXPOSURE INHALATION: Vapors are irritating at low concentrations and prolonged exposure can result in headache, dizziness, nausea, vomiting and confusion. Higher concentrations produce labored breathing, pulmonary edema, convulsions and death. EYES: Contact with the liquid or vapor can cause irritation. SKIN: Prolonged or repeated contact can cause skin irritation and toxic amounts can be absorbed through the intact skin.

EMERGENCY AND FIRST-AID PROCEDURES

INHALATION: Remove to fresh air. If not breathing, give artificial respiration. If breathing is difficult, give oxygen. Call a physician. EYES AND SKIN: Immediately flush areas of contact with plenty of water for 15 minutes and get medical attention. Remove contaminated clothing.

SECTION VI   REACTIVITY DATA

STABILITY	UNSTABLE		CONDITIONS TO AVOID
	STABLE	X	

INCOMPATIBILITY (Materials to avoid)	Strong oxidizers

HAZARDOUS DECOMPOSITION PRODUCTS	Thermal decomposition or burning may produce carbon monoxide and/or carbon dioxide, toxic fumes, HBr

HAZARDOUS POLYMERIZATION		CONDITIONS TO AVOID
May Occur	Will Not Occur	
	X	

SECTION VII   SPILL OR LEAK PROCEDURES

STEPS TO BE TAKEN IN CASE MATERIAL IS RELEASED OR SPILLED   Wearing suitable protective clothing, sweep material onto paper. Place in fiber carton. Incinerate. Wash spill area well with soap and water. Or, cover spill with calcium hypochlorite and flush to sewer.

WASTE DISPOSAL METHOD   Wearing suitable protective clothing, make up small packages in paper or other flammable materials. Incinerate. State and local laws take precedence.

SECTION VIII   SPECIAL PROTECTION INFORMATION

RESPIRATORY PROTECTION (Specify Type)   An approved self-contained breathing apparatus or air-line respirator, if necessary

VENTILATION	LOCAL EXHAUST	Yes	SPECIAL	No
	MECHANICAL (general)	Yes	OTHER	No

PROTECTIVE GLOVES		EYE PROTECTION	
	Yes		Yes

OTHER PROTECTIVE EQUIPMENT   As necessary to prevent skin contact.

SECTION IX   SPECIAL PRECAUTIONS

PRECAUTIONS TO BE TAKEN IN HANDLING AND STORING   Due to its high volatility, care should be taken when opening bottles of cyanogen bromide or when storing near other chemicals. Keep refrigerated.

OTHER PRECAUTIONS   INGESTION: Large doses may produce sudden loss of consciousness, respiratory arrest, and death. If ingested, induce vomiting immediately by giving two glasses of water and touching back of throat with finger or blunt object. Call a physician. Never give anything by mouth to a unconscious person.

The information herein is believed to be correct as of the date hereof, but is provided without warranty of any kind.

FIG 1. *Continued*.

CONTROLLED CHEMICAL CARCINOGENS
AND
ULTRATOXINS INVENTORY TRANSACTION FORM

1. Type of Entry:

   ☒ 1   Initial Inventory
   ☐ 2   Update
   ☐ 3   Inventory Additions
   ☐ 4   Deletions

2. Date Inventory Prepared: $\underline{0}$ $\underline{8}$ $\underline{0}$ $\underline{1}$ $\underline{8}$ $\underline{0}$
                             M   M   D   D   Y   Y

3. Standard Administrative Codes: HCL436

4. Controlled Chemical CAS #: $\underline{0}$ $\underline{0}$ $\underline{0}$ $\underline{0}$ $\underline{6}$ $\underline{7}$ $\underline{6}$ $\underline{6}$ $\underline{3}$

                NAME   :   Chloroform

5. Quantity Units: 6. Quantity: $\underline{0}$ $\underline{0}$ $\underline{0}$ $\underline{4}$ $\underline{0}$ $\underline{0}$ $\underline{0}$

   ☐ 1 mg
   ☒ 2 mL

7. Use of Chemical: $\underline{1}$ $\underline{A}$ 8. Frequency of Use: ☐ 1 Daily
                                               ☒ 2 Weekly
                                               ☐ 3 Monthly
                                               ☐ 4 Semi-annually

9. Location of Use: $\underline{A}$ $\underline{4}$ $\underline{0}$ $\underline{8}$ $\underline{1}$ $\underline{0}$ $\underline{0}$ $\underline{6}$
                           $\underline{A}$ $\underline{4}$ $\underline{0}$ $\underline{8}$ $\underline{1}$ $\underline{0}$ $\underline{0}$ $\underline{8}$
                           ─ ─   ─ ─   ─   ─ ─ ─

10. Location of Storage: $\underline{A}$ $\underline{4}$ $\underline{0}$ $\underline{8}$ $\underline{1}$ $\underline{0}$ $\underline{0}$ $\underline{7}$

11. Personnel Authorized to Use Carcinogens/Ultratoxins:

Last Name		I.D. #		Use Dates Initial		Ending
Gill ,	1 2 3	4 5	6 7 8 9	0 8	0 1	8 0	─ ─ ─
Bouchlas ,	2 2 4	4 5	6 7 8 9	0 8	0 1	8 0	─ ─ ─
Slayton ,	3 2 3	4 5	6 7 8 9	0 8	0 1	8 0	─ ─ ─

(Can enter up to 5 names)

Prepared by: _____ James Gill _____

FIG. 2. Sample of controlled chemical carcinogens/ultratoxins inventory transaction form.

office personnel will flag the purchase order with a symbol or notation at the end of the purchase order number. All Category A chemicals should be placed on separate purchase orders to facilitate special handling once these chemicals are received. The receiving department must be familiar with the special handling procedures required for these toxic substances. If a package displaying a Category A flagged purchase order number is received, it should be sent to the investigator for opening under appropriate containment. The receiving department will send a copy of the purchase order for Category A compounds to the Office of Biosafety to signal that additional or new quantities of Category A compounds have been purchased. The investigator will also submit an inventory transaction form reporting that additional quantities have been received.

Control of work practices, storage, and spills

Work practices to be followed in laboratories working with Category A ultratoxins and carcinogens must generally be those outlined in the Department of Health and Human Services guidelines (3). More stringent controls may be required for chemical carcinogens that may present unusual hazards because of their physical and chemical properties, their toxic effects, or their proposed use.

Laboratories which use Category A chemicals may contain special design features, including limited personnel access, filtration of exhaust air, personnel and emergency showers, primary containment equipment such as fume hoods and glove boxes, and eyewash facilities.

Category A compounds should be stored in a

specially designated area, which may be within the laboratory or in a separate room. Office of Biosafety personnel will periodically inspect the storage area and the laboratory facility in which the toxic compounds are used to ensure that safety features are operational. A sign bearing the legend "Caution— Chemical Carcinogen" or "Caution—Chemical Hazard" and a list of the specific toxic chemical agents being used must be posted in the work area. Only authorized users should have access to the ultratoxins/carcinogens. Working stock quantities outside the storage area should be limited to a 1-week supply.

Provision for control of spills should be a part of the laboratory's workplan. The SDS will often provide information on spill control. Each authorized user should be aware of the appropriate cleanup procedure.

CHEMICAL WASTE
FOR DISPOSAL

INSTRUCTIONS: Firmly attach this label to each container of chemical waste. Provide sufficient information to fully identify contents. Deliver to Waste Chemical Storage Room. Call Biosafety (3883) for additional information.

SOURCE: BLDG. _____ RM. _____

INVESTIGATOR: _____

TELEPHONE: _____

CHEMICAL NAME _____
(No trade names)

_____

COMPOSITION _____
(if mixture)

_____

HAZARDOUS PROPERTIES:

☐ Toxic   ☐ Corrosive

☐ Explosive   ☐ Caustic

☐ Flammable   ☐ Irritant

☐ Other _____

CDC 0.886 8/80

FIG. 3. "Chemical Waste for Disposal" label.

Safety monographs and safety plans are to be prepared by the laboratory supervisor and approved by the appropriate safety authority before work projects involving Category A compounds are started.

Disposal of carcinogens/ultratoxins

The appropriate disposal of chemical wastes is critically important. All chemical wastes, including those that contain a carcinogen/ultratoxin, must be identified by a "Chemical Waste for Disposal" label (Fig. 3) attached to the waste container, which is delivered to the Waste Chemical Storage Room. The Office of Biosafety must review and approve procedures for disposing of or deactivating chemical wastes within the laboratory. The disposal of any toxic waste, whether carcinogenic or not, must comply with current Environmental Protection Agency and Department of Transportation regulations (1, 2). This is coordinated through the Office of Biosafety and a private disposal contractor with a permit from the Environmental Protection Agency. A "notification of deletion of inventory" record will be sent to the Office of Biosafety to update the Controlled Chemical Inventory System.

SUMMARY

The system described was designed to meet an immediate need at CDC, that is, to control OSHA-regulated chemical carcinogens and certain other carcinogenic or extremely toxic chemicals. The system is flexible enough to permit flammable, explosive, radioactive, biological, and other hazardous chemicals to be included later. The weak point in the overall program may be the requirement for laboratorians to put complete and timely information into the system. During the initial inventory phase, however, this has not been a problem, and with adequate follow-up, we expect the system to produce the desired results.

LITERATURE CITED

1. **Department of Transportation.** 1980. Transport of hazardous wastes and hazardous substances, 49 CFR, Part 171, 172, 173, 174, 176, 177, May 22, 1980. U.S. Government Printing Office, Washington, D.C.
2. **Environmental Protection Agency.** 1980. Hazardous waste management system, 40 CFR, vol. 45-no. 98, book 2, p. 33063–33285, May 19, 1980. U.S. Government Printing Office, Washington, D.C.
3. **Laboratory Chemical Carcinogen Safety Standards Subcommittee of the Department of Health and Human Services Committee to Coordinate Toxicology and Related Programs.** 1980. Proposed guidelines for the laboratory use of chemical substances posing a potential occupational carcinogenic risk. U.S. Government Printing Office, Washington, D.C.

4. **Occupational Safety and Health Administration.** 1978. Occupational Safety and Health Standards, 29 CFR, subpart Z, 1910.1000–1910.1046. U.S. Government Printing Office, Washington, D.C.

5. **Occupational Safety and Health Administration.** 1980. Occupational Safety and Health Standards, 29 CFR 1990. Identification, classification and regulation of potential occupational carcinogens, January 22, 1980. U.S. Government Printing Office, Washington, D.C.

6. **Public Law 94-469.** Toxic Substances Control Act, October 11, 1976. U.S. Government Printing Office, Washington, D.C.

Chapter 15

Management and Codification of Risks in the Laboratory

JOSEPH R. SONGER

Assessment of risk is an essential element of safety. A thing is safe if the risks associated with it are judged to be acceptable at the time of use. While measuring risk is an objective but probabilistic exercise, judging the acceptability of that risk (judging safety) involves both personal and social value judgments. Lowrance (1) points out that it is false to expect that scientists can measure whether something is safe or not. Scientists are prepared principally to measure risk. While the level of a risk might remain the same, the acceptability of that risk is constantly subject to change; e.g., risk of exposure to 50 parts per million (ppm; μl/liter) of ethylene oxide, previously judged acceptable, is now unacceptable (the new standard is 1.0 ppm).

CODIFICATION OF RISK CONTROLS

Codification of risk controls in the form of guidelines, directives, and regulations is worthwhile. There have been many attempts to do this for the biomedical work environment. The latest and most comprehensive effort was a joint enterprise by the Centers for Disease Control (CDC) and the National Institutes of Health (NIH), resulting in the publication "Biosafety in Microbiological and Biomedical Laboratories," which appears in its entirety in Appendix 1 of this volume. Section II of those guidelines deals with the principles of biosafety, reinforcing both the physical barrier system concept and the biocontainment level concept. Four levels of biosafety containment and practices are detailed in section III. Concern is for the protection of personnel, the community, and the environment.

Because of the increased risk associated with diseases of animals, a wholly new set of guidelines for working with vertebrate animals was developed (Section IV). Recommended biosafety levels for infectious agents and infected animals were based on several factors, including the virulence, pathogenicity, route of spread, biological stability, and communicability of the agent; the procedures to be followed; the quantity and concentration of the agent; the endemic nature of the agent; and the availability of effective therapeutic measures and vaccines.

Summary statements regarding biosafety levels for infectious agents and infected animals are presented in section V, "Recommended Biosafety Levels for Infectious Agents and Infected Animals." Statements are presented for agents which meet one or more of the following criteria: (i) the agent is a proven hazard to laboratory personnel working with infectious materials; (ii) the potential for laboratory-associated infections is high even in the absence of previous documentation of such infections; or (iii) the consequences of infection are grave.

The risk assessments and biosafety levels recommended presuppose a population of immunocompetent individuals. People with altered immunocompetence may be at an increased risk when exposed to infectious agents.

Each agent summary statement contains recommendations for the use of vaccines and toxoids, either as licensed or investigational new drugs. Appropriate precautions should be taken in the administration of live attenuated virus vaccines to individuals with altered immunocompetence.

Because of the hazards peculiar to working with arboviruses, a section of the guidelines is devoted to this group of viruses. Biosafety Levels 2 through 4 are required for these organisms.

Activities involving large volumes or highly concentrated preparations of infectious agents, or manipulations which are likely to produce aerosols or are otherwise intrinsically hazardous, should be conducted at a higher control level than usually recommended for that agent. The laboratory director must make a risk assessment, and facilities must be appropriate to the risk involved.

Assessment of risk by cross-referencing agents and activities is a logical and straightforward approach. If proper values are entered, a valid assessment is made.

Official standards of biosafety for biological laboratories have been neither written nor authorized by any federal agency. A form of exception to this statement are federal or state regulations imposed to prohibit transportation into the country (or state or region) of certain animal and plant pathogens considered to be unacceptable, such as foot-and-mouth-diseased or -exposed animals. However, the guidelines set forth in "Biosafety in Microbiological and Biomedical Laboratories" could be considered as prestandards.

Safety standards have traditionally developed in

one of two ways. Either they are developed by a standards-writing organization such as the American National Standards Institute, or they may arise simply by consensus (consensus standards). Standards from either source can be made official legal standards. It seems most likely that as these CDC/NIH prestandards are followed, a consensus will be reached and official standards will evolve. It is very important, then, that as we follow these prestandards we seek to improve them before they become official.

HAZARDS OF CODIFYING RISKS

Regulations, guidelines, and directives are essential elements in a safety program; however, safety programs will be grossly inadequate if organizers depend on regulations alone. An ancient sage once said, "A regulation is something you rely on when you lose your ability to reason." People tend to reason differently, and hence regulation by consensus offers a basic, commonly understood framework for a safety program. But regulations are inadequate alone; they must be backed up by the actions of people.

Regulations are inadequate primarily because it is impossible to write enough to cover every possible situation. Furthermore, even if there were thorough enough regulations, we would be hard pressed in an emergency to recall which specific regulation to follow. Besides this, in a sense codification of hazards along with prescribed ritual antidotes may stifle further search for the best safety procedures. Once a practice has been established, especially over a long period of time, it goes unchallenged.

Risk management should include a continual search for better control methods. Safety managers examining codes or guidelines should continually attempt to translate them into work practices, keeping in mind that they are not intended to cover every detail of activity but need to be adapted to fit each immediate risk situation. One of the most difficult assumptions to overcome in the laboratory is that a well-qualified, highly educated researcher will also be well qualified to assess biohazard risks. For the most part, measurement of risk is not part of the college curriculum, and in specific courses it is treated only in a cursory manner.

JUDGING THE ACCEPTABILITY OF RISKS

Safety is not measured. Risks are measured. Only when those risks are weighted in a value judgment system can the degree of safety be judged. As previously indicated, a thing is safe if its attending risks are judged to be acceptable. Judging the acceptability of risk is a normative process. We must avoid any suggestion that safety is an intrinsic, absolute,

measurable property of things. Lowrance (1) provides some criteria for judging the acceptability of risks.

Reasonableness is by far the most commonly cited principle in safety judgments; this is often referred to as the "rule of reason." The problem is to decide what is reasonable. Tort liability charges which might occur in the laboratory are most likely to be the result of negligence. Negligence in the eyes of the law may be defined as conduct that falls below a standard of care established by law to protect others against an unreasonable risk of harm (2). If the standard of care has not been specifically established by statute, the actions or inactions of an individual will be measured against what a hypothetical reasonably prudent individual would have done under the same circumstances.

Several factors enter into the determination of reasonableness, as follows.

(i) **Custom of usage (or prevailing professional practice).** Many laboratory activities have been done for years and are generally recognized as safe; if a practice has been in common professional use it must be okay, since any adverse effects would have become evident, and a thing approved by custom is safer than one not tested at all. However, long-time usage does not automatically mean that an activity is safer. A prime example is mouth pipetting.

(ii) **Best available practice, highest practicable protection, and lowest practicable exposure.** In the laboratory we should seek to find the best available procedure with the highest level of protection and lowest level of exposure which is practicable.

(iii) **Degree of necessity of benefit.** Are the benefits worth the risk? Sometimes it is difficult to justify the risks taken in the laboratory on the basis of the benefits derived. Therefore, we should ask ourselves, is the procedure really necessary? It there an alternative, safer way?

(iv) **No detectable adverse effects.** This is a weak criterion since it is an admission of uncertainty or ignorance.

(v) **Toxicologically insignificant levels.** "There are no safe substances, only safe doses" is a maxim in common usage today. This approach could be criticized as being quite arbitrary. What is judged a safe dose today might not be judged safe tomorrow.

CONCLUSIONS

Codification of risks is an important part of a safety program. Regulations, guidelines, and directives are very important in the implementation of safety in the labortory; however, we must not rely on regulations alone. Utilizing the framework provided by regulations and guidelines, we must measure risks and make safety judgments on a day-to-day basis.

A task force under the direction of the NIH and the CDC has codified biohazards in the form of guidelines. These guidelines, titled "Biosafety in Microbiological and Biomedical Laboratories" (see Appendix 1 of this volume), form an excellent framework for a good biosafety program.

It is important to remember that safety is not measured. We can only measure risks and make safety judgments regarding the acceptability of those risks.

LITERATURE CITED

1. **Lowrance, W. W.** 1976. Of acceptable risk. Wm. Kaufmann, Inc., Los Altos, Calif.
2. **U.S. Department of Health, Education and Welfare.** 1977. Legal aspects of classroom safety, p. 250. *In* Safety in the school science laboratory. U.S. Department of Health, Education and Welfare, NIOSH, Cincinnati.

CONTAINMENT EQUIPMENT AND DESIGN

Chapter 16

Introduction

DONALD VESLEY

The following chapters are intended to provide the reader with a review of current thinking about facilities and equipment pertinent to safety in microbiological activities. It has been estimated that physical design and equipment factors contribute only 10% of the laboratory safety problem, while personal behavior factors contribute 90%. Nevertheless, appropriate choices in laboratory design and equipment acquisitions can greatly expedite safety programs while minimizing expenditures for unnecessary frills.

Chapter 17 provides a comprehensive outline of physical facility features appropriate to the various biosafety levels, while Chapter 18 concentrates on factors pertinent to remodeling of facilities rather than new construction. Chapter 19 addresses the various types of biological safety cabinet which have become so commonplace in microbiology facilities. These cabinets undoubtedly represent the single most important advance in biological safety equipment in the past two decades. In Chapter 20 the issues of personal protective equipment are addressed, stressing the option of direct worker protection even in the absence of biological safety cabinets. Finally, in Chapter 21 an array of miscellaneous safety devices is discussed. It should be kept in mind that new devices are continually being developed in response to specific safety needs. Thus, for that chapter and for the others as well, the reader should not construe the information presented as the last word in biosafety. Rather, this information should be considered as a basis from which new ideas and new developments should continually evolve.

Chapter 17

Design of Microbiological and Biomedical Research Facilities

DAVID L. WEST AND MARK A. CHATIGNY

Before the 1960s, there was little recorded experience in the design and utilization of laboratories specifically for microbiological and biomedical work. Over the past two decades, basic concepts of laboratory safety and environmental control have evolved, as have the engineering components of the control schemes. As a result, the design and use principles for such laboratories have become reasonably well defined.

This chapter deals with the architectural and engineering features that can be incorporated into the design of a biomedical laboratory facility to provide a safe environment and enhance productivity. The approaches described evolved during an era of emphasis on microbiology of infectious diseases and were further refined for viral oncology and recombinant DNA research. The text draws heavily on, and in a few instances is extracted from, works that have been published to promote awareness of an interest in biohazard safety and environmental control among laboratory users and designers. Among these are *Space Planning Principles for Biomedical Research Laboratories* (19), *Design Criteria for Viral Oncology Research Facilities* (12), *Proceedings of the Workshop on Cancer Research Safety* (14), *Design of Biomedical Research Facilities: Proceedings of a Cancer Research Safety Symposium* (6), *Laboratory Safety: Theory and Practice* (7), and *1982 Applications Volume of the ASHRAE Handbook* (1).

This chapter is written as a general guide for the laboratory user, and its major focus is on biological and chemical safety and environmental control aspects of facility design. The reader should be aware that there are numerous other architectural and engineering aspects of laboratory design, but those not pertaining specifically to safety and environmental control are beyond the scope of this chapter. For greater design detail, the reader should refer to the references acknowledged above and those cited in the text. Nothing included in this chapter is intended to supersede life safety and building codes, including seismic, electrical, and fire protection standards. The information provided is intended primarily to familiarize the reader with laboratory design problems; it is not intended to serve as a substitute for competent laboratory design consultants or architects.

Design goal

The goal in designing and constructing a laboratory facility is to furnish workspace that contributes to minimizing hazard potential for laboratory workers and the general environment, while providing an environment that facilitates research or clinical diagnostic laboratory work. Several objectives must be met to achieve an optimum design.

Design objectives

First, the design must be for a facility that is economically feasible to construct and operate. Because of rapidly rising construction and energy costs, this objective can be met only when great thought and effort are invested in meeting the functional design objectives.

Second, the safety and control features designed into the laboratory should match the assessed hazard potentials. This should be qualified by emphasizing that the facility must be adaptable to meet inevitable program changes. In practice, this will mean that the design must include provision for modification of the facility without excessive control measures being installed at the outset. This may require including in the design safety features not fully utilized at first. However, no safety systems should be installed and then circumvented.

Third, the facility must provide for the safe management of the various types of hazards usually encountered in a laboratory. In addition to the potential biohazards specific to the subject of research or clinical specimen, laboratory activities often expose workers to other less obvious yet significant hazards including toxic chemicals, radiation, fire or explosion, and traumatic injury.

Fourth, the facility must facilitate research or clinical support. Proper arrangement of functional spaces and zoning of mechanical support systems not only can help satisfy safety requirements, but also can promote efficient operations and contribute to productivity. A related goal, which can also be achieved by proper facility design and utilization, is the safeguarding of the experimental materials or clinical specimens from contamination or detrimental environmental factors.

Team effort: development of operational concepts

The design of a laboratory should be the product of a team effort. The team should include: (i) representatives of management who allocate capital investment and operating funds, coordinate operational support, and authorize and enforce policies of procedure relevant to safety practices; (ii) representatives of the users and direct support services, including the laboratory director and the various scientific disciplines, engineering support and maintenance, custodial service, and other relevant services (for example, shipping and receiving department); (iii) representatives of the health and safety organization that will be concerned with safety matters within the laboratory; and (iv) the architect and engineering staffs engaged to create the design and space.

The task for this team is to define the projected research activities and to develop operational concepts and a facility plan. Defining the research activities to be housed both immediately after construction and in the future is important. It is noteworthy that in many laboratory design and construction projects the original users and planning individuals often are not the only or final occupants. Wedum and Phillips (24) prepared an exhaustive set of questions which focus on information needed for cost-effective development of a research facility. It is desirable that the first two groups listed above take the lead in clarifying the research scope before the development of operational concepts and a facility plan is considered. McKinney (10) identified and discussed some basic elements and the time relationships to be considered in concept development and facility planning. The nature of the laboratory program(s) must be considered in context with applicable health and safety regulations and guidelines as well as good laboratory practice. The planners should guard against the tendency to focus too narrowly on research activity needs. Adequate attention should be given to elements that may be considered peripheral, but can have major impact on the program and facility design, including personnel and material movement, storage, maintenance and custodial services, and plans for managing emergency situations (11). Consideration of these elements, to be discussed in later sections of this chapter, is essential in development of a program of requirements for any research facility.

Operational concepts (e.g., work sequences and material flow patterns) should be developed concurrently with facility planning and design (10). Initial concepts should be formulated and incorporated into the program of requirements and facility plan at the schematic stage of design. The concepts should be refined to ensure that the final design (usually including many compromises) will provide a fully usable facility with an operating plan.

Health and safety approach to design

A safe laboratory facility can be developed by providing for the safety and health of the users and adding to this further safeguards to protect the surrounding community and general environment. The strategy most frequently used to protect workers from occupational exposure to a hazardous material is (i) to control the hazardous material at its source to prevent its release into the worker's environment, (ii) to protect the worker from contact with the material in the event the worker is in a contaminated environment, and (iii) to minimize the consequences of accidental release of material to the workplace environment. The first two elements of this strategy are usually effected by the use of engineering controls, process control practices, and personal protective equipment and clothing. The last element may be accomplished by staffing practices, immunizations, and administrative controls.

Laboratory safety equipment and features of facility construction constitute the engineering controls. These controls must be used to complement good laboratory practice; together they can control and confine hazardous materials. This complementary aspect underscores the need to coordinate the operational concepts and the facility design.

Engineering controls (primary and secondary barriers)

Aerosols of research materials can be major sources of contamination which can lead to infection of laboratory workers (26) and spread infections throughout a building (18). Small-particle, respirable aerosols can be generated by simple laboratory procedures (e.g., pipetting), mechanized operations (e.g., centrifugation), and overt accidents. It has been suggested that large-particle aerosol splatter and settling particles also produced by these activities can be as significant a source of laboratory contamination as the aerosol itself (20). Engineering controls, or "barrier systems" in laboratory vernacular, can be used to confine aerosols and spills, with the resulting contact contamination, at or near their sources.

A primary barrier, enclosing the work process and agent, is intended to confine and isolate the agent from the worker as well as to provide protection to other persons in the laboratory room (25, 26). Examples of primary barriers include ventilated fume hoods, biological safety cabinets, and glove box enclosures.

Architectural and engineering features of the laboratory can form a secondary barrier to protect personnel in other areas of the building and the environment from exposure to research materials released into a laboratory room (25, 26). Such a release of contamination may be due to the absence or failure of the primary barrier, to the barrier's being overwhelmed beyond design capability, or to a laboratory accident occurring outside the primary barrier. The secondary barrier is not intended to eliminate the risk of exposure for personnel inside the laboratory room where the release may occur.

Primary and secondary barriers serve to safeguard both laboratory personnel and the environment from hazardous research materials. Additionally, many of these systems are effective in maintaining the purity of research materials. Ubiquitous fungal or bacterial contamination from outdoors, indoors, or even laboratory personnel can be detrimental to the research. The potential for cross-contamination of research materials within the facility also can be reduced by barrier systems.

A thorough discussion of primary barriers is given in Chapter 19. The present chapter is devoted to secondary barriers. Many of the basic features described in this chapter are considered "good practice" in the design of any microbiology or biomedical laboratory. At the end of this chapter, guidance is provided on which features are recommended for various levels of containment.

CONTAINMENT LEVELS FOR SECONDARY BARRIERS

The following sections of this chapter describe the architectural and mechanical-electrical design features of laboratories, first in general terms and then more specifically with respect to engineering aspects. Accordingly, some repetition is necessary. It is expected that the reader will be concerned with specific facility needs. It is suggested that a review of the operational practices in the design development process and selection of the applicable containment level will enable the reader to identify material of immediate interest.

The concepts and approaches described in sections of this chapter can be used to design safe and productive laboratory facilities. However, not all of the features described are necessarily required for all laboratories. Indeed, as stated above, the safety and control features designed into a laboratory should closely match the assessed or anticipated hazard potential.

The National Cancer Institute has classified research facilities on the basis of the contamination control features incorporated into their design and construction (2, 12). Research facilities are classified as basic, containment, or maximum containment (2). The Centers for Disease Control and the National Institutes of Health, in their recent joint recommendations, use the same terms to classify laboratory facilities to achieve four levels of containment (23). These are Biosafety Levels 1, 2, 3, and 4, in ascending order of degree of containment, paralleling the former P1, P2, P3, and P4 designations (15) for recombinant DNA laboratories. The P1 to P4 designations of the recombinant DNA guidelines are being modified to adapt to the Biosafety Level designations (with some exceptions for large-scale fermentors).

Table 1 lists the facility classifications and the specific study areas in which each type of facility is applicable. With some adjustment, primarily concerning nomenclature for facilities used with the various Biosafety Levels, the National Cancer Institute definitions of basic, containment, and maximum containment facilities apply and are used as the basis for further discussion in this chapter. The table indicates approximate equivalences. Refer to current editions or revisions of the respective National Cancer Institute, National Institutes of Health, or Centers for Disease Control publications for technical requirements.

Biosafety Level 1 laboratories are as described for the basic facility except that no provision for biological safety cabinets is required. All work is normally done on the open bench, and special containment equipment is generally not required for manipulations of agents assigned to Biosafety Level 1. There are other operational considerations of access control, clothing, and medical monitoring that differentiate activity in the Biosafety Level 1 laboratory from that in the simple basic facility. Discussion here is limited to the physical facility.

THE LABORATORY MODULE: EXAMPLE OF A SECONDARY BARRIER

The secondary barrier to confine contamination can be thought of as an envelope consisting of the architectural and engineering features of the laboratory and its supporting mechanical systems. The elements of a secondary barrier may include (i) materials and methods of construction that facilitate cleaning and prevent accumulation of contamination, (ii) pest- and vector-proof construction, (iii) protection of utility distribution systems from contamination, (iv) treatment of liquid and air effluents to remove contaminants, and (v) air pressure gradients to maintain migration and infiltration of air from noncontaminated areas toward potentially contaminated ones (25, 26).

The simplest secondary barrier envelope is the laboratory module, which can be defined as the smallest repetitive unit of space within which are found all architectural, mechanical, electrical, and other functional requisites for the operation of a complete and environmentally self-sufficient laboratory unit (19). The individual laboratory or module can be considered alone or be viewed as one of several elemental or cellular units within a secondary barrier envelope. In either case, it is good practice to establish a barrier system for the smallest practical set of laboratory rooms or modules. The value of this practice is apparent when one considers the limited

TABLE 1. Facility classifications for laboratory activities and control levels[a]

Laboratory activity	Hazard or Biosafety Level	Facility classification[b]
Cancer viruses (11)	Low-risk viruses	Basic
	Moderate-risk viruses	Containment
	High-risk viruses	Maximum containment
Chemical carcinogens (13)	All	Containment
Organisms containing recombinant	Level 1	Basic[c]
DNA molecules (15); etiologic	Level 2	Basic
agents and other microorganisms	Level 3	Containment
(23)	Level 4	Maximum containment

[a]Refer to current editions of the respective National Cancer Institute, National Institutes of Health, or Centers for Disease Control publications for technical requirements. See also Appendix 1 of this volume.
[b]Facility classification and nomenclature per reference 2.
[c]Biosafety Level 1 does not require the primary barrier equipment called for in the description of basic facilities. Fume hoods for chemical or fire protection should be used as needed.

area contaminated in the event of a spill or the need to decontaminate a space after such a spill.

Envelope of the secondary barrier

Methods of construction should minimize uncontrolled infiltration or exfiltration of air to minimize migration of contamination into or out of the envelope (12). Depending on the degree of containment required (an issue discussed below), wall and ceiling panel joints, utility pipe and duct penetrations, electrical conduits, and other passages should be sealed or be capable of being sealed. Construction techniques that produce tight, durable joints between structural surfaces, without voids or open cracks, will also contribute to this end. Obviously, walls should extend from floor to ceiling. To help maintain balance of the facility ventilation system, doors to the access corridor should have self-closing devices (2). Such features can also aid in controlling pest and insect vectors.

Materials for work surfaces and structural finishes should be easy to clean and prevent accumulation of contamination. They should withstand detergent cleaning solutions and decontaminating agents (12). Many materials that designers use to make spaces attractive will conflict with criteria for laboratory sanitation and maintenance. Soft acoustical materials provide harborage for dust, fungi, and bacteria, and attractive nonglare finishes often do not stand up to heavy scrubbing. Surfaces that are most easily cleaned and sanitized are usually hard and shiny and contribute to glare and noise. Resolving this conflict will require the careful attention of the entire design team (4), and some design compromises are to be expected.

Surface finishes that have few seams are preferred. Coved-base floors of seamless sheet vinyl or monolithic membranes (e.g., troweled epoxy) are good examples. Wall and ceiling finishes of paint or special coatings that produce a monolithic finish on

plaster or wallboard are preferred over materials such as ceramic or glazed tile, with their troublesome mortar joints (4). Bench tops should be coved, impervious to liquids, and resistant to acids, alkalis, organic solvents, and moderate heat (2).

Dust- and harborage-free design

Utility distribution systems, furnishings, and equipment should be designed or selected to minimize dust-collecting areas and harborage areas for ubiquitous fungal or bacterial contamination. Exposed horizontal utility pipes and ductwork, animal cage framing, and open storage cabinets should be kept to a minimum to minimize dust-collecting surfaces. Spaces around furniture should be accessible for cleaning. If suspended ceilings are used to conceal utility pipes and ductwork, they should be of plaster or drywall construction. Moisture-collecting areas, such as condensate drip pans, should be avoided whenever possible. Where they are necessary above ceilings, service access panels should be provided.

Ventilation

The laboratory must be ventilated to maintain the directional airflow of the secondary barrier as well as to maintain temperature and humidity conditions and adequate fresh air. In recent years, natural ventilation via windows and vent stacks has been replaced by mechanical ventilation, which affords substantially better control of the air quality and quantities needed for temperature and pressure control in the laboratory. This ventilation usually includes heating and cooling as needed. In many cases, the rate at which air is moved through a laboratory, i.e., the ventilation rate, will depend primarily on the air quantities needed to pass through local exhaust systems (e.g., fume hoods and canopy hoods over autoclaves) and primary barrier systems (e.g., biological safety cabinets) for their proper functioning. These quantities may be larger than other ventilation needs.

Ventilation also may be required to maintain an area at a relative air pressure lower than adjacent areas (e.g., the access corridor). Air should migrate, or infiltrate, from noncontaminated areas toward areas that might be contaminated (25, 26). To maintain the negative relative pressure in a space, more air must be removed from the space than is provided. For a conventional laboratory module (e.g., 10 by 20 ft [ca. 3 by 6.1 m] with a 10-ft ceiling) a difference between supply and exhaust quantities of about 10%, but not less than 50 ft³/min (0.024 m³/s), will yield the desirable differential pressure. About 0.05 in. of water (12.4 Pa) pressure difference across a well-fitted door can be achieved without requiring undue door-operating force. This air volume is based on the assumption that most of the difference will infiltrate through rather controlled openings, such as the cracks around the normally closed access door to the corridor. The size of the space and the "porosity" of its construction can influence greatly the difference between supply and exhaust air quantities required to maintain the nominal pressure difference. The varying effects of outside wind pressure on a building should not be overlooked; this is considered further below.

Temperature control and maintenance of comfortable conditions are also functions of the ventilation system. The designer can estimate the cooling and heating requirements based on estimates of occupancy and equipment loading.

Depending on the degree of containment required, ventilation air may or may not be recirculated. As a general principle, it is desirable to limit recirculation to the one or more modules wherein a single laboratory operation is done. Such recirculation can minimize the space contaminated in the event of a spill, aid in vapor-phase decontamination, and provide good temperature control while conserving energy. Even in recirculating-type systems, a certain amount of air from the outdoors will be added to provide "freshness." Also depending on the degree of containment, exhaust air may have to be filtered before discharge to the outdoors. Further details of the ventilation system, including air filtration, will be discussed below under Engineering Support.

Utility systems

Another component of the secondary barrier is the utility system(s). In all cases, the domestic water entering the laboratory building from the service main (street) should be protected (5, 12). In biomedical research facilities there should be dual water systems, both having hot and cold water service. One system is for potable water uses exclusively. A second system (sometimes called institutionalized system) is considered for industrial uses exclusively, in ancillary support spaces and laboratories. Drinking fountains in laboratory areas should be on the potable (domestic) system and should be foot operated. The institutionalized or industrial water system is connected to laboratory fixtures and ice makers (ice makers should be labeled "NOT FOR HUMAN CONSUMPTION"). Laboratory fixtures should be installed with vacuum breaker backflow preventors. The institutionalized or industrial water system should be separated from the potable (domestic) water system by an approved reduced pressure backflow preventor or a break tank with an air gap.

Ideally, laboratory vacuum service should be provided by the use of small individual vacuum pumps, each fitted with a disinfective trap and absolute filter on the suction side (12, 16). Central house vacuum systems are not recommended because of problems associated with the maintenance of radiologically and biologically contaminated lines. If central vacuum service cannot be avoided, the system should include in-line absolute filters and liquid collection traps as near as practical to each use point or service cock. The filter installation should permit filter decontamination and replacement. The vacuum system exhaust air should be discharged directly outdoors. Design guidelines of the Compressed Gas Association, to facilitate system cleaning and maintenance, should be followed.

No special barrier precautions are warranted for the distribution of normal laboratory gases (i.e., compressed air and flammable gas). Specialty gases (e.g., carbon dioxide and nitrogen) are usually provided in the laboratory in compressed-gas cylinders, which should be secured with safety straps or chains. If specialty gases are used at a high rate, consideration should be given to installing a permanent rack for cylinders outside the laboratory. A permanent manifold and piping from the corridor or central supply room can convey the specialty gases to service outlets within the laboratory. This will reduce the frequency of having to handle the large and heavy cylinders in the laboratory.

Liquid waste treatment

The discussion in this section applies only to potentially infectious, biologically contaminated wastes. It does not cover toxic or flammable chemical wastes or radioactive materials.

Sinks and drains should not be used for disposal of infectious or otherwise contaminated materials. In keeping with the concept of the secondary barrier, contaminated liquid wastes leaving the laboratory should be either decontaminated before removal or transported under controlled conditions to a service area where they can be treated. Biologically contaminated liquid wastes generated in a laboratory can include spent cell culture media, fermentation broth, water contaminated with infectious materials, etc. Liquid wastes are easily sterilized in steam autoclaves. If the quantities are not too large, on-site sterilization is preferable. If quantities are too large for sterilization within the laboratory, liquid wastes can be carefully transported in closed break-proof containers to a central sterilization service area, where large autoclaves would be available. Chemical decontamination can also be used, but the process should be demonstrated to be effective and should not leave the user with the problem of the disposal of highly corrosive liquids. In any case, the laboratory design should accommodate the need for safe collection or storage areas for the quantities of contaminated liquid wastes anticipated.

Biological liquid waste sterilizer systems (often called biowaste treatment systems) are needed only for maximum containment laboratories. General engineering approaches to the design of these laboratories are described below under Engineering Support.

Solid waste treatment

Much of the solid waste from microbiology and biomedical laboratories will not be contaminated. The safest practice is to work with the assumption that if any of the waste leaving the laboratory is contaminated, all wastes should be controlled until they are decontaminated or destroyed.

Solid wastes can include used glassware, used disposable plastic ware, and refuse (e.g., paper, absorbent pads, animal carcasses, laboratory specimens, etc.). Glass and plastic ware should be placed in trays containing disinfectant solution. The filled trays can then be autoclaved, either within the laboratory or in a central service area. Refuse can be collected in plastic bags. If not incinerated, contaminated refuse should be autoclaved before disposal. Animal carcasses and clinical specimens should be incinerated to complete destruction before they leave the building. As with liquid wastes, the laboratory design should accommodate the need for safe collection or storage areas for the quantities of solid wastes anticipated.

ARRANGEMENT OF WORKSPACE WITHIN THE LABORATORY MODULE

Laboratory dimensions

Savings in construction costs can be realized by repeating, in various combinations, a two- and sometimes three-dimensional unit of space (19). Hence, modular design is often employed in laboratory facilities. Experience at the National Institutes of Health has led designers there to adopt the following as inside dimensions for a standard modular unit: approximately 10 ft (ca. 3 m) for width, 20 ft (ca. 6.1 m) for depth, and 9.5 ft (ca. 2.9 m) for height (4). The 10-ft width accommodates bench tops on both

side walls and provides optimum space between benches for two to three persons to work (19). The width can vary from 10 to 11 ft, but should not be much wider; a 12-ft (ca. 3.7-m) module has been found to waste space in the center of the module, which laboratory workers are tempted to fill with equipment (4). Laboratory modules are frequently 16 to 25 ft (4.9 io 7.7 m) in depth; National Institutes of Health designers has found 20 ft practical for standard design (4). A larger laboratory can be obtained by eliminating the wall between two adjacent modules. This could result in a double-width laboratory with two work aisles and peninsular as well as sidewall benchworks. A clear ceiling height of at least 9.5 ft accommodates fume hoods, primary barrier systems, and other laboratory equipment.

Relationship of equipment and furnishings to work

Space within a laboratory module can be adapted to a wide variety of functions including microbiology, cell culture, biochemistry, chemistry, physiology, animal holding, animal surgery, pathology, necropsy, and electron microscopy (4).

Arrangement of workspace within the module can be optimized by analyzing the needed placement of equipment and furnishings for anticipated laboratory operations. Individual steps in the operations should be analyzed with a view toward identifying the relationship between operational steps, personnel activity, and equipment and furnishings. Equipment related to one operation should be grouped in a logical arrangement and away from unrelated operations. Where multiple operations will be conducted concurrently in the laboratory, the activity and movement of one worker should be arranged to have minimal effect on the activity of others. Creation of functional working centers rather than worker property areas is usually most effective.

Fume hood and safety cabinet location

Fume hoods and biological safety cabinets should be placed at the far end of the module, usually along a side wall. This placement is to ensure that the hood or cabinet is away from doorways so the exit will not be blocked in the event of explosions or fire within the hood or cabinet. Also, the hood or cabinet will be away from swinging doors and passing traffic, which cause eddy air currents. These air currents adversely affect hood or cabinet containment performance by disturbing the airflow pattern at the work opening. For this same reason, air-conditioning supply outlets should be installed away from hoods or cabinets so that the associated turbulent air motion does not interfere with the hood or cabinet. Preferences for cabinet or hood location are shown in Fig. 1.

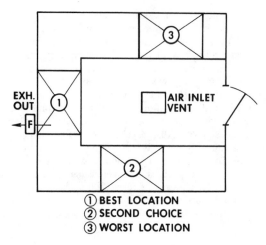

① BEST LOCATION
② SECOND CHOICE
③ WORST LOCATION

FIG. 1. Preferred location of safety cabinet or hood in the laboratory.

Laboratory storage

Though this is seldom achieved, storage of materials in the laboratory should be restricted to working quantities only. This practice does have its practical limitations. However, it is a safety rule that should be strictly followed for hazardous materials such as flammable solvents, toxic or hazardous chemicals, and radioactive isotopes. Adherence to this rule will require adequate storage with convenient access elsewhere in the facility (see Facility Plan, below). In any case, materials will inevitably be stored in the laboratory, and adequate and safe storage space meeting safety codes and standards should be provided in the design. Refer to the relevant chapters in this volume for the storage and handling requirements for flammable solvents, toxic or hazardous chemicals, and radioactive isotopes.

Provisions for waste collection or storage

Contaminated liquid and solid waste storage spaces should be out of the main stream of activity and sized to hold a limited amount of material. It is essential that users remove this material from the laboratory for decontamination at least at the end of each work day. Also, it may be desirable to have an area clearly marked for noncontaminated waste if the decontamination is done in the laboratory and a custodial service is expected to remove the waste. This latter space is best provided outside the laboratory module. In both cases, the space should be well lighted and simple with no fixed furnishings, to facilitate decontamination and cleaning.

Hand-washing facilities

Good laboratory practice in any biological laboratory dictates that laboratory workers wash their hands whenever they may have become contaminated. Further, washing one's hands whenever leaving the laboratory is a habit that is strongly recommended. Consequently, each laboratory should be furnished with a sink located near the access door.

Safety showers and eyewash fountains

A safety shower should be provided for diluting and removing chemicals splashed onto the body and for extinguishing clothing fires. The shower should be easily accessible, near fume hood areas. Sometimes showers may be considered for glass-washing areas if chromic acid is used. Floor drains for the showers are not needed and should be discouraged. After the shower has been used in an emergency, the water can be cleaned up by squeegee and absorbent material.

Eyewash fountains should be provided for emergency treatment (eye irrigation) after chemical splashes to the face. It is best to locate eyewash fountains adjacent to emergency showers.

FACILITY PLAN: ARRANGEMENT OF MODULES AND FUNCTIONAL AREAS WITHIN THE FACILITY

General principles in development of the facility plan

The design team will be striving to develop a plan that is functional and convenient for the users while promoting safety and environmental control. To achieve both functional convenience and safety while developing operational concepts and the facility plan, the following general principles should be applied.

(i) Establish personnel and material traffic flow patterns which clearly define the boundaries of clean and potentially contaminated areas.

(ii) Separate areas known to be sources of general contamination (e.g., animal rooms and waste staging areas) from areas where there are operations or materials sensitive to contamination (e.g., sterile media preparation, cell cultures).

(iii) Separate office areas from laboratory areas.

(iv) Provide visitor and service personnel access only through controlled areas.

Depending on the nature of the programs to be housed in the facility, it is usually desirable for a portion of the facility to have the capability of conversion from basic laboratory areas to ones of greater containment. (The reverse should always be possible.) The spaces selected for this conversion capability should be identified early in the planning so that architectural features and mechanical and electrical systems can be planned to facilitate future adaptation. If future requirements for greater laboratory containment are anticipated, and it is not feasible or desirable to provide complete conversion capability during the original design, then sufficient space and mechanical system capacity, or provisions for expanding capacity, should be incorporated in the design to permit future scale-up with a minimum of facility renovation and concomitant laboratory interruption.

Specific considerations

Management of personnel movement. The management of personnel movement in a laboratory facility should be considered for both public and nonpublic areas (10). Public areas usually include lobbies, office areas, libraries and reference rooms, and other areas where, other than for security, there is no concern for controlling access. Nonpublic areas usually include laboratories, animal rooms, laboratory support areas such as central sterilization service, storage, shipping and receiving, and mechanical equipment rooms. Access to nonpublic areas should be limited to persons who have been instructed in the operational concepts of the facility and the safety and environmental control procedures. Therefore, points of access to nonpublic areas should be few in number and should be supervised or otherwise controlled.

For the nonpublic areas, personnel movement should be managed to minimize personnel exposure to hazards (this includes biohazards, toxic chemicals, traumatic injury, etc.) and to minimize the potential for personnel to serve as vehicles for the transport of contamination. The facility plan should address (i) the need to limit access into specific hazard areas to specific workers, (ii) movement of personnel between program areas such as animal facilities and laboratories, and (iii) movement of both institutional and outside service personnel. Prescribed movement of personnel should extend from the entrance of the facility to the individual work station and should consider normal operations, emergency situations, and security (10).

Management of material movement. Movement of material into, out of, and within the laboratory facility should be planned at the same time as movement of personnel (10). To the extent practical, inbound materials should be kept separate from outbound materials. The movement of materials into and out of laboratories usually requires that they traverse a common pathway. The separation of inbound and outbound materials, to minimize potential for dispersion of contaminants within the facility, can be accomplished through timing (e.g., adjustment of the work cycle) and containerization. Sterilization of outbound wastes within the laboratory will help protect inbound materials from contamination. Provision must be made for unpacking, storing, and possibly repackaging incoming materials (e.g., disposable supplies) to minimize the spread of external contaminants in the controlled laboratory areas.

Storage. A central storage area should be provided for all materials used in the facility. Storage and staging areas will have to be provided for incoming material and for laboratory wastes that are to be removed for disposal or treatment. Certain types of materials (or wastes) may require special storage areas or conditions. Biological materials may require storage at specified conditions of temperature or humidity. Often chemicals also have specified storage conditions. In trying to consolidate storage of materials requiring the same environmental conditions, the potential for and consequences of cross-contamination should not be overlooked. Special attention to pest- and vector-proof construction is needed for animal feed and bedding storage. Refrigerated space may be needed for animal carcasses or chemical specimens awaiting incineration.

Safety should not be overlooked in storage planning. Flammable solvents, toxic or hazardous chemicals, and radioactive isotopes all have special storage requirements (17). Refer to the relevant chapters in this volume for the respective requirements.

For individual laboratories to function productively while maintaining only working quantities of materials, facility storage must be readily accessible. Access can be considered in terms of distance or the availability of storeroom personnel to make laboratory deliveries. In either case, the storage area should be considered a transition point between external and laboratory material flow. Ideally the staff should have control of the material flow both into and out of the laboratory.

Custodial and maintenance services. Planning for custodial and maintenance services must be considered early and reflected in the design. Attention must be given to the delineation of service responsibilities, staffing patterns, placement of service spaces (e.g., custodial closets, utility chases, etc.), and specific operational and maintenance procedures. These details can be developed for either institutional or contract service support or combinations. Operational procedures and design should permit performance of custodial or maintenance service without exposing service personnel to hazards, dispersing contaminants within the facility or outside the facility, compromising the integrity of the research materials, or unreasonably disrupting normal activities (10). Including custodial and maintenance supervisory personnel in the planning process is essential.

Potential emergency situations. Development of operational concepts and facility design should include consideration of emergency situations that may occur (10). Program and operating elements in the facility should be reviewed to identify particular potential emergencies and to devise emergency plans. In this phase of the project it is advisable to consider a range of emergency situations, up to worst-case spills and natural disasters.

Adaptability for future program changes or expansion. Laboratory buildings are often renovated or extended to accommodate changing or expanding programs. The original design should give consideration to architectural modification and future expansion of engineering systems and utility services. To reiterate a basic design objective, renovation of a laboratory area should not jeopardize safety or environmental control in other areas of the facility (see Chapter 18).

Consideration for special requirements for engineering support. Containment laboratories and those furnished with specialized equipment may require special engineering features. Specific requirements for specialized engineering support (e.g., access to process controllers or to air filters in an air duct) need to be identified early and considered in conjunction with physical plant maintenance and operating procedures.

ENGINEERING SUPPORT

Not all details of laboratory engineering design are discussed in this chapter. This section is intended to augment the earlier discussions of mechanical and electrical systems that have been described as part of the secondary barrier.

Mechanical

Heating, ventilating, and air conditioning

General. Heating, ventilating, and air-conditioning (HVAC) systems in laboratory facilities are employed to (i) provide make-up air for local exhaust systems, (ii) provide a means for cooling or heating a space to maintain conditions of temperature and humidity, (iii) provide gradients of relative air pressure to control the direction of contaminant migration, (iv) dilute and remove undesirable airborne contaminants (e.g., odors), and (v) maintain air quality (i.e., freshness).

Ventilation rates. Ventilation can be expressed quantitatively as the volumetric flow rate of air, measured in cubic feet per minute or cubic meters per second, supplied to an enclosed space. A ventilation rate for an enclosure or room may be defined as the number of air changes per unit time and in common practice has the units of changes per hour. An air change is a volume of air equal to the volume of the room. Although these quantitative expressions imply forced ventilation, the ventilation of air infiltrating from outdoors and other sources must also be considered. It is not uncommon to have 3 or more air changes per h from such leakages in buildings where close attention is not paid to sealing the secondary barrier. Such infiltration can bring into the laboratory spores, fungi, pollens, and air pollutants which may be detrimental to sensitive microbiological work.

The notion that removal of contamination generated in the workspace can be done successfully by normal or even high-change-rate ventilating systems has been overplayed. Dilution ventilation does not preclude personnel exposure, particularly from short-term burst exposures. Conventional ventilation (often in the range of 6 to 12 air changes per h) usually is not adequate to remove "burst source" contaminants before they disperse (3). It can be helpful in limiting the concentration of a continuously generated low-level contaminant to acceptable levels (e.g., humidity or odor in animal rooms). In designing ventilation systems, HVAC engineers strive for good air mixing in the room to minimize drafts and get good heat transfer. In most cases this turbulent mixing results in an aerosol removal of approximately 63% for each air change (8). However, in the laboratory there is no assurance that acceptable contaminant levels will be maintained in zones immediately surrounding the sources of contaminants or in personnel breathing zones. Local exhaust and primary barrier systems are preferred because they control the contaminants at their sources and thereby minimize personnel exposure.

The make-up air requirement for local exhaust and primary barrier systems will likely be the dominant factor in establishing the ventilation rate for a laboratory. Another factor that is likely to be dominant is the heating-cooling load requirement. This load comprises the burden of the make-up air and infiltration or exfiltration, solar radiation, occupants, equipment heat gain, and gain or loss of heat and moisture from the processes housed. Either or both of these factors establish the ventilation requirements to be calculated by the HVAC engineer. The ventilation rate thus derived should provide adequate air quality (freshness) and remove obnoxious odors. In no case should the user accept an arbitrarily specified ventilation rate as a "given" design criterion.

Primary barrier equipment effects. The installation of local exhaust and primary barrier systems in a laboratory building requires that air be supplied to rooms to replace that removed by the exhaust systems for the following reasons.

(i) To eliminate undesirable infiltration. An excessive negative pressure condition resulting from inadequate make-up air can cause infiltration of air into the room through door and window cracks and other openings found in normal construction. If the air comes from the outdoors or from uncontrolled corridors it may introduce undesirable contaminants into the laboratory.

(ii) To eliminate unintended differential pressures between adjacent spaces. The negative pressure condition may produce differential pressures that are at variance in magnitude, direction, or both, with the design intent. This may result in the migration of potentially contaminated air from or to nearby rooms and thus may promote the cross-contamination of experiment or product.

(iii) To ensure that the local exhaust system or primary barriers function properly. Lack of available air may cause a negative pressure condition which would increase the static pressure that the exhaust fan(s) must overcome. This may result in reduction of exhaust volume and jeopardize the system's protective capability.

Ventilation system balance. The ventilation system maintains air pressure gradients in the facility. In general, air should migrate, or infiltrate, from non-contaminated areas (such as public areas and clean storage areas) through access corridors to areas that might be contaminated (such as laboratories or animal rooms). Laboratories should have negative air pressure relative to the access corridor, and public areas should have positive air pressure relative to the access corridor. Generally, adjacent laboratories or laboratories on a common corridor should be at the same air pressure to minimize laboratory-to-laboratory migration of contamination. To achieve proper air balance within the facility, supply and exhaust ducts, each fitted with adjustable damper controls, should run to each ventilated space. The system design should be the least demanding of the operator, requiring the least sophisticated controls, maintenance, and operator attention for its reliable function. As the system ages, or should it fail, air balance should be maintained to the extent possible and drastic reversals of flow should be prohibited. It is essential that any designed balance of air flows in a system be maintained regardless of air filter loading, primary barrier operation, or door(s) being open. Exceptions to these generalities should be examined carefully.

The balance described above (i.e., air inflow to the laboratory) is not always achieved as simply as stated. Highly hazardous operations or operations particularly sensitive to contaminants from other spaces may require isolation in addition to the primary and secondary barriers suggested. Here, the use of air locks is sometimes necessary to ensure that the air pressure gradient and directional airflow are maintained. Figures 2 and 3 show some suggested airflow directions for laboratories accessed through air locks, to meet different requirements.

Ventilation air filtration. The extent of filtration for supply air will depend on the activity conducted in the laboratory. The supply air for most biomedical laboratories should pass through 85 to 95% efficient filters (American Society of Heating, Refrigerating and Air-Conditioning Engineers Standard 52-68 Test Method) (1). High-efficiency particulate air (HEPA) filters can be installed for the most critical applications such as environmental studies, specific-pathogen-free animals, or certain cell culture operations in which laboratory materials or animals are particularly susceptible to contamination from external sources (1).

Air recirculation and discharge. As stated above, ventilation air can be recirculated under certain conditions, depending on the nature of the laboratory work and the degree of containment required.

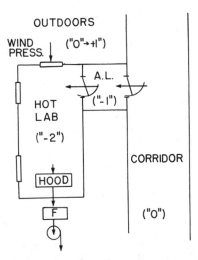

FIG. 2. Ventilating airflow direction for laboratory and air lock for work with highly hazardous materials (relative air pressures are illustrated with dimensionless, nominal values).

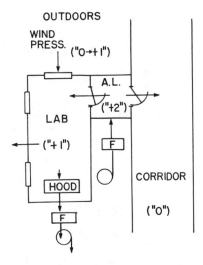

FIG. 3. Ventilating airflow direction for laboratory and air lock for work with materials susceptible to contamination from personnel traffic (relative air pressures are illustrated with dimensionless, nominal values).

Refer to Engineering Features, below, for guidance on this point. Any recirculated air should be filtered, and air from the outdoors must be added to provide freshness. The outdoor air intake should be as far as possible from exhaust discharges and activity that may contaminate the air supply (e.g., vehicle emissions from loading dock operations).

Exhaust air from the facility should be discharged to the outdoors and dispersed to prevent re-entry into the building. Whether or not the exhaust air needs to be filtered or otherwise treated before discharge depends on the containment level of the laboratory. Refer to Engineering Features, below, for guidance on this point.

Plumbing and utilities

General. General discussions on plumbing and utility systems are presented above. The only additional information to be presented here (next paragraph) will pertain to biological liquid waste sterilizer systems (often called biowaste treatment systems). Biowaste treatment systems are needed only for maximum containment laboratories; refer to Engineering Features, below, for further guidance.

Liquid waste (biowaste) treatment. Two basic types of biowaste treatment systems have been described (16). One system is designed and operated as a continuous-flow heat-exchange sterilization system. This system consists of pumps, heaters, retention tubes, and heat reclaim exchanges (18). Storage tanks are required to collect untreated liquid effluent and to ensure continuous operation of the system over a fixed period of time. To conserve energy, an efficient heat-exchange unit is necessary. This system is recommended when the effluent flow rate averages 15 to 25 gal (ca. 57 to 95 liters) per min or more. These systems are costly and difficult to build and operate; specialized engineering design is required (18). An alternative system is a pressure-rated batch sterilization process. The effluent is collected in a tank and heated by injecting steam into the tank or into an internal steam coil, by immersed electrical resistance coils, or by oil- or gas-fired burners. Two full-sized tanks are required so that one tank may be used to sterilize waste while the other is collecting effluent. Both batch and continuous-flow systems must be designed and equipped with a special sampling system to facilitate certification and operational testing.

The required capacity of the biowaste treatment system is determined by the size of the waste drainage system it serves. Microbiological facilities vary greatly in their daily water consumption; the U.S. Army found an average consumption rate to be one gal (ca. 3.8 liters)/day per ft^2 of building area (exclusive of cooling towers), with one-quarter to three-quarters of the water going to the drainage system (21). One basis for sizing the biowaste treatment system is to calculate the total working capacity of all chambers and safety cabinets connected to the biowaste system and add 25% of this capacity to allow for condensed steam and other waste (22). For economy, water usage should be minimized in areas served by biowaste treatment systems by use of spring-loaded faucets, demand-type watering devices for animals, and careful attention to wash-down procedures.

Electrical: Emergency Power

Emergency power should be provided to equipment required for personnel safety, for preservation of research material, and for the protection of the building and its contents. The following loads should be included as a minimum (9, 12): (i) exit and egress lighting; (ii) lighting in areas such as engine-generator rooms, electrical switchgear rooms, mechanical rooms; (iii) fire alarm system, fire pumps; (iv) lighting and ventilation in animal spaces; (v) exhaust fans for safety cabinets and fume hoods containing hazardous radiological, biological, or chemical materials; (vi) elevators (to operate one at a time), (vii) cold rooms, freezers, and refrigerators holding critical research material; (viii) critical research apparatus; and (ix) ventilation in containment areas. In most developed countries electrical outages will be of short duration. The need to maintain air-conditioning cooling should be examined carefully to minimize load. Similarly, the ventilation systems (no. ix above) can be run on a minimal basis (e.g., exhaust only).

Selection of internal loads to be served by the emergency power system should include consideration of the internal electrical distribution system. It may be economical to provide a substantial part of the laboratory area with emergency power rather than attempt to provide separate power supplies to individual loads.

Whether the system provides manual or automatic switching of the emergency generator/service feeder is a matter for the planning team to evaluate on the basis of the institution's operation and maintenance personnel procedures. With an adequate operating staff available 24 h a day, manual operation could be considered satisfactory. If automatic switching is planned, assurance should be had that there will be regular tests of switch gear operation.

The sources of emergency electrical power used most frequently (12) are (i) emergency service circuits and a propane gas- or diesel-driven generator, or (ii) an emergency feeder from a public utility substation which is independent of the network normally supplying electrical service. (This may require some variance from local electrical codes, depending on the site and method of connection.)

Considerations for Maintenance and Access

During facility design, attention should be given to providing adequate access to mechanical and electrical systems so that they can be safely maintained and serviced with minimum disruption to the laboratory program. Some mechanical systems such as exhaust air systems or drain systems may have to be decontaminated before servicing. These systems should be designed to permit decontamination (and verification of decontamination). Details of equipment installation relevant to servicing should be thoroughly worked out before construction, with particular attention given to the need for service workers to enter laboratory areas.

ENGINEERING FEATURES OF THE VARIOUS SECONDARY BARRIER SYSTEMS

The basic facility (2)

The basic facility provides general laboratory space for research where the potential hazards associated with the use of research materials can be controlled by ordinary laboratory practice. The basic facility is not characterized by special engineering features; conventional design approaches are adequate for this facility. The general features of the basic facility are: (i) organized space arrangement; (ii) washable surfaces; (iii) hand-washing facilities; (iv) self-closing doors; (v) availability of an autoclave (for infectious work); and (vi) provisions for ventilated biological safety cabinets. Work is commonly conducted on the open bench, but operations which create considerable aerosols may be confined to ventilated biological safety cabinets (not required in Biosafety Level 1 laboratories).

Ventilated safety cabinets should be available for work with potentially hazardous materials when open bench work practices do not provide adequate protection to the laboratory worker. Exhaust air from fume hoods and safety cabinets should be discharged to the outdoors. Although it is generally undesirable, filtered exhaust air from Class I and Class II biological safety cabinets (see Chapter 19) may be returned to the laboratory environment provided that the HEPA exhaust filters have been certified and the use of volatile, toxic, or radioactive materials in the cabinets is prohibited. Conditioned air for comfort ventilation can be provided in a conventional manner. General (room) ventilation air may be recirculated within the laboratory building without special treatment.

The containment facility (2)

The containment facility has special engineering features which make it possible for laboratory workers to handle moderately hazardous materials. The features that distinguish the containment facility from the basic facility include provisions for personnel access control and a specialized ventilation system. Ventilated safety cabinets should be used whenever the research materials are handled outside fully contained vessels. The general features of the containment laboratory are: (i) organized space arrangement; (ii) access control; (iii) surface finishes impervious to liquids; (iv) foot-, elbow-, or automatically operated hand-washing facilities; (v) self-closing doors; (vi) mechanical exhaust ventilation; (vii) directional airflow into the laboratory from access areas; (viii) autoclave available, preferably in the space (for infectious work); and (ix) ventilated safety cabinets.

The containment facility may be a single laboratory module or a complex of modules within a building. In either case, the containment facility is separated by a controlled-access zone from areas open to the public and other laboratory persons who do not work within the containment facility. The access zone may include clothing change and shower rooms ("clean" and "contaminated") and air locks to be used for the passage of equipment, materials, or supplies.

Considerable attention is given to the architectural features of a containment facility. The surface finishes of walls, floors, and ceilings should be impervious to liquids and readily cleanable. The laboratory modules are constructed so that they can be sealed to permit space decontamination. If windows are provided they should be sealed shut. If false ceilings are installed to conceal air ducts and utility distribution lines, they should be constructed of plaster or dry-wall. All ceiling joints should be taped and sealed, and the entire ceiling should be finished with epoxy, phenolic or polyurethane paint. Horizontal surfaces which may serve as dust collectors should be kept to a minimum.

A foot-, elbow-, or automatically operated handwashing facility should be provided near the exit area of each primary laboratory module. All doors of the containment facility should be self-closing.

An autoclave should be located within the containment facility if the facility is used for research involving hazardous microbiological materials. With appropriate procedural controls, it is possible to locate the autoclave outside of the containment facility, providing it is located within the same building.

The ventilation system supporting the containment facility is capable of controlling air movement. The system is balanced so that there is infiltration of air into each laboratory module from its access corridor. The containment facility may be served by the same ventilation system that serves areas outside the containment facility, provided the exhaust air is not recirculated to areas outside the containment facility and air balance can be maintained. The general (room) exhaust air from containment facilities is discharged to the outdoors without filtration or other treatment, clear of occupied buildings and supply air intakes. Room exhaust air may be recirculated provided it is HEPA filtered. It is recommended that recirculation be limited to the module (or group of modules) constituting a work unit in the containment laboratory.

The ventilation system should be capable of accommodating at least one ventilated safety cabinet in each laboratory module. The cabinet exhaust air may be discharged either directly to the outdoors through an individual duct and exhaust fan or through the general exhaust system of the containment facility. In the latter case it is important that the exhaust system be designed and operated in a manner that avoids interference with the air balance of the containment facility and the ventilated safety cabinet. Pressurization of the exhaust duct must be avoided.

The maximum containment facility (2)

The design objective of the maximum containment facility is to create a facility that will allow the safe conduct of research involving biological agents that are extremely hazardous to the laboratory worker or that may cause serious epidemic disease. The distinguishing characteristic of the maximum containment laboratory is the provision for secure secondary barriers to prevent the escape of any hazardous materials to the environment, particularly in the event of a failure in any primary barrier system (see Chapter 19). The secondary barriers serve to isolate the laboratory area from the surrounding environment. The primary protection for the laboratory worker within the maximum containment laboratory is provided by the use of Class III (glove box) biological safety cabinets (see Chapter 19) or Class I or II safety cabinets used in conjunction with special ventilated protective clothing.

The general features of the maximum containment facility are: (i) separate building or isolated area; (ii) change room and shower facilities; (iii) sealed structures; (iv) minimal horizontal surfaces; (v) special hand-washing facilities; (vi) separate ventilation system; (vii) negative air pressure; (viii) directional airflow; (ix) exhaust air treatment; (x) double-door autoclave; (xi) special facilities for liquid decontamination; and (xii) Class III (glove box) cabinet systems, or provisions for ventilated protective clothing to be used in conjunction with Class I or II biological safety cabinets.

Although the maximum containment facility is generally a separate building, it may be constructed as an isolated area within a building. The perimeter wall partitions of the facility should be installed the full height from finished floor to the under surface of the floor or roof above. If windows are installed in the perimeter partitions, they should be fixed shut and the frames should be thoroughly caulked with sealant. The window glass should be safety glass. Perimeter doors should be insect and rodent proof. Wall, floor, and ceiling construction joints, utility pipe and duct penetrations, and electrical conduits and other passages should be sealed to ensure isolation of the laboratory environment. The surface finishes should be selected on the basis of their ability to provide a monolithic surface barrier. Epoxy, phenolic, and polyurethane finishes have proved satisfactory for this purpose.

The clothing change and shower rooms are part of the perimeter of the facility. They are generally arranged so that the clean clothing change area is separated from the laboratory zone by an air lock or shower area. Personnel egress from the laboratory zone must be through the shower area to the clean clothing change room. Air locks for movement of materials, supplies, and equipment into the facility are also a part of the perimeter structure. The double-door autoclave is located so that either the interior or exterior door frame is sealed to the perimeter barrier wall. It is preferable to make the interior door frame contiguous with the barrier wall so that autoclave maintenance can be performed outside the laboratory zone.

The maximum containment facility is ventilated by its own supply and exhaust air mechanical ventilation system. The system is operated so that the air pressure within the facility can be maintained lower than the air pressure outside the perimeter walls. The air system is balanced so that airflow within the facility is from areas of least hazard potential toward areas of greatest hazard potential.

The air handling system should provide an air supply consisting of 100% outdoor air on a year-round basis. The system should provide separate branch supply and exhaust air ducts to each space to permit proper air balance. The supply and exhaust fans should be interlocked to prevent reversal of airflow in event of exhaust fan failure.

The general exhaust air is filtered by passage through HEPA filters before being discharged to the outdoors. The air filters should be located as near to the laboratory module as possible to minimize the length of potentially contaminated air ducts. The filter plenums should be designed to facilitate both testing of filters after installation and in-place decontamination before filter removal and replacement. General room air may be recirculated within a single room provided it is HEPA filtered.

As noted above, the maximum containment facility may contain specially designated areas where the primary barriers are Class I or II cabinets and in which all personnel are required to wear one-piece positive-pressure ventilated suits. The exhaust air from such a "suit area" must be filtered through two sets of HEPA filters in series. Duplicate filtration units and exhaust fans must be provided. This space must be airtight and have an air pressure lower than that in any adjacent area. A separate double-door autoclave must be provided for any "suit area" within the maximum containment laboratory. Standby (backup) systems must be provided for suit air, lighting, and room exhaust.

Mechanical systems should be designed so that maintenance of building machinery, piping, and controls can be performed from outside the laboratory environment.

Liquid effluents from the maximum containment facility should be collected and decontaminated before disposal into the sanitary sewers. Effluents from laboratory sinks, cabinets, floors, and autoclaves should be sterilized by heat treatment. Drain pipe vents should be fitted with HEPA filters. Liquid wastes from the shower room may be decontaminated with chemical disinfectants. The wastes from toilets may be discharged directly into the sanitary sewers.

LITERATURE CITED

1. **ASHRAE Technical Committee.** 1982. Applications volume of the ASHRAE Handbook, 9.2, Industrial air conditioning, laboratories, p. 15.1–15.14. American Society of Heating, Refrigerating and Air-Conditioning Engineers, Inc., New York.

2. **Barkley, W. E.** 1977. The research facility, p. 77–88. *In* Proceedings of the Workshop on Cancer Research Safety: 1977 September 26–29, Dulles Marriott Hotel, Washington, D.C. National Institutes of Health, Bethesda, Md.

3. **Chatigny, M. A., and D. L. West.** 1982. Ventilation rates: theoretical and practical considerations, p. 71–100. *In* M. S. Barbeito and D. L. West (ed.), Laboratory ventilation for hazard control: proceedings of a symposium 1976 October 21–22, Frederick Cancer Research Center, Frederick, Md. NIH Publication no. 82-1293. National Institutes of Health, Bethesda, Md.

4. **Dolan, D. C.** 1981. Design for biomedical research facilities: architectural features of biomedical design, p. 75–86. *In* D. G. Fox (ed.), Design of biomedical research facilities: proceedings of a cancer research safety symposium, 1979 October 18–19, Frederick Cancer Research Center, Frederick, Md. NIH Publication no. 81-2305. National Institutes of Health, Bethesda, Md.

5. **Duncan, W. E.** 1981. Mechanical design features for biomedical research facilities, p. 87–96. *In* D. G. Fox (ed.), Design of biomedical research facilities: proceedings of a cancer research safety symposium, 1979 October 18–19, Frederick Cancer Research Center, Frederick, Md. NIH Publication no. 81-2305. National Institutes of Health, Bethesda, Md.

6. **Fox, D. G. (ed.).** 1981. Design of biomedical research facilities: proceedings of a cancer research safety symposium, 1979 October 18–19, Frederick Cancer Research Center, Frederick, Md. NIH Publication no. 81-2305. National Institutes of Health, Bethesda, Md.

7. **Fuscaldo, A. A., B. J. Erlick, and B. Hindman (ed.).** 1980. Laboratory safety: theory and practice. Academic Press, Inc., New York.

8. **Kethley, T. W.** 1964. Air: its importance and control, p. 35–46. *In* G. S. Michaelsen (ed.), Proceedings of the National Conference on Institutionally Acquired Infections, 1963 September 4–6, Minneapolis, Minn. PHS Publication no. 1188. Communicable Disease Center, Atlanta, Ga.

9. **Marsteller, H. H.** 1981. Electrical design features for biomedical research facilities, p. 97–108. *In* D. G. Fox (ed.), Design of biomedical research facilities: proceedings of a cancer research safety symposium, 1979 October 18–19, Frederick Cancer Research Center, Frederick, Md. NIH Publication no. 81-2305. National Institutes of Health, Bethesda, Md.

10. **McKinney, R. W.** 1981. Design of biomedical research facilities: operational concepts and facility planning, p. 39–43. *In* D. G. Fox (ed.), Design of biomedical research facilities: proceedings of a cancer research safety symposium, 1979 October 18–19, Frederick Cancer Research Center, Frederick, Md. NIH Publication no. 81-2305. National Institutes of Health, Bethesda, Md.

11. **National Cancer Institute.** 1974. Safety standards for research involving oncogenic viruses. DHEW Publica-tion no. (NIH) 75-790. National Institutes of Health, Bethesda, Md.

12. **National Cancer Institute.** 1975. Design criteria for viral oncology research facilities. DHEW Publication no. (NIH) 76-891. National Institutes of Health, Bethesda, Md.

13. **National Cancer Institute.** 1975. Safety standards for research involving chemical carcinogens. DHEW Publication no. (NIH) 75-900. National Institutes of Health, Bethesda, Md.

14. **National Cancer Institute (ed.).** 1977. Proceedings of the Workshop on Cancer Research Safety, 1977 September 26–29, Dulles Marriott Hotel, Washington, D.C. National Institutes of Health, Bethesda, Md.

15. **National Cancer Institute.** 1978. Recombinant DNA research: guidelines. Fed. Reg. **43:**60108–60131.

16. **National Institutes of Health, Special Committee of Safety and Health Experts.** 1979. Laboratory safety monograph, a supplement to the NIH guidelines for recombinant DNA research. National Institutes of Health, Bethesda, Md.

17. **Phillips, G. B.** 1965. Safety in the chemical laboratory. Part 1. Control. J. Chem. Ed. **42:**A43–A44, A46–A48.

18. **Runkle, R. S., and G. B. Phillips (ed.).** 1969. Microbial contamination control facilities. Van Nostrand Reinhold Co., New York.

19. **Snow, D. L.** 1963. Space planning principles for biomedical research laboratories. Public Health Monogr. no. 71, PHS Publication no. 1025. Public Health Service, Washington, D.C.

20. **Stern, E. L., J. W. Johnson, D. Vesley, M. M. Halbert, L. E. Williams, and P. Blume.** 1974. Aerosol production associated with clinical laboratory procedures. Am. J. Clin. Pathol. **62:**591–600.

21. **U.S. Army.** 1970. Design criteria for microbiological facilities at Ft. Detrick, 2nd ed., vol. I, p. 19. U.S. Dept. of Defense, U.S. Army. Available from NTIS, Springfield, Va.

22. **U.S. Army.** 1970. Design criteria for microbiological facilities at Ft. Detrick, 2nd ed., vol. II, p. 3.30. U.S. Dept. of Defense, U.S. Army. Available from NTIS, Springfield, Va.

23. **U.S. Department of Health and Human Services.** 1984. Biosafety in microbiological and biomedical laboratories. HHS Publication no. (CDC) 84-8395. U.S. Government Printing Office, Washington, D.C.

24. **Wedum, A. G. and G. B. Phillips.** 1964. Criteria for design of a microbiological research laboratory. ASHRAE J. **6:**46–52.

25. **West, D. L.** 1981. Special design considerations: biohazard facilities for infectious microorganisms, recombinant DNA materials, and oncogenic viruses, p. 147–164. *In* D. G. Fox (ed.), Design of biomedical research facilities: proceedings of a cancer research safety symposium, 1979 October 18–19, Frederick Cancer Research Center, Frederick, Md. NIH Publication no. 81-2305. National Institutes of Health, Bethesda, Md.

26. **West, D. L., D. R. Twardzik, R. W. McKinney, W. E. Barkley, and A. Hellman.** 1981. Identification, analysis, and control of biohazards in viral cancer research, p. 167–223. *In* A. A. Fuscaldo, B. J. Erlick, and B. Hindman (ed.), Laboratory safety: theory and practice. Academic Press, Inc., New York.

Chapter 18

Designs to Simplify Laboratory Construction and Maintenance, Improve Safety, and Conserve Energy

HERBERT M. BLOOM

The renovation of existing facilities is generally recognized as being more difficult than the construction of new ones. Normally, there is greater freedom of design in the construction of new facilities, whereas the redesign and renovation of existing laboratory facilities will encounter major inherent obstacles, such as: (i) design constraints due to existing load-bearing walls, ceiling heights, and floor loadings, as well as the existing location of entrances for utilities (water, electricity, etc.); (ii) relocation of operations occupying the areas to be renovated; and (iii) interference with ongoing laboratory and other functions in the vicinity of the renovations.

Renovations are particularly challenging since one must attempt to utilize the existing utilities, if possible, in their present state. An existing electrical system which no longer meets the National Electrical Code and must be replaced might present a problem if the circuit breakers to be replaced or the electrical panels which must be increased in capacity are no longer manufactured. Another obstacle often encountered is the lack of up-to-date structural drawings. Changes made in a building over the years are not always accurately recorded, and it is difficult to trace utilities that disappear into walls or into concrete floors that have no open level below. Additionally, it is difficult to determine floor loadings in the absence of design drawings and design information.

New structures can be designed to satisfy known functional requirements. When a building is renovated, however, its functional requirements must be adapted to fit within the existing configuration. Both in constructing a new facility and in renovating an existing facility, utility systems, equipment locations, partitions, etc., should be so located that future changes can be accomplished at a reasonable cost and with minimum disruption to the occupants. The interior of framed buildings can be modified more easily and cheaply than that of buildings whose partitions are load bearing. Not all buildings are suitable for renovation.

The design of biological and chemical laboratories and animal facilities is complex, and their construction is expensive. Design of such facilities must take into consideration operating and maintenance costs which will increase with time. Accordingly, design flexibility is important for future changes. Of equal importance is the consideration of designing a facility that can be readily cleaned and maintained.

The proper design, operation, and maintenance of laboratory facilities is essential to achieving efficient operations. The first, essential step toward this end is the well-thought-out preparation of an Architect-Engineer Scope of Work that clearly indicates the purpose of the facility; the environmental conditions under which the facility is to operate; the flow of personnel and material; special equipment such as autoclaves, cage and rack washers, electron microscopes, darkrooms, etc.; and special safety features such as HEPA (high-efficiency particulate air) and charcoal filtration for hood systems as required, life safety systems, etc. A thorough Architect-Engineer Scope of Work should provide final drawings and specifications that will produce the desired facility. Care in its preparation will definitely result in fewer construction change orders.

LAYOUT AND DESIGN

A primary consideration in the arrangement of space is that the laboratory floor plan should suit the functional convenience of the research program. In containment-type laboratories, physical isolation from other areas must be provided.

In all cases, the traffic flow pattern should minimize the proximity of clean and contaminated areas or items. Laboratories must be provided with air, gas, vacuum, and hot-, cold-, and pure-water outlets. Quadriplex electrical outlets should be provided for each 4 feet (ca. 1.2 m) of linear bench space. Sinks should be provided, and sized, as required. Laboratory air compressors, pumps, etc., wherever possible, should not be located in the normal flow of laboratory traffic. Desk space for technicians' use should be provided within each laboratory. Walk-in cold and incubator rooms should be provided as required. Shared equipment rooms are desirable and will reduce capital equipment costs. Autoclaves/sterilizers should be provided as required.

The design of new or renovated structures must include provisions for handicapped personnel such as entrance/exit ramps, toilet and washing facilities, elevator service, etc.

Thought should be given to maintenance, repair, and replacement during engineering design and layout. Avoid locating any equipment at excessive heights, in limited-access areas, or in areas subject to high temperature and humidity or excessive vibration. Thought should be given to relamping of lighting installations. Fixtures must not be placed in locations where bulb replacement is difficult or impossible. Piping systems should be valved to isolate sections in the event of a failure. Provision should be made for telephone locations, raceways, and readily accessible central panelboards. Adequate ventilation of utility areas reduces heat load and permits motors to operate more efficiently. Transformers located indoors should be well ventilated to prevent overheating.

It is absolutely essential that adequate space be provided for utility equipment such as pumps, air compressors, electrical panels and motor control centers, refrigeration equipment, etc. Failure to do so results in higher maintenance costs and the inability of maintenance personnel to service the equipment adequately. Space must be provided for the removal and maintenance of installed equipment, the replacement of air filters, periodic servicing of installed equipment, and ready access to equipment and its controls.

Not to be overlooked are janitorial closets, laboratory supply storage, maintenance storage areas, window locations, power sources for cleaning equipment (particularly in hallways and lobbies), exterior building water faucets, lockers for maintenance personnel, the ability to introduce and rearrange equipment within the structure, stairways and landings of adequate size for moving furniture from floor to floor in the absence of an elevator, etc.

For proper facility operations, equipment access doors of adequate size and location must be provided. If the facility has a below-ground-level utility room, the entrance well must be of sufficient size for large equipment to be safely lowered and maneuvered through the doorway. If the facility has utility equipment installed in the attic area, hoists in fixed locations and with adequate lifting capacity are required to permit the safe movement of equipment.

A design aspect often overlooked is the provision of safe exits from a congested utility area, particularly attic utility areas. The design should include adequate lighting and at least two means of safe egress.

CONSTRUCTION CONSIDERATIONS

The construction contractor should be required to prepare and maintain a set of complete "as-built" drawings on the job. These should be updated as construction progresses and not after completion of the job. A frequent failure is an attempt by the contractor to modify the drawings months after the changes have been completed. Inaccurate or incomplete as-built drawings are of little value or worse than none at all.

Equipment specifications should be standardized as much as possible. Where possible, only equipment for which replacement parts and service are widely available should be selected. Equipment that can be obtained from only one source must be avoided. If replacement parts and service are available, repairs or maintenance can be accomplished in a shorter time. Frequently, entire systems or equipment items must be replaced because the replacement parts needed for simple repairs are not available. Selection of equipment which is overly proprietary makes the owner captive to the single source of parts or service.

ARCHITECTURAL CONSIDERATIONS

Architectural arrangements should be approved before the start of design. Changes made during the design phase are usually costly, particularly if they are major in scope and the plans are well under way.

Walls and ceilings

Laboratory walls and ceilings should be smooth, and all joints and pipe, conduit, light, and duct penetrations should be sealed to ensure integrity between laboratories and permit a high degree of cleanliness. Ceilings in laboratories may be hung acoustical tile; animal rooms should have plaster ceilings. Walls should be constructed with impact-resistant materials. Exterior walls should be insulated to reduce energy consumption; this is particularly important in animal rooms to maintain constant temperature. Ground floors should also be insulated. Concrete-block walls of fine texture are often utilized in construction of both laboratories and animal rooms. The report of an Investigation of Animal Room Finishes for Co-Carcinogenesis Research Laboratory, Building 9211, and Animal Health Laboratory, Oak Ridge National Laboratory, prepared by A. M. Kinney, Inc., Consulting Engineers, recommended that animal rooms be constructed with light-weight aggregate concrete blocks sealed with a sprayed-on epoxy coating. This type of construction will not only resist quaternary ammonium solutions, phenolic compounds (Lysol, Kretol, etc.), chlorinated compounds, detergents, urine, and steam, but will also bridge hairline cracks. However, such construction will not resist larger cracks caused by building settlement or other structural movement. The report concluded that glazed tile block is not satisfactory for animal room construction. A curb suitable for protecting the wall from impact of wheeled vehicles is recommended for animal rooms and corridors within animal facilities. Doors to laboratories should be fitted with a window and a self-closer. Doors to animal rooms should be at least 42 in. (106.7 cm) wide and have a viewing panel with speaking diaphragms, armor plate on both sides of the door, recessed pull handles, and door sills that are not an obstacle to movement of racks. Animal room doors should be equipped with self-closers, and their bottoms should be mouse proofed.

Walls in administrative areas may be plasterboard, paneling, or whatever the client desires and can afford. The ceilings may be plasterboard or lay-in type with recessed lighting. Whatever type of wall or ceiling is utilized, it should be easy to clean. A chair rail should be installed to protect corridor walls, particularly in the laboratory area.

Flooring

Laboratory floors should be covered, preferably with seamless sheet vinyl (smooth surface, non-dirt catching) including a 4-in. (ca. 10.2-cm) cove base sealed at top and bottom. Administrative areas may be floored with vinyl tile and 4-in. cove base or carpeting. Lunchroom areas should not be carpeted; lunchrooms with carpeting have proved to be difficult to clean and maintain. Animal rooms and animal area corridors should be finished with a troweled-on epoxy material with a carefully integrated 4-in. cove base that is non-dirt catching. Service area floors should be finished in the same manner.

Painting

The selection of paint colors should be carefully considered. The policy of permitting building occupants to select colors for their particular area may be a great morale builder, but it creates a maintenance nightmare. The use of several different colors within a facility not only is more expensive than the use of a few basic colors, but it also does not allow "touch-up" with any degree of overall satisfaction because the same color from the same manufacturer will often vary from one lot to another. A high-quality paint properly applied is adequate for most laboratory areas.

Laboratory furniture

Laboratory furniture should be fabricated from steel with an appropriate factory-applied finish that will not be adversely affected by decontaminants and will resist impact from lab carts and other objects. Work benches may be surfaced with Formica or a solvent-resistant surface. These surfaces permit easy cleaning and are presentable in appearance. A water-tight seal at floor and wall surfaces is required for workbenches, cabinets, and other pieces of installed equipment, unless the items are easily moved to permit cleaning. It is recommended that all laboratory storage cabinets, lockers, wall-hung cabinets,

and similar furniture that does not extend to the ceiling have tops that slope forward to prevent build-up of dust and to facilitate drainage after washdown with decontaminating solutions.

Rooms for handling radioactive materials (1)

It is most important to determine the amount of radiobiology anticipated before the facility is designed, since the design of laboratories handling radioactive materials is significantly different, involving safety regulations and higher construction costs. It may be difficult to meet radiological safety standards after the building is constructed. For example, if gamma-ray emitters are to be used, appropriate lead shielding must be provided, and the weight of the shielding may exceed the live load for which the floor was designed.

All surfaces in laboratories and animal rooms where radioactive materials are to be used must be monolithic, nonporous, and washable. Cracks, crevices, and joints must be sealed. Vinyl, rubber, or linoleum sheets can be applied over a concrete floor to provide protection, since these materials are nonporous and can be removed if necessary for radiological decontamination. Epoxy resin paints or polyurethane coatings will seal plaster walls effectively if properly applied. Stainless-steel sinks, bench tops, animal cages, and other items that might be contaminated with radioactive materials are recommended since they can easily be decontaminated.

Provisions should be made to collect, store, and monitor all liquid and solid wastes from radiological areas. All drains from radiological work areas should empty into buckets or other suitable containers. All radioactive waste should be disposed of in accordance with applicable Nuclear Regulatory Commission requirements.

A chemical fume hood is required for all radiological procedures in which there is any chance of air pollution by radioisotopes. The airflow rate into the hood opening should be at least 100 linear ft/min. The hood must be equipped with an ultra-high-efficiency filter for exhaust air. A charcoal filter is also required.

In the microbiological containment facility the design engineer may find that he must provide facilities for infectious microbial agents tagged with radioactive materials. Selection of the primary barrier system (e.g., chemical fume hood, Class I or II total exhaust biological safety cabinet, or Class III biological safety cabinet) should be based on the need to provide protection to personnel, the environment, and the experiment (see Chapter 19 and Appendix 1).

In some instances when highly radioactive materials are to be used, the design will have to include remote-handling equipment such as tongs, forceps, clamps, or remote mechanical systems. Such systems are expensive and ordinarily will require large amounts of shielding.

Rooms for electron microscopes (1)

Many containment facilities will require an electron microscope. A complete electron microscopy laboratory must provide space and facilities for operation of the microscope and its accessories, as well as laboratory facilities for specimen preparation and photographic facilities for developing and printing. The room should be located to avoid excessive vibration, defined as $\geq 10^{-4}$ g. Basements or first floors are usually more stable than upper floors. Not only can elevators, fans, pumps, and other motors cause excessive vibration, but also electrical equipment, such as telephone switchboards, may generate troublesome stray magnetic fields. The electron microscopy room should have at least high-efficiency filtration of supply air to control dust. Ordinarily the room should be at a positive air pressure to prevent entrance of contaminants. Large fluctuations in temperature and relative humidity should be avoided. The heat load imposed by the microscope and its power supply should be considered when planning temperature control. The usual required utilities are water, electricity, and compressed air.

Walk-in refrigerator and incubator rooms (1)

Refrigeration usually requires more space than incubation. Commercial prefabricated units are usually suitable for most needs.

Provisions should be made in the design of the building for condensate drip lines to run from the cooling coils through the floor rather than on top of the floor. All crevices between the refrigerator or incubator and the walls and floor of the room should be sealed unless the unit is readily movable. Walk-in refrigerators, and usually walk-in incubators, should be provided with floor drains to facilitate cleaning. The floor drains should be placed as near the back wall as possible. Traps should be deep enough to prevent their being emptied by positive pressure when the doors are closed. Doors to walk-in refrigerators and incubators should have a sealed, double-glass viewing panel in the center with the bottom of the viewing panel located ca. 50 in. (ca. 1.25 m) above the floor. The hardware for walk-in refrigerator and incubator doors should be corrosion resistant and should be equipped with a padlocking handle and an interior automatic release that operates whether the outside handles are padlocked or free. Control panels should be located outside the rooms.

Solvent storage room (1)

One of the major causes of fires in research facilities is the improper storage and handling of flammable solvents. Most facilities should have a special storage room or cubicle for flammable materials. The flammable-material storage room should conform to the National Board of Fire Underwriters requirements for type B inside storage or mixing rooms and should be protected with a fixed carbon dioxide extinguisher system, or equal, installed according to the requirements of the National Board of Fire Underwriters or the National Fire Protection Association. Approximately 1 lb of CO_2 is required for each 15 ft^3 of space. The air exhaust fan for the flammable-material storage room should be spark resistant (AMCA type B) and should have an explosion-proof motor.

MECHANICAL CONSIDERATIONS

Heating, ventilation, and air conditioning systems

The type and number of heating, ventilation, and air conditioning (HVAC) systems selected to meet facility needs will have an important bearing on energy consumption. Systems that can be easily controlled to meet changing requirements will improve energy efficiency. There is nothing wrong with having different types of systems within one building, especially if each system is selected to provide the necessary conditions at maximum efficiency. It may complicate the building and the design but will frequently be worthwhile in terms of operating costs.

Laboratory facilities by their very nature are large energy users, consuming 5 to 25 times as much energy per square foot of floor area as typical commercial buildings. The concern over energy consumption must be made known before engineering design commences. Facility design must take into account the building's use, equipment, temperature and humidity requirements, etc. Construction or renovation designs must consider the use of energy recovery devices and equipment such as the energy wheel, run-around coil, and thermal heat pipe, which can recover between 50 and 70% of the heat from exhaust air, thus reducing operating costs.

The design and location of air louvers often receive insufficient attention. Needless to say, the supply air intake should be as far from the building air exhaust as possible and should have adequate air intake area. Inadequate area will cause high air velocity and can carry in rain and snow, which will clog the primary filters. Snow-clogged air supply filters, for example, will cause animal rooms to alarm due to reduced airflow, yet to remove these filters will leave the incoming air unfiltered.

Roughing filters, preferably curtain type, should be installed ahead of the main supply filters; these will extend the life of the more expensive main filters. The main supply filters should be designed to be readily installed and removed by maintenance personnel. The exhaust filter system should also be designed for easy servicing. Exhaust filters serving chemical and radiological hoods and bioassay animal areas should be a bag-in-bag-out type, readily removable with minimum exposure to maintenance personnel. During these filter changes, maintenance personnel must be masked and wear disposable protective clothing.

No single utility system can be more frustrating to maintenance personnel than an undersized HVAC duct system, improperly installed, and with insufficient dampers for proper air balancing. The design must expressly specify the proper size ducts to the contractor. Generally, the duct ratio of width to depth should not exceed 3 or 3.5 to 1. All duct joints should be sealed, even in low-pressure duct systems. Supply ducts should be insulated. Dampers should be easily accessible for balancing operations. In animal facilities, a roughing filter should be installed at the exhaust inlet in the animal room to prevent animal hair and dandruff from entering the ducts.

Fans and cooling and heating coils should be adequately sized to provide for future expansion. In today's market, equipment such as heating coils is designed for a given maximum performance. In the event that laboratory or animal areas are renovated and the airflow must be increased, new coils and fans are often required because of the inadequacy of existing equipment.

Temperature and humidity sensor locations must be chosen with care. If sensors are located in the exhaust ducts, the exhaust ducts must be equipped with a room exhaust filter and the sensors should be the capillary type. The best type of control for animal room temperature and humidity is aspirators located in the animal room. Such aspirators are equipped with a filter and a small fan which mixes the inlet air being monitored.

Freeze stats should be installed on preheat coils to avoid freezing of coils. This will save down time and eliminate costly repairs and inconvenience to facility occupants.

Autoclaves and rack and cage washers must be separately vented to remove excessive heat loads.

Each chemical fume hood and Class II Type B biological safety cabinet must have its own spark-proof exhaust fan and be exhausted through the roof and not through the building ventilation exhaust system. All hoods are to be equipped with a pressure switch and alarm to indicate lack of safe level exhaust air flow. The exhaust from these hoods should be filtered through absolute filters. Chemical fume exhaust systems should be equipped with a charcoal filter after the absolute filter. The filter housings should be the bag-out type. Heavy-gauge galvanized-steel welded ducts should be utilitized on all chemical fume hood exhaust systems. The design hood flow rate must be 100 linear ft/min with the sash at full opening for chemical fume and Class II Type B hoods.

Too much is at stake not to have backup chilled-water equipment in animal facilities and backup chilled- and condensing water pumps for continuously operating laboratories.

Install oil-free instrument air compressors with dryers. Oil-free compressors are highly recommended to protect sensitive instrumentation. Duplex equipment will meet these requirements. A backup capability is absolutely mandatory if pneumatic HVAC controls are installed.

Piping

The incoming water line should have a bag-type, self-flushing water filter system. This equipment will provide higher-quality water for laboratory and general use and reduce filter maintenance costs on pure-water systems.

Cold water to drinking fountains should be the first water line from the supply header, just after the water filter and with no other connections to this line. A reduced-pressure backflow preventer with an interrupted bypass should be installed immediately after the building water supply filter system. This will prevent contamination of the incoming water supply.

Safety showers with an eyewash station should be provided where required.

In areas where contamination control is necessary, foot-, elbow-, or knee-operated levers should be provided to operate faucets or valves. Drinking fountains should have foot pedals only.

Drains with adequate flow are required in all utility areas, particularly upper levels. Flooding within the basement utility rooms can cause severe damage to equipment. Flooding in attic and upper areas without proper drains will cause severe damage to the rooms and equipment below. Even concrete floors, with their many penetrations, just cannot be made flood proof.

Other considerations

Utility systems should be installed in such a manner that they can be readily isolated by valves in the event repairs are required on a portion of the system. This is particularly true for preheat and reheat coils, cooling coils, air systems, etc. This procedure will reduce building down time and complaints by the building inhabitants.

All utility lines should be identified clearly and visibly, as to content and direction of flow. It is also an excellent idea to tag similarly HVAC ducts, HVAC blowers serving individual areas, and particularly blowers and filters connected to fume and biological hoods.

ELECTRICAL

Conductors, especially feeders, should allow capacity for future growth in electrical power requirements. The future potential of the building or space served should be assessed, and conductors should be sized accordingly.

Raceway systems should be made accessible. Concealed raceways should be kept to a minimum, and long runs without pull boxes, or other means of access, should be avoided. Where work must be run concealed, such as in slabs, sufficient empty spare conduits should be installed to allow for future needs. Similarly, telephone raceway systems should include sufficient spares and be designed to permit expansion as needed.

All conductors, circuits, control functions, etc., must be identified in construction documents and specifications. These requirements should include:

1. Color coding schemes for conductors
2. Requirements to tag, mark, and otherwise identify conductors
3. Requirements to mark lighting switches, receptacle outlets, control enclosures, disconnect switches, etc., with the circuit from which the device receives its supply
4. Requirements for complete and accurate circuit directories, switchboard legend plates, etc.

Alarms for sensitive areas (animal rooms) and equipment will reduce equipment and product losses. The alarm will alert personnel that a malfunction has occurred, thus providing a chance to relocate the product (media, cultures, animals, etc.) and possibly enabling maintenance staff to respond quickly enough to prevent greater damage to equipment or products.

Fire alarm systems should be a supervised, zoned, coded or noncoded, annunciated type incorporating manual stations, automatic detectors, alarm bells, trouble alarm, and code transmitter.

Lighting fixtures should be moderately priced commercial grade, with acrylic diffusers and high power factor, low-noise ballasts. They may be either recessed or nonrecessed.

Spare circuits should be provided and terminated at disconnect switches.

Motor control centers ought to be installed in a well-ventilated area, away from water pipes and moisture. Their design should include connection and control wiring for all motor loads, including motor protection, interlocking automatic control, etc. A manual fire shutdown switch should be provided, as well as programmed timing for automatic restart of motors. A quantity of spares should be provided for future expansion.

Life safety systems must include emergency circuit panels in accordance with Article 700 of the National Electric Code, exit signs, and emergency light sources such as 12 V DC wall-mounted units, etc.

CONCLUSIONS

Not all of the opportunities to simplify maintenance have been outlined here. Furthermore, some of the points covered not only simplify maintenance but are absolutely vital to the continued operation of critical facilities. Backup chilled-water equipment, chilled-water and condensing water pumps, and instrument air compressors are absolutely vital to the continued operation of critical facilities. Careful architectural considerations will not only reduce initial costs but also provide for affordable maintenance. Electrical installations that provide a modest capacity for growth beyond immediate needs should be standard. System designs must be dependable, readily understood, and reasonably easy to maintain. Exotic, unproven system designs should be avoided as much as possible.

It behooves the designer to discuss his concepts with the maintenance supervisor. Maintenance personnel have a wealth of knowledge that can be put to good practical use. A good policy is to have the mechanical and electrical drawings, in particular, reviewed by maintenance supervisors even before construction bids are solicited.

Any design that improves ease of maintenance, decreases maintenance costs, and enhances reliability will produce a facility that not only is a credit to the designer and the engineering profession but will, in the long run, result in a better product—whether the making of steel or the operation of a research facility.

LITERATURE CITED

1. **Runkle, R. S., and G. B. Phillips.** 1969. Microbial contamination control facilities. Van Nostrand Reinhold Co., New York.

Chapter 19

Primary Barriers

MARK A. CHATIGNY

The objective of this chapter is to provide the laboratory worker with information and advice regarding selection, use, and efficacy of various barrier systems. The design features of importance will be discussed only as they affect the selection and use of the devices. The term "primary barriers" is defined as the physical containment methods and equipment used to isolate the operator or laboratory worker from the chemical or biological agent in use. The reverse may be preferred, although sometimes with differing methodology.

Primary barriers, in one form or another, have been used in microbiology since Louis Pasteur demonstrated that sealed or plugged vessels did not become contaminated with airborne microbes. As microbiology technology advanced, so did that of containment. Attention to containment lagged somewhat behind laboratory worker infection experience. The latter must be considered a historically strong driving force. The high incidence of infection in scrub typhus laboratory workers led to fabrication and use of a safety cabinet described by Van den Ende (29) in 1943. Later, work with the rickettsiae of Q fever was the source of many infections. An open-front cabinet was used (26) to contain work with this agent. The evolution of the microbiological containment cabinet has been reviewed by Wedum (31) and Chatigny (9). Only the infection experiences were unique to the microbiology laboratory, because parallel efforts were in progress to protect workers from chemical and radiological hazards in laboratory work in those disciplines. In recent years, the problem of protecting the work or product from the worker and the environment has gained importance, particularly in the areas of microelectronics, cell culture, and preparation of materials for parenteral use.

One of the original objectives in containment of a process or operation in a primary enclosure was to control aerosols. The intuition of Pasteur was followed by the experiences cited above and described more fully elsewhere in this text. Workers at the U.S. Army Biological Laboratories contributed strongly in demonstrating that surface contamination and aerosols were produced whenever microbial suspensions were manipulated (31). On this basis, the Army Biological Laboratories Safety Group and Engineering Department did much of the basic design, fabrication, and testing of physical protection devices

in the 1940s and 1950s. The current view, based on these data and further quantitative tests (17), is that the containment should include control of contact, spatter, and small-particle aerosol spread of the agents and that the containment should be a "system" encompassing clothing, mechanical devices, laboratory design, and work practices.

The general principles applicable to containment within a primary barrier system include the following.

1. Minimize the volume to be contained.
2. Provide safe (i.e., noncontaminating) transfer of material into and out of the container without destroying the barrier.
3. Provide means for decontamination of the enclosure and effluents.

Primary barriers include physical barriers of steel, plastic, glass, or other similar materials; air barriers or air "curtains"; and exhaust and supply barriers, such as high-efficiency particulate air (HEPA) filters, charcoal filters, or air incinerators. The choice of materials depends on the procedures to be housed, the contaminating agents of concern, the methods of decontamination, and the need for access to the work.

TYPICAL SIMPLE-PROCESS CONTAINMENT TECHNIQUES

One of the readily recognizable hazards has been addressed for a number of years by a simple containment procedure, that is, the construction of safety cups for the microbiological centrifuge. These containers range from individual sealed tubes (Fig. 1) to larger screw-capped buckets and sealed rotors. These devices can prevent leakage of small-batch materials under low-, medium-, or even high-speed centrifugation. Other sealing measures are used for large batch rotors. Blenders have also been described as "heavy" aerosol producers. Without special sealing design, they can, like centrifuges, rapidly contaminate spaces and spread high levels of surface contamination. Workers at the U.S. Army Biological Laboratories, many years ago, devised a safety blender cup, improvements of which are now marketed commercially (Waring Products Division, New Hartford, Conn.).

In the development of simple, laboratory-made

144

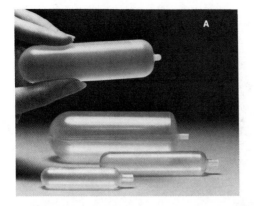

FIG. 1. (A) Heat-sealed centrifuge tube. (B) Centrifuge tube sealer. (Courtesy Beckman-Spinco Co., Palo Alto, Calif.)

enclosures for a variety of essentially hand-performed operations, LaManna (19) enclosed individual operations, e.g., a Ten Broek grinder, a blender, and a mixing operation, in heat-sealed plastic film (Fig. 2). The "Stomacher" (Colworth Stomacher Lab-Blender; Cook Laboratory Products Division, Dynatech Lab Products, Inc., Alexandria, Va.), a plastic-bag blender, is a more recent example of this technique. Other very simple primary barriers include sealed, ventilated flasks for making shake cultures, safety waste discard containers, and, of course, the well-described containment procedures used for shipping etiologic agents described in the

Code of Federal Regulations (5). Although only a few examples of simple primary closures have been described, there is literally no limit to the application of such techniques. Many of these "primary barrier" units are described further in other sections. A very simple primary barrier, applied carefully, can often be extremely effective. Although there are many manufactured products available, there is still ample room for exercise of ingenuity and development of ad hoc containment techniques.

The blender and the sonic homogenizer are known to be two of the worst offenders with respect to aerosol generation. On the other hand, the centrifuge appears to be of less hazard. This may appear to be somewhat at odds with the history of laboratory-caused epidemics. As noted above, major accidents with centrifuges are rare, but when they occur the consequences can be severe. This is a direct indication that the centrifuge imparts a great deal of energy to the microbial suspension, and one of the major factors controlling the quantity and particle size of aerosol output is the energy input to the process. Because the primary containers used in a centrifuge may be subject to extremely high stresses, careful attention must be paid to the quality of their sealing. Some early tube closures depended on expansion of O-rings. These were not satisfactory. Today, most manufacturers produce effective closures, but some must be considered only a partial answer to the containment of the tube or cluster of tubes, and in such cases appropriately filtered chamber evacuation procedures should be used. Large bulk or zonal rotors and continuous-flow centrifuges are particularly difficult to seal, and extreme care should be taken in their use. The simple primary barriers described above can be effective, but one must also consider the possibility of major accident (e.g., rotor rupture). Work with large volumes of infectious agents may merit putting the entire centrifuge in a ventilated enclosure (11).

Regardless of the effectiveness of individual containment devices, it is usually necessary to have some kind of cabinet or enclosure in which to load and unload centrifuge rotors and containers. The safety hood or biosafety cabinet is most widely accepted for this purpose and is available in a variety of sizes and designs. It is perhaps the single most useful safety device in the microbiology laboratory and is second in importance only to the safe worker in maintaining control of the environment. Although the discussion below will deal primarily with three recognized "classes" of cabinets, many of the special cabinet enclosures available, particularly in plastic fabrications, can play a useful role in the laboratory. These include simple plastic bags with built-in glove access, flexible-film isolators, "static" enclosures often used on a bench top, filter-bonnet animal cages, and many other special application techniques that can often be used to supplement the simple devices described above.

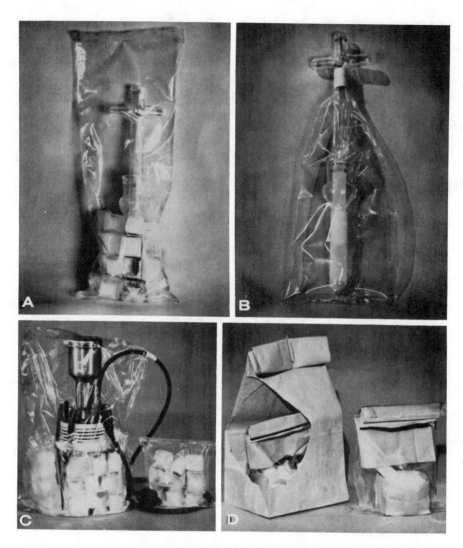

FIG. 2. Flexible, autoclavable plastic film enclosures for blender and grinders (A, B, C). (D) Autoclaving pack. (Courtesy U.S. Navy.)

BIOLOGICAL SAFETY CABINETS

Some of the early versions of the biological safety cabinets were variants of the familiar chemical fume hood. In the course of some years of use, it was observed by various workers that some of the requirements for the biological safety cabinet that differed from those for the chemical fume hood were: the need for a fixed view screen that would protect against splatter of large droplets; a requirement to close the cabinet for decontamination; and the ability to provide a flow of air through the cabinet with effective decontamination of the effluent airstream. The need for decontamination required minimizing the quantity of airflow through the cabinet, and a front access opening of 10 in. (25 cm) or less height was accepted. By 1960, there was broad agreement that a hinged, sloped-window cabinet with air drawn in the front and exhausted out the top through high-efficiency filters was almost a necessary adjunct to every laboratory working with pathogens. Several reviews have described the types of cabinets and rationales for their use (9, 13, 31). An early development in clean-air biological safety cabinets was reported in 1969 (1).

At present, three general classes of these cabinets are defined. These are: Class I, the open-front air inflow cabinet, usually with a fixed height opening and sloped view window; Class II, open-front, vertical airflow cabinets, of which there are several subtypes; and Class III, cabinets hermetically sealed with access through gas-tight air locks and work access through fixed, heavy-duty, arm-length rubber gloves. Table 1 provides a summary description of the essential characteristics of each of these classes of cabinet. A descriptive review has been published (27).

Biological safety cabinets are heavily dependent on the availability of the high-efficiency filter designated the high-efficiency particulate air (HEPA) filter, developed by the U.S. Army Chemical Corps, Atomic Energy Commission, and other workers in the World War II period. No less important is the instrumentation for measurement of the very high efficiency of these filters. Light-scatter photometers are used to demonstrate 99.97 + % collection of near monodisperse airborne particles of di-octyl-*n*-pthalate approximately 0.3 μm in diameter. Particles in this size range are stated to be the most difficult to arrest. The filters collect larger particules by simple interception and smaller particles, such as bacteriophage and animal viruses, by virtue of both impaction and diffusion collection.

In evaluating the parameters affecting containment characteristics of safety cabinets, Barkley (W. E. Barkley, Ph.D. thesis, University of Minnesota, Minneapolis, 1972) used a design "containment fac-

tor" of 10^5, as does the British Standard 5726 (7). It is convenient to use the reciprocal (e.g., 10^{-5}) as a "leakage factor." This is an expression of the fraction of aerosol in the cabinet that is expected to leak out. The factor 10^5 (or 10^{-5}) is a model number on which to base a cabinet design analysis, as done by Barkley, but it is also a practical estimate of expected cabinet performance. In fact, a range of leakage factors should be expected when there is activity in the cabinet (14, 30). The position of the operator, how far into the cabinet he performs his operations, and the rapidity of his movements all affect this leakage. Our opinion is that one should expect a range of 10^{-4} to 10^{-7} for open-front Class I cabinet leakage factors. The highest of these (10^{-4}) shows the effect of poor practice, e.g., work within 4 in. (10 cm) of the front and moving arms out of the cabinet, and the lowest (10^{-7}) requires good opening design and little activity at the workspace. Class II cabinets would have approximately the same range, whereas Class III cabinets should have factors of 10^{-8} or less.

CLASS I BIOSAFETY CABINETS

The earliest cabinets, as stated above, were Class I cabinets as diagrammed in Fig. 3. These cabinets

TABLE 1. General characteristics of biological safety cabinets

Cabinet type	Face velocity in lfpm (m/s)	Airflow pattern	Negative pressure, in in. of water static pressure (cm)	Exhaust efficiency
Class I, open front	75 + [a] (0.38 +)	In at front; out rear and top through filter	NA[b]	99.97% + for 0.3-μm particles (HEPA)
Class I, front panel without gloves	150+ (0.75+)	In at front; out rear and top	NA	Same as Class I open front
Class I, front panel with gloves	NA	In through openings and makeup HEPA filter	$\Delta P^c > 0.5$ (>1.25)	Same as Class I open front
Class IIA	Avg 75 (0.38)	±30% into front; ±70% recirculated through top filter; exit via HEPA	NA	Same as Class I open front; recirculating filter also HEPA
Class IIB1	Avg 100[a] (0.5)	±70% into front; ±30% recirculated through top filter	NA	Same as Class I open front
Class IIB2	Avg 100[a] (0.5)	±30 to 70% into front; ±70 to 30% in at top; exhaust through HEPA; no recirculation	NA	Same as Class I open front; top inlet air HEPA filtered
Class IIB3	Avg 100[a] (0.5)	Adjustable to provide B1 or B2 performance	NA	Exhaust and recirculation or makeup HEPA filtered
Class III	NA	Air inlets and exhaust through HEPA filters	$\Delta P > 0.5$ (1.25)	2 HEPA filters in series, 99.997±%

[a]Minimum, may be higher depending on design and use. These velocities are usually average for the opening and are measured by determining quantity of air exhausted and supplied.
[b]NA, Not applicable.
[c]ΔP, Differential pressure.

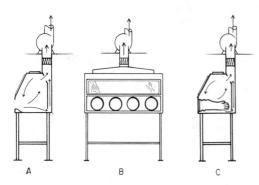

FIG. 3. Diagram of Class I open-front biological safety cabinet: (A) conventional use; (B) with armhole plate attached; (C) with rubber gloves attached to armhole ports.

depend on a flow of air into the front work opening, across the work surface, and out through a de-contamination device, usually a high-efficiency filter, and an exhaust blower. The cabinet can provide good protection of the operator from the work and allows use of burners, small centrifuges, and other equipment without seriously degrading the containment effectiveness. The cabinets may be constructed of stainless steel or fire-resistant reinforced plastic, with glass or clear optical-grade plastic for view windows. They are usually available in lengths from 3 ft (0.9 m) to about 6 ft (1.8 m). Materials and equipment may be moved in and out through the front opening, through a hinged view window, or via air lock doors added to the cabinet end. There have been extensive tests and discussions of the required input air velocity needed for safe operation of these cabinets. Inlet air velocity is a grossly inadequate measure of containment effectiveness; nonetheless it is quoted in the commercial literature. The current consensus is that inlet air velocities from 75 to 125 linear feet per minute (lfpm) (0.38 to 0.63 m/s), depending on inlet design, provide optimal operator protection. Such a cabinet can facilitate day-to-day operations because work access is excellent. However, this ease of access tends to encourage operators to make rapid motions at the front of the unit and to move their arms in and out of the cabinet; these kinds of motions, plus those of room-air currents and drafts at 60 lfpm (0.25 m/s) or higher (many room ventilating systems provide from 75 to 150 lfpm [0.38 to 0.76 m/s] air velocity in the workspaces), can degrade the containment provided by the inflowing air.

There is general agreement that the cabinet should have an interior rear baffle to provide a smooth airflow across the work surface while permitting some air to be removed from the upper section. The front opening design is also important, and the user should ensure that this aspect of design has been resolved satisfactorily. The addition of an air foil or suction slots at the opening, as shown in Fig. 4, can be effective. The Class I cabinet is an example of the

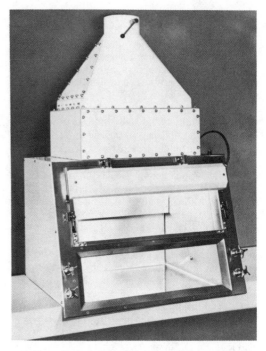

FIG. 4. Class I biological safety cabinet with airfoil formed front entry. (Courtesy Baker Co., Sanford, Maine.)

use of directional airflow, compared to the popular concept of "negative pressure," for aerosol capture and containment, since the degree of negative pressure immediately inside such a cabinet is probably less than 0.010 in. (0.025 cm) of water negative to the room atmosphere. In fact, the thermal pressure head caused by a large-sized laboratory burner can overcome this differential pressure and cause backflow out the front opening if a portion of the ventilating air is not removed near the top of the cabinet.

Performance of this cabinet can be substantially improved by the addition of a closure panel bearing circular ports for arm holes, as shown in Fig. 3B. The panel need not be airtight. Since the blower is not changed, the resultant air velocity through the arm-hole ports may be increased to 150 to 250 lfpm (0.74 to 1.25 m/s). Because the area of the opening (and periphery) is substantially reduced, the turbulence created by this high-velocity airflow is not critical, and substantially less leakage usually occurs. Additional containment can be afforded with this cabinet by use of arm-length rubber gloves added to this closure panel (Fig. 3C). In this case, the airflow is now restricted only to the leakage points of the cabinet and the cabinet may become substantially negative in pressure (ca. 2 in. [5 cm] of water static pressure) with very little exchange rate of air unless inlet air pressure relief (usually through HEPA filters) is provided.

Class I cabinets offer no protection of the work from the operator or the environment, and in a laboratory that does not supply clean air, or in a cell culture operation wherein the contamination from the worker himself may affect the work product, they may be unsatisfactory. However, they are simple and economical, are easily installed, can be used with radioisotopes and some toxic chemicals, and can be adapted in various forms to meet unique needs of any special process. For example, it is not at all uncommon to see modifications of these devices added to the tops of centrifuges (6), as enclosures around fermentor devices, or in work involving both biological and chemical agents of low to moderate risk.

Most Class I cabinet installations incorporate a high-efficiency filter in the cabinet or in a sealed housing immediately before the exhaust fan which provides discharge (usually to the outdoors). An air incinerator may supplement or be substituted for the air filter. However, the use of an incinerator for exhausting this type of cabinet is not usually warranted. Activated carbon or other chemical treatment filters may be used in conjunction with the HEPA filters if the work involves significant quantities of carcinogenic or toxic materials.

The installation requirements for this type of cabinet are similar to those of the Class II cabinets described below and will be discussed subsequently, as will decontamination and general use procedures, many of which are the same.

CLASS II BIOSAFETY CABINETS

The major difference between the Class I and Class II cabinets, of consequence to the user, is the fact that the Class II (vertical laminar-flow biological safety) cabinet can afford protection both for the operator and for the work. The practicality of construction of this type of cabinet stems from the observation that a stream of air at approximately 100 lfpm (0.5 m/s) forced through a HEPA filter provided a particle-free environment for several feet downstream of the filter if there were no obstructions. This has been termed a "laminar-flow" (i.e., essentially nonmixing) airstream and is used in "clean rooms" and in "clean benches." Laminar airflows are widely used in the preparation of precision microelectronics and machine parts needed for the U.S. space program.

The horizontal-flow clean bench (Fig. 5) has been used for some years by workers in cell culture laboratories to provide near-sterile work area conditions. However, it does not provide protection for the operator, and in fact can expose the worker to aerosols of proteinaceous materials. This type of cabinet can be useful in microelectronics fabrication, in the hospital laboratory for final preparation of parenteral solutions, or other applications where the product is unlikely to have any ill effect on the cabinet user. However, it is considered unsuitable for microbiological laboratory work. The simple vertical-flow

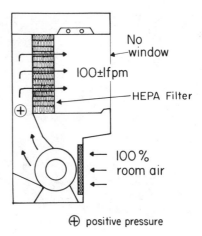

FIG. 5. Diagram of horizontal laminar-flow clean bench.

cabinet, utilizing that same basic principle of "piston" flow of clean air, overcomes this difficulty while maintaining good protection for the work. The complexity of airflow for the fully enclosed vertical-flow cabinet often makes laminar airflow impossible or even unnecessary. However, this terminology is widely used in descriptions of Class II cabinets. In this section it will be interpreted to indicate an airflow of clean filtered air over the work surface with minimal mixing with the airstream coming into the cabinet via the work opening.

Two general subtypes of Class II cabinets are available. They are designated IIA and IIB in the United States National Sanitation Foundation (NSF) Standard no. 49 (NSF-49) (22) and the British Standard no. 5726 (7). In general, the Class IIA cabinet maintains a minimum of 75 ft/min (0.4 m/s) inflow velocity through the work opening and recirculates a major part of the air traversing the work surface (ca. 70%), whereas the Class IIB cabinet maintains an inlet flow velocity of 100 ft/min (0.5 m/s) and exhausts most or all of the air traversing the work surface.

Class IIA biosafety cabinets. Figure 6 shows a typical Class IIA cabinet. The airflow characteristics of these cabinets are generally as shown in Fig. 7. The air drawn into and over the blower and then, under pressure, up to the recirculating or exhaust filters and through the exit filter is contaminated both from the work and from the room. Therefore this air plenum must be airtight and leak proof. NSF-49 calls for a halide gas (e.g., R-12 refrigerant) test of this plenum, allowing less that 10^{-7} ml/s of mixed air and gas to escape while the plenum is under a positive presure of 2 in. (5 cm) of water static pressure. Because a substantial fraction of the air in the cabinet (up to 70%) is recirculated through the filters, the type A cabinet is generally not considered

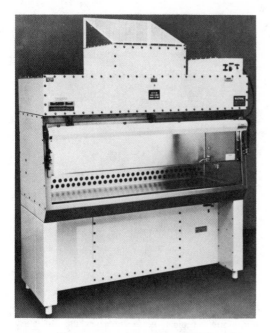

FIG. 6. Class IIA biological safety cabinet. Exhaust arranged to discharge to laboratory. (Courtesy Baker Co., Sanford, Maine.)

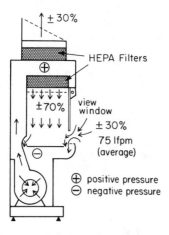

FIG. 7. Simplified diagram of Class IIA biological safety cabinet airflow scheme.

blower in the cabinet forces the air both through the recirculating air filter and the exhaust air filter, and a careful balance must be achieved to adjust the work opening inlet air to the desired velocity.

The concept of sealing the type A positive-pressure plenum Freon tight is good, but testing it is a difficult task. An alternative to guaranteeing that the positive-pressure plenum is leak tight is to surround the plenum with a negative-pressure area, as shown in Fig. 8. The performance characteristics of this type of cabinet are equal to those of the conventional Class IIA cabinet. At the confluence of the down-flowing stream of air over the workplace and the air entering the work opening, there is a joining of two airstreams that may be of substantially different velocities. This offers some difficult aerodynamic problems, and this cabinet can suffer from the same problems of turbulence around the work opening as the simple Class I cabinet has (12). Its containment characteristics are approximately the same, but if the cabinet is vented directly to the room, an overall leakage factor of 10^{-5} (exhaust filter penetration)

suitable for use with high-activity radioactive materials or with toxic or carcinogenic chemicals. The essential elements of Class IIA cabinets are HEPA-filtered laminar-flow recirculated air, traveling downward over the work surface, air inlet into the front with immediate conveyance away from the work surface, and discharge of excess air from the cabinet via a HEPA filter to the room or outdoors (Fig. 7 and 8). As the simplified diagram shows, the

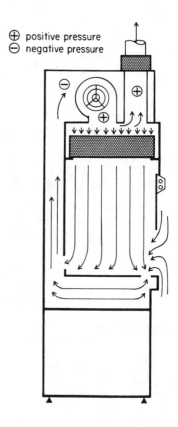

FIG. 8. Simplified diagram of Class IIA biological safety cabinet with positive-pressure plenum enclosed within negative pressure space.

may be expected. The cabinet is particularly sensitive to interruption of the vertical airflow by gas burners. Use of disposable sterile transfer loops or electric heater sterilizers is recommended (2).

Class IIB biosafety cabinets. At this time there are three subtypes of these cabinets, designated B1, B2, and B3. The Class IIB1 cabinet (Fig. 9) has filtered downflow air that is composed largely of uncontaminated recirculated inflow air, exhausts this air directly out of the cabinet after HEPA filtration, has all contaminated ducts and plena under negative pressure, and has an inflow air velocity averaging 10 lfpm (0.5 m/s). The cabinet can be useful for microbiological work and for work with low-level radioisotopes and limited amounts of toxic chemicals. However, the degree of air mixing and recirculation in the cabinet requires that use of such materials be restricted to levels not considered toxic to the work product. Further, this class of cabinet will not usually meet the air inflow standards for work with carcinogens. The type IIB2 "total exhaust" cabinet (Fig. 10) is similar in design, but has all air entering the cabinet making only one traverse through the cabinet before being discharged through a HEPA filter. The work opening inlet air velocity averages 100 lfpm (0.5 m/s) or higher. This air is prevented from contaminating the work by a protective flow of HEPA-filtered room air entering the top of the cabinet. The diagram shows enclosure of the positive-pressure exhaust duct within the negative-pressure volume established by the exhaust fan suction. Class IIB2 cabinets are designed to be used for work with limited quantities of toxic chemi-

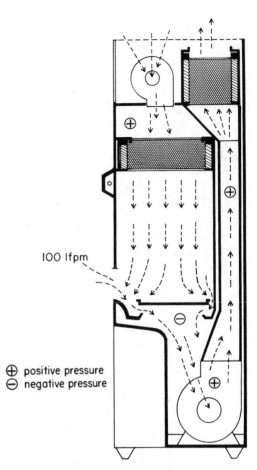

FIG. 10. Simplified diagram of Class IIB2 biological safety cabinet. (Adapted from Germfree Laboratories, Miami, Fla.)

cals or radionuclides required in microbiological studies. At least one manufacturer states that his total exhaust cabinet will meet Environmental Protection Agency and Occupational Safety and Health Administration Standards (125 to 150 lfpm [0.6 to 0.75 m/s] inlet velocity) for work with chemical carcinogens. Cabinets of this design meet NSF-49 standards for biocontainment and product protection. Biological leakage factors have not been established, and data on containment of volatile chemicals or of inlet air velocity profiles are not available as of this writing. If air velocities (downward and inward) are maintained similar to those in the IIB1 configuration, the containment performance should be equal. The Class IIB3 cabinet definition by NSF is essentially that of a type IIB2 cabinet convertible to a B1 cabinet by arrangement of ducts and dampers. In use it must be considered equivalent to a IIB2 cabinet (total exhaust) or a IIB1 (some recirculation) cabinet. Comments in reference to these two types are also applicable to the IIB3 cabinet.

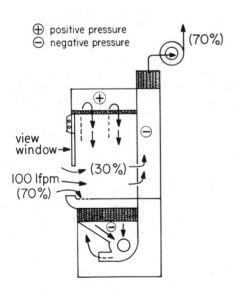

FIG. 9. Simplified diagram of Class IIB1 biological safety cabinet showing airflow directions. Note requirement for external exhaust blower.

A major drawback to the Class IIB cabinet is the requirement for pass-through of relatively large quantities of room air and subsequent discharge to the atmosphere. As is done with fume hoods, some unconditioned air may be supplied directly from outdoors by separate ducting. This can be expensive and difficult to accomplish. Another drawback common to most Class IIB cabinets is that they must have at least two fans (supply and exhaust) operating in balance, that is, with the exhaust always exceeding the input to provide the necessary work opening inflow and negative pressure within the cabinet. Considering that the exhaust and supply filters are subject to differing rates of dirt loading, airflow at the inlet can vary with usage. This added complication in installation and set-up should be examined by the prospective users of these cabinets.

It is essential that there be little or no leakage through the recirculating or exhaust air filters. Many of the recirculating filters are large (as big as 24 by 72 by 8 in. deep [61 by 183 by 20 cm]) and are extremely fragile. For many years the incidence of failure due to damage to these filters after shipping was quite high. This problem has been reduced by improving handling and design of the filter mounting, but the possibility of such leakage after movement of a cabinet persists. Figure 11 shows details of typical HEPA filter construction. The filter medium is installed in the frame in a pleated manner, with corrugated metal or fiber separators. It is fragile, being very thin, and is subject to puncture or cracking by handling or shock in shipping. The paper is usually attached to the edges of the frame by a cement that hardens. The paper becomes highly stressed at the point of attachment, and dropping the filter, or handling it with any degree of roughness, causes cracks at this juncture. In many cases, these small cracks are not readily visible and cannot be located by simple inspection. The procedure for testing is to generate a fine-particle (0.1- to 3.0-μm) di-octyl-n-phthalate aerosol upstream of the filter and to measure penetration of particulates immediately downstream by using a light-scatter type aerosol penetrometer to scan the face of the filter in a pattern ensuring coverages of all edges and pleats. The edge gasket, on which the filter is seated within the frame in the cabinet, has also been a source of many leaks and can be tested in a similar manner. Leakage greater than 0.01% at any one point, measured by the di-octyl-n-phthalate/penetrometer method, is unacceptable. If the leaks in the filter or gasket are small, they may be repaired readily by the use of room temperature-vulcanizing silicone sealant (Dow-Corning). The edge groove seal shown (Flanders Filters Inc., Washington, N.C.) is used with the groove filled with a silicone grease meeting the on-edge mating flange. It can be very effective, particularly in large installations, but should be used with caution in locations where the grease can become contaminated.

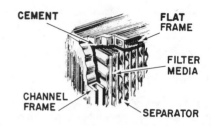

FIG. 11. Detail of typical HEPA filter construction. Two types of edge seals are shown.

Both the circulating air velocities and the inflow/exhaust air rates are important in the Class II cabinets. For the most part, the recirculating air velocity is affected by the pressure drop across the HEPA supply filter at the top or bottom of the cabinet and by dampers or deflectors. The velocity of the vertical downflow and the incoming airstreams may be adjusted by setting the correct speed of the exhaust and recirculating blowers or resetting dampers. The measurements are made with an air velocity meter (anemometer) with suitable sensitivity. It should be noted that the front inlet air velocities cannot usually be measured directly in the Class II type cabinets; this is customarily done by measuring the quantity of air exhausted out of the cabinet and then, by calculation, determining the average inflow air velocity through the front opening. The velocity profile of downflow air throughout the cabinet should be within ±10% of the manufacturer's stated performance, but it is often a function of the cabinet design and cannot be adjusted readily in the field. Exceptions would occur if a tester or operator were to permit an object to block part of a filter or blower. (Paper towels have been retrieved from blower inlets on more than one occasion.)

It has been our observation that one other criterion, often not tested in this particular set, can be of equal importance, and that is the degree of turbulence around the work opening. It is reasonably simple to use a titanium tetrachloride smoke stick (Mine Safety Appliance Co., Pittsburgh, Pa.) moved slowly around the periphery of the front opening to determine and ensure that the air is flowing smoothly into the cabinet. The smoke stick may be held and traversed a few inches above the work surface so that, again, the desired direction of air is observed. There are other tests, e.g., vibration and noise, illumination intensity, etc. The tests described here and above, including the details of the certification test used by the manufacturer, are fully described in NSF-49 (22). Field testing will be described further under Selection and Use Procedures, below.

SELECTION OF CLASS I AND II SAFETY CABINETS

In selection of one of the Class I or Class II safety cabinets, there are several major factors to be considered. These include the operations to be conducted in the cabinet; the classification of the etiologic agents to be used; the protection required for the work product; the possibility of use of radioisotopes or toxic or carcinogenic material in the course of the work; the funds available; and the need to use cabinets that meet accepted standards for housing the work to be performed. While cost is very often a strong factor in cabinet selection, the purchaser should be aware of the "marginal cost" of improvements or additions to the cabinet that may provide increased flexibility in its use and ensure a longer life and wider applicability in the laboratory. For example, the 4-ft (1.22 m)-long cabinet appears to be a fairly standard length for many laboratories. It is suitable for a single operator workplace; however, a recommended procedure in the use of a cabinet is to minimize the movement of the operator's arms in and out of the cabinet. This can only be done by loading in as much of the work material as possible, e.g., dilution blanks, petri plates, test tubes, etc., at the start of the use period, and using it in a reasonably well-laid-out work pattern (2). This frequently requires some extra space so that the material does not interrupt the airflow over the work area. Therefore, it is recommended that if space and funds permit, a 6-ft (1.8-m) cabinet should be acquired. The cost for the increased size is not usually proportional to the increase in space available, and a 6-ft cabinet may be big enough to provide workspace for two people.

Cabinet designs can be varied to meet special needs. One of the rather specialized kinds of use that can be met by cabinets is that of small-animal necropsy. Cabinets are available with two-sided openings, facing each other. These are often very helpful when one operator can prepare the animal and the other can retrieve the desired samples.

In general, the criteria shown in Table 2, and in a National Institutes of Health slide-cassette package (5), provide a fair guide for a rational selection of cabinets. It is perhaps unfortunate that the Class I

cabinet has not been receiving as much attention from manufacturers and distributors as the Class IIA and B, because it is intrinsically a low-cost, simple device and affords adequate operator protection for workers using moderate-risk microbial agents and for animal work. Furthermore, it can be useful for toxic or carcinogenic chemicals and radioisotopes with the proper decontamination of exhaust air. The disadvantage that it permits contamination of the work by room air may not always be of major concern. Newsome (23, 24) has shown that a well-engineered cabinet gives some degree of product protection, and the "Code of Practice" of the Howie Working Party in England (16) strongly recommends the Class I cabinet for routine bacteriological work. Nonetheless, the Class II cabinets have achieved rapid acceptance in the laboratory community in the United States, and a good variety of designs and competitive pricing structures are currently available. With the considerations of space, as described above, having been made, the choice between a type A and a type B cabinet can usually be made on the basis of the need for use of radioisotopes, toxic chemicals, or carcinogens in the course of the work, the type B being the desired cabinet to cope with these hazards. It should be noted that the original type IIB cabinet is useful only for very low-level, short-lived radioisotopes and limited quantities of toxic or carcinogenic material. These limitations can be relaxed substantially by use of the Class IIB2 cabinet, which provides a single pass of 100% room or outside air through the workspace. The IIA and the IIB cabinets offer nearly equal protection for the work. One factor that must not be overlooked in selection of safety cabinets is the requirement for exhaust air. It is apparent that if a 6-ft (1.8-m)-long Class I or Class IIB2 cabinet is installed in a room, approximately 200 to 900 ft³/min (0.1 to 0.5 m³/s) of air will need to be exhausted from the room. Smaller quantities are needed for some Class II cabinets, but the volume of air to be exhausted can be substantial. This is often amelio-

TABLE 2. Exhaust requirements for biological safety cabinets

Type of cabinet	Average inlet velocity in lfpm (m/s)	Opening height in in. (cm)	Exhaust air, ft³/min (m³/s)	
			4-ft cabinet (1.2 m)	6-ft cabinet (1.8 m)
Class I	75–100 (0.4–0.5)	9 (23)	225–300 (0.1–0.15)	350–450 (0.17–0.23)
Class II-A	75 (0.4)	8–10 (20–25)	250± (0.1+)	400± (0.15+)
IIB1	100 (minimum) (0.5)	8 (20)	275 (0.13)	400± (0.2)
IIB2	100 (minimum) (0.5)	8 (20)	700 (0.35)[a]	1,050 (0.53)
IIB3	100 (0.5)	9 (20)	260 (0.13)[b] 700 (0.35)	400 (0.2) 1,050 (0.53)
Class III			1 to 10 ft³/min[c] per ft of cabinet	

[a]Total, including front inflow and filtered downflow air. Quantities will vary with actual dimensions of cabinet and work opening.

[b]Adjusted.

[c]May increase to minimum of 100 ft³/min (0.05 m³/s) in the event of glove or air lock failure.

rated with the type IIA by exhausting directly into the room; this will be discussed under Installation Procedures, below.

The use of a statement that any Class II cabinet acquired for work with low- to moderate-risk agents should, as a minimum, meet the standards of NSF-49 (22) is perhaps the simplest way to ensure that a purchaser will acquire a cabinet with some assurance of its performance and reliability. This may be particularly important when one considers that the purchaser is often the responsible individual in the laboratory, buying the equipment for use by co-workers. In so doing, he accepts some responsibility for their occupational safety. This liability can press heavily if other than fully certified equipment is employed. However, the NSF Standard is a minimum set of requirements and should not be used as a purchase specification. In the acquisition process, the proposed use of the cabinet should be described clearly to the suppliers or manufacturer. Most manufacturers are anxious to ensure that they are furnishing equipment adequate to meet the needs of the user. It is not unreasonable to ask for test data demonstrating performance claims and clear statements that the equipment proposed can provide the needed product and operator protection. The purchaser should also require that the supplier of the cabinet will ensure that the cabinet will be put into place and demonstrated to be operating in accordance with the stated performance specifications before the purchase commitment is completed. Most manufacturers do not provide full installation services but can make arrangements with local servicing organizations to provide installation assistance and in-place testing. This is considered an essential part of any purchase order unless the user has the full mechanical installation and test capabilities in hand and is willing to accept responsibility for performance of the cabinet "as received" at the loading dock. Damage in shipping to these units is not infrequent, components are often missing, and controls or dampers have become misadjusted during shipment more than once. On-site inspection and initial "field" certification are essential. Similarly, provision for periodic retest should be discussed with the supplier.

In summary, the selection and acquisition procedure for Class I and II cabinets should include the steps listed below.

(i) Establish the class and type of cabinet to be used, commensurate with the operator and product protection required. The classification of microbial agents used and definition of the tasks to be done should be described in a "proposed application" statement in the purchase document. The supplier should provide statements and test data, if possible, in substantiation of the suitability of the equipment proposed.

(ii) Establish the size of the cabinet based on the number of workers using it, the space available, and the complexity of the work process (i.e., amount of supplies and material being loaded for a given operation).

(iii) Evaluate both current and future needs. Recognize that early stages of exploratory work frequently require high levels of contamination protection due to lack of knowledge of the toxicity or pathogenicity of the work material. In later phases, it may be more desirable to ensure a lack of contamination of the work material, in which case a cabinet that affords good worker protection and product protection can be of good value and continuing benefit to the user.

(iv) Ensure that the size of cabinet selected can be brought through the various doorways to its final use point and that there is adequate headspace for installation of filter boxes, exhaust ducts, fans, etc. Exhaust ducts, fans, and other connection needs should be considered.

(v) If the work is to be done with moderate- or higher risk material, it is well to consider a cabinet certified to national standards (NSF-49 for biological; Occupational Safety and Health Administration and Environmental Protection Agency for chemical) for work with potentially or known hazardous materials.

(vi) Specify that the cabinet is to be purchased with the condition that acceptance will be based on satisfactory acceptance tests in place, in its final location. (This often entails an increased cost, but there is little choice in the matter if one is to meet the criteria of no. v above.) Assistance in interlocking room ventilation blowers and providing vent failure alarms may also be needed.

(vii) The manufacturer should provide complete circuit diagrams, operating instructions, maintenance instructions, spare parts listing, and a listing of expendable replaceable parts, e.g., filters, that are available on a continuing basis or from other recognized specialty manufacturers.

(viii) Stay within a budget. This is included as the last item because if one cannot meet the technical requirements of the operation within the budget, then either the operation must be changed or the budget must be changed. The biological safety cabinet is a basic device, useful in Biosafety Level 2, 3, and 4 or equivalent laboratories. Its cost should be considered an essential part of the cost of the laboratory project.

CLASS III CABINETS (GLOVE BOXES)

For the most part, the Class III cabinet system comprises a hermetically sealed cabinet system, suitable for extremely hazardous work, e.g., in containment laboratories meeting Centers for Disease Control Biosafety Level 3 and 4 requirements (28). The cabinets are hermetically sealed, and all operations within the cabinet are conducted through attached arm-length rubber gloves. Entry into the cabinet is usually through a sealed air lock, and exit of material may be through an autoclave, a decontamination-type air lock, or a "dunk tank" filled with liquid disinfectant. Figure 12 shows a Class III cabinet system installation. These cabinets are often built in a modular basis and assembled into specialty lines or systems encompassing a full set of operations in the laboratory (18). Ideally, one should be able to put all of the necessary raw materials into the cabinet system, conduct the work, and remove only waste products and hard data. For example, some cabinets have been made for such uses as animal inoculation by syringe or aerosol challenge; others may accommodate centrifuges, incubators, refrigerators, and other equipment. Most such cabinet assemblies are made of stainless steel, although some are of plastic. The latter are often used for controlled-atmosphere protective systems, e.g., anaerobic chambers, germfree animal isolators, etc. Class III cabinets or "glove boxes" may be provided with strippable or removable liners and additional shielding if the work involves the use of high-activity or long-life radioisotopes. Class III cabinet installations are expensive. It is difficult to become accustomed to using these cabinets because of both the preparatory work required and the difficulty in working through arm-length gloves. However, when a very high level of protection is required for the operator and the environment, these cabinets can be extremely useful, and with proper training an operator rapidly becomes accustomed to the limitations afforded by working through fixed gloves. The gloves provide both aerosol containment and protection from hand and arm

contamination, which can be a major source of contamination release from Class I or II cabinets.

Although there are commercially available modular units and a variety of plastic special-purpose chambers, many of the Class III cabinet installations will be fabricated or assembled to order, and considerable special engineering is involved. Some of the design factors that must be considered in selection and use of these cabinets are as follows.

(i) Ventilation. The ventilation rate of these cabinets may be varied to minimize cross contamination inside, but is usually minimal, although at least one manufacturer produces a unit offering near-laminar flow, similar to a Class II cabinet, within the Class III system (Baker Co., Sanford, Maine). Air enters these cabinets through at least a single HEPA filter and is exhausted through specially tested ultra-high-efficiency HEPA filters (99.999%) in tandem, an exhaust incinerator, or both. It is usual to maintain approximately 0.5 to 0.75 in. (1.3 to 1.9 cm) of water negative pressure inside the cabinet. More pressure tends to make the gloves distend and lose feel. Any blowers selected should be arranged so that they provide inlet air velocities of at least 100 ft/min (0.5 m/s) through any glove port in the event of glove rupture, or through any air lock entry in the event both doors of an air lock become opened at the same time. The air exhaust blowers from these cabinets are often equipped with dual fans, each of which is adequate to maintain the necessary pressure in the system. The fans may be electrically interlocked to

FIG. 12. Class III biological safety cabinet line.

provide increased airflow in the event of glove failure or other inadvertent opening to the system.

(ii) Accessories. A Class III cabinet is rarely used by itself, but should incorporate the necessary equipment for conduct of the work. If a Class III cabinet is used in conjunction with a centrifuge, it should enclose the whole centrifuge if the material to be used warrants Class III containment (11). A cabinet system mounted only on the top of a centrifuge, refrigerator, or other device is subject to leakage of the barrier construction of the centrifuge to which it is attached or possible failure of the centrifuge chamber or connecting piping (vacuum system).

(iii) Access. The cabinet system should have entry and exit locks, a double-door autoclave (preferably with hydraulic or electrically operated doors), and chemical dunk tanks or fumigation chambers. A series of very effective entry/exit air lock chambers designed to exchange with a standard glove port hole is available (Central Research Laboratories, Inc., Redwing, Minn.).

(iv) Containment requirements. If the cabinets are to be used for work on Centers for Disease Control Class 3 or 4 etiologic agents (8), they should be tested to be gas tight under slightly positive pressure. The customary test is to use R-12 halogen refrigerant (1 oz per 30 ft^3; 33.4 g/m^3) at 3 in. (7.5 cm) of water static pressure and requires that no leak equivalent to more than 10^{-5} ml of halogen occurs per s at any point in the system (21) (10^{-7} ml of mixed air and R-12 per s). It is a difficult standard to meet, but the equipment is available and if the operations proposed warrant use of the Class III cabinet, the test must be made. If tandem or special entry/exit HEPA filters are used, the exhaust filter leakage factors can be expected to be 10^{-8} to 10^{-9}. Air exhausts should be directed clear of occupied spaces. Although Class III cabinets are usable in both containment and high-containment facilities, their use in a containment (Biosafety Level 3) facility must be given careful consideration because it is apparent that a Biosafety Level 4 agent should be confined entirely within the cabinet system or a secure container. Thus, in a Biosafety Level 3 laboratory it is desirable to conduct as much preparation as possible outside the cabinet under less highly contained conditions and to conduct all open manipulations with the pathogens inside the cabinet. The possibility and consequences of a major failure of the system or transport containers should be considered. This factor alone may make consideration of Biosafety Level 4 facilities desirable.

Because the selection and installation of Class III cabinets is a highly specialized procedure, the generalized material provided above is only an indication of important factors in selection of equipment. However, the reader is cautioned to scan the cited literature well and to define the work requirements very carefully before embarking on development of such a cabinet installation. At that time, consideration

should be given to a "suit protection" area as described below.

BIOSAFETY CABINET INSTALLATION RECOMMENDATIONS

The installation of Class III cabinets is highly specialized. While it is possible to use only a single element of Class III modular cabinetry, such equipment is usually installed as a "system," and specialized design requirements often include use of continuous spaces for animal holding and other activities, as described below. It should be noted that the space within which a Class III cabinet system is used must be suitable for containment in event of failure of the cabinet. This is discussed in Chapter 17.

Class I and II open-front cabinets are more frequently used in Biosafety Level 2 and 3 laboratories under a variety of conditions. They are basic tools for use in the microbiology laboratory. As discussed in Chapter 17, the best location for such cabinets is at the end of a U-configuration, where there will be a minimum of cross-traffic in front of the work surface to interrupt the airflow or to disrupt the operation, and at the same time work bench space will be available at either end for materials. Positioning the workspace against an outside wall permits ready installation of duct work to the outside. An inside wall adjacent to service chases can permit connection to ventilation exhausts or a duct to the roof. It is important to avoid high-velocity air drafts from room air inlets, either in the ceiling or walls or from air conditioners in windows, or even such items as swinging doors which create transient, high-velocity airflow across the face of the cabinet. Leakage, both into and out of the cabinets, has been shown to be proportional to the velocity of air crossing in front of the cabinet (25). Class I cabinets can be exhausted through a HEPA filter to the laboratory. However, a direct outdoor duct will permit use of the cabinets for chemicals and radioisotopes and is the preferred installation.

Class IIA cabinets do not specifically require ventilation of the exhaust directly to the outside. These cabinets, if they have blowers and exhaust filters incorporated within them, may be exhausted through the filter into the laboratory room. However, this is not considered the best practice. Exhaust filters occasionally develop leakage. Furthermore, the user may wish to decontaminate the cabinet with formaldehyde vapor. Discharge and ventilation of the waste formaldehyde vapor from the cabinet is considerably easier when the cabinet has an exhaust to the outdoors. The cabinet should have its own exhaust blower (in IIA this may be the main blower) and be exhausted directly outside with an anti-backflow damper in the exhaust to ensure that there is no flow back through the filter and hood into the "negative-pressure" laboratory when the hood is shut off. The exhaust should be run to an area clear of and at least 7 ft (ca. 2.1 m) above the roof line, so that workmen do not come near the outlet. It should not discharge out into a courtyard or where it may be drawn into other parts of the building. In some cases it may be possible to exhaust the cabinet into a building exhaust system that does not recirculate to other parts of the building. This is frequently done with a loose connection to the exhaust called a "thimble piece" (Fig. 13), or variants thereof, as shown in NSF-49 (22). The user is cautioned to recall that decontamination will require that the cabinet exhaust be substantially blocked during decontamination gassing. If necessary, even a "hard" connection can be used if the duct-work system is dedicated to a limited number of cabinets or exhaust systems and suitable provisions are made to prevent backflow and to shut down the cabinet in the event no exhaust flow is present.

With space at a premium in most microbiology laboratories, it is often difficult to ensure that there is adequate room for removal and exchange of the in-cabinet and exhaust air filters from the Class I and II cabinets. Provision must be made for ready access by maintenance men and recertification personnel who will be required to inspect the cabinet periodically.

Class IIB cabinets may require connection to a separate exhaust system because many such cabinets are not furnished with their own exhaust blowers. Even if the cabinet has its own exhauster, the fact that some radioactive or toxic chemicals may be used requires that these agents be discharged clear of occupied spaces. Most toxic chemical collectors that are available for use within the cabinet (carbon, molecular sieve, etc.) permit penetration as they load up, in contrast to the HEPA filter used for purely particulate microbial burdens. The chemical filters are difficult to test to determine effective use life remaining. Pyrolysis or other destruction may be required for some toxic chemicals. This caveat is the most obvious installation problem for IIB series cabinets. The requirement to provide sufficient inflow of air and lack of cross-drafts is much like that of the Class I cabinet except that the cabinet may use more makeup air. The effect on room ventilation

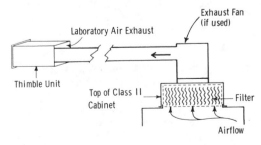

FIG. 13. Suggested Class II biological safety cabinet venting connection.

balance is similar to that described for Class I cabinets.

BIOSAFETY CABINET USE PROCEDURES

The installation of a Class I or Class II cabinet within a laboratory is itself an indication that careful work practices are needed. The cabinets are not substitutes for good practice. They can only complement a careful worker. Correct operating procedures are described in a National Cancer Institute filmstrip (2) slide-cassette set. The operator should wear a closed-front overgarment, e.g., surgical gown with full-length sleeves and surgeon's rubber gloves. Use of bare hands and a button-front laboratory coat in the cabinet for typical operations is not advised. For many procedures that are considered only minor aerosol-producing operations, there still can be hand, arm, and surface contamination. The cabinet will remove aerosols rapidly, but it is up to the user to dispose of spilled contamination.

It is imperative that most of the material for a given operation be placed in the cabinet before the work is initiated, both to minimize in-and-out motions and to permit working in an efficient manner. Inoculation of small animals is an excellent case in point. Figure 14 shows a mouse inoculation procedure being done in a Class I cabinet. Several points are worthy of note. The operator is wearing a surgical gown and gloves; there is a disinfectant-soaked towel on the work surface of the cabinet; pipettes and syringes are all available within the cabinet, as are shallow discard trays containing disinfectant for receipt of the waste syringes and material. It is apparent that it may be necessary to bring in more cages of animals. This procedure is typical of many done in the cabinet, and unavoidable in-out motions should be minimized and done with extreme care. A nearby cart or bench (and suitably clothed helper) are most useful.

In Class II cabinets, the operator should work well within the cabinet and not out close to the front. Substantial leakage from the cabinet can occur when the work is being done within 4 in. (ca. 10 cm) of the cabinet opening. The condition is almost unique to Class II cabinets because of the relatively low inlet air velocities at the top edge of the work opening and the air removal slot or grill immediately along the front of the work surface of some cabinets.

Provision must be made for decontamination of waste materials, discard of pipettes and supplies, and disposal of excess material. In addition, the cabinet itself will need periodic decontamination as described below. The cabinet, in any case, should be decontaminated to the degree possible with an effective liquid disinfectant at the end of each work operation, or at least at the end of each work day. Most cabinets of stainless steel or durable plastic will withstand use of 500 to 5,000 ppm sodium hypochlo-

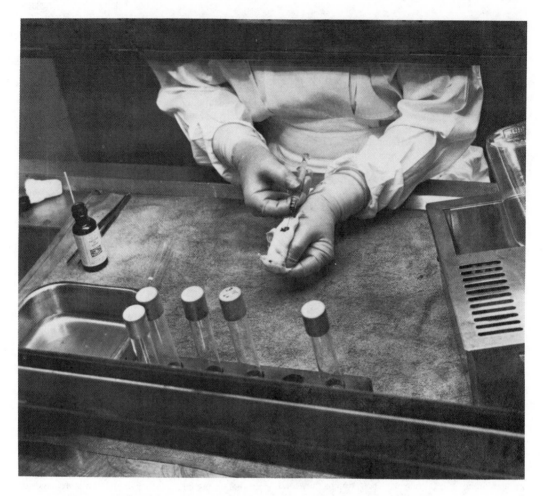

FIG. 14. Inoculation of a mouse in a Class I cabinet. (U.S. Navy photo.)

rite (10:1 to 100:1; Clorox) periodically. However, this material is corrosive, and other disinfectants, e.g., 70% ethanol, quaternary ammonium compounds, etc., may also be satisfactory. Many disinfectants have an effective residual decontaminating action. The subject of decontaminant selection and use is treated in Chapter 22 of this volume.

UV irradiation may also be used inside the cabinet. Its benefit is primarily for decontamination of physically "cleaned" surfaces after the work use. UV is not an effective mode of purification of the air within a cabinet because the time of air contact with the UV tube is usually low. It is not recommended for most installations (8), but it can be useful in disinfecting susceptible bacteria and viruses from surfaces. If the cabinet has a stainless-steel interior, a light placed at almost any location within the upper part of the cabinet will bounce 2,537-A (ca. 254-nm) illumination around the inside (and out the opening) of the cabinet quite thoroughly. Care should be taken to ensure that the materials of the view window will

withstand the UV and ozone and that the operator and others in the room are adequately protected against UV illumination because it can cause painful eye burns.

When a cabinet is in use, particularly in a small laboratory room, the entry door to the laboratory should be posted to minimize traffic through the room, vigorous swinging of any entry doors (which causes sudden drafts), and distraction of the worker. This, of course, is in addition to the required biohazards sign.

BIOSAFETY CABINET DECONTAMINATION AND RECERTIFICATION

It has been stated categorically that any cabinet that is purchased should be certified in place initially by the manufacturer or his representative. The cabinet should be recertified annually or at approximately 1,000 h of service. If the filters develop excessive pressure loss, or if the cabinet is moved, even within the same room, recertification involves testing the specified air velocities and readjusting them to the correct values, leak-testing filters and cabinets, repairing or replacing the filters as required, and replacing or repairing any other elements of the cabinet. Since the personnel performing these tasks may not be fully familiar with the laboratory operation or immunized against the agents employed, it is desirable that cabinets be decontaminated before any repairs or certification testing is conducted. Procedures for decontaminating cabinets with formaldehyde vapor are described in a National Cancer Institute slide-cassette presentation (3) and Chapter 22 of this volume. Basically, this comprises sealing off the cabinets (including both inlets and exhausts) and vaporizing dry paraformaldehyde to provide a concentration from 250 to 300 mg/ft^3 (7 to 8.5 mg/m^3). The overall volume of the cabinet should be used to allow for takeup in the supply and exhaust filters. The vapor is then held in the cabinet for approximately 4 h. It is usually necessary to ensure that the temperature remains in the 20 to 25°C range and humidity is at least 70%. After sufficient contact time with the formaldehyde gas has been achieved, the gas may be discharged through the exhaust filter to the outdoors. If the cabinet is not connected directly to an exhaust duct, a temporary hose may be attached from the exhaust, out the nearest window, or to the room exhaust system provided it does not recirculate into any laboratory spaces. This procedure, although effective, can create a problem with deposition of paraformaldehyde on surfaces and on the extended surface of the HEPA filters. Therefore, it is recommended that 300 mg/ft^3 (8.5 mg/m^3) be used sparingly, and that 250+ mg/ft^3 (7+ mg/m^3) may be adequate for most uses. Effectiveness can be estimated by placing *Bacillus subtilis* spore strips (10^7 to 10^8 per strip) in the cabinet. After decontamination and airing, these spore strips are incubated on Trypticase soy agar. The residual formaldehyde may be of concern to people working with cell cultures or viruses that may be highly sensitive to formaldehyde vapors. In our own work we have not seen the level of residual formaldehyde in cabinets above 30 μg/liter after 8 h or more of ventilation. We have determined that the persisting trace of paraformaldehyde on the HEPA filter is usually not sufficient to affect cell culture growth.

Although formaldehyde is the best choice for most cabinet space decontamination procedures, it must be emphasized that the effectiveness of any vapor-phase decontaminant against the specific agents used in the cabinet should be ensured. For example, formaldehyde is not effective against many of the so-called "slow viruses," such as the agents of scrapie, Creutzfeldt-Jakob disease, etc. In such cases, vigorously applied liquid decontaminants may be required. It may be desirable to wet down, remove, and autoclave filters in such cases. Considering the relatively poor penetrating capability of formaldehyde vapor, it is prudent to autoclave or incinerate HEPA filters after use in infectious disease laboratories.

Considering the Class I and II hood in-place test requirements described previously, the following is the minimum test and certification that should be done in the laboratory on a periodic basis. It is illustrated in a slide-cassette package (4) from the National Audiovisual Center, Washington, D.C.

1. Decontaminate as required.
2. Test filters and cabinets for leakage; repair as required.
3. Replace filters as required.
4. Set inlet and downflow air velocities as specified by the manufacturer.
5. For testing procedures, see NSF-49 or manufacturer's instructions.
6. Test at least yearly, at 1,000 h of use, or whenever the unit is moved. Post the unit with the date tested, the name of the person doing the testing, and the list of tests passed satisfactorily.

Note. Testing for "clean bench" operation does not require a challenge of the filter with aerosol. "Clean" certification is not an acceptable substitute for safety certification. The filters must be challenged with a test aerosol.

THE VENTILATED SUIT AS A PRIMARY BARRIER

In previous sections, we have discussed the philosophy of barrier protection as being "containment of the work process," first in minimal-volume packages or housings, and second, in work cabinets that allow for the housing of whole operations. A third choice may be needed in cases where high-hazard materials are used under conditions, such as handling primates, where the physical manipulations within cabinets or within small containers may be extremely difficult. This is a method wherein the operators are packaged, rather than the work or the work materials. The accepted method for this containment is to provide the operator with a one- or two-piece ventilated suit, usually of reinforced vinyl plastic, with fresh air supplied from a breathing-air supply (21). Such a suit arrangement requires that there be very little leakage into the suit, which is maintained at approximately 0.5 in. (1.25 cm) of water, positive pressure to the room. The suited worker must have clean, cool air supplied at approximately 25 to 50 ft^3/min (ca. 0.12 to 0.24 m^3/min). The air supply should be oil free, of fairly low humidity, and have a back-up system, usually compressed air in cylinders, so that in the event of compressor or pump failure the operator will have some time to leave the hazardous space. Supplementary air cooling (e.g., by Vortex coolers) may be required for heavy work or in warm spaces. Because the noise level within such suits tends to be very high due to the rushing of air, it is desirable that the suit be provided with an audiocommunications system and a full alarm system so that in the event of failure of any of the supporting elements the operator is quickly notified. The protection factor offered by such suits can be high, and they have been used for work with microorganisms requiring Biosafety Level 4 containment. Simple head masks, supplied-air head masks, or full-face cover respirators may also be used under these conditions for less hazardous operations. Such attire can be helpful in handling animals being held within primary barrier units.

The room in which the suit is used must, in effect, be equivalent to a large Class III cabinet. Accordingly, it must be operated under a negative pressure; have an airtight internal shell all around, provisions for discard of waste material through an autoclave or other system, an entry air lock, and change room and shower access and egress; and be capable of full decontamination. Often a paracetic acid shower is used in the entry/exit. All air exhausted from these spaces should be passed through tandem HEPA filters or incinerated. Duct work connected to the suit room from supply or exhaust filters should be Freon tight like that for Class III cabinets.

The internal finish of such spaces must permit ready decontamination. Finishes such as epoxies or polyester resins have been quite successful. It is usual to construct such spaces with recessed lighting and utilities that may be serviced from outside the space and with provisions for remote control of paraformaldehyde-vaporizing heaters.

The similarity of suit rooms to a Class III hood system is quite close, and the space and all connecting conduits, duct works, and cabinets should be tested to the same containment standards as the Class III system. The guidelines for work defining Biosafety Level 4 laboratories (21) describe most of the requirements of this type of facility.

Suit rooms may be used for housing animals that have been exposed to pathogens within the Class III cabinet system. It should be noted that both the cabinet system and the suit room are considered Biosafety Level 4 laboratories, as defined by National Institutes of Health guidelines. The animals may be held in individually ventilated cabinet-cages within the space, or held in filter-topped type cages that are opened only within a Class I or Class II cabinet within this suit room, or returned to within the Class III cabinet system for servicing.

The suit room, like the Class III cabinet system, is a highly specialized installation. The brief description above is provided only to introduce the potential user to some of the requirements for such spaces and the uses to which they can be put. As for the Class III cabinet systems, detailed engineering and test procedures are necessary for installation and use of such space. Conversion of existing space of standard construction to such use is difficult and costly and should be approached with care. The containment equipment described can function reliably only when adequate electrical and mechanical support services are available. For example, redundancy in electrical services, exhaust fans, supply fans, heater systems, communication systems, breathing air, liquid waste decontamination, etc., must be provided to ensure adequate protection both to the worker and to the environment. Maintenance and operation of such equipment should be considered. There is a need for highly trained and well-motivated support staff who can provide the necessary mechanical and instrumentation services.

APPLICATION OF PRIMARY BARRIERS TO MICROBIOLOGY LABORATORY RESEARCH

We have defined three classes of cabinets and an equivalent ventilated-suit containment method. To the degree possible, considering the rapid changes in this field, we have outlined the physical characteristics of each of the primary barrier systems described. With knowledge of the containment systems available, the user's earliest major problem is usually in matching the containment equipment to the requirements. Table 3 summarizes application data for cabinets.

(i) **Biosafety Level 1.** For materials designated by the Centers for Disease Control (CDC) (8, 28) as Biosafety Level 1 agents, there are no generalized requirements for containment systems.

(ii) **Biosafety Level 2.** For materials classified by CDC as Biosafety Level 2 etiologic agents, by the National Cancer Institute as low- to moderate-risk viruses, or by the National Institutes of Health Recombinant DNA Guidelines (21) as requiring BL2 containment, the primary barriers should be applied to those operations known to produce appreciable aerosols. Class I or II biological safety cabinets can be used to contain aerosol-producing equipment such as blenders, lyophilizers, sonicators, table-top centrifuges, and open-vessel devices used for growth or propagation of microorganisms by aeration or shaking. The safety cabinet may not be required with aerosol-producing equipment as contained by other primary barriers; for example, the centrifuge can be operated in the open if a sealed head or effective safety cup is used.

(iii) **Biosafety Level 3.** The Class I or II safety cabinet is required in the Biosafety Level 3 (BL3) laboratory, as defined in the National Institutes of Health Recombinant DNA Guidelines (21), and when working with Biosafety Level 3 (8, 28) etiologic agents. In addition, oncogenic viral agents designated moderate-risk agents should be handled within cabinets. In addition to the operations described above, as suggested for cabinet enclosures, one should also consider housing such operations as

TABLE 3. Applications of biological safety cabinets (27)

Cabinet			Applications			
Type	Work opening	Face air velocity, ft/min (m/s)	Oncogenic viruses (20)	Chemical carcinogens	Etiologic agents (8)	Biosafety Level (28)
Class I	Open front	$\geq$ 75 (0.4)	Low and moderate risk	No	CDC 1–3	2, 3
	Glove panel, no gloves	150 (0.75)	Low and moderate risk	Yes[a]	CDC 1–3	2, 3
	Glove panel, with gloves	N/A	Low and moderate risk	Yes	CDC 1–3	2, 3
Class II						
Type A	8–10 in.	75 (0.4)	Low and moderate risk	No	CDC 1–3	2, 3
Type B1	Sliding sash 8–10 in.; use 8-in. opening	100 (0.5)	Low and moderate risk	Yes in low dilution and volatility	CDC 1–3	2, 3
Type B2	Same as above	100 (0.5)	Low and moderate risk	Yes with approved[a] air inflow and exhaust	CDC 1–3	2, 3
Type B3	Same as above	100 (0.5)	Low and moderate risk	Yes with approved[a] air inflow and exhaust	CDC 1–3	2, 3
Class III	NA	NA	Low and moderate risk	Yes	CDC 3–4	3, 4

[a]Meeting local standards and Environmental Protection Agency and Occupational Safety and Health Administration standards specific to agents used. Usually requires hard connection to direct exhaust outdoors.

pipetting, serial dilutions, culture transfer operations, plating, flaming, grinding cultures, blending dry materials, and virtually any operation in which high concentrations of microbial agents may be used or, alternatively, there is a potential for contamination of the product or even a xenotropic contaminant within the work product, e.g., as is found in some cell lines. Although laboratory animal holding will be discussed separately, animals infected with agents in these categories can be housed in partial-containment caging systems, such as open cages placed in ventilated enclosures, solid-wall and -bottom cages covered by filter bonnets, and solid-wall and -bottom cages placed on holding racks equipped with UV lamps and reflectors. The last should be used only with etiologic agents shown to be inactivated readily by UV irradiation.

The use of extensive Class III cabinet systems or suit rooms is substantially as defined for Biosafety Level 4 physical containment facilities (28) for work with high-hazard agents and procedures. Laboratory animals held in the facility should be housed either in cages contained within Class III cabinets or in partial containment caging systems such as open cages placed in ventilated enclosures, or solid-wall and -bottom cages covered by filter bonnets and located in a specially designed suit room. A Class III cabinet system may be used within a Biosafety Level 3 or 4 laboratory to enclose an operation that can be a heavy aerosol producer or to house biological agents that are very poorly defined with respect to routes of infection, infectious dose, or persistence.

FINAL COMMENTS

The primary barriers described in this section are basic tools in microbiology laboratory safety. It is of value to reiterate a few facts, conclusions and admonitions.

(i) Class I and Class II cabinets are usually used in an open-front configuration. Accordingly, they must be installed and used carefully to get full benefit of the protection available.

(ii) Hands, arms, the front of the worker's body, and the interior work areas will probably be contaminated whenever work with a liquid suspension of microbial agents or lyophilized powder is conducted in Class I or Class II cabinets.

(iii) Whenever microbial agents are used, effective decontamination procedures must be available.

(iv) The biohazard from microbial agents is often combined with lower-level chemical and radiological hazards. Equipment and operating procedures should be chosen to meet both containment needs.

(v) Primary barriers are aids to good practice, not substitutes; therefore, the best definition of good practice is "quality laboratory work done in a safe manner."

(vi) There are no short cuts in safe operations of primary barriers. Install quality equipment and insist that its performance be fully demonstrated before it is put into use.

(vii) The laboratory is a continually changing environment, and equipment often wears out, is used for other than the original purposes, or is treated as a piece of furniture. Regular inspection, test, and repair programs are essential if primary barrier equipment is to provide the product and worker protection desired by workers in the laboratory and suppliers of the equipment. The lab operator can do some simple tests himself, but validation and retest of equipment should be done by qualified testers recommended by public agencies or by manufacturers.

GLOSSARY

Back-draft damper. Airflow-actuated plate in a duct that allows airflow in only one direction.

Barrier. Physical object or directional air stream (or combination thereof) that serves to prevent or control passage of microorganisms.

Biosafety Level. As defined in reference 28. Comparable rating system to BL1, BL2, BL3, and BL4 of the NIH Recombinant DNA guidelines.

Face velocity. Measured or calculated velocity of air flowing into the work opening of a cabinet. With some exceptions, it is not uniform across the opening in the vertical direction but should be uniform in the horizontal direction (exception: horizontal laminar-flow clean bench).

HEPA filter. Air filter tested to collect at least 99.97% of all 0.3-μm spherical particles in aerosol with which it is challenged.

Laminar airflow. Airflow in which the entire body of air flows at uniform velocity with minimal mixing.

Penetrometer. An optical device employing a light source and phototube and electronics to measure light scatter from aerosols. It measures the mass concentration of the aerosol.

Plenum. When not otherwise specified, refers to filter chambers or enclosures between fan and filter in the biological safety cabinet.

Turbulence. Airflow in which flow directions vary within the body of air and there is substantial mixing both within the air mass and from the surrounding space.

Ventilated suit. Pressurized, impermeable outer garment fully enclosing the wearer and provided with conditioned air by hose or pressure tank.

The invaluable assistance of Caro Hopper in the preparation of this manuscript is gratefully acknowledged.

This work was supported by the Office of Naval Research (Code 400), Washington, D.C.

LITERATURE CITED

1. **Akers, R. L., R. J. Walker, F. L. Sabel, and J. J. McDade.** 1969. Development of a laminar flow biological cabinet. J. Am. Ind. Hyg. Assoc. **30:**177–185.
2. **Anonymous.** 1976. Effective use of the laminar flow biological safety cabinet. (Slide-cassette.) NAC no. 00971 and no. 003087. National Audiovisual Center (GSA), Washington, D.C.
3. **Anonymous.** 1976. Formaldehyde decontamination of the laminar flow biological safety cabinet. (Slide-cassette.) NAC no. 005137 and no. 003148. National Audiovisual Center (GSA), Washington, D.C.
4. **Anonymous.** 1976. Certification of Class II (laminar flow) biological safety cabinets. (Slide-cassette.) NAC no. 003134 and no. 009771. National Audiovisual Center (GSA), Washington, D.C.
5. **Anonymous.** 1976. Selecting a biological safety cabinet. (Slide-cassette.) NAC no. 00709 and no. 01006. National Audiovisual Center (GSA), Washington, D.C.
6. **Baldwin, C. L. (ed.).** 1973. Centrifuge biohazards: proceedings of National Cancer Institute Symposium on Centrifuge Biohazards. DHEW Publication no. (NIH)78–373. National Institutes of Health, Bethesda, Md.
7. **British Standards Institution.** 1979. Specifications for microbiological safety cabinets, Brit. Std. 5726. British Standards Institution, London.
8. **Center for Disease Control.** 1976. Classification of etiologic agents on the basis of hazard. Center for Disease Control, Atlanta, Ga.
9. **Chatigny, M. A.** 1961. Protection against infection in the microbiology laboratory: devices and procedures. Adv. Appl. Microbiol. **3:**131–192.
10. **Chatigny, M. A., and D. I. Clinger.** 1969. Contamination control in aerobiology, p. 194. *In* R. L. Dimmick and A. B. Akers (ed.), An introduction to experimental aerobiology. John Wiley and Sons, New York.
11. **Chatigny, M. A., S. Dunn, K. Ishimaru, J. A. Eagleson, and S. B. Prusiner.** 1979. Evaluation of a Class III biological safety cabinet for enclosure of an ultracentrifuge. Appl. Environ. Microbiol. **38:**934–939.
12. **Clark, R. P., and B. J. Mullan.** 1978. Airflows in and around downflow "safety" cabinets. J. Appl. Bacteriol. **48:**131–135.
13. **Darlow, H. M.** 1969. Safety in the microbiological laboratory, p. 169–204. *In* J. R. Norris and D. W. Ribbons (ed.), Methods in microbiology, vol. 1. Academic Press, Inc., New York.
14. **Dimmick, R. L., W. F. Vogl, and M. A. Chatigny.** 1973. Potential for accidental microbial transmission in the biological laboratory, p. 246–266. *In* A. Hellman, M. N. Oxman, and R. Pollack (ed.), Biohazards in biological research. Cold Spring Harbor Laboratory, Cold Spring Harbor, N.Y.
15. **Federal Register.** 1971. Amendment of Para. 72.25, Part 72, Title 42, Code of Federal Regulations. 36 F.R. 8815. U.S. Government Printing Office, Washington, D.C.
16. **Howie, Sir J.** 1978. A code of practice for the prevention of infection in clinical laboratories (report of the Working Party). Her Majesty's Stationery Office, London.
17. **Kenny, M. T., and F. L. Sabel.** 1968. Particle size

distribution of *Serratia marcescens* aerosols created during common laboratory procedures and simulated laboratory accidents. Appl. Microbiol. **16:**1146–1150.

18. **Kuehne, R. W.** 1973. Biological containment facility for studying infectious disease. Appl. Microbiol. **26:**239–243.

19. **LaManna, C.** 1960. The use of autoclavable plastic material for do-it-yourself solutions to laboratory problems of safety. J. Lab. Clin. Med. **55:**319–321.

20. **National Cancer Institute.** 1974. Safety standards for research involving oncogenic viruses. DHEW Publication no. (NIH)78–790. National Cancer Institute, Bethesda, Md.

21. **National Institutes of Health.** 1978. NIH guidelines for recombinant DNA research supplement: laboratory safety monograph. National Institutes of Health, Bethesda, Md.

22. **National Sanitation Foundation.** 1976 (revised 1983). NSF Standard no. 49 for Class II (laminar flow) biohazard cabinetry. National Sanitation Foundation, Ann Arbor, Mich.

23. **Newsome, S. W. B.** 1979. Class II (laminar flow) biological safety cabinet. J. Clin. Pathol. **32:**505–513.

24. **Newsome, S. W. B.** 1979. Performance of exhaust protective (Class I) biological "safety" cabinets. J. Clin. Pathol. **32:**576–583.

25. **Rake, B. W.** 1978. Influence of cross drafts on the performance of a biological safety cabinet. Appl. Environ. Microbiol. **36:**278–283.

26. **Sheppard, C. C., C. W. May, and N. H. Topping.** 1945. A protective cabinet for infectious disease laboratories. J. Lab. Clin. Med. **30:**712–716.

27. **Stuart, D. G., J. J. Grenier, R. A. Rumery, and J. M. Eagleson.** 1982. Survey, performance and use of biological safety cabinets. Am. Ind. Hyg. J. **43:**265–270.

28. **U.S. Department of Health and Human Services.** 1984. Biosafety in microbiological and biomedical laboratories. HHS Publication no. (CDC)84–8395. U.S. Government Printing Office, Washington, D.C.

29. **Van den Ende, M.** 1943. An apparatus for the safe inoculation of animals with dangerous pathogens. J. Hyg. **43:**189–194.

30. **Vogl, W. F., and M. A. Chatigny.** 1973. A simplified nomograph system for estimation of risk in the microbiological laboratory. *In* Reinraumtechnik I: Berichte des international Symposium Reinraumtechnik gehallen Zurich. Schweiz 18–20 Oct. 1972.

31. **Wedum, A. G.** 1953. Bacteriological safety. Am. J. Public Health **43:**1428–1437.

Chapter 20

Personal Protective Equipment

BRUCE W. STAINBROOK AND ROBERT S. RUNKLE

The laboratory environment encompasses several hazards, known and unknown, which pose potential risks to personnel. Many of the hazards, which include physical, biological, chemical, and radiation exposure, can be reduced or minimized through the proper use of personal protective equipment. Such equipment includes gloves, respirators, eye and ear protection, garments, and anything else worn by the employee to guard against the hazard he or she may encounter in the laboratory environment. The personal protective equipment may be required to reduce the risk of infection or to protect an employee from contact, inhalation, or ingestion of a chemical solution or radioactive material involved in the work at hand.

Occupational Safety and Health Administration (OSHA) legislation makes it mandatory for an employer to furnish employees with a working environment free from the recognized hazards that could cause death, injury, or illness to the employees. (Certain state agencies have been given precedence over OSHA as long as their requirements equal the OSHA standards.) OSHA requires that the hazard be controlled through engineering means when economically feasible; that is, the facility must provide some device which removes the hazard from the employees' work area and thereby eliminates the necessity for personal protective equipment. In many cases, however, it may not be feasible to eliminate the hazardous situation. Both OSHA and state agencies recognize that some hazards require a cost-benefit risk analysis both of the process and of the laboratory procedure to determine the practicality of eliminating the problem. Protective clothing is a legitimate solution in many instances, particularly when elimination of the hazard is impractical or not cost effective.

Frequently, common safety hazards have been neglected in the laboratory environment, which presents many of the same hazards found in the industrial environment: employees can encounter flying particles which could cause physical injury and can face exposure to dangerous chemicals and solvents, whether vapor or fumes. In addition, many laboratories pose more serious problems to the health and welfare of the persons working in the facility (1, 14, 23). Therefore, in weighing the hazards, personal protective equipment may be not only necessary but

also the most practical, cost-effective means available to prevent injury or illness to the laboratory employee.

Selection of proper type of personal protective equipment

Once the hazard has been recognized, it is necessary to identify the type of equipment the worker should wear in the laboratory. In selecting such equipment, two criteria should be used: (i) the degree of protection which a particular piece of equipment affords under varying conditions, and (ii) the ease with which it may be used.

Considerable time and risk can be saved by checking with manufacturers, appropriate governmental agencies, and private testing laboratories for solutions to similar problems. Information should be available with the product or upon request from the manufacturer; any questionable data should be verified by contacting OSHA or the appropriate state agency directly. (See Table 1 for a current listing of appropriate agencies and institutions with active relevant programs [1, 14].)

Using the proper equipment to control the hazard

Once the need for personal protective equipment has been determined, the next problem is to motivate the employee to use the equipment properly. Employees are now required by law to wear appropriate personal protective equipment once hazards have been identified according to government standards (14, 23). Some factors that will simplify compliance with requirements are as follows.

(i) The employee must understand the procedure or process that is hazardous and, therefore, the need for such equipment.

(ii) The employee must realize that the job can be performed with minimum interference with his or her normal work procedures.

(iii) The employee should have maximum familiarity with the personal protective equipment; it should fit properly and be easy to wear.

(iv) Simple instructions for the care of such equipment should be provided, and supervisory personnel should be responsible for the initial

TABLE 1. U.S. federal and other agencies active in testing and evaluating protective equipment

Department of Health and Human Services, Public Health Service

Centers for Disease Control
1600 Clifton Road, N.E.
Atlanta, GA 30333

National Institute for Occupational Safety and Health (NIOSH)
1600 Clifton Road, N.E.
Atlanta, GA 30333

Also: Appalachian Laboratory for Occupational Safety and Health
944 Chestnut Ridge Road
Morgantown, WV 26505

Robert A. Taft Laboratories
4676 Columbia Parkway
Cincinnati, OH 45226

Food and Drug Administration
Bureau of Radiological Health[a]
12720 Twinbrook Parkway
Rockville, MD 20852

Office of Health Research, Statistics and Technology
National Center for Health Statistics
Center Building #2
3700 East-West Highway
Hyattsville, MD 20782

National Institutes of Health
Division of Safety
9000 Rockville Pike
Bethesda, MD 20892

National Institute of Environmental Health Sciences
P.O. Box 12233
Research Triangle Park, NC 27709

Department of the Interior

Bureau of Mines
2401 E Street, N.W. (Columbia Plaza)
Washington, DC 20506

Department of Labor

Mine Safety and Health Administration[b]
Ballston Tower #3
4015 Wilson Blvd.
Arlington, VA 22203

Occupational Safety and Health Administration
200 Constitution Avenue, N.W.
Washington, DC 20210

Nuclear Regulatory Commission
1717 H Street, N.W.
Washington, DC 20555

U.S. Government Printing Office
Washington, DC 20402

Nonprofit organization

The National Council on Radiation Protection and Measurements
7910 Woodmont Avenue
Bethesda, MD 20814

[a]This bureau is being merged with the Bureau of Medical Devices. New names and locations have not been announced.
[b]This succeeds the Mining Enforcement and Safety Administration at the Interior Department.

demonstration and periodic follow-up of proper maintenance.

The following sections cover in some detail specific types of personal protective equipment. More detailed studies will be found in the referenced technical source material.

PERSONAL PROTECTIVE EQUIPMENT

Lab coat

The most common and cost-effective type of personal protective equipment used in the laboratory today is the lab coat. This clothing is often overlooked, and the degree of protection that it offers is frequently underestimated. The lab coat can be used to protect clothing against biological or chemical spills as well as to provide some additional body protection (7).

The specific hazard(s) and the needed degree of protection should be identified when selecting coats for laboratory personnel. It must be determined whether or not the lab coat is actually sufficient to protect the person from the immediate danger. For instance, is the material thick enough to protect the employee from the spill? Can the spill be removed from the lab coat after an accident? Another important consideration is the coat's material. For example, where there is a potential for fire exposure in the laboratory, the lab coat should be made of some type of flame-resistant material. A polyester-cotton blend material is flammable and will melt on the skin when in contact with a spark, heat source, or some corrosives. Therefore, it may be better to select a 100% cotton lab coat, which is nonreactive to many chemicals as well as flame resistant (12, 16, 17). The lab coat itself should cover the arms as well as the majority of the middle body. It is good laboratory practice that the coat be buttoned at all times in the laboratory.

Hearing protection

Noise exposure is a frequent hazard in the laboratory. OSHA regulations stipulate that a worker not be exposed to more than 85 dB (A scale) for an 8-h working day without the use of hearing protection. (That standard is, however, still under litigation, with OSHA trying to lower the exposure even further [1, 4, 14].) In a laboratory environment, extreme noise exposure is rare. However, devices such as ultrasonic cleaners and specialized centrifuges or separators can cause personnel much discomfort due to high noise levels. Since the OSHA requirements are based on timed exposure, ultrasonic cleaners and other noisy laboratory devices such as compressors may actually exceed the permissible limits, depending upon how long the worker is in the equipment space. It is important, therefore, that such hazards be

identified and that the proper hearing protection be selected.

All hearing protection equipment must be approved by the National Institute of Occupational Safety and Health Administration, the Mine Safety and Health Administration or both. Hearing protection devices range from ear plugs to Canadian wool. Either disposable or reusable ear plugs can be purchased. A hygiene program for cleaning reusable ear plugs daily or before use with soap and water or alcohol must be implemented and strictly enforced. Though ear muffs do not usually come in contact with the inner ear, they should also be cleaned regularly as part of a good hygiene program.

There are many problems affiliated with hearing protection. Comfort is probably the first consideration. The employees should be given an opportunity to select a particular type of hearing protection from an approved listing. The second consideration is that the established rules for hearing protection must be enforced. Employees should be adequately instructed in the proper method of wearing the hearing protection and the hazards involved in its operation (1, 8, 14, 23). This is usually done in conjunction with a medical department in a large organization. In addition, pre-employment and subsequent annual hearing tests should be given to those individuals working in laboratories with known noise hazards (23). Hearing tests should be analyzed by a qualified professional who can determine whether there has been any hearing loss since the previous exam and whether additional medical treatment or reassignment of the employee to a less hazardous area is required.

A key provision of the OSHA guidelines is the requirement that an effort be made to eliminate the noise problem through engineering or design changes to the facilities (23). In many cases, this can be accomplished by isolating the noise-producing equipment by use of enclosure guards or baffles. This will allow the elimination of personal protective equipment, and personnel exposure will be brought within acceptable limits. As long as the noise exposure does exist, however, it is imperative that regular follow-up noise studies of the area be completed and documented to determine whether there has been an increase or decrease in the noise exposure, based on the allowable limits.

Eye protection

Protection of eyesight is a critical responsibility of management in laboratory safety (14). To protect laboratory personnel from potential eye damage, attention should first be focused on elimination of the problem. For example, personnel must be trained in the proper use of a fume hood sash or laser light sources. Protective guards should be provided over equipment to prevent splashes or aerosol particles from being thrown back at the worker. Housekeeping is also essential; a clean laboratory can prevent materials from being spilled or dust being blown around by motorized equipment.

If the hazardous procedure cannot be modified to eliminate the exposure risk to the eye, appropriate personal protective equipment must be utilized. In the area of eye safety the following groups of protective measures can be recommended (2).

(i) Safety glasses and side shields. "Safety" glasses must meet more rigid standards for strength and endurance than regular prescription glasses. Most safety glasses can easily provide side shield protection against particles and physical splashes. Often safety glasses are worn in conjunction with face shields to ensure maximum protection.

(ii) Face shields. Protection against a chemical or biological splash is best provided by face shields, which are commercially available in many sizes and shapes to cover the entire face and neck. Applications for this type of equipment include autoclave loading and unloading, grinding operations, and disposing of acids or solvents.

(iii) Goggles. Goggles, another alternative, provide both side and top protection to the eye. Goggles are available in many materials which can provide impact, radiation, and chemical splash protection. They usually have a flexible membrane for sealing against the face. Some goggles are vented to provide air movement through the goggle while preventing a spill or splash from seeping through the rim.

Other available eye protection is usually a modification of the three types listed above. Each type can be purchased with specific lenses which can provide protection against lasers, radiation, UV light, or other damaging energy sources (21). Obviously, safety glasses and goggles can also be provided with prescription lenses to meet individual needs.

Eye protection equipment should be cleaned regularly, and a specific place must be provided for the storage of cleaning supplies. Supervisors must enforce eye protection programs and stress frequently the need for the equipment. The laboratory director must decide whether the eye protection program is limited to only one area of a laboratory module or whether the risk is such that an entire section, wing, or floor must be isolated to protect all employees properly.

A discussion on eye protection would not be complete without some comments and cautions on contact lenses. Contact lenses provide no protection and should not be substituted for required eye protection (15). Although folklore "documents" instances in which contact lenses have supposedly saved someone's sight, there are no definitive studies to support these claims. In fact, the use of contact lenses may increase risk of eye damage because fluids may be trapped between the lenses and the eye. In the laboratory environment work may involve not only biohazards but the use of chemicals and solvents as well. Should any of these come in contact with the eye,

they will usually migrate between the lens and the eye. The damage will continue despite the proper use of eyewash. In some cases, the lens will actually adhere to the eye and not the let the water cleanse the eye. Some chemicals are solvents for plastic lenses and will cause the lens to melt to the eyeball surface, creating worse problems. In an impact accident, the contact lens may be driven into the eye, causing damage or loss of sight. Therefore, contact lenses must not be substituted for good eye proection. In most cases, contact lenses should not be allowed in the laboratory at any time; rather, safety glasses or equivalent devices should be worn instead. Laboratory management should identify which employees wear contact lenses and maintain an up-to-date listing (15).

Respiratory protection

Respiratory protection may be one of the least understood and poorest examples of adequate protection in the average laboratory. (See also Chapter 21.) The risk of inhalation of toxic or caustic substances, whether biohazard, radioactive, or chemical, frequently exists in the laboratory environment and poses a significant potential health hazard to the employee (1, 5, 14).

The first effective approach is to attempt to eliminate the problem through design or equipment selection. This is usually accomplished by providing adequate rates of ventilation or by redesigning the ventilation process to a completely closed system. Engineering controls are designed to eliminate employee contact with the hazardous agent. When engineering controls are not feasible, then personal protective equipment becomes mandatory. The respirator is the primary choice (3).

There are many commercially available kinds of respirators or self-contained breathing apparatus. The most common and simplest kind of respirator is the mask, which is available in either a disposable or a reusable version. Selection of a mask usually is based on a hazard study performed by a qualified safety professional. The mask selected should be approved by the National Institute of Occupational Safety and Health or the Mine Safety and Health Administration and identified for specific hazards. Disposable masks have the advantage of being lightweight and can also be discarded at the end of the allowable working period (10, 14). Disposable masks have many applications, such as the surgical mask, which chiefly provides product protection, or masks specifically designed to be used in environments with exposure to mercury, toxic dusts, spray paints, and odors.

The reusable mask has both advantages and disadvantages. These masks do have a capacity for protecting against a larger variety of substances and may therefore be the protection of choice in certain applications. Their disadvantages are that they are heavy and more cumbersome than disposable masks and require a maintenance and cleaning program, as well as frequent replacement of the filter cartridge.

Another type of respiratory equipment is the air flow hood (Fig. 1), which is most often used in lieu of the respirator to provide clean filtered air. These hoods are usually portable and are equipped with a battery power source. Variations include units which provide air from an air compressor through a hose and CO_2 filter, emergency escape hoods (Fig. 2), or the self-contained apparatus (Fig. 3) used for longer durations. These hoods can be used where certain toxic dusts, biohazards, and radioisotopes are utilized. These types are discussed in more detail below.

Foot protection

An area of laboratory safety that is frequently omitted is foot protection. The laboratory can be a wilderness of foot injuries from spilled chemicals (acids, bases, solvents, etc.), broken glass, hot liquids, and dropped heavy objects. Most foot exposure problems can be solved very simply by requiring that all employees wear low-heel, fully enclosed, leather shoes (1, 14). Cloth or plastic shoes should be prohibited because these can absorb chemicals and liquids. Moving heavy equipment in a pilot laboratory or in a storeroom can present a significant hazard to unprotected feet. Steel-toed shoes are recommended to prevent crushed or broken feet; steel caps

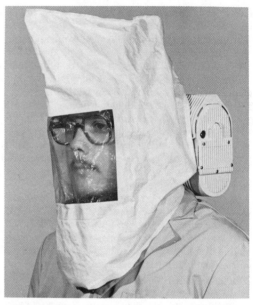

FIG. 2. Emergency escape hood: Scott SCRAM solid-state oxygen, 15-min emergency escape device (courtesy Scott Aviation, Div. A.T.O., Inc., Lancaster, N.Y.).

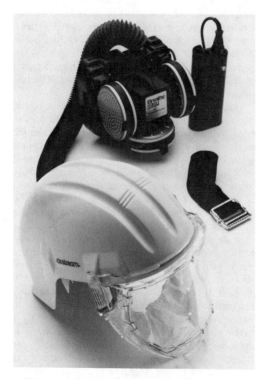

FIG. 3. Self-contained breathing apparatus: Scott Air-Pak IIa SCBA, 30-min duration, 2,216-lb/in^2 system (courtesy Scott Aviation).

FIG. 1. Airflow hood (courtesy Racal Airstream, Inc., Frederick, Md.).

are also available which can be attached to leather shoes to provide this type of protection. The laboratory director must decide which areas of the laboratory require restriction and make sure that appropriate signs are posted and that restrictions are continually enforced.

Gloves

Industrial use of gloves is very widespread, not only for product or process protection but also for personnel protection. Gloves can provide protection from biohazard exposure, acid or caustic burns, chemical burns, and excessive heat or cold, as well as bites, cuts, and other physical hazards (6). Broader use in the laboratory could provide opportunity for reduced accident/injury rates.

The primary reason for gloves is to guard against the immediate hazard which can occur, such as the chemical burn or extreme heat or cold from laboratory procedures. Gloves also provide a long-term protection for prevention of dermatitis (1, 13, 14, 20). Repeated exposure to low doses of many compounds can eventually create an adverse skin reaction. Recently a dermatitis reaction to "barrier creams" was noted in our laboratories; those individuals who consistently wore gloves avoided the problem.

Selection of the proper glove to provide maximum protection in known exposure situations is most important (12). Unlike other types of protective clothing or equipment, gloves do not usually carry National Institute of Occupational Safety and Health, Mine Safety and Health Administration, or OSHA approval. Therefore, the laboratory supervisor or manager must rely on the vendor or manufacturer for technical specifications and test data to decide whether or not the glove will protect against a specific hazard.

One simple test to verify whether a glove is able to perform its assigned task is to fill the glove partially with the hazardous substance and let it hang for 24 to 48 h in a ventilated enclosure or fume hood. After this exposure, if the substance has not eaten through the glove, then adequate protection should be provided to the employee for short-duration exposures. This is, however, an unscientific test and should only be used to verify the claims made by the manufacturer.

The most frequently used type of glove in the laboratory is the surgical glove. Applications include prevention of contact with biohazards, minimal radiation hazards, and some chemical hazards. Protection from acids, caustics, and solvents often requires different degrees of protection, such as that available from gloves made of thick rubber or vinyl, neoprene, or polyvinyl alcohol. Most of these types of gloves are specifically designed for unique environments (22).

Leather or asbestos gloves with liners are often used for heat exposures such as unloading an autoclave or kiln. Many vendors are also offering alternative gloves made of the newer fabrics to provide such protection. Usually a wool/leather glove with a heavy liner made of cotton or wool will provide good protection when working in cold environments, such as removal of cultures from the freezer. The protection for either hot or cold exposure situations must be selected based upon total exposure time.

As with masks and earplugs, gloves are available as either disposable or reusable. Disposable gloves offer the convenience of single use and should be used when the risk of high contamination to the work or the employee exists, such as in biohazard or radiation procedures. Reusable gloves must be cleaned regularly or after each use so that cross-contamination does not occur, either with the experiment or with worker contact. Reusable gloves must also be checked before each use for tears, holes, etc. Any defects should result in the immediate discarding and replacement of the gloves.

Table 2 indicates the kinds of gloves available and the substances they can guard against (12). This table is a useful tool in the development of good glove protection practices in all laboratory situations.

Air-supplied suits

Aerospace technology has brought to industry and the laboratory the air-supplied suit (Fig. 4), many of which are lightweight, comfortable to wear, flexible enough for most tasks, and disposable. These pressurized or ventilated suits provide total body protec-

FIG. 4. Ventilated suit (courtesy Durafab, Inc., Cleburne, Tex.).

TABLE 2. Matching gloves to job hazards[a]

Substance	Neoprene	PVC	Paracril/PVC	Polyurethane	SBR	Special PVC	Substance	Neoprene	PVC	Paracril/PVC	Polyurethane	SBR	Special PVC
Hydrocarbons (oils and solvents)							Organic acids						
Castor oil	E	E	E	G	NR	E	Acetic acid	E	E	E	G	E	E
Coconut oil	E	E	E	G	F	E	Citric acid	E	E	E	G	E	E
Cottonseed oil	E	E	E	G	F	E	Formic acid	E	E	E	G	E	E
Mineral oil	E	E	E	E	NR	E	Lactic acid	E	E	E	G	E	E
Petroleum oils	E	E	E	E	NR	E	Oleic acid	E	E	E	F	G	E
Gasoline (SR)	E	G	G	E	NR	G	Malic acid	E	E	E	F	G	E
Gasoline (cracked)	E	G	G	E	NR	G	Steric acid	E	E	E	G	E	E
Coal tar solvents	G	G	G	E	NR	G	Tannic acid	E	E	E	G	E	E
Naphtha	E	G	E	E	NR	G	Carbolic acid	F	E	E	G	E	E
Grease (all kinds)	G	E	E	G	NR	E	Inorganic acids						
Kerosene (pet)	G	G	E	E	NR	G	Carbonic acid	E	E	E	F	F	E
Kerosene (c-t)	F	G	E	E	NR	G	Chlorine water	E	E	E	G	G	E
Turpentine	E	G	G	E	NR	G	Hydrobromic acid	E	E	E	F	G	E
Propane	G	G	G	E	F	G	Hydrochloric acid	G	E	G	NR	F	E
Carbon tetrachloride	G	G	G	F	NR	G	Hydrogen sulfide	G	E	G	NR	F	E
Trichloroethylene	G	G	G	F	F	G	Phosphoric acid	E	E	E	F	F	E
Perchloroethylene	G	F	G	F	NR	F	Sulfuric acid	G	E	G	NR	F	E
Benzene (benzol)	F	F	F	G	NR	F	Nitric acid	G	E	G	NR	F	E
Toluol	F	F	G	G	NR	F	Perchloric acid	G	E	G	F	F	E
Toluene	F	F	G	G	NR	F	Hydrofluoric acid	NR	E	F	NR	NR	E
Butane	G	G	G	E	F	G	Salts, alkalis						
Hexane	E	G	G	F	F	G	Ammonium hydroxide	E	E	E	G	G	E
Hylene	G	G	F	G	NR	G	Potassium hydroxide	E	E	E	F	G	E
Cyclohexane	G	F	NR	G	NR	F	Sodium hydroxide	E	E	E	F	G	E
Benzyl chloride	F	G	E	F	F	G	Calcium hypochlorite	E	E	E	G	G	E
Methyl chloride	E	G	E	F	F	G	Ammonium sulfate	E	E	E	E	E	E
Chloroform	F	G	G	F	F	G	Caustic potash	E	E	E	E	G	E
							Copper chloride	G	E	E	G	F	E
Ketones and aldehydes							Copper sulfate	G	E	E	G	F	E
Acetaldehyde	E	E	E	F	F	E	Potassium dichromate	G	E	E	G	G	E
Butyraldehyde	E	E	E	F	G	E	Calcium chloride	E	E	E	E	E	E
Formaldehyde	E	E	E	F	G	E	Ferric chloride	E	E	E	E	G	E
Furfural	G	E	G	F	F	E	Tin chloride	E	E	E	E	G	E
Benzaldehyde	G	G	G	F	F	G	Miscellaneous						
Acetone	G	F	NR	F	F	F	Ethyl ether	E	E	E	F	G	E
Methyl ethyl ketone	G	F	F	F	F	F	Methyl Cellosolve	E	G	E	F	F	E
Chloroacetone	G	F	F	F	F	F	Dibenzyl ether	G	E	E	F	F	E
							Soaps	E	E	E	G	F	E
Organic esters							Aniline	E	E	E	G	F	E
Amyl acetate	F	E	G	F	F	E	Acrylonitrile	G	E	E	G	F	E
Butyl acetate	F	E	G	F	F	E	Nitrobenzene	F	G	E	F	NR	G
Ethyl acetate	F	G	F	F	F	G	Trinitrotoluene	F	E	G	F	NR	E
Methyl acetate	G	G	F	F	F	G	Morpholine	E	E	E	F	NR	E
Propyl acetate	G	G	F	F	F	G	Diethanolamine	G	E	E	F	F	E
Ethyl formate	G	G	G	F	G	G	Triethanolamine	G	E	E	F	F	E
Dibutyl phthalate	G	E	E	F	G	E	Monoethanolamine	G	E	E	F	F	E
Tricresol phosphate	G	E	E	F	G	E	Carbon disulfide	F	E	E	F	F	E
							Chlorox	F	E	E	G	E	E
Alcohols							Hydrogen peroxide	G	E	E	G	E	E
Amyl alcohol (fusel oil)	E	E	E	F	G	E	Butter	E	E	E	G	F	E
Butyl alcohol	E	E	E	F	NR	E	Buttermilk	E	E	E	G	F	E
Ethyl alcohol	E	E	E	F	G	E	Milk	E	E	E	F	F	E
Methyl alcohol	E	E	E	F	G	E	Lard oil	E	E	E	F	F	E
Propyl alcohol	E	E	E	F	G	E	Linseed oil	E	E	E	F	F	E
Benzyl alcohol	E	E	E	F	G	E	Olive oil	E	E	E	F	F	E
Octyl alcohol	E	E	E	F	G	E	Pine oil	E	E	E	F	NR	E
Glycerine	E	E	E	E	E	E	Cutting oil	G	G	G	G	NR	E
Ethylene glycol	E	E	E	F	E	E	Paint remover	G	F	G	F	F	G
Diacetone alcohol	E	E	E	F	G	E	Petroleum solvent	G	F	G	E	NR	G
Triethanolamine	G	E	E	F	F	E	Battery acid	F	E	E	F	NR	E

[a]Note: these ratings are intended as a general guide only. Employers should run their own on-the-job tests. Reprinted by permission from *Occupational Health and Safety*.

[b]E, Excellent; G, good; F, fair; NR, not recommended.

tion of the worker from a hazardous environment and still permit effective performance of the assigned work. Most of these suits are made of vinyl plastics. They may be considered an alternative to conventional respiratory equipment, but their basic purpose is to protect the wearer from exposure to radiation or contaminants (chemical or biological) which could be absorbed through the skin, ingested, or inhaled. They provide a safe environment, for example, for the employee handling hazardous materials outside a biosafety cabinet or working in a maximum-containment closed room.

Several factors should be considered in wearing any type of air-supplied suit. Since noise produced by the incoming airflow can be high enough to cause damage to the inner ear (14), it may be necessary to install a silencer in the hood to reduce noise levels. Most workers, however, confess to a greater feeling of safety if they constantly hear the hiss of incoming air (14).

Thermal comfort of the employee is also very important and is achieved by providing sufficient airflow of both respirable and cooling air. Since suits are usually impervious to vapor flow, they can add significant heat stress to the individual in a work situation, resulting in as much as a 25% loss of efficiency (5, 19). The suit should function over a range of 20 to 50°C ambient temperature and should be equipped with variable air supplies ranging from 1.52 to 10.66 m^3/min, depending on the type of work in which the worker will be involved (5, 19). Special precautions have been proposed for prolonged use of the self-contained breathing apparatus in freezing or near-freezing environments such as a cold room used to contain an experiment (5, 14).

The supply air source may need to be cleaned with high-efficiency particulate air filters, particularly in areas of radioactive particulates or laboratories that employ highly infectious agents (11). In most cases, it is economical to purchase the air suit as a disposable garment. An advantage of disposable clothing is that it can be specifically selected for a particular situation, thereby avoiding reactions between the suit material and various hazardous agents. In addition, the suit can be disposed of without transmitting the hazard to other areas in the building.

Suits are available with a self-contained battery-powered air supply or with an umbilical hose system to a fixed supply point. The forced filtered air creates positive pressure inside the suit, thereby preventing contaminated air or material from entering the suit (18). The head is protected with a rigid "bubble" or helmet device; gloves and shoe or foot protection are also provided. The suits themselves are flexible and either elasticized or capable of being tightened in some fashion around the wrist and leg openings.

Special suits have been designed for the handling of radioactive materials. Most of these suits, however, will not protect against the external hazard of penetrating radiation; shielding lead, control over the time of exposure, or remote handling techniques such as glove boxes or mechanical arms are required to provide maximum protection. The ventilated suits do, however, very effectively prevent the dangers of skin contact, inhalation, and ingestion of airborne radioactive materials.

Bunny suits

Another type of outer protection is the "bunny suit," a front-zippered coverall with integral booties and head cover. These suits are often used in areas where sterile or "clean" environments must be maintained and should be worn in Biosafety Level 3 and 4 biological containment or radioactive facilities. These suits provide protection to the underclothing and skin from spills or aerosols; the zippered front provides a good seal so that spills are less likely to enter. They are more effective than the traditional button-down lab coat because of their minimum amount of openings and greater coverage. In a radioactive facility, appropriate shielding from penetrating radiation should be provided via conventional means.

LITERATURE CITED

1. **Allen, R. W., M. D. Ellis, and A. W. Hart.** 1976. Industrial hygiene. Prentice-Hall, Inc., Englewood Cliffs, N.J.
2. **Anonymous.** 1975. Care of eyes in 1975. Occupat. Health (London) **27:**253–255.
3. **Aucion, T. A., Jr.** 1975. A successful respirator program. Am. Ind. Hyg. Assoc. J. **36:**752–754.
4. **Beitter, R. C.** 1982. Effective hearing protection—a four-part program. Natl. Safety News **September:**32–35.
5. **Dionne, E. D.** 1982. Help your respirator help your respiration. Natl. Safety News **September:**39–44.
6. **Dionne, E. D.** 1982. A glove affair. Natl. Safety News **September:**359–363.
7. **Dowsett, E. G., and J. F. Heggie.** 1972. Protective pathology laboratory coat. Lancet **i:**1271.
8. **Else, D.** 1973. A note on the protection afforded by hearing protectors—implications of the energy principle. Ann. Occupat. Hyg. **16:**81–83.
9. **Else, D.** 1975. Personal protection. Occupat. Health (London) **27:**71–74.
10. **Goldstein, L.** 1980. Disposable items are here to stay. Natl. Safety News **October:**50–51.
11. **Goldstein, L.** 1980. How to protect against radiation. Natl. Safety News **October:**46–47.
12. **Lynch, P.** 1982. Matching protective clothing to job hazards. Occupat. Health Safety **January:**30–34.
13. **Moursiden, H. T., and O. Faber.** 1973. Penetration of protective gloves by allergens and irritants. Trans. St. Johns Hosp. Dermatol. Soc. **59:**230–234.
14. **National Safety Council.** 1974. Accident prevention manual for industrial operations, 7th ed. National Safety Council, Chicago.
15. **Nejmeh, G., Jr.** 1982. Contact lens. Part II. An analysis. Natl. Safety News **September:**38.

16. **Penland, W. Z., and A. S. Levine.** 1980. Protection from flame and heat. Natl. Safety News **October:**48–49.

17. **Peraldi, D. M.** 1971. Role of work clothes in the extension of burns. Arch. Mal. Prof., p. 407–410.

18. **Poplack, D. G., W. Z. Penland, and A. S. Levine.** 1974. Bacteriological isolation garment. Lancet **i:**1261–1262.

19. **Sansone, E. B., and M. W. Slein.** 1976. Application of the microbiological safety experience to work with chemical carcinogens. Am. Ind. Hyg. Assoc. J. **37:**711–720.

20. **Sansone, E. B., and Y. B. Tewari.** 1978. The permeability of laboratory gloves to selected solvents. Am. Ind. Hyg. Assoc. J. **39:**169–174.

21. **Staffanou, R. S., D. L. Ditmars, C. Drucker, and G. W. Middleton.** 1976. Eye protection from light radiation. J. Prosthet. Dent. **35:**682–688.

22. **Thomas, S. M. G.** 1970. The use of protective gloves. Occupat. Health **22:**281–284.

23. **U.S. Department of Health, Education and Welfare, Public Health Service, Center For Disease Control, and National Institute for Occupational Safety and Health.** 1973. The industrial environment—its evaluation and control. U.S. Government Printing Office, Washington, D.C.

Chapter 21

Miscellaneous Safety Devices

JOSEPH R. SONGER

Major pieces of safety equipment designed for specific laboratory activities are covered in other chapters. Devices which aid in performing laboratory work in a safe and expeditious manner and are not covered in other areas are presented in this chapter.

Blending equipment

Hazards associated with the use of high-speed blenders are well documented (1, 2, 6) (see also Chapter 19). A safety blender container, or "bowl," described by Reitman et al. (4) and commercially available, is shown in Fig. 1. This bowl (Waring AS-1) is constructed of stainless steel and can be sterilized by autoclaving. It has an "O" ring gasketed lid, Teflon bearings, and inlet and outlet fittings to the chamber to allow continuous flow. It is also jacketed so that cooling brine can be circulated to extend the blending time and avoid biological inactivation due to excessive heat. The present cost of the AS-1 safety blender bowl exceeds $500, which might be prohibitive because of budget constraints. A possible alternative is to use less expensive blender containers and to conduct the work in a safety cabinet. Even if the AS-1 bowl is used, it should be opened in a safety cabinet. Lamanna (3) describes a flexible-film plastic cover for controlling aerosols generated by a high-speed blender. This adaptation provides adequate protection during the blending process; however, the bag should be opened in a safety cabinet.

Centrifugation equipment

The potential for multiple infections from a single centrifuge accident is great, especially when fluid escapes from the rotor or cup while the centrifuge is operating at high speed. Tubes breaking during centrifugation pose the greatest hazard when a swinging-cup centrifuge is used.

Care must be exercised when selecting centrifuge tubes. They should not be subjected to a greater centrifugal force than they are designed for. Periodic inspection with a polariscope will detect stress lines in glass that are not visible with ordinary light. Safety trunion cups (Fig. 2) should be used to prevent the escape of aerosols should the centrifuge tube break or any other type of leakage occur.

Extra-strength polypropylene centrifuge tubes (Fig. 3) are available in 250-ml and 125-ml sizes. They are fitted with leakprooof polyethylene screw

FIG. 1. Safety blender container.

FIG. 2. Safety trunion cups.

caps for safety when centrifuging radioactive or pathogenic fluids. Each lot of tubes is tested for accuracy, sterility, and leakage.

Lyophilization ampoules

Accidental exposure to lyophilized microorganisms can occur when ampoules sealed under vacuum implode or are dropped. If glass ampoules are used, they should be thick walled (Fig. 4A). Implosions can be prevented if ampoules are filled with nitrogen gas rather than being evacuated before sealing. This also minimizes the escape of microorganisms which occurs when evacuated ampoules are opened.

Ampoules for liquid nitrogen freezer storage

Ampoules used to store microorganisms in liquid nitrogen have exploded, causing eye injuries. Polypropylene tubes (Fig. 4B) eliminate this hazard. These tubes are available dust-free or presterilized and are fitted with HD polyethylene caps with sili-

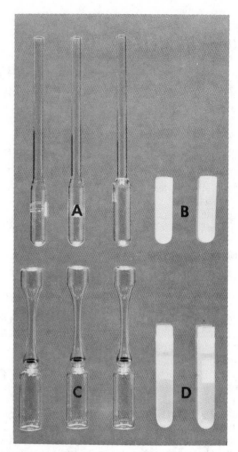

FIG. 4. Lyophilization and nitrogen freezer storage ampoules. (A) Thick-walled glass lyophilization ampoules. (B) Polypropylene tubes fitted with HD polyethylene caps with silicone washers, available in 2- and 5-ml sizes; explosion proof. (C) Thin-wall ampoules for serum freezing in conventional freezers; not for pressure vacuum. (D) Polypropylene serum storage tubes for use with nitrogen freezers.

cone washers. They are available in 2- and 5-ml sizes. Heat-sealable polypropylene tubes are also available.

Hand protection during necropsy procedures

Knife cuts are a common problem during necropsy. Protective gloves should be worn during these procedures (Fig. 5). The woven inner glove is made of fabric-covered stainless-steel wire and protects against cuts; the neoprene outer glove minimizes direct skin contact with contaminated tissue and fluids.

Electric culture loop sterilizer

Sterilizing culture loops in an open flame generates small-particle aerosols which can contain viable

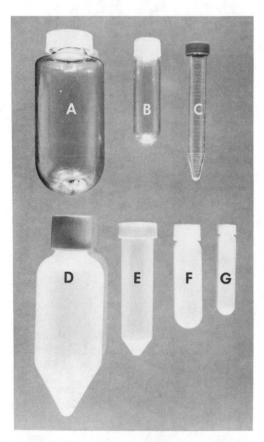

FIG. 3. Centrifuge tubes. (A through C) Transparent polycarbonate tubes: (A) 250 ml; (B) 30 ml; (C) 25 ml. (D through G) Opaque, extra-strength polypropylene tubes: (D) 250 ml; (E) 50 ml; (F) 30 ml; (G) 10 ml.

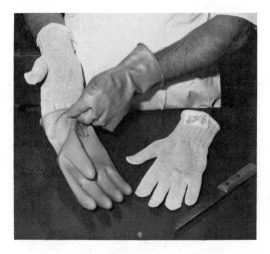

FIG. 5. Double-thickness necropsy gloves.

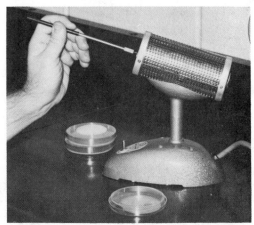

FIG. 6. Electric incinerator loop sterilizer.

microorganisms. The culture loop sterilizer shown in Fig. 6 is an electric incinerator, shielded to contain the aerosol generated by boiling liquid until the organisms are inactivated.

Flameless electric burner

The flameless electric burner shown in Fig. 7 can be used where an open flame is not desirable or in atmospheres which will not support a flame. It operates on 120-V current and requires a 6- to 35-lb/in^2 air supply. It can be placed on a counter top or held in the hand.

Disposable plastic inoculating instruments

Disposable plastic loops and needles (Fig. 8) can be used for culture work where high temperatures or open flames for sterilizing metal instruments are not desirable. They have a working temperature range of -40 to $+70°C$. The loops are semiquantitative and can be used for counting bacteria.

Hypodermic syringes and needles

Syringes and needles have been described as the single most hazardous item of equipment in the biological laboratory (5). One of the most common accidents is the needle popping off the syringe when pressure is applied, which can result in eye as well as respiratory exposure. Disposable glass syringes with permanently attached needles (Fig. 9A) and needle-locking disposable plastic syringes (Fig. 9B) will eliminate this risk. Reusable glass syringes with needle-locking connections are available in capacities from 1 to 100 ml (Fig. 9C and D). Reusable glass syringes sometimes ''freeze'' during sterilization.

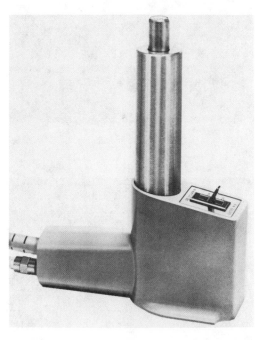

FIG. 7. Flameless electric burner. Develops a heat stream up to 1,100°F (ca. 600°C).

Attempts to open these syringes sometimes result in breakage and hand cuts. Syringe openers such as the one shown in Fig. 10 should be used.

Contaminated-glassware sterilizer pans

Stainless-steel steam table pans available in a variety of sizes (Fig. 11) can be used for sterilizing equipment in the autoclave. Glassware should be submerged in water to facilitate sterilization. These pans are self-stacking and have lids.

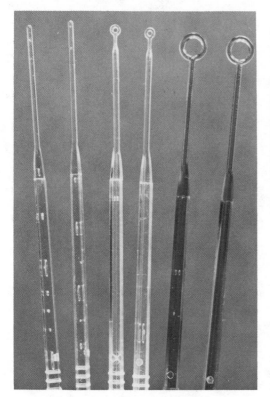

FIG. 8. Disposable plastic inoculating loops and needles.

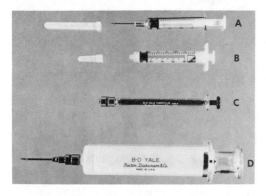

FIG. 9. Hypodermic syringes and needles.

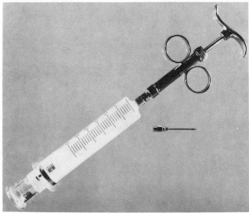

FIG. 10. Syringe opener. Fill syringe opener with water and attach to syringe. Pressure developed by the syringe opener will force the syringe open.

FIG. 11. Contaminated-glassware sterilization pans.

Vacuum guard

Safe operation of laboratory vacuum devices requires a vacuum guard (Fig. 12) which provides constant air filtration and an automatic vacuum break. The guard is installed between the vacuum stopcock and the vacuum device. If a biological solution is accidentally drawn into the vacuum guard, the guard's float will move and a mercury switch will deenergize a valve in series with the vacuum line, switching off the device.

Personnel safety shields

A personnel safety shield should be placed between a potentially dangerous apparatus setup and the personnel working with it. The shield shown in Fig. 13 is made of polycarbonate plastic and has a heavily weighted base for stability.

Electric heating mantle

An electric heating mantle will protect flasks from thermal shock and prevent damage or injury when dangerous solutions are being heated (Fig. 14). It can be used for any heating operation including distillation, fractionation, extraction, etc. The heating element is embedded in layers of glass fabric and

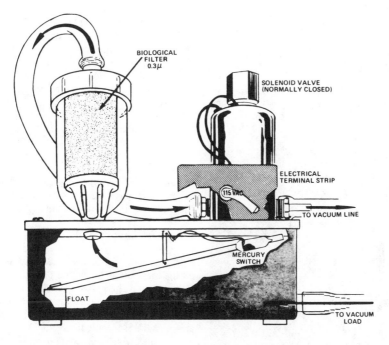

FIG. 12. Vacuum guard to protect the vacuum system from contaminated air and liquid.

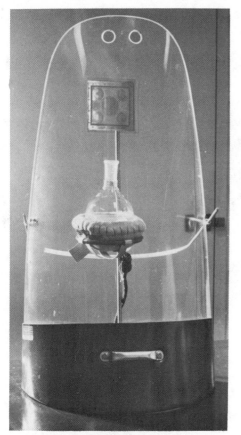

FIG. 13. Personnel safety shield.

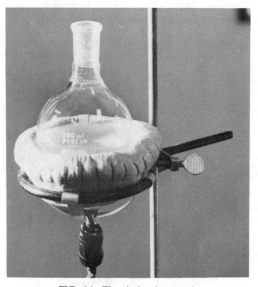

FIG. 14. Electric heating mantle.

covered with glass insulation. This covering protects the vessel wall from thermal strain.

Glassware cuts: protective equipment

Glassware cuts are the most common laboratory accident. A number of devices can be used to minimize these accidents. Safety glasses, goggles, and face shields (Fig. 15) should be worn where warranted. Leather gloves should be worn when handling glass tubing. Large glass vessels, especially those used under pressure or vacuum, should be checked for cracks, stars, or stress lines with a polariscope (Fig. 16).

To insert glass tubing through a rubber stopper, first insert the next larger size cork bore, and then place the tubing inside the bore and withdraw the

FIG. 15. Eye and face protective equipment. (A and B) Goggles for eye protection against broken glass, chemical splashes, eye exposure to infectious microorganisms, and low-energy impacts. If high-energy impacts are possible, safety glasses should be worn under goggles (A). (C) Safety glasses provide protection against high-energy impacts as well as frontal exposure to chemicals and biological hazards. (D) Face shield provides protection against splashes with chemicals and infectious microorganisms but not against high-energy impacts.

bore, leaving the tubing in place (Fig. 17).

Glass stopcocks sometimes become frozen, and attempts to force them open can result in serious cuts. The stopcock remover shown in Fig. 18 greatly simplifies this problem.

Glassware should never be mixed with regular trash for disposal. A specially marked disposal or puncture-proof bag should be used for broken glassware. Broken glass should be picked up with a brush and dust pan, never with the bare hands.

Glass is generally to be avoided in work with pressure and vacuum. If it must be used, it should be shielded as shown in Fig. 13 or contained in a suitable cabinet system. Glass vacuum desiccators should be covered with a shield (Fig. 19).

Hazardous liquids in glass bottles

Plastic bottle carriers (Fig. 20) provide protection during transport and handling of hazardous liquids in glass bottles. The plastic container labeled B in the figure has the advantage that it can remain in use during dispensing. Several chemical manufacturers are marketing hazardous chemicals in vinyl-coated glass bottles which will contain the liquid even if the glass is broken.

Waste organic solvent: storage and transport cans

Waste flammable solvents should be stored in safety cans which are leak tight and have a spring-loaded, self-closing cap (Fig. 21). The cap also acts as a relief vent when pressure builds up in the can.

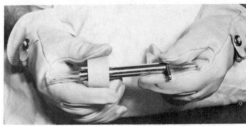

FIG. 17. Inserting glass tubing in a rubber stopper.

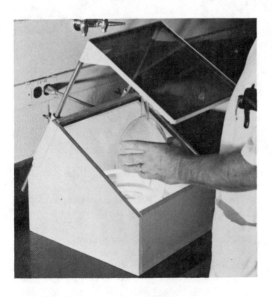

FIG. 16. Polariscope.

FIG. 18. Stopcock remover.

FIG. 19. Glass vacuum desiccator with protective shield.

FIG. 20. Protective plastic bottle carriers. (A) Protects jugs during storage and transport, but the jug must be removed to pour contents. (B) Protects jugs during storage, transport, and use; contents can be poured from the jug with the jug still in the carrier.

Flash arrestor screens inside the spout prevent fire flashback to the can's contents.

Disposal cans for needles and scalpel blades

Mixing needles, scalpel blades, and other sharps with trash is dangerous. Disposal cans (Fig. 22) can be purchased in pint, quart, half-gallon, and gallon sizes. A 0.5-in.-diameter hole is punched in the lid to accommodate needles and blades. When the cans are full, they can be autoclaved before disposal. Hypodermic needles are a special problem around landfills and other disposal sites because they create a permanent slow leak when they puncture pneumatic tires.

Biohazard transport containers

When carrying biohazardous material from one area to another, a liquid-tight transport container

FIG. 21. Waste organic solvent storage and transport safety can.

FIG. 22. Disposal can for needle and scalpel blades.

should be used. One-gallon and one-half-gallon paint pails (Fig. 23) are relatively inexpensive and work

FIG. 23. Biohazard transport container (paint pail with biohazard label).

quite well for this purpose. A biohazard label should be placed on the container.

Safety chains or straps for compressed-gas cylinders

When compressed-gas cylinders are used in the laboratory, they should be secured to a bench or some other strong support by means of a chain or belt fitted with an adjustable clamp (Fig. 24).

Biological sterility indicator systems

Biological sterility indicator systems (Fig. 25) are designed to monitor the effectiveness of sterilization of material and equipment. *Bacillus subtilis* spore strips are used for ethylene oxide sterilization, and *Bacillus stearothermophilus* spore strips are used for steam sterilization. The systems marked A and B in Fig. 25 consist of a small plastic vial (labeled 1) with a filter top which contains an ampoule of nutrient medium with an indicator dye and a spore strip. This vial is placed in the load to be sterilized. After exposure to steam or ethylene oxide, the vial is removed and the ampoule of nutrient medium is ruptured by firm pressure on the sides of the vial. The vial is then placed in the appropriate small fixed-temperature incubator (labeled 2 in Fig. 25) (37°C for *B. subtilis* and 55°C for *B. stearothermophilus*) and incubated for 24 h. If spores have survived, the indicator dye will change from purple to yellow. Other commercially available spore strips can also be used but require some experience in microbiology (Fig. 25C). If sufficient quantities are needed to justify the effort, spore strips can be prepared in the laboratory (Fig. 25D).

FIG. 24. Safety clamp and strap for compressed-gas cylinders.

Ventilation smoke tubes

Smoke tubes can be used to determine the direction and approximate velocity of airstream (Fig. 26). The glass ampoule contains titanium tetrachloride (TiCl$_4$) absorbed on a substrate. The ends of the tube are broken, and air is blown through the tube with the aspirator bulb. As the TiCl$_4$ reacts with moisture in the air, a cloud of white smoke is formed. The smoke is irritating to the respiratory system, so care should be exercised to minimize respiratory exposure. These tubes are especially useful for quick checks on cabinet openings and negative-pressure rooms.

Respiratory protective equipment

Respiratory protective equipment should be available and used as needed. Some types of portable respiratory protective equipment are shown in Fig. 27. Full-face gas masks which provide eye as well as respiratory protection can be fitted with a variety of cartridges to protect against low concentrations of toxic gases for short periods of time. Respirators (Fig. 27C) are for respiratory protection only and can also be fitted with cartridges to protect against low concentrations of toxic gases for short periods of time. Both full-face gas masks and respirators can be fitted with a particle arrestance filter which is

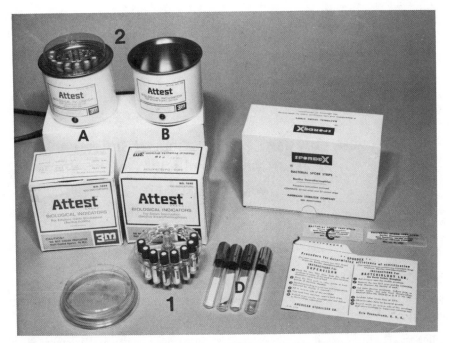

FIG. 25. Biological sterility indicator systems. (A) *B. subtilis* spore strip system. (B) *B. stearothermophilus* spore strip system: (1) spore strips in incubator rack: (2) incubators. (C) Commercial spore strips. (D) Prepared spore strips.

FIG. 26. Ventilation smoke tube for testing air flow directions.

99.97% efficient on particles 0.3 μm in diameter. Dust masks (Fig. 27D and E) are for the most part effective in removing large dust particles. The mask labeled E is effective against dust particles and, being impregnated with charcoal, is also effective in odor control. In areas of extreme hazard, ventilated whole-body suits and hoods can be used. (See also Chapter 20.)

LITERATURE CITED

1. **Anderson, R. E., L. Stein, M. L. Moss, and N. H. Gross.** 1952. Potential infectious hazards of common bacteriological techniques. J. Bacteriol. **64:**473–481.

2. **Kenny, M. T., and F. L. Sabel.** 1968. Particle size distribution of *Serratia marcescens* aerosols created during common laboratory procedures and simulated laboratory accidents. Appl. Microbiol. **16:**1146–1150.

3. **Lamanna, C.** 1960. Use of autoclavable plastic material for do-it-yourself solutions to laboratory problems of safety. J. Lab. Clin. Med. **55:**319-321.

4. **Reitman, M., M. A. Frank, Sr., R. Alg, and W. S. Miller.** 1954. Modifications of the high speed safety blender. Appl. Microbiol. **2:**173.

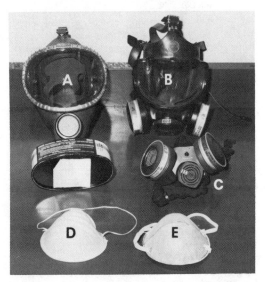

FIG. 27. Respiratory protective equipment. See the text for details.

5. **Wedum, A. G.** 1974. Biohazard control, p. 191–210. *In* E. C. Melby and N. H. Altman (ed.), Handbook of laboratory animal science, vol.1. CRC Press, Inc., Cleveland.

6. **Winkler, W. G., T. R. Fashinell, L. Leffingwell, P. Howard, and J. R. Conomy.** 1973. Airborne rabies transmission in the laboratory worker. J. Am. Vet. Med. Assoc. **226:**1219–1221.

CONTAINMENT TECHNIQUES

Chapter 22

Decontamination, Sterilization, Disinfection, and Antisepsis in the Microbiology Laboratory

DONALD VESLEY AND JAMES LAUER

The array of microbicidal tasks routinely performed in laboratories can be ascribed either to safety purposes (which we will refer to as decontamination) or to quality control purposes, and sometimes to both. The decontamination of cultures and items contaminated by biohazardous agents is a vital step toward protection of laboratory workers from infectious disease. The decontamination process is also necessary to prevent release of such agents into the community at large, an objective which has been largely attained as attested by the small number of reported incidents involving secondary spread. Those which have been reported primarily involve person-to-person rather than environmental transmission (4). Conversely, sterilization of media and equipment is vital to the integrity of any microbiological manipulation as a component of standard quality control practices.

These microbicidal processes are taken for granted by most laboratory workers, yet considerable confusion remains over terminology and over specific techniques to be utilized, and in some instances there is a dearth of specific knowledge about the effectiveness of a given microbicidal process for its intended purpose.

It is our objective in this chapter to propose workable terminology and to spell out recommendations which may be of practical use over a wide range of laboratory practices. Many publications are available which delve into the theoretical aspects of microbicidal action and into the intricacies of specific microbicidal agents. The reader is referred particularly to the third edition of *Disinfection, Sterilization, and Preservation*, edited by Seymour S. Block (5), for a comprehensive treatment of this subject. It is not our intention to reiterate the technical details of that very useful text. Instead, we focus on the practical aspects of everyday microbicidal tasks which must be carried out in the microbiology laboratory. In so doing, we will identify the various categories or objectives of these tasks, highlight pitfalls, limitations, and hazards associated with the use of microbicidal agents, and then catalog suggested procedures for many specific tasks.

TERMINOLOGY OF MICROBICIDAL ACTION

Discussion of microbicidal terminology has occupied the attention of microbiologists since the inception of the science. Such discussions frequently degenerate into semantic dilemmas in which everyone apparently agrees in principle but then goes right on misusing terms due to ingrained habits. We do not expect to end this confusion definitively, but nonetheless we feel compelled to lay down ground rules which will at least enable the reader to interpret the information in this chapter consistently.

Kelsey brought up this dilemma in his article, "The Myth of Surgical Sterility" (14). Sterility testing is impractical due to its destructive nature, the large number of samples needed to detect even a relatively high percentage of nonsterile items, and the ever-present possibility of false-positives due to lab contamination. The alternative, of basing destruction cycles on experimentally determined D values (the time required for a 90% reduction in number of surviving microbes under specific test conditions), also has pitfalls in that the D values themselves are subject to variation. This criterion also has the built-in difficulty of requiring assignment of a probability of sterility due to the logarithmic decline in survivors which implies an infinite time to reach zero, the only accepted definition of sterility. Brown and Gilbert (6) have proposed abandonment of the term "sterility" in favor of "safe for its intended use," provided that an "acceptable" probability of sterility can be demonstrated. Campbell (7) has proposed that this probability be expressed in terms of a "microbiological safety index," a number representing the logarithm of the reciprocal of the theoretical number of survivors. Thus, an autoclave pack with an assumed initial bioload of 10^6 most resistant spores, subjected to a cycle transversing 12 logs (e.g., a 24-min cycle based on a D value of 2 min), would end up with 10^{-6} survivors, for a microbiological safety index of 6. It is obvious that this concept also depends on assumptions which cannot be verified easily. However, in practice, autoclave sterilization cycles probably grossly overestimate the initial spore load and the D value, thus creating a large margin of safety if the steam does in fact penetrate to all organisms in or on the item to be sterilized.

The difficulties inherent in assuming sterility are compounded when the objective is decontamination for safety purposes. Decontamination is usually defined as the destruction or removal of microorganisms to some lower level, but not necessarily total destruction. Logically the safety objective can only be met by a sertilization process when a high-risk etiological agent (Biosafety Level 3 or 4) is present, but it may be met by a lesser microbicidal treatment for lower-risk agents. Thus, decontamination may variably be achieved by application of sterilization, disinfection, or antisepsis treatments as the situation dictates. The situation is further complicated by the nature of decontamination treatments, which often involve the presence of large quantities of organic matter such as microbiological media, a factor which can be overcome only by increasing cycle time or increasing the concentration of the microbicidal agent. Verification of results under these assumptions is even more nebulous than for conventional quality control sterilization processes carried out on "clean" items.

We are usually left depending on chemical and biological indicators to determine our decision whether to accept the assumption of sterility. An excellent monograph on this subject has been prepared by the Health Industry Manufacturers Association (13). Chemical indicators can confirm exposure of the indicator to minimum time and temperature conditions. Recent technical advances have produced indicators which integrate time and temperature progression in a steam sterilizer cycle, providing confirmation of theoretically adequate microbicidal conditions at a given chamber location.

The biological indicator comes closest to a legitimate measure of spore kill because a negative result actually requires the total destruction (or at least nonrecoverability) of a given number of resistant spores (*Bacillus stearothermophilus* for steam and *Bacillus subtilis* subsp. *globigii* for ethylene oxide or dry heat). Even biological indicators, however, have limitations (e.g., variable resistance levels and a variable number of spores inoculated on the strip). In addition, they do not compensate for the high organic loads inherent in decontamination tasks and thus are of more questionable value for those tasks. However, they remain the most popular indicators of total microbial destruction. For both chemical and biological indicators it is essential that the location of the indicator be chosen carefully for meaningful in-

terpretation, preferably at a point most resistant to steam penetration and in the correct chamber location. All of this does not even address the difficulty of determining the effectiveness of nonsporicidal chemical germicides. The success of such compounds can only be assumed if the manufacturer's recommended dilution of a liquid antimicrobial agent is used for the recommended contact period. Even assuming that the Environmental Protection Agency registration process for such products is meaningful (requiring passage by the Association of Official Analytical Chemists [use-dilution test]), we are left with the difficult assignment of attempting to quantitate the degree of destruction achieved under given circumstances. Microbiological sampling of surfaces subjected to germicidal treatment may be useful in the evaluation of specific treatments, but is generally not cost-effective for routine monitoring.

In fact, then, we do not have consistently reliable methods for confirming the degree of destruction actually achieved in day-to-day microbicidal activities. We suggest that the objectives of decontamination can best be reached by selecting empirically acceptable techniques and applying them diligently.

For the purposes of this chapter we propose the following terms.

Decontamination

Decontamination is probably the most common and significant category of antimicrobial action in microbiological laboratories. The laboratory environment and the objects within it are subjected to contamination with infectious agents, often in the presence of extraneous organic matter (culture media, etc.). Decontamination, therefore, implies the application of microbicidal steam, gas, or liquid chemical agents in situations where microbes may be protected from contact by extraneous matter. It is not usually possible to assume total destruction of either spores or vegetative microbes. The objective of decontamination is simply to render the material safe for further handling, whether it is an instrument to be subsequently washed and sterilized, or liquid or solid waste to be disposed of by regular means. (For high-risk agents, however, total destruction may well be required for adequate safety.)

Sterilization, disinfection, and antisepsis are all forms of decontamination; decontamination, however, may or may not involve any of these. Liquid cultures are decontaminated before being discharged to the sewer; instruments are decontaminated (sometimes by both a liquid chemical soak and an autoclave cycle) before being washed and sterilized for reuse. Even routine day-end floor cleaning can be considered decontamination, as the process involves a contaminated surface which is treated with a germicide (detergent-germicide) to make it suitable for further use, without an assumption of the specific degree of microbe destruction.

It is important to emphasize that appreciably greater concentrations and times are often recommended for decontamination treatments than for sterilization or disinfection. The steam (121°) cycle for sterilization is 30 min, but that for decontamination is 1 h; disinfection techniques call for 100 mg of chlorine per liter, but 10,000 mg/liter is recommended for decontamination (emergency spill cleanup). This reflects the necessity of reaching the microbes through the possible protection of extraneous matter, as discussed above.

Sterilization

The term "sterilization" is reserved for the following uses. (i) It applies to the processing of clean, prewrapped items in steam sterilizers or ethylene oxide autoclaves using cycles which consistently produce negative results in chemical and biological indicators. It is assumed that the packaging and wrapping will permit penetration of the steam or gas to all areas of interest and then will protect against posttreatment recontamination. (ii) Liquids such as microbiological culture media are sterilized in steam autoclaves to ensure the integrity of the experiment. (iii) Sterilization also can be the objective when items are decontaminated by steam or gas with increased cycle times to compensate for heavy organic loads.

Disinfection

The term "disinfection" implies the use of liquid antimicrobial chemicals on inanimate objects (work surfaces, equipment items, etc.) with the objective of destroying all nonsporeforming organisms of potential hazard to humans or which could compromise the integrity of the experiment. Thus disinfection may be carried out for safety or quality control purposes and frequently for both simultaneously.

The same objective applied to heat treatment of a liquid or to subboiling wet-heat treatment of objects is termed pasteurization.

Antisepsis

The term "antisepsis" concerns the application of a liquid antimicrobial chemical to living tissue (human or animal). The objective is to prevent sepsis; this may be achieved either by destroying potentially infectious organisms or by preventing their growth and multiplication. Again, no sporicidal activity is implied. This category includes such diverse applications as swabbing of an injection site on a research animal and hand washing with germicidal solutions for laboratory personnel. In the case of hand washing, the objective is obviously broadened to include preventing the spread of infectious or contaminating agents. This, again, implies that both safety and quality control objectives may be pertinent.

Sterilants, disinfectants, and antiseptics

One further note on terminology involves the use of words such as "sterilant," "disinfectant," or "antiseptic." These terms should be used to denote products or processes capable of achieving sterilization, disinfection, or antisepsis, respectively, under ideal conditions.

PRECAUTIONS, LIMITATIONS, AND TOXIC HAZARDS OF MICROBICIDAL AGENTS

The application of microbicidal treatments in many laboratory activities has long been considered to be necessary and routine. Less attention, however, has been focused on the opposite interpretation of this decontamination activity; namely, that the capacity for DNA disruption which makes such agents useful may also make them dangerous to personnel, disruptive to research microbes, or both.

The more obvious of these undesirable properties can be mitigated by common-sense precautions, but some hazards are more subtle, and greater attention to potential dangers should be included in recommended procedures. For example, the hazard of burns from escaping steam can be overcome by good autoclave design and maintenance, together with thorough training of personnel, but the risk of exposure to ethylene oxide remains controversial. In 1984 the Occupational Safety and Health Administration standard was reduced to 1.0 mg/liter (time weighted average), with an action level of 0.5 mg/liter. These levels replaced a previous standard of 50 mg/liter and are near the limit which can accurately be measured by currently available monitoring instruments. The National Institute of Occupational Safety and Health has published a comprehensive review of ethylene oxide hazards (12).

Indeed, the literature is replete with reports of acute toxic effects of one kind or another for virtually every class of microbicidal agent used in laboratories. Chronic exposure effects have been more difficult to document, and projection of mutagenic, carcinogenic, or teratogenic consequences remains primarily speculative. Effects on animals have frequently been more convincingly documented than those on humans, for the obvious reason that animal models are necessarily used for toxicity evaluations. The complication for microbiology laboratories is obvious, however, in that the effect of microbicidal agents on research animals becomes an important quality control consideration.

Table 1 summarizes some of the harmful effects reported for the various types of microbicidal agents used in laboratories. A few representative literature citations are included.

In many instances the toxic hazards described are related to careless overuse of microbicidal agents at

TABLE 1. Human hazards associated with selected antimicrobial agents

Antimicrobial agent	Indicated hazard	Reference
Steam (autoclave)	Burns from escaping steam or hot liquids; cuts from exploding bottles	Many incidents reported
	Aerosols and chemical vapors from improper vacuum exhaust	Barbeito and Brookey (1)
UV light	Corneal and skin burns from direct or deflected light	Many incidents reported
Ethylene oxide gas	Eye and respiratory irritant, skin desiccant, mutagen, potential carcinogen	Glaser (12)
Formaldehyde gas	Highly irritating; toxicity and hypersensitivity	National Institute of Occupational Safety and Health (18)
Alcohol (isopropyl)	Acute toxicity	Freireich et al. (11)
	Contact dermatitis	Fregert et al. (10)
Chlorine	Gaseous form highly toxic; liquid Cl_2 not toxic at active dilutions	
Glutaraldehyde	Contact dermatitis	Sanderson and Cronin (22)
Hexachlorophene (bisphenol)	Acute neurotoxin	Mullick (17)
Iodine (iodophors)	Skin irritation	Zinner et al. (27)
Phenols	Occupational leukoderma	Bentley-Phillips (3)
	Depigmentation	Odom and Stein (19)
	Idiopathic neonatal hyperbilirubinemia	Center for Disease Control (8)
Quaternary ammonium compounds	Minor contact dermatitis	Shmunes and Levy (23)
Chloroform	Weak carcinogen (animal models only)	

concentrations well above the recommended use dilution. In fact, deaths from hexachlorophene and the neonatal hyperbilirubinemia problem are examples of that type of careless mistake (8). There are also a variety of precautions related to microbicidal treatments which stem from theoretical knowledge of potential hazards. Barbeito and Brookey (1) have described the potential for aerosols to escape from a vacuum autoclave with air exhausted before the steam cycle. Methods are described to ensure that autoclave design will not allow such escape. Another example is the precaution of limiting the quantity of ethanol used in wiping down laminar-flow cabinets. The explosion potential of ethanol should restrict that quantity to 400 ml for a 4-ft cabinet and 600 ml for a 6-ft cabinet. These quantities are significantly above the amount needed for surface decontamination, but should be kept in mind relative to total quantities contained in the cabinet. Another theoretical potential for hazards exists in the practice of autoclaving pans of items which are soaked in a chlorine solution. It is theoretically possible for poisonous gaseous

chlorine to be generated during autoclaving. Generally the high organic load in such a situation will serve to neutralize the chlorine, but a worthwhile precaution would be to add sodium thiosulfate (1 ml of 5% sodium thiosulfate per ml of 5% hypochlorite ion). This would also help to prevent corrosive action of Cl_2 on susceptible autoclave components.

The damaging effect of microbicidal treatments on containers and other surfaces is a legitimate concern. Table 2 summarizes the characteristics of various plastics relative to steam, ethylene oxide, and dry-heat treatments. Table 3 indicates the relative resistance of these plastics to various categories of liquid chemicals. Centrifuge rotors and accessories are particularly susceptible to damage from chemicals which may be used for decontamination purposes. Manufacturer's instructions should always be followed for such practices. Generally, titanium rotors are more resistant than aluminum, and stainless-steel tubes and caps are more resistant than most plastics. Many rotors, including O-rings and gaskets, can be autoclaved at 121°C for 1 h without damage (2).

TABLE 2. Physical properties of commonly used laboratory plastics relative to decontamination or sterilization[a]

Plastic	Maximum use temp (°C)	Transparency	Steam autoclave	ETO[b] or formaldehyde gas	Dry heat
Low-density polyethylene	80	Translucent	No	Yes	No
High-density polyethylene	120	Translucent	—[c]	Yes	No
Polypropylene	135	Translucent	Yes	Yes	No
Polyallomer	130	Translucent	Yes	Yes	Yes
Polymethylpentene	175	Clear	Yes	Yes	No
Teflon	205	Translucent	Yes	Yes	Yes
Polycarbonate	135	Clear	Yes[d]	Yes	No
Polyvinyl chloride	70[e]	Clear	No[e]	Yes	No
Polysulfone	165	Clear	Yes	Yes	Yes

[a] By permission of Nalgene Co., Division of Sybron Corp., Rochester, N.Y.
[b] ETO, Ethylene oxide.
[c] Can be autoclaved at 120°C for 20 min if containers are empty and uncovered.
[d] Mechanical strength may be reduced by autoclaving.
[e] Polyvinyl chloride in tubing can withstand 121°C.

TABLE 3. Resistance of commonly used laboratory plastics to a variety of substances[a]

Plastic	Weak acids	Strong acids	Alcohols	Aldehydes	Bases	Halogenated hydrocarbons	Strong oxidizing agents
Low-density polyethylene	a	a	a	b	a	d	c
High-density polyethylene	a	a	a	b	a	c	c
Polypropylene	a	a	a	b	a	c	c
Polyallomer	a	a	a	b	a	c	c
Polymethylpentene	a	a	a	b	a	d	c
Teflon	a	a	a	a	a	a	a
Polycarbonate	a	d	b	c	d	d	d
Polyvinyl chloride	a	a	a	d	a	d	b
Polysulfone	a	b	b	c	a	d	c

[a] By permission of Nalgene Co., Division of Sybron Corp., Rochester, N.Y.
[b] a, No damage after 30 days of exposure. b, No damage after 7 days; some effect in 30 days. c, Some damage after 7 days of exposure. d, Not recommended for continuous use.

It should be pointed out again that frequently the concentrations of microbicidal agents recommended for decontamination tasks are significantly higher than the conventional use dilution recommended by the manufacturer. Therefore, precautions are all the more important when considering the risk versus benefit for a particular activity.

GENERAL CONSIDERATIONS

Before embarking on a catalog of specific recommendations for handling microbicidal tasks in a laboratory, we wish to point out several general considerations which we consider important and which experience tells us are frequently overlooked.

The first of these has to do with the general clutter which so frequently pervades the laboratory. Many investigators seem to believe that every piece of equipment which conceivably could be required in a lifetime career must be within arm's length at all times. The result is a working environment incompatible with decontamination needs. A require-

ment even at the basic laboratory level is for daily wipe-down of bench tops with a suitable disinfectant. In fact, even this task is frequently ignored or is complicated by clutter. Correction of a problem such as this requires an attitude change on the part of the worker and as such may become one of the most difficult solutions to implement despite its technical simplicity. We emphasize again how much the reality of laboratory safety depends on individual willingness and desire to comply rather than on technical considerations.

Another reason for minimizing the amount of extraneous material in the work environment has to do with accidental spills. The procedures spelled out in the section below for emergency cleanup become extremely cumbersome in the presence of clutter. Examples of obstacles to efficient cleanup include venetian blinds on windows and cardboard boxes stored on the floor. Thus, limiting the laboratory to those items specifically to be utilized in current work will expedite decontamination tasks of several types and, incidentally, contribute to quality control and to fire safety.

Another general consideration involves the separation and clear identification of items to be decontaminated. A "dirty" cart clearly labeled "To Be Autoclaved" is recommended for all laboratories where infectious disease agents are handled. Once more, the diligence and training of all workers is essential to ensure that appropriate items are promptly placed on the cart, but such a system must be instituted to avoid having culture plates, flasks, etc., lying around with no indication of their status or future. Autoclave cycles must then be scheduled with sufficient frequency to avoid overloading of the cart and the consequent misplacement of contaminated items. Additionally, separate containers must be provided for reusable and disposable items to avoid problems with melting plastic.

RECOMMENDED PROTOCOLS

A list of various decontaminants, their effectiveness against different microbial groups, their important characteristics, and their most appropriate application in research and clinical laboratories is given in Table 4. However, for the effective use of these decontaminants, specific protocols must be followed.

Emergency Spill Protocols

It is recommended that every laboratory should prepare a specific protocol to be followed in the event of a biohazardous spill. The protocol should include designation of responsibilities as to who will carry out each of the specified steps. In most instances this should be the person who perpetrated the accident, but some institutions may prefer to assign the Biosafety Officer to this task, particularly when high-risk or even moderate-risk agents are involved.

To make the cleanup protocol effective, it is essential that all materials designated in the written procedure be readily available and functional on a continuous basis. At a minimum, the materials available for emergency cleanup should include suitable disinfectants, paper towels, forceps, an autoclavable squeegee and dust pan, sponges, and autoclavable plastic bags. All laboratory personnel should be familiar with the procedures and should have immediate access to the necessary equipment. Periodic emergency drills should be scheduled to ascertain that this state of readiness indeed exists.

In a properly designed and equipped laboratory, spill cleanup will be enhanced by easily cleanable surfaces and properly installed equipment (either built in or free standing to permit cleaning access under and around all components). Minimizing unnecessary storage of supplies and removing items and equipment not needed in the day's activities will serve to simplify cleanup should a spill occur.

Accessories such as venetian blinds or decorations should be minimized for the same reason.

Specific protocols should be developed for various potential emergency spill situations. The following common accidents will be discussed in this section: (i) spill in a laminar flow biological safety cabinet; (ii) spill on a laboratory bench or floor (distinction will also be made for containment versus noncontainment laboratories); and (iii) spill outside of a laboratory (such as during transport of biohazardous agents).

Spill in a laminar-flow biological safety cabinet

If a spill should occur within a laminar-flow biological safety cabinet, the cabinet should continue to operate, as it is designed specifically to contain aerosols and all exhaust air is filtered. Cleanup should be initiated as soon as possible, utilizing a suitable disinfectant. A germicidal detergent such as a phenolic or iodophor should be used in wiping all reachable cabinet surfaces.

If the cabinet incorporates a catch basin beneath the work surface, this should be flooded with the disinfectant. Alcohol is not recommended due to the explosive potential of large quantities.

At least 20 min of contact time with the germicide should be allowed before items are removed from the cabinet. All items within the cabinet should be packaged for transport to the autoclave or, for items which cannot be autoclaved, such as viable cultures to be retained, wiped carefully with the disinfectant. The operator's gloves and long-sleeved gown should be considered contaminated and should be placed in a container for autoclaving after the cleanup process is completed. The cabinet should be run for 10 min or so after this process and before activity is resumed.

This procedure is recommended for all spills of moderate- or high-risk agents, but may be modified for low-risk agents (i.e., conventional cabinet wipedown may be utilized without flooding of the catch basin if a small quantity of an agent being handled under Biosafety Level 2 is spilled).

A decision must be made after spill cleanup as to whether or not formaldehyde decontamination of the cabinet, including filters, is necessary. In general this should be done for any high-risk agent (although Biosafety Level 4 mandates the use only of Class III cabinets) and for a major spill of a moderate-risk agent.

The procedure recommended by the National Institutes of Health (see below, Protocols for Laminar-Flow Biological Safety Cabinets) for formaldehyde decontamination should be followed. This procedure utilizes either flake paraformaldehyde, at 0.3 g/ft^3, or 37% Formalin. The humidity in the cabinet should be raised (by vaporizing water) to 70% or more, and the cabinet must be sealed with plastic sheets and tape to contain the formaldehyde gas. If the cabinet does not exhaust directly to the outdoors, a flexible hose connection must be utilized to remove the gas after decontamination. During the decontamination process the exhaust must be sealed to achieve adequate contact time (1 to 3 h is recommended). After this period, the exhaust damper is opened and the cabinet is turned on to exhaust all residual formaldehyde. It is recommended that the cabinet be run at least overnight before normal activity is resumed. Formaldehyde is highly irritating and toxic, with an Occupational Safety and Health Administration-established permissible exposure level of 2 ppm.

Spills in a laboratory

The procedure recommended assumes a spill of a moderate-risk agent and therefore also assumes that the agent is being handled in a containment laboratory with negative pressure (Biosafety Level 3). If that is the case, the initial response should be to warn others in the room (or suite of rooms) to leave as quickly as possible. The worker involved in the spill should remove contaminated clothing (the containment lab situation also suggests that a wrap-around garment is being worn) and place it in an autoclavable plastic bag for decontamination immediately upon leaving the laboratory.

It is important to note that if the laboratory is not under negative pressure, then it may be necessary to begin cleanup immediately after others are warned to leave. The advantage of the negative pressure is that it permits a suggested 30-min waiting period for any aerosol resulting from the spill to settle before cleanup is begun.

During the waiting period the cleanup materials can be assembled, and the person who will carry out the procedure can don suitable attire. For moderate-risk agents, disposable booties, a long-sleeved gown, single-use mask, and rubber gloves are suggested. If a high-risk agent is involved, a jump suit with tight-fitting wrists and a full-face respirator with high-efficiency particulate air filter cartridge should be considered.

The cleanup process should utilize a concentrated germicide such as 5% bleach (50,000 ppm Cl_2). The germicide should be poured around the edges of the spill to avoid further aerosolization, or, alternatively, paper towels soaked in the germicide can be placed over the spill area. The extent of the spill area must be determined to define the scope of the cleanup activity. It is at this point that the value of having

TABLE 4. Decontaminants and their use in research and clinical laboratories[a]

Decontaminant	Use parameters				Effective against:					Important characteristics						
	Concn of active ingredient	Temp (°C)	Relative humidity (%)	Contact time (min)	Vegetative bacteria	Lipo viruses	Tubercle bacilli	Hydrophilic viruses	Bacterial spores	Inactivated by organic matter	Residual	Corrosive	Skin irritant	Eye irritant	Respiratory irritant	Toxic (absorbed or ingested)
Autoclave (15 lb/in²)	Saturated steam	121		50–90	+	+	+	+	+							
Autoclave (27 lb/in²)	Saturated steam	132		10–20	+	+	+	+	+							
Dry-heat oven		160–180		180–240	+	+	+	+	+							
Incinerator	Heat	649–926		1–60+	+	+	+	+	+							
UV radiation (253.7 μm)	40 μW/cm²			10–30	+	+	+	±		+[c]						
Ethylene oxide	400–800 mg/liter	35–60	30–60	105–240	+	+	+	+	+		+				+	+
Paraformaldehyde (gas)	0.3 g/ft³	>23	>60	60–180	+	+	+	+	+		+				+	+
Quaternary ammonium compounds	0.1–2%			10–30	+	+				+						+
Phenolic compounds	0.2–3%			10–30	+	+	+	±		±	+	+	+	+	±	+
Chlorine compounds	0.01–5%			10–30	+	+	+	+	±	+	±	+	+	+	+	+
Iodophor compounds	0.47%			10–30	+	+	+	±		+	+	+	+	+		+
Alcohol (ethyl or isopropyl)	70–85%			10–30	+	+	±								+	+
Formaldehyde (liquid)	4–8%			10–30	+	+	+	+	±		+		+	+	±	+
Glutaraldehyde	2%			10–600	+	+	+	+	+		+		+	+	+	+

[a] +, Very positive response; ±, less positive response. A blank denotes a negative response or not applicable.
[b] Contact manufacturer of instruments.
[c] Soil and other materials are not penetrated by UV radiation.
[d] The pungent and irritating characteristics of formaldehyde preclude its use for biohazard spills.

removed extraneous items will be apparent, as all objects within the spill area must be decontaminated. Approximately 20 min of contact time should be allowed to ensure adequate germicidal action. All spill materials are then gathered with the autoclavable squeegee and dust pan and transferred to an autoclavable plastic bag which may itself be put in a stainless-steel bucket or pan. Larger items which can be autoclaved may be separately bagged for transport to the autoclave. A cart with side rails to prevent materials from sliding off should be used for transport. A direct route avoiding populated areas should be selected. At this point it becomes obvious why a steam autoclave should be located within a containment lab or very nearby.

The specific precautions and parameters for use of the steam autoclave will be described in a later section.

Spills outside the laboratory: during transport

If a biohazardous agent is spilled during transport outside a laboratory, the major difference from the previous procedure would be to initiate the cleanup immediately, as in a noncontainment laboratory. Because it would already be too late to prevent aerosolization with its unpredictable consequences, the major emphasis should be on preventing spills during transport. All cultures which must be removed from the lab for incubation, refrigeration, or any other purpose should be placed in an unbreakable container which would prevent escape of any liquid or aerosol if it should be dropped. As described in Chapter 21, one-gallon or half-gallon paint pails are an inexpensive means of accomplishing this objective. A biohazard symbol should be affixed to containers used for this purpose. Protective plastic jug carriers are also commercially available.

Combination spills (biohazardous and radioactive materials)

In the event that a spill of biohazardous materials should also involve radioactive substances, the cleanup procedure becomes somewhat more complex. The institution's Radiation Protection officer should be consulted before cleanup begins, to determine what, if any, modification may have to be instituted. The most likely type of radiation hazard in a microbiological laboratory will be either ^{14}C or ^{3}H,

TABLE 4. continued

	Applications																					
Work surface maintenance	Floor maintenance	Biohazard spill, floor surfaces	Safety cabinet surface maintenance	Safety cabinet biohazard spill	Safety cabinet total decontamination	Large area and air systems decontamination	Water baths	Contaminated liquid—discard	Infectious laboratory waste	Contaminated glassware	Contaminated instruments	Equipment surfaces	Equipment total decontamination	Floor drains	Animal injection site	Contaminated animal bedding	Infected animal carcass	Microbial transfer loop	Books, paper, shoes	Electrical instruments	Lensed instruments[b]
								+	+	+	+					+	±				
								+	+	+	+					+	±				
										+	+				±	+	+				
		±																			
										±	±		±					+	+		
						+	+						+							+	
+	+		+					±		+		+									
+	+	±	+	±				+		+		+	+		+						
		+	+		+			±		+	+	+		+							
+	+	±	+	±				±		+		+									
+	+		+									+				+					
		±d			+							±	±	±							
										+	+	±									±

neither of which presents an external hazard. The spill should be surveyed to determine whether there is a need to protect against hand and body radiation exposures from more energetic beta and gamma emitters. Under the joint guidance of the Biosafety Officer and the Radiation Protection Officer, the cleanup can proceed along the lines previously specified for laboratory biohazard spills. The Radiation Protection Officer should evaluate the potential hazard of radioactivity release before the steam autoclave is used. Again, usually ^{14}C and ^{3}H can be autoclaved without hazard. However, ^{125}I may be of sufficient activity to preclude that step. In that case the biohazardous agent may have to be inactivated by use of a compatible liquid chemical germicide before the shipment is packaged as a radioactive waste.

Do not use Cl_2 as a liquid germicide; there is the potential for release of I_2 through the chemical reaction. Either Formalin or glutaraldehyde would be a suitable substitute.

After the area of the spill is cleaned up, a final radiation survey should be made. If the radiation level exceeds twice the background level, radiation decontamination is needed and should be carried out under the supervision of the Radiation Protection Officer, using an approved detergent for radiation cleanup.

Further information on radiation problems in a microbiological laboratory is presented in Chapter 27.

Protocols for Decontaminating Infectious Waste

All laboratories handling moderate- to high-risk agents should have a steam autoclave within a restricted area which serves a number of laboratories. For laboratories handling low-risk agents, an autoclave should also be available, preferably on the same floor and in the general vicinity of the laboratory.

Infectious laboratory waste (petri dishes, culture tubes, animal bedding, contaminated liquid, etc.) and contaminated instruments and glassware can be decontaminated in either a gravity displacement (most common type) or a prevacuum autoclave operating at 121 to 132°C. However, the processing time necessary to achieve decontamination will depend on specific loading factors. These factors include the type of waste container used (metal versus polypropylene), the use of autoclavable waste bags, the amount of water added to the waste, and the weight of the waste load (15).

Figures 1 through 5 show the average time-temperature profiles in waste relative to these waste loading factors. Figure 1 shows that when 1 liter of water was added to waste in a steel container and to waste in an autoclavable waste bag in the steel container, the temperature of the waste increased rapidly with processing time, reaching 120 to 121°C

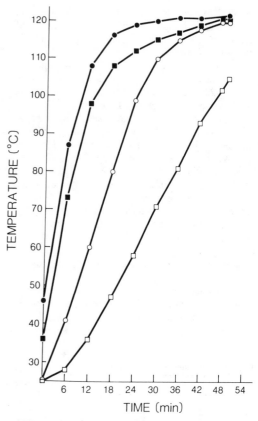

FIG. 1. Average time-temperature profiles of waste loads (1,750 g) consisting of petri dishes (100 by 15 mm) containing agar. The loads were autoclaved in a steel container with (●) or without (○) 1 liter of water and in a steel container with an autoclavable plastic waste bag with (■) or without (□) 1 liter of water. Averages are based on three separate trials conducted in a gravity displacement autoclave that reached 121 to 122°C within 3 min after the cycle was initiated.

at 50 min. When water was not added to the steel container or to the autoclavable bag in the steel container, the temperature increase was significantly slower. Note that the worst time-temperature profile occurred when water was not added to the waste in an autoclavable bag.

Figure 2 depicts the average temperature of waste being processed in polypropylene containers. The data indicate that the addition of 1 liter of water to the container and to the autoclavable bag in the container did not significantly increase the temperature in the waste as compared to the trials with no additional water. In addition, note that the time-temperature profiles of the waste in polypropylene containers are much lower than those in steel containers (Fig. 1).

The effect of the amount of water added to the waste in an autoclavable bag that is processed in a steel container is illustrated in Fig. 3. Note that 1 liter

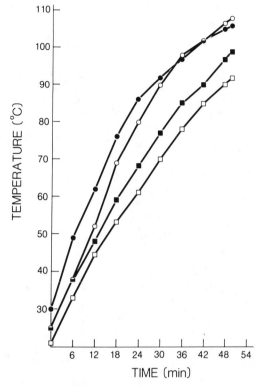

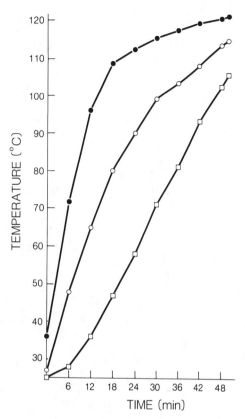

FIG. 2. Average time-temperature profiles of waste loads (1,750 g) consisting of petri dishes (100 by 15 mm) containing agar. The loads were autoclaved in a polypropylene container with (●) or without (○) 1 liter of water and in a polypropylene container with an autoclavable plastic waste bag with (■) or without (□) 1 liter of water. Based on three separate trials conducted in a gravity displacement autoclave that reached 121 to 122°C within 3 min after the cycle was initiated.

FIG. 3. Average time-temperature profiles of waste loads (1,750 g) consisting of petri dishes (100 by 15 mm) containing agar. The loads were autoclaved in autoclavable plastic waste bags with 1 liter (●), 100 ml (○), or no (□) water and then autoclaved in a steel container. Based on three separate trials conducted in a gravity displacement autoclave that reached 121 to 122°C within 3 min after the cycle was initiated.

of water was better than 100 ml of water, which was better than no additional water.

Figure 4 shows that the average time-temperature profile was depressed (after 24 min of processing) by about 3 to 8°C when the amount of waste being processed was doubled. Figure 5 depicts the average time-temperature profiles when 1 liter and 500 ml of water were added to 10 pounds of waste in an autoclavable bag which was then processed in a steel pan. Note that the time-temperature profile was higher when 1 liter of water was used. These results are in contrast with the interpretation by Rutala et al. (21) that the addition of water does not significantly affect the average time-temperature profile of waste.

These data are based on studies using a gravity displacement autoclave. An inherent problem with this type of autoclave is that air entrapment will occur in upright containers. The temperature within an air pocket is much lower than that of the saturated steam surrounding it. Thus, materials in an air pocket will take longer to reach an adequate decontamination

temperature (5). A prevacuum autoclave removes the air from the chamber and upright containers by pulling a vacuum before the saturated steam enters. This resolves the air entrapment problem, except when the prevacuum autoclave is operated on the liquid cycle. In the liquid cycle, air must be removed by gravity displacement since the mechanical vacuum is bypassed (liquid in containers could be forcefully drawn from the containers if a vacuum were applied).

We suggest that each laboratory review its present protocol to ensure that the processing time used produces a temperature in waste of at least 115°C for 20 min. This may necessitate a total processing time of 60 to 90 min or more, depending on the loading conditions used. Also, extreme caution must be used when adding water to infectious waste so that aerosols containing infectious agents are not generated. Water, if added, should be trickled gently down the sides of the container rather than poured in directly,

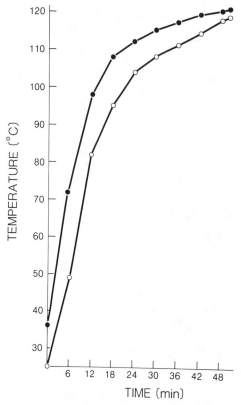

FIG. 4. Average time-temperature profiles of waste loads 1,570 g (●) or 3,500 g (○) consisting of petri dishes (100 by 15 mm) containing agar. The loads were placed in autoclavable plastic waste bags with 1 liter of water and then autoclaved in a steel container. Based on three separate trials conducted in a gravity displacement autoclave that reached 121 to 122°C within 3 min after the cycle was initiated.

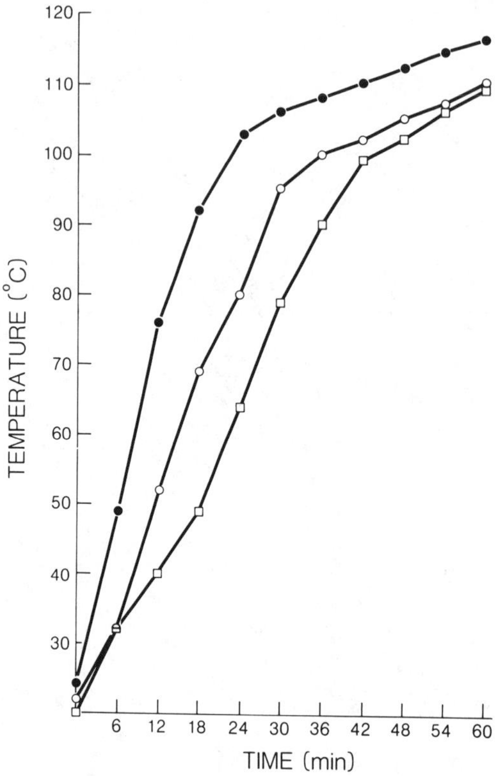

FIG. 5. Average time-temperature profiles of waste loads (4.54 kg) consisting of petri dishes (100 by 15 mm) containing agar. The loads were placed in autoclavable plastic waste bags with 1 liter (●), 500 ml (○), or no (□) water and then autoclaved in a steel tray (53 by 36 by 13 cm). Based on three separate trials conducted in a gravity displacement autoclave that reached 121 to 122°C within 3 min after the cycle was initiated. '

and any items added to the container after the water should be handled gently to avoid splashing.

Incineration can also be used to destroy infectious laboratory waste, contaminated animal bedding (bedding waste from animals infected with pathogenic agents), infected animal carcasses, and human pathological waste. However, as for any other process, specific design and operating criteria must be adhered to (5). Inadequate or incomplete incineration may well be accompanied by inadequate decontamination. These criteria will not be covered here but should be reviewed by persons responsible for operating the incinerator.

Incineration is the process of choice for the decontamination and destruction of infected and noninfected animal carcasses and for human pathological waste (limbs, organs, etc.). It is recommended that such waste be placed in a designated labeled waste bag, preferably within a metal or reinforced cardboard container (example, "Infectious Waste for Incineration—DO NOT PLACE SHARP OBJECTS, NORMAL WASTE, ETC. IN THIS BAG"). This labeling is important to designate the contents of the bag and to alert custodial personnel to use greater care in handling it. The waste bag should be of durable material that is not easily ripped and can hold small amounts of liquid (3-mm polyethylene bags, plastic-lined paper bags, etc.). Laboratory staff should then transport this bag to a secured, refrigerated storage area and place it in a covered metal container. Custodial personnel trained in the handling of infectious waste should transport the covered metal containers to the incinerator, or to the loading dock where they will be picked up and delivered to the incinerator.

The emptied metal container should be decontaminated and washed at a site next to either the incinerator or the loading dock. The metal containers should be cleaned with a disinfectant-detergent, such as a phenolic or iodophor compound, followed by steam. An alternative to the metal containers would be to use leakproof cardboard boxes or drums. These could be placed directly in the incinerator.

Rendering of noninfected animal carcasses and infected animal carcasses containing low-risk agents is also a potentially adequate decontamination-disposal process. It should be noted that rendering plants may not accept small animals such as rats, mice, etc. In addition, one should contact the state agency responsible for regulating rendering plants to ascertain the requirements that must be complied with, especially as related to infected animal carcasses.

The decontamination or sterilization of glassware, instruments, and other materials can also be achieved by dry heat. Dry-heat ovens are used for glassware, instruments, and anhydrous materials such as oils, greases, and powder. The longer time necessary for sterilization is due to the fact that dry heat is less effective in destroying microbes than is moist heat (steam) (5). However, for the sterilization of anhydrous materials and closed containers, dry heat is recommended over steam autoclaving. The reason for this is that the moisture component of saturated steam at 121 to 132°C, which is responsible for the rapid destruction of microbes, will not penetrate anhydrous materials and closed containers. Thus, the highest dry-heat temperature that materials and container interiors will reach in an autoclave is 121 to 132°C. Obviously, the highest temperature reached in a dry-heat oven will depend on the temperature setting (i.e., 160 to 180°C). The contact time given in Table 4 takes into account the lag time for certain materials to reach the temperature in the oven and may or may not be applicable to all situations. Good examples of the times required to decontaminate various materials are presented by Reddish (20).

Contaminated glassware and instruments, regardless of the microbial agent present, can be decontaminated with a 2% alkaline glutaraldehyde solution. This method is a possible alternative for items which cannot tolerate steam autoclaving. Glutaraldehyde is commercially available as a 2% alkaline solution and should be used undiluted. Its use is mainly restricted to items that can be submerged and soaked in a covered container of the solution. The soak or contact time will depend on a number of factors, including the contaminating microbe. Items contaminated with a large concentration of bacterial spores should be soaked for 10 to 12 h. Lightly contaminated items with less resistant microbes can be soaked for 10 to 30 min. The lumen in various instruments and equipment can also be decontaminated with this solution. If it is used in this manner, great care must be taken to ensure that the solution has reached all areas in the lumen. Also, if decontaminating a lensed instrument, make sure that the components of the instrument are compatible

with glutaradehyde. After decontamination, the items must be rinsed with sterile water to remove all residuals.

Formaldehyde can also be used for much the same applications as glutaraldehyde, with a few exceptions. An 8 to 10% solution of formaldehyde applied for several hours is necessary to decontaminate items containing bacterial spores and some strains of atypical mycobacteria. Formaldehyde vapors from the solution are very pungent and irritating, making it extremely difficult to work with this decontaminant. Iodine, phenolic, and chlorine solutions are usually not recommended to be used in this manner because they either are too corrosive, are rapidly inactivated by organic matter, or are not sporicidal. However, a chlorine or iodine solution (0.5 to 1.0%) may be used if the items are glass or plastic, the organic load is low, and a sporicidal activity is not needed. Phenolic solutions are not recommended because some phenolics corrode plastics and very high concentrations are needed to inactivate hydrophilic viruses.

Floor Care Protocols

There are numerous chemical disinfectant-detergent compounds commercially available for the general maintenance of laboratory and animal care area floors. Selection of a suitable compound depends upon the microbicidal effectiveness needed, the corrosiveness of the compound, cost, and its residual and irritating properties.

Phenolic-detergent and quaternary ammonium compounds are the most commonly used decontaminants for floor maintenance. Phenolic-detergent-based compounds have a broad microbicidal spectrum and are less affected by organic matter than are the quaternary ammonium-based compounds. However, phenolic-detergent compounds are more toxic, and many can be absorbed through intact skin. Both types of compounds are sold as a concentrated solution and must be diluted per manufacturer's directions (1 to 3 oz [ca. 30 to 89 ml] per 128 oz [ca. 3,785 ml] of water) before use.

A two-bucket protocol is recommended for wet mopping of laboratory floors. The principle of this protocol is that fresh disinfectant-detergent solution is always applied to the floor from one bucket, while all spent solution removed from the floor is wrung from the mop and collected in the second bucket. In practice, the mop is saturated with the disinfectant-detergent solution from the first bucket, very lightly wrung into the second bucket, and applied to the floor using a side-to-side motion while slowly rotating the mop head. An area of 100 to 150 ft^2 is covered in this manner. Then after allowing at least a 5-min contact period, the solution is removed with a wrung-out mop. This procedure is continued until the total floor area is treated.

An alternative protocol would be to flood the floor area with the disinfectant-detergent solution and then, after 5 min, either pick it up with a wet vacuum or push it down a floor drain with a squeegee. This protocol requires that floors be completely sealed or of monolithic construction so that liquid cannot leak to adjacent areas. This protocol is more suitable to animal care areas where there is an absence of permanently placed benches and fixed equipment. The disinfectant-detergent solution is applied at a rate of about 1 gal/144 ft^2. If tank or wall sprayers are used to apply the solution, the setting on the nozzle should be adjusted so that the solution flows freely (not a spray) and is applied close to the floor.

When wet mopping, a freshly laundered cotton mop head should be used daily. Before a used mop head is sent to the laundry, it should be soaked in a freshly prepared disinfectant-detergent solution for 30 min or autoclaved. For laboratories handling moderate- to high-risk agents, the mop head should be autoclaved at 121°C for 20 to 30 min.

Sweeping and vacuum cleaning for general floor maintenance should not be allowed in laboratories handling infectious agents. Such cleaning techniques will readily aerosolize microbes present on the floor.

Protocol for Floor Drains

If a floor drain(s) is present in a laboratory, a sufficient amount of liquid must be present in the trap to prevent the backup of sewer gases. In addition, floor drains can serve as a reservoir for microbial agents. Thus, a disinfectant solution should be poured into the drain each month. A half-gallon (ca. 1.9 liter) of either a 1:100 dilution of household bleach (chlorine) or a phenolic-detergent solution is recommended.

Protocol for Water Baths

The addition of a disinfectant to the water in baths is recommended to resolve two potential concerns. First, inapparent contamination of the water during the incubation of pathogenic agents in the bath is not unlikely. Such contamination could lead to the transmission of this agent to susceptible staff. Second, water, including distilled, deionized, and tap water, usually contains nonpathogenic bacteria. Thus, laboratory experiments requiring a water bath could become contaminated with these bacteria, causing erroneous results. Because of this it is recommended that either 0.1 oz (ca. 2.9 ml) of bleach (chlorine) or 1 oz (ca. 29 ml) of phenolic detergent be added to each gallon (3.8 liter) of water used in a bath. The phenolic disinfectant is preferred over chlorine because it is more stable, less corrosive, and less inactivated by organic matter. The water in the bath should be changed weekly. Propylene glycol has been used effectively as an alternative in cold-water

baths, and if the bath can be turned up to temperatures of 90°C or more, that should be done as a decontaminating process for about 30 min every week.

Protocols for Laminar-Flow Biological Safety Cabinets

Surfaces in the work area of laminar-flow biological safety cabinets should be wiped down with a germicide both before and after each use. Before work is started the cabinet should be operated (fan on) for 5 to 10 min. The surface of the work area is then decontaminated by wiping the area with a clean cloth soaked with a 70 to 85% solution of ethanol. If moderate-risk agents or hepatitis B virus have been handled in the cabinet, a wipe-down with a 2% iodophor solution, followed by alcohol to remove the iodine, would be more appropriate. After the completion of work the same wipe-down procedure should be used, while the cabinet is operating. Gloves and a gown should be worn during the wipe-down procedure. The cloth used should be placed in a container that will be subsequently autoclaved. The gloves and gown worn during the work may continue to be worn while cleaning up afterward, but after the final wipe-down they should also be autoclaved.

UV radiation produced by germicidal lamps is also used within cabinets to decontaminate the work surfaces. UV radiation is effective against most microbes. Its effective use, however, requires an understanding of its abilities and limitations. The 253.7-nm wavelength emitted by a germicidal lamp has limited penetrating power. It is thus primarily effective against unprotected microbes on exposed surfaces or in the air (will not penetrate soil, dust, etc.). The intensity or destructive power decreases by the square of the distance from the lamp (ca. $I = 1/d^2$). Thus exposure time is always related to distance. The intensity of the lamp will also decrease with time. Therefore it should be checked with a UV meter at least yearly. The intensity from the lamp is drastically affected by the accumulation of dust and dirt on it. Therefore, unless the lamp is cleaned weekly or biweekly the intensity from it may be inadequate. Eye protection against direct or indirect (reflected from surfaces or clothing) UV exposure is required. The UV lamp should *never* be on while an operator is working in the cabinet. On the basis of these points, UV cannot be recommended as a decontaminant unless the lamp is properly maintained (checked yearly and cleaned often) and UV safety glasses are worn when there is a potential for exposure.

The decontamination of the total cabinet, which includes the work surfaces, the supply and exhaust filters, the surfaces of the air plenums, and the fan unit, is best achieved by exposing these areas to formaldehyde gas. Such decontamination is recommended (i) before maintenance work on cabinet; (ii) before filter changes; (iii) before the cabinet is moved; (iv) before performance tests which require access to the sealed portion of the cabinet; (v) after a gross spill of infectious material in the cabinet; or (vi) before the cabinet activity is changed from work with moderate- or high-risk infectious materials to work with noninfectious materials.

Formaldehyde gas is most easily generated by heat-accelerated depolymerization of flake paraformaldehyde, which can be purchased commercially. The recommended protocol for the use of this decontaminant is as follows.

1. Calculate the total cabinet volume in cubic feet by multiplying in feet the height, width, and depth of the cabinet.

2. The amount of paraformaldehyde to be used is determined by multiplying the volume of cabinet in cubic feet by 0.3 g of paraformaldehyde. This amount will provide an equivalent concentration of 0.8% by weight or 10,000 ppm by volume in air. *Caution*: Be careful not to use a greater amount of paraformaldehyde than is required. Concentrations of formaldehyde over 8% in air are explosive! Also, an excess amount can cause the formaldehyde to polymerize on surfaces as a white powder, which will necessitate a longer aeration period before the cabinet can be used.

3. Weigh out the amount of paraformaldehyde calculated and place in a plastic bag until it is needed. Flake paraformaldehyde is recommended because it has less tendency to splatter when heated.

4. Assemble the following equipment and materials:
   (a) Electric frying pan with a thermostatic control that can be set at 450°F
   (b) Polyethlene film (thickness, 3 to 4 mm)
   (c) Duct tape
   (d) Heavy-duty electrical extension cords (two)
   (e) Hot plate
   (f) Beaker of water (500 ml)
   (g) Temperature- and humidity-indicating device(s)
   (h) Flexible ventilating duct
   (i) A sign labeled "Keep Out—decontamination with formaldehyde gas in process"
   (j) Safety glasses, face shield, or goggles
   (k) Industrial canister-type gas mask or a self-contained breathing apparatus
   (l) Coveralls and gloves
   (m) Knife or scissors

5. Set-up of equipment and materials and preparation of cabinet and exhaust air system.
   (a) Attach the sign, "Keep Out—etc.," to cabinet.
   (b) Put on coveralls, eye protection, and gloves.
   (c) Spread the preweighed paraformaldehyde evenly in the bottom ofthe frying pan, which is preset at 450°F, and place the pan in the cabinet. Then attach the electrical cord of the pan to an extension cord and extend it out of the front of the cabinet. *Caution*: Do not connect the extension cord to the electric supply at this time.
   (d) Place the hot plate in the cabinet and put the beaker of water on it. Attach the plate's electric cord to the second extension cord and extend it out the front of cabinet. Do not connect the electric supply at this time.
   (e) Place temperature- and humidity-indicating device(s) in cabinet so that the indicators can be seen through the front glass portion of the cabinet.
   (f) If the cabinet exhaust air port is not connected directly to the building's exhaust ventilation system, attach the flexible hose to the cabinet's exhaust port (seal and tape around this connection). Then extend the hose to a room air exhaust grill or an exterior window. Seal the opening at the end of the hose with plastic film and tape. *Caution*: If the room exhaust air is recirculated within the building, the hose must be extended to a window. Also, if the volume of air exhausted through the room grill is less than the volume of air being exhausted by the cabinet, extend hose to window.
   (g) If the cabinet exhaust air port is connected directly to the building's exhaust ventilation system, close the exhaust damper. *Caution*: The building's exhaust ventilation system should not in this case be connected to the building's air recirculation system. However, check to make sure that the air is not recirculated. If it is, extend hose to window and seal as in step (f) above.
   (h) Now tape all cracks and seams of the cabinet to make it as gas tight as possible.
   (i) Then, using the plastic film and the duct tape, seal the work opening of the cabinet. Tape around the protruding electric cords should produce a seal that is also gas tight.
   (j) Plug in the extension cord attached to the hot plate until the temperature is about 74°F and the relative humidity is between 60 and 80%.
   (k) Now plug in the electric fry pan and allow it to heat until about half of the paraformaldehyde has depolymerized (about 10 min). Then turn on the cabinet fan for 3 to 5 s to disseminate the gas to inaccessible areas.
   (l) Depolymerize the remaining paraformaldehyde and then turn on the fan again for 3 to 5 s. Allow the cabinet to stand for a minimum of 1 and preferably 3 h.

(m) After 1 to 3 h of exposure, begin the ventilation process by removing the film from the flexible hose or by opening the exhaust damper. Then, slit the plastic covering the front opening of the cabinet and immediately turn the cabinet fan on.

(n) Ventilate the cabinet overnight or for about 16 h to remove formaldehyde gas and polymer that has adsorbed to surface area and filters.

6. Now work (changing filters, etc.) can be performed. *Caution*: Formaldehyde gas has a very pungent and irritating odor and is a toxic substance having a permissible exposure limit (maximum amount a worker can be exposed to, averaged over an 8-h work day) of 2 ppm. Repeated exposure to formaldehyde is known to produce a hypersensitive condition in certain individuals. Because of this and the potential for exposure to formaldehyde during cabinet decontamination, it is essential that a self-contained breathing apparatus or an industrial-type gas mask is immediately available. In addition, it is strongly recommended that formaldehyde gas decontamination be done only by individuals with sufficient experience and training.

Large areas, air handling systems, laboratory equipment, and electrical instruments can also be decontaminated with formaldehyde gas. The protocol for equipment and instrument decontamination would be similar to that used in cabinet decontamination, except that large polyethylene bags made by heat-sealing polyethylene film would be used to enclose the equipment or instruments. Also, an exhaust fan and duct would have to be attached to the bag to disseminate the gas throughout the bag and then to exhaust the gas from the bag to the outside atmosphere.

Protocol for Work Surface Maintenance

Research and clinical laboratories that deal with microorganisms or clinical specimens from animals and human beings should decontaminate work surfaces. It is recommended that work surfaces be decontaminated each morning before starting work, before restarting work after a break, and at the end of the work day.

A quaternary ammonium compound diluted according to manufacturer's instructions, or a 70 to 85% alcohol solution, is suitable for laboratories dealing with ordinary noninfectious microbes. These solutions could be used in other laboratories dealing with specific low-risk microbes as long as these germicides are known to be effective against the agent(s). For laboratories working at Biosafety Level 3 or 4, agents should not be handled on the open bench except under extenuating circumstances. In

that case, a phenolic or iodine solution should be used, followed by an alcohol wipe (alcohol will remove the residual phenolic compound that can be absorbed through intact skin and will remove the sticky iodine-detergent residual from iodophors); manufacturer's directions should be followed in preparing these solutions. The disinfectant solution should be applied to the surface with a clean cloth or paper towel dampened with the solution (spray or pour on cloth). Light equipment and materials should be moved so that the entire surface is dampened. Allow at least 5 min of contact before wiping the surface with a different cloth or paper towel which has been dampened with alcohol. The cloth or towel should be discarded in a container that will be autoclaved or incinerated. The individual applying the solution should wear a glove on the hand that is in contact with the cloth or towel.

The same procedure is followed at the end of the day and after breaks. It is also a good practice to wipe down all vertical surfaces, at least on a monthly basis, with a disinfectant.

Hand Washing Protocol

Hand washing is an extremely important procedure for preventing exposure to and dissemination of infectious agents. Hand contamination with transient microbes (pathogenic and saprophytic microbes which are not normal components of the resident skin flora) can readily occur during manipulation of specimens, equipment, and supplies and during contact with work surfaces (16, 24). Unless microbial contamination is routinely removed, exposure via contact with mucous membranes, inoculation through skin, ingestion, etc., becomes inevitable.

A nonantiseptic soap can be used for hand washing in laboratories that do not handle human and animal specimens and in laboratories handling noninfectious agents. The use of an antiseptic hand-washing compound is recommended for all other laboratories. The antiseptic compound used should be effective against most microbes and should not cause excessive skin dryness, cracking, or dermatitis upon repeated use. In addition, compounds which can be absorbed through intact skin and cause systemic poisoning should never be used.

Two compounds which are widely used as antiseptic handwashing agents are chlorhexidine gluconate and povidone-iodine. An alternative would be to use a nonantiseptic soap followed by a rinse with 70% ethanol. Liquid dispensers rather than bar soap are recommended for all hand-washing products.

Laboratory personnel should wash their hands:
—When coming on duty
—On leaving the laboratory for whatever reason
—When hands are obviously soiled
—Before and after completion of a task in a biological safety cabinet, even if gloves are worn

—Before contact around one's face or mouth
—On completion of duty

A protocol for hand washing is as follows.

1. Turn on faucets and wet hands with tepid water.
2. Dispense nonantiseptic soap or antiseptic compound into a cupped hand.
3. Spread soap or compound around both hands and between fingers. If needed, add a little more water to facilitate spread and lathering.
4. Wash hands for about 30 s. Vigorously rub both sides of hands starting from a few inches above the wrist, extending downward between the fingers, and around and under the fingernails.
5. Rinse thoroughly under the tepid running water. Rinsing should start above the wrist area and proceed to the tips of the fingers. *Note*: If faucets are not knee or foot operated, do not turn off water (touch faucet handles) yet.
6. Dry hands thoroughly with paper towel(s). If faucets are hand operated, turn them off now, using a dry paper towel to protect clean hand.

Hand-washing sinks should be of adequate size and conveniently located throughout the laboratory and at each entry/exit door to encourage hand washing. The faucet should be operated by a knee- or foot-control device, especially for laboratories handling materials containing moderate- to high-risk microbial agents. Dispensers for the hand-washing compound and paper towels should be available and maintained at each of the hand-washing sinks.

GLOSSARY

Antisepsis. Application of a chemical to living tissue for the purpose of preventing infection.

Decontamination. Destruction or removal of microorganisms to some lower level, not necessarily zero. (Also applies to removal or neutralization of toxic agents and, in microbiology laboratories, implies microbicidal action for safety purposes.)

Disinfection. Chemical or physical treatment which destroys the most resistant vegetative microbes (or viruses), but not spores, in or on inanimate substances.

Dunk tank. Device containing liquid germicide, used for the passage of materials into or out of a gas-tight chamber without the transfer of microorganisms.

D value. Time required, in a given treatment, to reduce the number of surviving organisms by one logarithm (90%).

Microbiological safety index. Logarithm of the reciprocal of number of survivors after a sterilization treatment.

Pasteurization. Heat treatment of a liquid medium to destroy the most resistant vegetative organisms of interest.

Sanitization. Reduction of microbial load on an inanimate surface to a "safe" public health level.

Sterilization. Unequivocal total destruction of all living organisms. In practice, sterility is very difficult to determine. Therefore, the term is applied to a very low probability (e.g., usually 10^{-6}) that even one organism has survived the process.

Thermal death time. Time required to achieve sterility under a given set of circumstances.

LITERATURE CITED

1. **Barbeito, M. S., and E. A. Brookey.** 1976. Microbiological hazard from the exhaustion of a high-vacuum sterilizer. Appl. Environ. Microbiol. **32:**671–678.
2. **Beckman Instruments.** 1980. Rotors and tubes for preparation ultracentrifuges: an operator's manual LR-1M-7. Spinco Division of Beckman Instruments, Palo Alto, Calif.
3. **Bentley-Phillips, B.** 1974. Occupational leucoderma following misuse of a disinfectant. S. African Med. J. **48:**810.
4. **Blaser, M. J., and J. P. Lofgren.** 1981. Fatal salmonellosis originating in a clinical microbiology laboratory. J. Clin. Microbiol. **13:**855–858.
5. **Block, S. (ed.).** 1983. Disinfection, sterilization and preservation, 3rd ed. Lea and Febiger, Philadelphia.
6. **Brown, M. R. W., and P. Gilbert.** 1977. Increasing the probability of sterility of medicinal products. J. Pharm. Pharmaceut. **29:**517–523.
7. **Campbell, R. W.** 1980. Sterile is a sterile word (The Microbiological Safety Index). Radiat. Phys. Chem. **15:**121–124.
8. **Center for Disease Control.** 1975. Neonatal hyperbilirubinemia. Epidemiol. Notes Rep. **24:**35.
9. **Favero, M. S.** 1982. Iodine—champagne in a tin cup. Infect. Control **3:**30–32.
10. **Fregert, S., O. Groth, N. Hjorth, B. Magnusson, H. Korsman, and P. Ovium.** 1969. Alcohol dermatitis. Acta Dermatol. Venerol. **49:**493–497.
11. **Freireich, A. W., J. J. Cinque, and G. Xanthaky.** 1967. Hemodialysis for isopropanol poisoning. N. Engl. J. Med. **277:**699–700.
12. **Glaser, Z. R.** 1977. Special occupational hazard review with control recommendation for the use of ethylene oxide as a sterilant in medical facilities. Department of Health and Human Services publication no. 77–200. U. S. Government Printing Office, Washington, D.C.
13. **Health Industry Manufacturers Association.** 1978. Biological and chemical indicators. Medical Device Sterilization Monograph Series. Health Industry Manufacturers Association, Washington, D. C.
14. **Kelsey, J. C.** 1972. The myth of surgical sterility. Lancet **ii:**1301–1303.
15. **Lauer, J. L., D. R. Battles, and D. Vesley.** 1982. Decontaminating infectious laboratory waste by autoclaving. Appl. Environ. Microbiol. **44:**690–694.
16. **Lauer, J. L., N. A. VanDrunen, J. W. Washburn, and H. H. Balfour, Jr.** 1979. Transmission of hepatitis B virus in clinical laboratory areas. J. Infect. Dis. **140:**513–516.

17. **Mullick, F. G.** 1972. Hexachlorophene toxicity—human experience at the Armed Forces Institute of Pathology. Pediatrics **51:**395–399.

18. **National Institute for Occupational Health and Safety.** 1976. Criteria for a recommended standard-occupational exposure to formaldehyde. Department of Health and Human Services publication no. 76–142. U.S. Government Printing Office, Washington, D.C.

19. **Odom, R. B., and K. M. Stein.** 1973. Depigmentation caused by a phenolic detergent-germicide. Arch. Dermatol. **108:**848.

20. **Reddish, G. F. (ed.).** 1975. Antiseptics, disinfectants, fungicides, and chemical and physical sterilization. Lea and Febiger, Philadelphia.

21. **Rutala, W. A., M. M. Stiegel, and F. A. Smith, Jr.** 1982. Decontamination of laboratory microbiological waste by steam sterilization. Appl. Environ. Microbiol. **43:**1311–1316.

22. **Sanderson, K. E., and D. Cronin.** 1968. Glutaraldehyde and contact dermatitis. Br. J. Med. **3:**802.

23. **Shmunes, E., and E. J. Levy.** 1972. Quaternary ammonium compound contact dermatitis from a deodorant. Arch. Dermatol. **105:**91–95.

24. **Steere, A. C., and G. F. Mallison.** 1975. Handwashing practices for the prevention of nosocomial infections. Ann. Intern. Med. **83:**683–690.

25. **U.S. Department of Health and Human Services.** 1979. Laboratory safety monograph: a supplement to National Institute of Health Guidelines for Recombinant DNA Research. National Institutes of Health, Bethesda, Md.

26. **Wedum, A. G.** 1953. Bacteriological safety. Am. J. Public Health **43:**1428–1437.

27. **Zinner, D. D., J. M. Jablon, and M. S. Saslow.** 1961. Bactericidal properties of povidone-iodine and its effectiveness as an oral antiseptic. Oral Surg. Oral Med. Oral Pathol. **14:**1377–1382.

Chapter 23

Housekeeping in the Laboratory

EDWARD A. SCHMIDT

The housekeeper who performs duties in the laboratory is faced with exposure to potential hazards not often seen in other housekeeping areas. Because of the general safety hazards inherent in housekeeping activities, as well as the additional hazards found in the laboratory, safety must be of great concern to all persons who deal with the housekeeping department and its employees (2, 3). Housekeeping personnel should be included in employee health programs (see Chapter 3). The housekeeping department also bears a large part of the responsibility for providing a continuously safe environment for all personnel within the facility served (1–3).

Recently there has been a trend toward obtaining contract cleaning services, or contract management of the internal housekeeping departments, from outside the institution (2). Whether housekeeping services are provided in-house or by contract, the considerations discussed in this chapter should be kept in mind.

Housekeeping in the laboratory makes use of relatively standard products, equipment, and procedures but will vary due to unique exposures and types of laboratories (2, 12). There are basically three categories of laboratory: research, industrial, and clinical. Research laboratories are primarily those involved in industrial research and development and those found in academic institutions. Industrial laboratories include those involved with production of biological, chemical, and radiological products, and quality control laboratories found throughout industry. Clinical laboratories include hospital, outpatient facility, health department, and reference laboratories. Regardless of the category of laboratory, there are some basic hazards associated with each laboratory which are present to some degree in any category.

The population at greatest risk from the safety hazards of the laboratory is naturally the laboratory workers themselves. However, any laboratory safety program must include personnel who do not work directly with the hazardous material but who nevertheless are at risk of exposure (3, 4). Personnel assigned to laboratory housekeeping duties have probably the second greatest job-related exposure to laboratory hazards. A third group at risk are maintenance and engineering personnel responsible for maintenance, control, and testing of laboratory and building equipment. At less, but still significant, risk are visitors to the laboratory. Thus the laboratory safety program must protect not only laboratory workers but housekeeping and maintenance personnel and visitors as well. This chapter deals with the laboratory responsibilities to and of the housekeeping staff.

A well-organized laboratory safety program will include several basic criteria (3–5, 13–15). In areas of high potential risk, housekeeping duties should be designated by the laboratory director. Hazardous areas in the laboratory must be designated. Each hazardous area must then be properly identified by appropriate warnings signs, barriers, etc. Access to hazardous areas must be limited to authorized persons. All hazardous materials must be identified and properly labeled; where appropriate, hazard and first aid information should be provided on the label. All safety aids in the laboratory must also be properly identified and readily accessible, including fire extinguishers, fire blankets, deluge showers, eye wash fountains, alarm systems, first aid supplies and equipment, and emergency telephone numbers.

All of this safety information must then be relayed to the housekeeping staff by means of a training program (1–4). This program may include an initial orientation to hazards, hazardous areas, safety aids, and safety precautions, followed up by periodic review, retraining and updating when necessary. All housekeepers potentially exposed to laboratory hazards, including supervisors and workers who fill in during vacations, absences, and weekends, should undergo the complete and continuing training program.

Only qualified personnel with the necessary technical knowledge should be authorized to handle laboratory equipment and materials (14, 15). For this reason the laboratory personnel themselves should provide housekeeping services related to the equipment and materials. Laboratory benches and equipment should be cleaned by laboratory personnel, not housekeeping staff. The need for technical knowledge dictates that laboratory personnel also carry the primary responsibility for cleanup of spills or accidents (3, 14).

The housekeeping department has the primary responsibility for training housekeeping personnel, assisted by technical support from the laboratory as described above. The specific housekeeping duties and the specific procedures required in each area of the laboratory must be taught to the person or people responsible for them. Most of the housekeeping duties in the laboratory will require much the same procedures, equipment, and products as those used elsewhere in the facility (1, 2, 4, 6, 7, 9, 10, 12). Where these do differ, it may well be for safety reasons and thus must be thoroughly explained by housekeeping supervisors and thoroughly understood by housekeepers. All hazardous areas and materials and all safety and protective measures must be covered in the housekeeper's initial and continuing training. Personal hygiene must also be stressed in the training program, with special emphasis placed on proper hand washing, especially in biological areas of the laboratory (1, 8, 10).

The laboratory personnel should be responsible for cleaning laboratory benches and equipment and any other laboratory areas which require specialized technical knowledge (3, 14, 15). The specific duties of housekeepers will generally include floor maintenance, removal of uncontaminated dry waste, rest room and locker room cleaning, and maintenance of administrative areas. High and low dusting, wall washing, window cleaning, etc., should be done under supervision. Specific areas of responsibility must be designated in departmental policies and understood by both housekeeping and laboratory personnel. A list of areas of responsibility should be posted permanently in the laboratory area as a reminder to everyone (3). Cleanup of spills and accidents should be the primary responsibility of the laboratory. In laboratories where housekeepers share this responsibility, the specific situations in which they are asked to participate must be clearly designated. Misunderstandings could lead to unqualified cleaning personnel working under hazardous conditions. When there is any doubt, laboratory personnel should take charge of spill cleanup.

In many institutions the housekeeping department has the responsibility for handling solid waste, including that from the laboratory. Housekeeping may handle general laboratory wastes, but it is advisable that special hazardous wastes be dealt with by laboratory or other designated qualified personnel (3, 14). This also applies to handling of hazardous linen and laboratory garments (14).

HOUSEKEEPING PROCEDURES

Cleaning frequencies

Laboratories where potentially infectious materials are handled should be cleaned according to the recommendations listed in Appendix 1. Other laboratories may be cleaned daily, or less frequently if daily cleaning is not necessary. Rest rooms should be cleaned at least daily. Administrative areas may be cleaned as often as desirable. Noninfectious and decontaminated solid waste should be removed at least daily. Laboratories should be cleaned in the evenings or during periods of minimal activity to provide less intrusion and disruption and to permit more thorough cleaning. A sample laboratory cleaning plan is shown in Table 1.

Floor maintenance

Most laboratory floors are hard surfaces. Floor finishes should be slip resistant (2–4).

Dusting. Hard-surface floors should be dusted daily. Dusting with a chemically treated reusable or disposable dust mop saves time and removes dust and debris efficiently. Vacuum dusting does not significantly improve dust removal and is more time consuming. If a vacuum cleaner is used in a biological laboratory, it should be equipped with a microbiological filter (1, 4, 9). Use of chemically treated dust mops does not significantly affect air quality (13a).

Wet cleaning. Floors should be wet mopped daily using a single-bucket or double-bucket procedure, (Table 2). Safety warning signs should be set up and

TABLE 1. Sample laboratory cleaning plan

Supplies and equipment
 1. Plastic waste container liners
 2. Paper towels and hand soap supply
 3. Wiping cloths or rags
 4. Spray bottle containing use dilution of disinfectant detergent
 5. Chemically treated high duster, frame, and handle
 6. Chemically treated dust mop, frame, and handle
 7. Clean wet mop and handle
 8. Mop wringer and bucket containing use dilution of disinfectant detergent
 9. Wet-floor signs

Note: Use an EPA-registered disinfectant detergent diluted according to label directions.

Procedures
 1. Empty waste containers. Double-bag and identify hazardous wastes. Replace plastic liners in waste containers.
 2. Dust overhead lights, ledges, ventilation ducts, etc., using chemically treated high duster.
 3. Damp wipe horizontal surfaces using spray bottle and wiping cloths.
 4. Clean sinks and handwashing facilities using spray bottle and wiping cloths. Replenish paper towels and hand soap supply, if needed.
 5. Spot clean walls, if needed, using spray bottle and wiping cloths.
 6. Dust floor using chemically treated dustmop.
 7. Post wet-floor signs.
 8. Wet mop floor using disinfectant-detergent solution and clean wetmop.
 9. Inspect work.
10. Remove wet-floor signs when floor is dry.

TABLE 2. Comparison of efficiency of
floor-cleaning methods

Procedure	Avg % bacterial reduction
Single-bucket mopping...................	89.8[a]
Double-bucket mopping..................	93.9
Flooding and wet-vacuum pickup..........	84.8
Machine scrubbing and wet-vacuum pickup ..	94.1

[a] The single-bucket procedure produces results statistically equivalent to those of the other floor-cleaning procedures ($P = <0.05$, Student's t test) (unpublished data).

alternative pathways should be provided when floors are wet. Flooding and wet-vacuum pickup is effective, but it is time consuming and s not generally suited to cleaning floors in laboratories. Floors in biological laboratories and rest rooms should be cleaned with a disinfectant-detergent solution; quaternary ammonium compounds, phenolic compounds, and, to a lesser degree, iodophors are suitable, if registered with the U.S. Environmental Protection Agency and used in accordance with label directions (4, 8, 9). Floors in areas where biological hazards are not of concern may be cleaned with a detergent. Good frictional cleaning and physical removal are most important in achieving good cleaning results (9). Reusable materials and equipment, e.g., sponges, mops, buckets, vacuum cleaners, etc., should be decontaminated after use.

Buffing. Periodic buffing of floors will maintain gloss and aesthetic appeal and improve antimicrobial cleaning efficiency (13a). Floors should be spray buffed. Spray buffing does not significantly affect air quality, whereas dry buffing can disseminate dust and microorganisms into the air and is not suitable for laboratories (1, 13a).

High and low dusting

Surfaces within normal reach may be dusted or cleaned by damp wiping with a disinfectant-detergent or detergent solution. Applying the solution with a pistol-grip sprayer and wiping with a clean reusable or disposable cloth is the best procedure for this purpose. Surfaces beyond reach may be dusted with a pole-mounted chemically treated duster or by vacuuming. Surfaces in biological areas should be cleaned daily.

Wall washing

Walls in general laboratory areas with satisfactory ventilation systems should be cleaned once or twice a year and with the same procedures as those in other areas. In animal areas, operating rooms, and special laboratories, walls may be cleaned routinely every month. In all cases the cleaning frequency must be determined by the laboratory director, the biosafety

officer, and the housekeeping director. Gross soil or contamination should be promptly spot cleaned. Biological contamination should be removed with a disinfectant-detergent solution. Areas out of reach may be cleaned with pole-mounted devices.

Window cleaning

Windows can be cleaned as required for other areas. Gross soil or can be spot cleaned as for wall washing. Interior window cleaning beyond normal reach should be performed with extension poles or authorized stools or ladders. Window washers should not stand or walk on laboratory benches.

Other areas

Housekeeping procedures for rest rooms, locker rooms, corridors, and administrative areas in laboratory facilities do not differ from those used in ordinary buildings.

Biological spills

Spills of biological materials must be cleaned with a disinfectant by laboratory personnel, as discussed above. For details see Chapter 22.

HOUSEKEEPING MANAGEMENT

Housekeepers must maintain equipment and supplies in clean and safe order (2, 3). Supplies must be properly labeled. Electrical equipment such as floor buffers and dry or wet vacuum cleaners must be stored properly and kept in good working order. Defective equipment should be repaired promptly. Three-prong electrical plugs frequently have damaged ground prongs and and thus should be checked periodically, preferably before each use. Highly visible electric cords should be used to prevent tripping. Only equipment that has undergone a safety check should be used in the laboratory.

Housekeepers must not enter areas they are not authorized to enter or do work they have not been specifically trained and assigned to do, since they may thus be exposed to unknown hazards. They must not handle laboratory animals other than as their duties require. They must not smell or handle any unknown materials or objects, or empty or handle unspecified or unidentified containers. The director of each laboratory should clearly define any other activities which might be hazardous and should be prohibited to the housekeeping staff.

A number of areas present management problems within housekeeping departments, whether in laboratories or elsewhere (2). It is not unusual to have a high turnover rate within the department. If new personnel are continually assigned to the laboratory, they may never reach the desired level of training.

Even with workers of adequate seniority, motivation can be a problem. Housekeeping personnel frequently come from lower socioeconomic levels and have limited education. Both training and motivation of these people may be difficult if they lack an appreciation of the importance of the job at hand and the necessity for proper safety precautions. A further difficulty in both training and motivation is posed by the increasing number of people entering the work force with a limited or no understanding of English. Bilingual supervisors and training aids can be of invaluable assistance in such situations (3). Finally, most housekeeping work is performed in the evening or on weekends or at other times when the workers will interfere as little as possible with laboratory operations. As a result, housekeepers frequently work with minimal supervision by either housekeeping supervisors or laboratory personnel. Housekeepers assigned to the laboratory should be adequately trained, properly motivated, stable in attendance, and thoroughly dependable.

The housekeeping director must carefully screen new job applicants for these desired qualities (2). Proven good workers should be promoted. Intensive specialized training should be provided regularly for all staff, including, as discussed above, any personnel who may be asked to fill in in case of vacations, illness, absence, etc.

The overall operation of the housekeeping department can be affected by a number of influences both internal and external to the institution. The institution's administration naturally has the greatest influence, setting the scope of housekeeping operations, establishing required quality standards, and administering the budget. Housekeeping management will need to keep in close contact with administration, as well as with the laboratory director, with regard to the continuing quality of the housekeeping staff's performance, and to keep abreast of possible changes in the laboratory operations which would affect housekeeping duties. Also, the institution's biosafety officer or safety committee will designate hazardous areas and practices in the laboratory, recommend precautions or protective measures, and follow up on compliance. In health care facilities the institution's infection control committee will have still further effect on housekeeping activities in line with its responsibility for sanitation and control of nosocomial infection (1, 2, 11).

External influences on the housekeeping department are varied. The Occupational Safety and Health Administration maintains standards for safety in the environment and working conditions of employees, as set out in the Occupational Safety and Health Standards for General Industry of 1974. Failure to meet these standards may result in penalties, and the housekeeping director should be aware of all the applicable requirements. The National Fire Protection Association, as well as state and local fire jurisdictions, also maintains regulatory standards.

State and local public health offices have an impact, especially in the case of health care facilities. Health care facilities are also influenced by the Joint Commission on Accreditation of Hospitals, which sets standards and conducts periodic surveys (11). Both the laboratory and the housekeeping department have definite safety responsibilities under this accreditation program. Worker right-to-know laws mandate worker awareness of hazardous exposures. Trade unions, civil liberties organizations, and federal, state, and local civil service commissions also wield influence, primarily through personnel practices such as hiring, promotion, termination, etc.

With relation to laboratory housekeeping, the goal of all of these organizations, as well as the various levels of administration within the institution, is safe and healthy employees and a clean and smoothly running laboratory program. Five components are essential to this goal: good equipment, good chemical products, good procedures, good training and education, and good supervision. With these five requirements met, the quality of housekeeping will be maintained at the desired level.

LITERATURE CITED

1. **American Hospital Association.** 1979. Infection control in the hospital, 4th ed. American Hospital Association, Chicago.
2. **American Hospital Association.** 1979. Hospital housekeeping handbook. American Hospital Association, Chicago.
3. **American Hospital Association and National Safety Council.** 1983. Safety guide for health care institutions, 3rd ed. American Hospital Association and National Safety Council, Chicago.
4. **Bond, R. G., G. S. Michaelsen, and R. L. DeRoos (ed.).** 1973. Environmental health and safety in health care facilities. The Macmillan Co., New York.
5. **Center for Disease Control.** 1974. Lab safety at the Center for Disease Control. Department of Health, Education and Walfare. Publication no. (CDC) 76–8118. Center for Disease Control, Atlanta, Ga.
6. **Center for Disease Control.** 1977. Control measures for hepatitis B in dialysis centers. Viral hepatitis: investigations and control series, November 1977. Center for Disease Control, Atlanta, Ga.
7. **Center for Disease Control.** 1979. Hepatitis surveillance, report no. 44. October 1979. Center for Disease Control, Atlanta, Ga.
8. **Centers for Disease Control.** 1981. Guidelines on infection control. Guidelines for hospital environmental control. Infect. Control 2:131–146.
9. **Centers for Disease Control.** 1982. Guidelines on infection control. Guidelines for hospital environmental control (cont.). Infect. Control 3:52–60.
10. **Centers for Disease Control.** 1983. Guideline for isolation precautions in hospitals. Infect. Control 4:245–325.
11. **Joint Commission on Accreditation of Hospitals.** 1984. Accreditation manual for hospitals, 1985. Joint Commission on Accreditation of Hospitals, Chicago.
12. **National Institutes of Health.** 1972. Cleaning pro-

cedure manual, clinical center. Department of Health, Education and Welfare Publication no. (NIH)72–106. U.S. Government Printing Office, Washington, D.C.

13. **National Institutes of Health.** 1974. Biohazards safety guide. U.S. Government Printing Office, Washington, D.C.

13a. **Schmidt, E. A., D. L. Coleman, and G. F. Mallison.** 1984. Improved system for floor cleaning in health care facilities. Appl. Environ. Microbiol. **47**:942–946.

14. **Steere, N.V. (ed.).** 1971. CRC handbook of laboratory safety, 2nd ed. CRC Press, Boca Raton, Fla.

15. **U.S. Department of Health and Human Services.** 1984. Biosafety in microbiological and biomedical laboratories. HHS Publication no. (CDC) 84–8395. U.S. Government Printing Office, Washington, D.C.

Chapter 24

Pipetting

EVERETT HANEL, JR., AND MARY M. HALBERT

The pipette is a universal laboratory device for the volumetric measurement and transfer of fluids from one container to another. Straws and tubes have been used for centuries to draw liquids from a container to the mouth, but it was not until the establishment of the science of chemistry that glass laboratory apparatus including pipettes was developed. In the late 18th and early 19th centuries, F. A. H. Descroizilles (1745–1825), C. L. Berthollet (1748–1822), and J. L. Gay-Lussac (1778–1880) developed calibrated glass burettes for delivering variable amounts of solution and pipettes for delivering fixed amounts of solution (18a, 31). These first glass pipettes were crude and varied in size and shape. They were calibrated by weighing a given volume of water and were marked at the delivery level (1.0, 5.0, 10 cc [ml]) with a gunflint or file (28).

During the period 1830 to 1880, when the germ theory of disease was established by Schwann, Pasteur, Koch, and many other renowned scientists, the new science of bacteriology borrowed much of its equipment, including the pipette, from the older science of chemistry. The pipette became an integral part of bacteriological equipment and rapidly evolved into the modern graduated liquid measuring device. An 1893 chemical apparatus catalogue of Queen and Co. of Philadelphia listed an array of modern-appearing serological, volumetric, and Mohr pipettes. The Mohr pipettes were graduated in divisions of 0.01 ml in the 1-ml size and 0.1 ml in the 50-ml size.

The isolation, production, and identification of an infectious agent that causes human disease is often followed by a laboratory-acquired infection. In 1884, Gaffky isolated *Salmonella typhi*, and in 1885 the first case of laboratory-acquired typhoid fever was noted. The mode of exposure for this first case of typhoid fever was not known, but pipetting was listed as the cause of a number of subsequent cases. In 1915, Kisskalt (12), in Germany, sent a questionnaire to colleagues in Europe and collected information on 50 cases of laboratory-acquired typhoid fever dating back to the first case in 1885. There were six deaths, giving a mortality rate of 12%. The mode of infection was known in 23 cases, and in 16 of these the pipette was the cause. By 1929, there were 130 reported laboratory-acquired infections due to *S. typhi, Salmonella paratyphi, Shigella* sp., and *Vibrio*

cholerae (7). Ingestion through a pipette was the most common means of infection.

The frequency of typhoid acquired by pipetting accidents stimulated the development of mechanical pipettors. Descriptions of mechanical pipettors using rubber bulbs attached to the pipette appeared in German scientific journals as early as 1907 (15). A mechanical pipettor very similar to several modern pipetting aids was developed by the Medical Research Laboratories of Parke Davis Co. in 1908 (14). This pipettor (Fig. 1) had a rubber diaphragm that was raised and lowered by a thumb screw to draw or expel liquids.

In 1918, Reinhardt (24), an Austrian physician, described 21 different pipetting devices. He stated that with the aid of the devices described, it was possible to work more quickly than with the oral pipette.

The large number of pipetting incidents resulting in laboratory-acquired infections stimulated governmental concern in Germany. In 1956, the government of the Federal Republic of Germany published a regulation (4) that prohibited all mouth pipetting in medical, dental, and veterinary laboratories even if infectious agents were not used. However, visits to representative German laboratories in 1960 showed that this regulation was frequently disregarded (19).

In spite of the obvious hazards, the promulgation of regulations requiring the use of mechanical devices to pipette hazardous biological materials has been slow. The Naval Bioscience Research Laboratory, University of California, has required the use of a rubber bulb for all pipetting since 1943, and this regulation is still in force. At the U.S. Army Biological Research Laboratories at Fort Detrick, Md., the initial safety regulations in 1945 required the use of cotton-plugged pipettes, while only recommending the use of pipetting aids. It was not until 1957 that regulations were issued forbidding mouth pipetting of infectious or toxic materials. The various safety regulations issued by the Centers for Disease Control over the years recommended bulb pipetting when working with infectious or toxic materials, until 1975, when new regulation (*Lab Safety at the CDC*) listed 107 agents that required bulb pipetting. Biological safety regulations of the National Institutes of Health recommended the use of mechanical pipetting aids when working with infectious materials, until

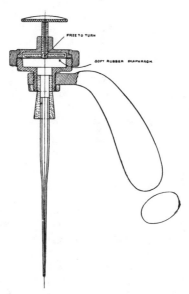

FIG. 1. (A) Mechanical pipettor developed by Parke Davis Medical Research Laboratories in 1908. (B) Cut-away view of pipettor.

1974, when the National Cancer Institute publication *Safety Standards for Research Involving Oncogenic Viruses* stated: "Mechanical pipetting aids shall be used for all pipetting procedures." This safety standard covered all pipetting procedures in both low- and moderate-risk oncogenic virus work. The National Institutes of Health warehouse has carried a variety of mechanical pipetting devices since 1958.

The *NIH Guidelines for Research Involving Recombinant DNA Molecules*, issued in 1976, stated: "Mechanical pipetting devices shall be used; pipetting by mouth is prohibited." This statement includes all work at biosafety levels 1 through 4.

In 1969, Phillips (20) found that aspiration through a pipette was the most frequent accident among 3,700 laboratory-acquired infections. Pike and co-workers, in their studies on laboratory-associated infections in 1951 (27), 1965 (23), and 1976 (22), found that aspiration through a pipette accounted for 15.4, 11.3, and 13.1%, respectively, of the total accidents reported.

Many laboratories now require the use of pipetting aids, and the frequency of pipetting accidents has decreased; however, the fact that exposures to hazardous materials through pipetting accidents still occur is somewhat discouraging. The major hazards of mouth pipetting are the production of aerosols, accidental aspiration of the fluid in the pipette, and contamination of the mouthpiece from the operator's contaminated finger (2, 18, 21, 25, 30).

Aerosols are created by the accidental escape of one or more drops from a pipette, which fall to the laboratory tabletop or floor. The amount of aerosol generated varies with the height of the fall, the type of surface on which the drop falls, and other factors. Hard surfaces such as stainless steel and painted wood produce the greatest aerosol concentration, but no surface tested completely prevented aerosol production (25). Cloth toweling dampened with a disinfectant will significantly reduce aerosol generation from a falling drop (Table 1). Pipetting of hazardous and potentially hazardous materials should be carried out in a ventilated biological safety cabinet.

The procedure used by many technicians for mixing a bacterial suspension in a test tube is that of

TABLE 1. Aerosol created by a falling drop[a]

Drop of bacterial suspension falling 12 in. (ca. 30.5 cm)	Avg no. of CFP recovered from air
Stainless steel	49.0
Painted wood	43.0
Dry cotton towel	28.0
Kemrock	23.0
Hand towel wet with 5% phenol ...	4.0
Dry wrapping paper	3.0
Dry paper towel	1.0
Pan of 5% phenol	0.1

[a] *S. indica* was the test organism. Results summarized in this table are primarily from Anderson (2) and Wedum (30). CFP, Colony-forming particles. Each procedure was repeated 10 times and the average was calculated. Air sampling was conducted by using seven sieve samplers (340 sieve holes) that surrounded the work area and were operated at a flow rate of 1 ft³/min each.

alternately sucking and blowing through a pipette. This procedure produces turbulence, and if air is passed through the suspension bubbles are produced, which upon bursting will generate an aerosol.

Whenever a pipette is used to transfer hazardous fluids, the pipette should be drained by holding the tip against the side of a tube or bottle and allowing the contents to drain gently without bubbling. Blowout type pipettes in which contents are forcibly expelled should not be used for hazardous fluids. This procedure was studied by Anderson et al. (2), using *Serratia indica* as the test organism. They found that when air was blown through a culture to produce bubbling, 0.8 colony-forming particle of *S. indica* per unit operation was recovered, whereas only 0.2 colony-forming particle of *S. indica* was recovered when the pipette contents were forcefully ejected without producing bubbles. Forceful ejection can result in turbulence and frothing in the receiver and the creation of droplet nuclei at the pipette tip, which can release aerosols into the surrounding environment. When multiple pipetting procedures such as serial dilutions are being carried out with infectious materials, it is prudent to conduct such tasks in a ventilated cabinet.

High-speed photography demonstrated that when the last drop of fluid in the tip of a pipette is blown out with moderate force, an aerosol of approximately 15,000 droplets is produced (11). The droplet sizes are mostly under 10 μm (Fig. 2). A study of cotton-plugged versus unplugged and blunt versus tapered-tip pipettes showed no appreciable differences in the concentrations or behavior of the aerosols formed. Pipettes calibrated mark to mark (Mohr type) are preferable to types requiring forceful expulsion of the last drop.

The aspiration of vapors or aerosols from liquids when mouth pipetting with unplugged pipettes was demonstrated by Bloom (5) and studied further by Phillips and Bailey (21). Both studies made use of a hypodermic syringe fastened by rubber tubing to the pipette to simulate oral pipetting. Bloom showed that orally pipetting tritium oxide resulted in the aspiration of 5 to 70 nl of the solution per operation. Constant pipetting can result in significant transfer of vapors from the solutions being used. The studies carried out by Phillips and Bailey (21) using broth cultures of either *Serratia marcesens* or *Bacillus subtilis* subsp. *niger* showed that aerosols of the test bacteria were present in the syringe in 34% of the trials with *S. marcesens* and in 58% of the trials with *B. subtilis*. These studies by Phillips and Bailey show that if unplugged pipettes are used in mouth pipetting there is a significant probability of oral contamination even if fluid is not accidentally aspirated. Cotton plugs will reduce or eliminate the passage of airborne particulates; when fluid is drawn into the plug, however, the flow of fluid may be slowed, but the plug can be easily sucked into the mouth followed by a gush of the liquid.

FIG. 2. Blowing the last drop from a pipette. A drop of liquid is retained because of capillary attraction. After gravity draining, pipette was blown out with moderate force. Ten-milliliter graduated pipette. From Johansson and Ferris (11).

The direct aspiration of infectious material during pipetting has resulted in more laboratory-acquired infections than any other accident except those involving hypodermic syringes and needles (20, 22, 23, 27). The aspiration of fluid through a pipette occurs so easily that it is not surprising that it occurs frequently. Most technicians who carry out a large number of pipetting operations by mouth, such as making serial dilutions, will aspirate fluid sooner or later. Habit and reflex action may result in excessive suction being applied to small pipettes after switching from large pipettes, which will result in the fluid rising in the small pipette with almost explosive velocity. It is obvious that infectious or toxic material should not be pipetted by mouth. Strict prohibition of mouth pipetting will eliminate the need for individuals to evaluate potential hazards and it will establish the habit of using pipetting aids.

Oral contamination can result from finger contamination of the upper end of the pipette. Workers would not think of placing a finger in the mouth when working with infectious or toxic materials, but an individual may be accomplishing the same result by placing a finger over the end of the pipette and then placing the end of the pipette in the mouth.

Contaminated pipettes should be placed horizontally in a pan containing enough suitable disinfectant, such as a solution of hypochlorite (Clorox diluted 1:100), to allow complete immersion of the pipette. The pipette discard tray should be placed in the biosafety cabinet. Pipettes should not be placed vertically in a cylinder that, because of its height, must be placed on the floor outside the cabinet. Removing contaminated pipettes from the safety cabinet and placing them vertically in a cylinder provides an opportunity for dripping from the pipette onto the floor, the rim of the cylinder, and the top of the pipettes protruding above the level of the disinfectant. The pipettes are often removed from the vertical cylinder and placed in another container for autoclaving. On the other hand, the tray containing pipettes within the safety cabinet can be covered and then removed and placed directly in the steam autoclave (Fig. 3). Autoclaving is the only procedure that will ensure sterilization of discarded pipettes.

Contamination of cell cultures occurs in many laboratories employing tissue and cell cultures for experimental or diagnostic purposes. Although contaminated cell cultures may or may not be a direct biohazard, the contaminated cultures can severely affect research results and be the cause of incorrect interpretations and diagnoses. Fogh et al. (9) stated in their review of cell culture contaminations: "Mycoplasmas can be disseminated from the oropharynx via droplets, saliva, mouth pipetting or the contaminated fingers of technicians." As to prevention procedures, Fogh et al. state: "Rather simple methods of prevention of mycoplasma contamination are available. Among the recommended procedures are: rigid aseptic techniques, including avoidance of the use of mouth pipetting." Other investigators who have studied the problem of cell culture contamination have concluded that mouth pipetting is a frequent cause of contamination and should be avoided (3, 6, 10, 26).

The modern laboratory is designed and equipped to provide an optimal level of safety. While directional airflows and biological safety cabinets are very desirable mechanisms for providing particulate containment, these systems are costly. The elimination of pipetting hazards will provide increased safety and yet add little, if any, cost to the operation. The safety regulations of many research and clinical laboratories still recommend rather than require the use of pipetting aids when pipetting hazardous materials. The widespread use of biological safety cabinets has promoted the use of pipetting aids since it is impossible to pipette directly by mouth within the cabinet. However, more and more regulations and guidelines, such as the *Guidelines for Research Involving Recombinant DNA Molecules* (17), state that all pipetting (biosafety levels 1 through 4) must be carried out using a mechanical pipetting device; pipetting by mouth is prohibited.

The hazards associated with mouth pipetting can be avoided by the use of one of the many commerically available pipetting aids. When selecting a pipetting aid, the following points should be considered:

Volume of liquid to be manipulated

Accuracy required

Ease of manipulation

Use of standard pipettes, "shortie" pipettes, or pipette tips

Viscosity of the liquid to be measured and transferred

Compatibility of equipment with the liquid if solvents or corrosive reagents are to be used

Ease of sterilization, cleaning, repair, and recalibration

Versatility of use

Ease of changing pipettes

Ease of controlling flow of fluid

Desirability of having the fourth and fifth fingers free to manipulate stoppers and bottle caps

Need for the equipment to be lightweight and comfortable to hold and for the operator to use it continuously without excessive fatigue

The School of Public Health, University of Minnesota, lists 71 different commercially available pipetting devices in their course, "Fundamentals for Safe Microbiological Research" (29). A particular type of pipetting aid that may be satisfactory to one individual or kind of operation may not meet the requirements of others; therefore, different types of pipettors should be tried in each situation. Care should be taken in selecting a device so that its design or use does not contribute to exposure to hazardous materials, that exposed parts (other than delivery tips) remain free of hazardous substances, and that the unit can be easily sterilized and cleaned after use. Note that many pipetting aids currently available are not designed to be steam sterilized.

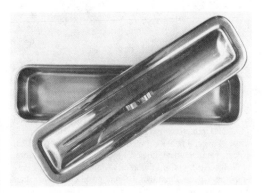

FIG. 3. Volrath instrument or pipette tray.

Mouth pipetting devices that use a hydrophobic membrane filter (usually 0.20 μm) in a mouthpiece attached to plastic tubing that extends to and fits over the end of a pipette will prevent the passage of liquid and will remove particulates. The operator controls delivery by finger pressure on a soft delivery tube mounted between the filter and the plastic tube. Fingers do not touch the portion of the mouthpiece entering the mouth. However, this reusable pipetting device requires mouth pipetting and the storage of the pipettor on the bench or other location between usage. Therefore, this mouth pipetting device is not recommended for pipetting operations.

Nonautomatic pipetting aids are available in a variety of bulb-actuated suction devices as well as piston or syringe types. Many pipetting aids that are in use are shown in Fig. 4, 5, and 6. Others are available, and new ones are developed and put on the market each year.

The pipetting aids used in biomedical and chemical laboratories can be classified into three categories: ultramicro (<0.1 ml), micro (0.1 to 1 ml), and macro (>1 ml). Advances in protein chemistry, enzymology, and immunochemistry, together with the development of methods for antibody, antigen, and enzyme purification, have led to the development of immunoassay methods such as fluorescence immunoassay, radioimmunoassay, voltammetric immunoassay, and enzyme-linked immunosorbent assay which use microtitrations. The ultramicro

pipettors are available in dispensing volumes from 0.001 to 0.1 ml, but most serial dilutions made in the course of microtitrations use disposable microtitration plates with wells of 0.25- to 3.0-ml capacity. Dilutions are made using ultramicropipettes or microdiluters with disposable plastic or reusable, autoclavable polypropylene pipette tips. Microdispensers made of stainless steel or autoclavable plastic designed for serial transfer of 0.025-, 0.05-, or 0.1-ml volumes are also available. These microdispensers are used in automatic titrating machines or in hand-held microdispensing devices. From 8 to 12 serial transfers are made simultaneously.

Pipette tips are available in either nonsterile bulk-packed cartons of 1,000 or as sterile tips in 8- or 12-place holders sealed with tape. The tape is peeled off and the tips are inserted simultaneously into the pipettor; liquid is dispensed in 8 or 12 wells of microtitration plate at each delivery stroke. The tips are ejected by activating an ejection button (Fig. 7).

The use of power-assisted vacuum sources is helpful in reducing fatigue when large numbers of pipetting operations are being conducted. A battery-powered pipetting aid, equipped with rechargeable batteries and a filter to prevent contamination of the unit, is available that will accept any glass or plastic pipette from 0.5- to 75-ml capacity (Fig. 8). Other pipetting devices use 115 V alternating current and may be equipped with high-efficiency particulate air

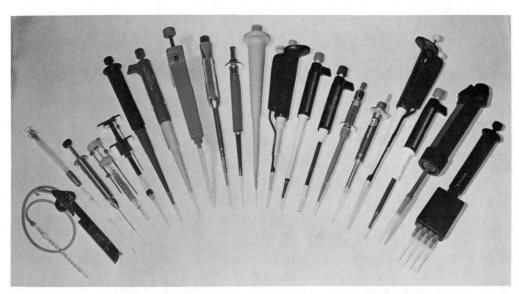

FIG. 4. Ultramicro pipetting aids. From left to right: Clinac safety pipettor; Sargent-Welch pipette syringe; Drummond Dialamatic micro dispenser; Hamilton Ultra Micro syringe pipette; Unimetrics micropipettor; Finnpipette (Finnish manufactured); Helena Quickpette; Clay Adams Selectapette pipette; Lancer Precision pipettor; MLA pipettor (Medical Laboratory Automation, Inc.); Eppendorf Microliter pipette (German manufactured); Gilson Pipetman Fixed Pushbutton microliter pipette (French manufactured); Oxford Sampler pipette (800 series); Oxford Sampler Ultramicro pipette; Elkay "Socorex" micropipette (Swiss manufactured); Centaur micropipette; Gilson Pipetman Digital pushbutton microliter pipette; Oxford sampler pipette, model Q; Oxford Micro-Doser Repetitive pipette; Flow Laboratories Titer-Tek Multi-channel pipette (French manufactured).

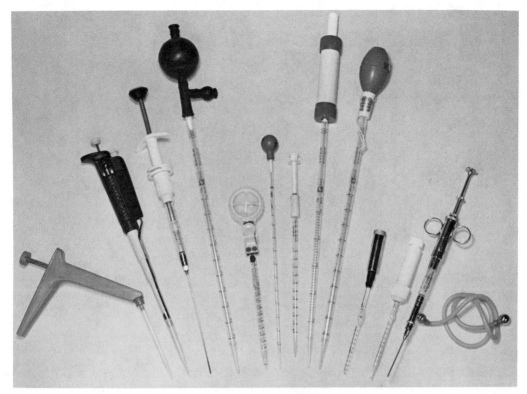

FIG. 5. Micro pipetting aids. From left to right: Clay Adams Selectapette pipette; Gilson Pipetman Digital Pushbutton microliter pipette (French manufactured); Manostat Vari-Pet; Spectroline pipette filler; Pumpett 18 (Swedish manufactured); Curtin Matheson rubber pipette bulb; Demuth safety pipette; Nalgene pipetting aid; Analytic Products safety bulb; Clay Adams pipette suction apparatus; Manostat Accropet Filler, micro; Cornwall continuous pipetting outfit.

filters on the pump. A membrane filter can be used as an auxiliary attachment to prevent aerosol passage and perform as a check valve to prevent the passage of liquid into the pipette holder (Fig. 9). It is desirable to protect the pipette holder from contamination because cross-contamination is a particularly sensitive problem in cell culture work.

The wide variety of available models with various special features should make it relatively easy to select the most suitable pipetting aid for the individual and the intended use. The almost universal availability and use of laminar-flow biological safety cabinets physically requires the use of pipetting aids, but the 8- to 10-in. (ca. 20- to 25-cm) open workspace across the front of the biological safety cabinet makes it awkward to use long pipettes with tall pipetting aids. Some workers find it more convenient and less tiring to use short pipettes (shorties) ranging in length from 5 to 9 in. (ca. 13 to 23 cm).

The following definitions are applicable to pipettes used in the biomedical laboratory.

To-deliver pipettes. To-deliver (TD) pipettes are calibrated to deliver a given volume after allowing for residual fluid remaining on the walls and tip. There are several types of pipettes in this category.

(i) Blow-out pipettes (indicated by a double ring or frosted band near the top) are designed for forceful expulsion of tip contents (and, therefore, are contraindicated for use with hazardous fluids, due to aerosol potential).

(ii) Drain-type pipettes are designed to drain slowly until the last drop is extracted by gently touching the tip to a surface of the receiving container.

(iii) Volumetric (or transfer) pipettes are designed to deliver a given volume by unrestricted draining from a vertical position with the tip held to the wall of the receiving container, without extracting the last drop.

(iv) Measuring (Mohr) pipettes are designed to deliver a given volume by unrestricted draining from a vertical position to a given line. The tip is then touched to the container surface to release any adhering drops (also referred to as mark-to-mark).

(v) Dual-purpose pipettes (with negative graduations) are designed to be used either in the blow-out mode or the measuring (mark-to-mark) mode. These pipettes are marked with two calibration lines. The upper line is marked "TD," and the lower graduation is marked "TC." This pipette carries the double band characteristic of blow-out pipettes. Dual-

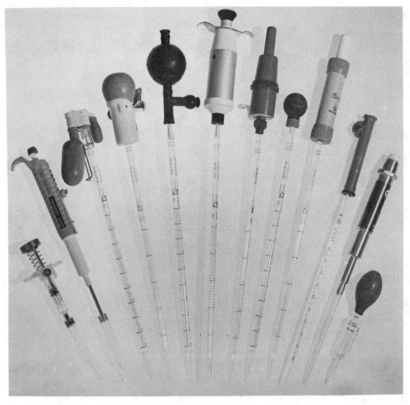

FIG. 6. Macro pipetting aids. From left to right: Lab Industries Repipet Sampler; Oxford Macro Set Transfer pipetting system; National Instrument Micropet pipettor; Pumpett 25, Original (Swedish manufactured); Spectroline pipette filler; Manostat Accropet Filler (Macro); Volac Universal Pipette Controller (British manufactured); Interex rubber transfer bulb; Nalgene pipetting aid; Pi Pump (West German manufactured); Centaur Macropipette; Lab Industries Micro/Mac Pipet.

purpose pipettes are not recommended because of the possibility of mix-up between the TD/blow-out and TC markings, with consequent misuse and improper delivery.

To-contain pipettes. To-contain (TC) pipettes are calibrated to contain a given volume within the pipette. The total volume can be recovered only by subsequent multiple rinsings.

Considerable error can be introduced if these definitions are misunderstood and the pipette is not used according to design specifications. Mechanical pipetting aids should be selected to be consistent with the desired mode of delivery. Note that unrestricted draining may be compromised by the forceful expulsion of pipette contents inherent in many aids.

Modern pipettes are constructed with excellent accuracy, usually ±2.0% or greater. This is generally acceptable, considering the many variables inherent in bacteriological procedures. However, both federal and state accrediting programs require laboratories performing certain procedures to document that volumetric equipment has undergone periodic maintenance, inspection, and testing. The Code of Federal Regulations, Title 42, Public Health, Part

74, requires pipettes, pipettors, pipettor microdispensers, and dilutors used in certain procedures to be tested periodically to comply with good quality control practices. The checklist of the College of American Pathologists states: "Automatic pipettes and diluting devices used for measurement of sample volumes must be checked for accuracy and reproducibility before being placed into service."

If a high degree of accuracy is required, pipettes meeting Class A requirements for volume accuracy and Class B flow time requirements of Federal Specification NNN-P-395 can be purchased. These pipettes bear the letter "A" near the top of the mouthpiece and comply with requirements for laboratory certification by the College of American Pathologists. Because the Code of Federal Regulations requiring the periodic testing of volumetric equipment does not define the time interval of "periodic," laboratory supervisors may wish to establish their own schedule for testing. In general, a pipette's accuracy will not vary over time, but automatic dilutors and dispensers may show some variations. Of course, accuracy is affected by the viscosity, surface tension, density, and vapor pressure of the

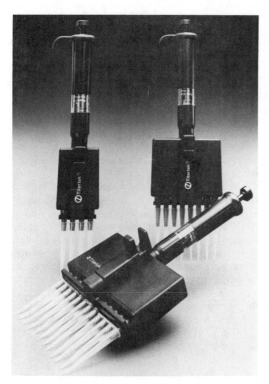

FIG. 7. Four-, 8-, or 12-tip pipettor with tip ejector for use with multiwell microplates or tissue culture plates. Picture courtesy of Flow Laboratories, Inc.

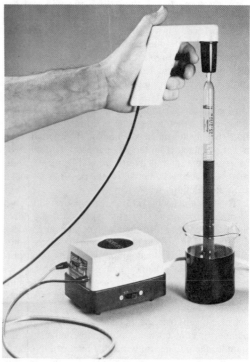

FIG. 9. Pipette aid activated by a 115 V alternating-current pump. Picture courtesy of Drummond Scientific Co.

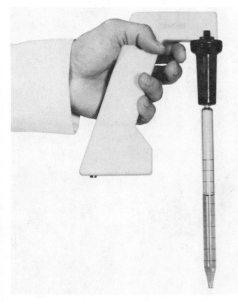

FIG. 8. Battery-powered pipetting aid with rechargeable batteries. Picture courtesy of Drummond Scientific Co.

measured reagent. Therefore, a change in the fluid being measured and transferred may require a check for accuracy of the volumetric equipment being used. Pipettes used in water analysis, particularly in work relative to enforcement of regulations of water quality, must be calibrated and marked. These pipettes, as well as milk dilution pipettes, should conform to the American Public Health Association standards given in the latest edition of *Standard Methods for the Examination of Dairy Products* (1).

There are several methods available for determining the accuracy and reproducibility (precision) of pipettes. The most common method has been to weigh the fluid delivered with an analytical balance. The volume of any liquid whose specific gravity is known can be accurately determined by comparing its weight with that of water on an analytical balance. A number of commercially available kits are available for the volume calibration of pipettes. Some kits employ a color dilution principle with the color intensity determined with a spectrophotometer. A standard curve is prepared, and a quantity of calibration reagent is pipetted into a cuvette. The color intensity is read in the spectrophotometer and compared with the standard curve to determine the volume delivered by the pipette (8). The accuracy is printed on most brands of pipettes.

Accuracy and reproducibility (precision) of TC pipettes can be measured by adding a radioisotope of known activity to the fluid being pipetted and determining the radioactive counts on a particle counter (13). At least four repetitive pipetting measurements should be made for each pipette when checking accuracy, and 10 replicate samples should be measured when checking reproducibility (precision). (For further information on pipette performance and testing, see reference 16.)

Manufacturers calibrate TD products with an overfill, so that when the pipette is drained, the film of fluid that always remains behind on the internal walls is not deducted from the amount actually delivered. Measurement and delivery from TD pipettes are relatively accurate, but it is not possible to make accurate dilutions with them.

There is a class of TD transfer pipettes which are positive-displacement syringe-type pipettes. These pipettes were first designed with a plunger inside a needle and are used to introduce small-volume samples into a gas chromatograph or a mass spectrometer. They give accuracy and precision with practically no sample loss or carry-over, and they require only one calibration for all samples, as the procedure is unaffected by surface tension, vapor pressure, density, or viscosity of the sample. Only the Teflon plunger tip contacts the sample; upon delivery of the sample, the plunger tip wipes the bore of the pipette tip clean. This forces out all of even the most viscous sample. Positive-displacement pipettes are used primarily in the microliter range, but 1-ml and larger sizes also are available.

Some procedures in the research or clinical laboratory are so time dependent or highly repetitive that a dilutor-dispenser or bottle dispenser is preferred to using the slower manual pipetting method. The electrically powered dilutor-dispenser uses syringes of various sizes and can be set within a limited range to dilute and dispense the desired volumes. A complete cycle of dilution and dispensing takes 6 s (Fig. 10). The bottle or reagent dispenser can be obtained in a range of sizes from 0.05 to 0.2 ml in the microliter range (Fig. 11) and 5 to 10 ml in the larger dispensing size (Fig. 12). Dispensing bottles range in size from 70 to 2,000 ml. Accuracy of the reagent dispenser is usually 1%, and reproducibility is usually 0.2%. Aqueous solutions, dilute acids, unhydrolyzed proteins, and other liquids can be quickly, conveniently, and accurately transferred when using a bottle or reagent dispenser.

The types of pipettes and pipetting aids to be used in the laboratory should be established by the supervisor and made known to all personnel in the laboratory. Mouth pipetting should not be permitted. In the modern laboratory, the use of a pipetting aid is safer, faster, and more accurate than mouth pipetting. When large numbers of dilutions and transfers are to be made, as in serological testing, then one of the variable-volume diluting-dispensing machines is fast, convenient, and accurate for the job. Where accuracy and precision are critical for volumes of <1.0 ml, TC pipettes or capillary tubes or TD positive-displacement manual samplers should be used.

SUMMARY OF THE USE OF PIPETTES AND PIPETTING AIDS

Never use mouth pipetting. Always use some type of a pipetting aid.

When working with infectious or toxic fluids,

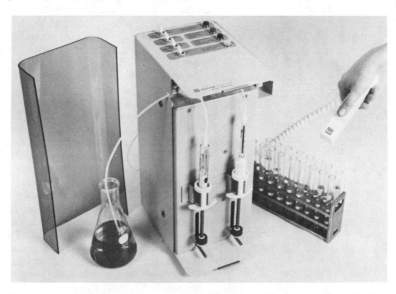

FIG. 10. Electric-powered dilutor-dispenser can be used within a wide range to dilute and dispense desired volumes. Picture courtesy of Manostat Corp.

FIG. 11. Bottle dispenser quipped with a microliter dispenser. Picture courtesy of Markson Science.

FIG. 12. Bottle dispensers equipped with dispensers in a range of sizes. Picture courtesy of Cole-Parmer Instrument Co.

pipetting operations should be confined to a ventilated biosafety cabinet or hood.

Infectious or toxic materials should never be expelled from a pipette.

Mark-to-mark pipettes are preferable to other types, since they do not require expulsion of the last drop.

Infectious or toxic fluids should never be mixed by bubbling air from a pipette through the fluid.

Infectious or toxic material should not be mixed by alternate suction and expulsion through a pipette.

Discharge from pipettes should be as close as possible to the fluid or agar level, and the contents should be allowed to run down the wall of the tube or bottle whenever possible, not dropped from a height.

Pipettes used for transferring infectious or toxic materials should always be plugged with cotton, even when safety pipetting aids are used.

When pipettes are used, avoid accidentally dropping infectious or toxic material from the pipette. Place a disinfectant-dampened towel on the work surface, and autoclave the towel after use. Disposable absorbent paper bonded to polyethylene to absorb drops of fluid is a suitable cover for the work surface and provides a waterproof lining.

Contaminated pipettes should be placed horizontally in a pan or tray containing enough suitable disinfectant to allow complete immersion. Pipettes should not be placed vertically in a cylinder.

Discard pans for used pipettes are to be housed within a biological safety cabinet.

The pan and used pipettes should be autoclaved as a unit and replaced by a clean pan with fresh disinfectant.

LITERATURE CITED

1. **American Public Health Association.** 1978. Standard methods for the examination of dairy products, 14th ed. American Public Health Association, Washington, D.C.
2. **Anderson, R. E., L. Stein, M. L. Moss, and N. H. Gross.** 1952. Potential infectious hazards of common bacteriological techniques. J. Bacteriol. **64**:473–481.
3. **Barile, M. F.** 1968. Mycoplasma and cell cultures. Natl. Cancer Inst. Monogr. **29**:201–204.
4. **Berufsgenossenschaft für Gesundheitsdienst und Wohfohrtspflege.** 1956. Accident prevention regulations for medical laboratories. Berufsgenossenchaft für Gesundheitsdienst und Wohfohrtspflege, Hamburg 36, Federal Republic of Germany.
5. **Bloom, B.** 1960. The hazard of orally pipetting tritium oxide. J. Lab. Clin. Med. **55**:164.
6. **Carski, T. R., and M. C. Shepard.** 1961. Pleuropneumonia-like (mycoplasma) infections of tissue culture. J. Bacteriol. **81**:626–635.
7. **Draese, K. D.** 1939. Über Laboratoriumsinfektionen mit Typhusbazillen und anderen Bakterien. Arch. Hyg. Bakteriol. **121**:232–291.
8. **Ellerbrook, L. D.** 1954. A simple colorimetric method for calibration of pipettes. Am. J. Clin. Pathol. **24**:868–874.
9. **Fogh, N. J. B. Holmgren, and P. P. Ludovici.** 1971. A review of cell culture contaminations. In Vitro **7**:26–41.
10. **Hayflick, L.** 1965. Tissue cultures and mycoplasmas. Texas Rep. Biol. Med. **23**:285–303.
11. **Johansson, K. R., and D. H. Ferris.** 1946. Photography of airborne particles during bacteriological plating operations. J. Infect. Dis. **78**:238–252.
12. **Kisskalt, K.** 1915. Laboratoriumsinfektionen mit Typhusbazillen. Z. Hyg. Infektionskr. **50**:145–162.
13. **Lopez, R. E.** 1972. Evaluation of a precision pipetting system. Am. Lab. **4**:28–29.
14. **Miller, E. C. L.** 1908. Some simple laboratory devices. Zentralbl. Bakteriol. **46**:728–732.
15. **Müller, R., and H. Gräf.** 1907. Wert der Blutuntersuchung für die Typhusdiagnose. Zentralbl. Bakteriol. **43**:856–879.
16. **National Committee for Clinical Laboratory Standards.** 1984. Determining performance of volumetric equipment: proposed guidelines (18-P). National Committee for Clinical Laboratory Standards, Villanova, Pa.
17. **National Institutes of Health.** 1984. Guidelines for

research involving recombinant DNA molecules. November 23, 1984. Fed. Reg. **48**:46266–46291.

18. **Paneth, L.** 1915. The prevention of laboratory infections. Med. Klin. **11**:1398–1399.

18a. **Partington, J. R.** 1964. A history of chemistry, vol. 4. The Macmillan Co., London.

19. **Phillips, G. B.** 1961. Microbiological safety in U.S. and foreign laboratories. Technical Study 35. U.S. Army Biological Laboratories, Fort Detrick, Md. (A.D. 268–635).

20. **Phillips, G. B.** 1969. Control of microbiological hazards in the laboratory. Am. Ind. Hyg. Assoc. J. **30**:170–176.

21. **Phillips, G. B., and S. P. Bailey.** 1966. Hazards of mouth pipetting. Am. J. Med. Technol. **32**:127–129.

22. **Pike, R. M.** 1976. Laboratory-associated infections: summary and analysis of 3,921 cases. Health Lab. Sci. **13**:105–114.

23. **Pike, R. M., S. E. Sulkin, and M. L. Schulze.** 1965. Continuing importance of laboratory-acquired infections. Am. J. Public Health **55**:190–199.

24. **Reinhardt, F.** 1918. Zur Verhütung von Laboratoriumsinfecktion. Zentralbl. Bakteriol. Abt. 1 Orig. Bd. 80 Heft 7 **55**:456–465.

25. **Reitman, M., and G. B. Phillips.** 1955. Biological hazards of common laboratory procedures. I. The pipette. Am. J. Med. Technol. **21**:338–342.

26. **Rothblat, G. H.** 1960. PPLO contamination in tissue cultures. Ann. N.Y. Acad. Sci. **79**:430–432.

27. **Sulkin, S. E., and R. M. Pike.** 1951. Survey of laboratory-acquired infections. Am. J. Public Health **41**:769–781.

28. **Szent-Györgyi, A.** 1969. Pharmaceutical laboratory glassware in the 18th century. Ther. Hung. **17**:93–98.

29. **University of Minnesota School of Public Health and American Society for Microbiology.** 1982. Fundamentals for safe microbiological research. Available from The National Audiovisual Center (GSA)-AO9753, Washington, D.C.

30. **Wedum, A. G.** 1953. Bacteriological safety. Am. J. Public Health **43**:1428–1437.

31. **Welter [sic] and J. L. Gay-Lussac.** 1828. Burette, pipette, and measuring flask. Ann. Chim. **39**:337–368.

Chapter 25

Packaging and Shipping Biological, Radioactive, and Chemical Materials

JOHN E. JAUGSTETTER AND WILLIAM M. WAGNER

Federal regulations prescribe procedures for packaging and shipping hazardous materials; unfortunately, these regulations are often difficult to obtain or to follow. Most of the requirements specified in the regulations are based on the physical characteristics of the material and just plain common sense.

In the safe transportation of hazardous material, proper packaging is the primary consideration. The prospective shipper should ask three basic questions:

(i) What material is being shipped?
(ii) What is the form of the material?
(iii) What is the quantity of the material?

Answers to these questions will determine the container required, the specific warning label to be attached, and the proper mode of transportation.

The basic requirements of the various federal and international regulations which govern the packaging and shipping of hazardous material are summarized below. Even the most conscientious and up-to-date shipper may, however, run into difficulty with the constantly changing requirements of different regulatory organizations.

The Department of Transportation (DOT) has the regulatory responsibility for the safe transportation of hazardous materials between states and to foreign countries. The DOT regulations govern all means of transportation except shipping by mail, which is covered by the U.S. Postal Service (USPS). The DOT regulations are found in the Code of Federal Regulations (CFR), Title 49, Parts 100–199 (12); USPS regulations are found in 39 CFR, Part 123. Most states have adopted these regulations for the movement of hazardous materials within a state.

The 91-page Hazardous Materials Table in 49 CFR, Part 172.101, lists most hazardous materials and includes a description, classification, regulation reference, and warning for each material. Hazardous biological materials are classified as "Etiological Agents," and most radioactive substances are classified as "Radioactive Material, n.o.s. [not otherwise specified]." In addition, other federal agencies, namely, the Public Health Service (PHS) and Nuclear Regulatory Commission, have responsibility for these materials.

Examples of the data in the Hazardous Materials

Table are presented in Table 1. The descriptions of the material and the proper shipping names are given in column 2; the hazard class is given in column 3. For example, in Table 1, "Nitric acid, fuming" is the description and proper shipping name for a material in the "Oxidizer" hazard class. If a material involves more than one hazard, such as radioactive phosphoric acid, it must be packaged according to the highest ranking hazard. Hazardous materials are ranked by degree of hazard in Table 2; radioactive materials head the list. If, however, the quantity of radioactivity is so small that the material is exempt from the regulations, then the chemical hazard of phosphoric acid would have the highest priority.

BIOLOGICAL MATERIALS

Developing Perspectives

In 1971, the Food and Drug Administration's authority for governing the interstate shipment of etiologic agents was delegated to the Center for Disease Control. The PHS Interstate Quarantine regulation (42 CFR, Section 72.25) was revised, effective 31 July 1972, to provide packaging requirements, a hazard warning label, and a system for receiving telephone reports of damaged or leaking shipments of etiologic agents and for responding to these reports. The regulation, "Interstate Shipment of Etiologic Agents" (42 CFR, Part 72), was later revised. The revised regulation, effective 20 August 1980 (23), did the following:

(i) Corrected the scientific nomenclature and expanded and updated the list of etiologic agents subject to the requirement

(ii) Provided for future revisions of the list of agents by publication of a notice in the Federal Register

(iii) Increased the maximum permissible volume of etiologic agent in a single primary container from 500 to 1,000 ml

(iv) Clarified the packaging, labeling, and shipping requirements for materials such as cultures, clinical specimens, and vaccines which contain or may contain an etiologic agent

TABLE 1. Hazardous Materials Table[a]

(1) Notes[b]	(2) Hazardous materials descriptions and proper shipping names	(3) Hazard class	(4) Label(s) required (if not excepted)	(5) Packaging		(6) Maximum net quantity in one package	
				(a) Exceptions	(b) Specific requirements	(a) Passenger-carrying aircraft or railcar	(b) Cargo-only aircraft
	Acetone	Flammable liquid	Flammable liquid	173.118	173.119	1 qt	10 gal
	Alcohol, n.o.s.	Flammable liquid	Flammable liquid	173.118	173.125	1 qt	10 gal
	Alcohol, n.o.s.	Combustible liquid	Combustible liquid	173.118a	None	No limit	No limit
	Bromine	Corrosive material	Corrosive	None	173.252	Forbidden	1 qt
A,W	Carbon dioxide, solid, or dry ice, or carbonice	ORM-A	None	None	173.615	440 lb	440 lb
	Corrosive, liquid, n.o.s.	Corrosive material	Corrosive	173.244	173.245 173.245a	1 qt	1 qt
	Ether (ethyl)	Flammable liquid	Flammable liquid	None	173.119	Forbidden	10 gal
	Etiologic agent	Etiologic agent	Etiologic agent	173.386	173.387	173.386	4 liters
*	Hydrochloric (muriatic) acid	Corrosive material	Corrosive	173.244	173.263	1 qt	1 gal
	Lead arsenate, solid	Poison B	Poison	173.364	173.367	50 lb	200 lb
	Nitric acid, fuming	Oxidizer	Oxidizer and poison	None	173.268	Forbidden	5 pints
*	Radioactive material, L.S.A., n.o.s.	Radioactive material	Radioactive	173.392	173.393		
*	Radioactive material, n.o.s.	Radioactive material	Radioactive	173.393	173.395		
	Succinic acid peroxide	Organic peroxide	Organic peroxide	173.153	173.157 173.158	Forbidden	25 lb
	Uranyl nitrate, solid	Radioactive material	Radioactive and oxidizer	173.392	173.393 173.396	Forbidden	

[a] Adapted from CFR Title 49, Part 172.101.
[b] *, Material may be regulated under another class; A, regulation for aircraft only; W, regulation for watercraft only. n.o.s., Not otherwise specified; L.S.A., low specific activity.

TABLE 2. Classification of hazardous materials[a]

1. Radioactive material
2. Poison A
3. Flammable gas
4. Nonflammable gas
5. Flammable liquid
6. Oxidizer
7. Flammable solid
8. Corrosive material (liquid)
9. Poison B
10. Corrosive material (solid)
11. Irritating material
12. Combustible liquid (110 gal or more)
13. ORM-B
14. ORM-A
15. Combustible liquid (110 gal or less)
16. Etiologic agent [classified in CFR Title 42, Part 72.30(c)]

[a]Adapted from CFR Title 49, Part 173.2.

The PHS Foreign Quarantine regulation (42 CFR, Part 71, Section 71.156) governs the importation and any subsequent receipt by transfer within the United States of both etiologic agents and vectors of human diseases. This regulation incorporates a permit system in which the Centers for Disease Control issues a permit to the recipient. The permit may be issued subject to certain conditions of issuance indicated on the permit.

Animal health authorities of the U.S. Department of Agriculture (USDA) have a similar permit system for animal host-specific agents and certain zoonotic agents. The USDA has both the responsibility and the authority to regulate transportation of such material. Shippers of these agents or of plant pathogens are referred to appropriate officials of the Veterinary Service or Plant Protection and Quarantine Program, USDA.

The shipment of etiologic agents, diagnostic specimens, and biological products by all modes of transportation—air, motor, rail, and water carriers— is governed by DOT regulation 49 CFR, Section 173.386–388. The DOT regulations differ from those of the PHS in two principal areas, as follows.

(i) Section 173.387 specifies environmental conditions, test conditions (water spray, free drop, and penetration), and testing procedures applicable to packages containing etiologic agents.
(ii) Authority of the DOT expands the application of its regulations to include the transport of etiologic agents by carriers licensed in interstate commerce regardless of whether material is shipped intrastate or interstate.

The USPS has requirements applicable to the mailing of "diseased tissues, blood, serum, and cultures of pathogenic microorganisms." These requirements, listed in the *Postal Service Manual* (17), are compatible with those of the PHS and DOT.

The Air Transport Association (ATA), an organization of domestic airline carriers, publishes and periodically revises Restricted Articles Tariff 6–D (1), which further specifies conditions for shipment of etiologic agents, diagnostic specimens, and biological products being transported by air. Petitions for revisions in Tariff 6–D are submitted to the Civil Aeronautics Board for consideration and approval. The ATA tariff incorporates the pertinent requirements of the PHS and DOT regulations. It also requires documentation in the form of a Shipper's Certification for Restricted Articles to further identify etiologic agents which are transported by air and specifies that volumes of etiologic agents of less than 50 ml may be transported on passenger-carrying aircraft. Volumes exceeding 50 ml are relegated to cargo-only aircraft.

The Universal Postal Union (UPU) has revised the regulations governing the mailing of hazardous or perishable substances via international mail. These regulations are based on recommendations jointly developed by the UPU and the International Air Transport Association. The changes were effective 22 May 1980, and are outlined below.

(i) Perishable biological materials, including infectious substances, are admissible as UPU (international) mail only if they are sent in the form of a registered airmail letter package. The package must not weight over 4 lb. (ca. 1.8 kg). Such packages require the Customs Declaration for Postal Mail (POD Form 2976) (Fig. 1). This form must be signed by the shipper.
(ii) An "Infectious Substance" label (Fig. 2) has been adopted by the UPU.

Other requirements for international mail that remain unchanged include the following.

FIG. 1. Customs Declaration for Postal Mail for international mailing.

FIG. 2. Shipping label for infectious substance.

(i) Etiologic Agents/Biomedical Material label. This label must be affixed to the outer shipping container.
(ii) Shipper's Certification for Restricted Articles. Two completed and signed copies of this form must be presented to the carrier.
(iii) Packaging.

Packaging Requirements and Instructions

Diagnostic specimens and biological products

Diagnostic specimens and biological products are subject to minimum packaging requirements described in 42 CFR, Part 72.2, of the PHS Interstate Shipment of Etiologic Agents regulation (23), which specifies that such material must be packaged "to withstand leakage of contents, shocks, pressure changes and other conditions incident to ordinary handling in transportation." It is the sender's responsibility to provide packaging that meets these conditions.

One packaging system that has proven satisfactory for shipping diagnostic specimens comprises commercially available molded polystyrene containers taped or enclosed in a cardboard sleeve or container. These molded containers are available in various sizes and shapes for single or multiple specimens. Several other packaging systems are equally effective for shipping diagnostic specimens. Some of these systems are described in references 5 and 24.

Diagnostic specimens and biological products are not subject to volume limitations of the primary specimen containers or to aggregate volume limitations in single shipping containers regardless of the mode of shipment.

Etiologic agents

Materials known or presumed to contain a viable microorganism or its toxin which causes, or may cause, human disease are subject to additional packaging and shipping requirements. Section 72.3 of the PHS Interstate Shipment of Etiologic Agents regulations regulates etiologic agents and describes packaging, labeling, and shipping requirements. Packaging requirements are based on the volume of materials in the primary container (that is, volumes less than 50 ml and volumes 50 ml or greater), as follows.

Quantities less than 50 ml. Place the material to be shipped in a securely closed, watertight primary container (test tube, glass ampoule, vial, etc.). Tighten the screw cap of the container and tape it to prevent it from loosening during shipment. Place this primary container in a durable, watertight secondary container. Several primary containers may be enclosed in a single secondary container, provided the total volume of all the enclosed primary containers does not exceed 50 ml. Fill the space at the top, bottom, and sides between the primary and secondary containers with sufficient nonparticulate absorbent material to absorb the entire contents of the primary container(s) in case of breakage or leakage. Next, enclose each set of primary and secondary containers in an outer shipping container constructed of corrugated fiberboard, cardboard, wood, or other material of equivalent strength. Patient and laboratory forms should be placed between the secondary container and the outer shipping container. A diagrammatic summary of packaging requirements can be found in Appendix 1 at the end of this volume.

If dry ice is used as a refrigerant, it must be placed outside the secondary container.

Quantities 50 ml or greater. Place the material in a securely closed, watertight primary container. Place this primary container in a durable, watertight secondary container. Single primary containers must not contain more than 1,000 ml of material. Two or more primary containers whose combined volumes do not exceed 1,000 ml may, however, be placed in a single secondary container. Fill the space at the top, bottom, and sides between the primary and secondary containers with sufficient nonparticulate absorbent material to absorb the entire contents of the primary container(s) in case of breakage or leakage. Then, enclose each set of primary and secondary containers in an outer shipping container constructed of corrugated fiberboard, wood, or other material of equivalent strength. Place a shock-absorbent material, equal in volume at least to that of the absorbent material, between the primary and secondary containers at the top, bottom, and sides, and between the secondary container and the outer shipping container. The maximum amount of material that may be enclosed in a single outer shipping container is 4,000 ml.

If dry ice is used as a refrigerant, it must be placed outside the secondary container(s), and the shock-absorbent material must be placed so that the secondary container does not become loose inside the outer shipping container as the dry ice sublimates.

Labeling Requirements and Instructions

Diagnostic specimens and biological products

Diagnostic specimens and biological products require no special markings, and hazard warning labels are not required. The shipper should not place "Etiologic Agents/Biomedical Material" labels on any shipment of diagnostic specimens or biological products.

Etiologic agents

All information identifying or describing etiologic material (laboratory data forms, patient information, etc.) should be placed around the outside of the secondary container. Never place this information between primary and secondary containers.

A label indicating the names and addresses of the sender and the addressee should be affixed to the outer shipping container.

The label for Etiologic Agents/Biomedical Material as illustrated and described in 42 CFR, Part 72.3, of the PHS Interstate Shipment of Etiologic Agents regulation (see Appendix 1 of this volume) should be placed on the outer shipping container of all packages containing etiologic agents which are to be transported in interstate traffic.

When plant pathogens, as defined by the USDA, are shipped, the outer shipping container must have the appropriate USDA shipping label affixed to it.

Transportation Instructions for Etiologic Agents

Domestic transportation

Quantities less than 50 ml. Specimens sent by air mail should have a "Notice to Carrier" (Fig. 3) affixed to the outer shipping container. These specimens may be transported on passenger-carrying aircraft.

Quantities 50 ml or greater. Shipments of etiologic agents exceeding 50 ml in volume are restricted to transport by cargo-only aircraft in accordance with ATA Restricted Articles Tariff 6–D (1). Such shipments must have two copies of the completed and signed Shipper's Certification for Restricted Articles, as specified in Tariff 6–D, affixed to or accompanying the shipment.

When dry ice is used as a refrigerant, an appropriate label (Fig. 4) should be affixed to the outer shipping container. The amount of dry ice and the date the shipment was packed should be indicated on the label.

International transportation

Perishable biological materials. Perishable biological substances, including infectious substances (etiologic agents), when offered for international shipment, must be sent as a registered airmail package. The package must not weigh over 4 lb (1.8 kg).

Required labels. Required labels, to be affixed to the outer shipping container, include the following:

   (i) Name and address of sender and receiver
   (ii) Etiologic Agents/Biomedical Material label (Fig. 3)
   (iii) Infectious Substance label (Fig. 2)
   (iv) Customs Declaration for Postal Mail (PS

NOTICE TO CARRIER

This package contains LESS THAN 50 ml OF AN **ETIOLOGIC AGENT**, N.O.S., is packaged and labeled in accordance with the U.S. Public Health Service Interstate Quarantine Regulations (42 CFR, Section 72.25 (c) (1) and (4)), and MEETS ALL REQUIREMENTS FOR SHIPMENT BY MAIL AND ON PASSENGER AIRCRAFT.

This shipment is EXEMPTED FROM ATA RESTRICTED ARTICLES TARIFF 6-D (see General Requirements 386 (d) (1)) and from DOT HAZARDOUS MATERIALS REGULATIONS (see 49 CFR, Section 173.386 (d) (3)). SHIPPER'S CERTIFICATES, SHIPPING PAPERS, AND OTHER DOCUMENTATION OR LABELING ARE NOT REQUIRED.

Date

Signature of Shipper

Address

FIG. 3. Notice-to-carrier label for use in sending etiologic agents by air mail.

KEEP FROZEN

ORA – Group A (IATA)   ORM – Group A (DOT)

DRY ICE

PUT IN –75° C FREEZER IMMEDIATELY ON ARRIVAL

Amount_____ lbs.   Date Packed _____

WARNING – DO NOT HANDLE DRY ICE WITH BARE
HANDS OR STORE IT IN CLOSED OR UNVENTILATED
SPACES.   HHS, PHS, CDC

FIG. 4. Shipping label for specimen in dry ice.

Form 2976; Fig. 1); this green form must be
signed by the shipper
(v) Shipper's Certification for Restricted Articles;
two completed and signed copies must be
affixed to or must accompany the shipment
(vi) Labels required in specific countries are listed
in "International Postage Rates and Fees"
(18)

RADIOACTIVE MATERIALS

Regulatory Responsibility

Radioactive material is probably the most regulated of the hazardous materials. An entire government agency, the Nuclear Regulatory Commission,
has the responsibility for the production, possession,
transportation, and use of most radionuclides. Some
important regulations are spelled out in CFR Title 10,
notably, Parts 19, 20, 31, and 71 (13). A license
issued by the Commission must be obtained to receive, possess, or use such material. Small specific
quantities, such as the quantities in prepackaged
radioimmunoassay kits, however, may be exempted
from many of the requirements. The transportation of
radioactive material is also the responsibility of the
DOT. Most radionuclides are not listed individually
in the Hazardous Materials Table, but are specified in
general as "Radioactive Material, n.o.s." or
"Radioactive Material, L.S.A. [low specific activ-
ity]."

Packaging Requirements and Instructions

Material classification

Within the United States, most shipments of
radioactive materials consist of small quantities used
in the practice of medicine or in research. These
radionuclides present different shipping hazards, depending on the radiation characteristics and the form
of the material. Radioactive materials in solid, metal
configuration, or which are tightly encapsulated,
are designated "special-form" nuclear material.
Special-form radionuclides present only an external

radiation hazard. All other forms of radionuclides,
that is, gas, liquid, or solid, are called "normal
form." In addition to being an external radiation
hazard, normal-form radionuclides may leak or release contamination. Normal-form radionuclides are
divided into seven transport groups (I through VII),
depending on the radiation toxicity of the material.
The most radiotoxic radionuclides are in group I; the
least radiotoxic are in group VII. The transport
groups for some of the radionuclides commonly
found in the laboratory are shown in Table 3. The
more radiotoxic the material, the more restrictive the
limits. Except for iodine-125 and iodine-131, which
are in group III, most radionuclides used in routine
laboratory procedures are in group IV.

The quantity of material used in research or medical practice is termed the "tracer level" and is
relatively low in concentration compared to that
found in the nuclear industry. These lesser quantities
are called "exempt," "small," or "type A" quantities. The activity levels for these three groups are
shown in Table 4. Small and exempt quantities,
sometimes referred to as L.S.A., can be shipped in a
strong, tightly packed container, but type A quantities must be shipped in a container designed under
DOT specifications.

Containers

Radioactive materials must be packaged in a
strong, tightly packed inner container. Liquids must
be packaged within a leak- and corrosion-resistant
inner container. Should the package be subjected to a
stress equal to a drop of 30 ft (ca. 9 m), this container
will prevent the material from leaking. The inner
container must contain enough absorbent materials to
absorb at least twice the volume of the liquid, posi-
tioned so that in the event of leakage they will readily
absorb the liquid. For liquids shipped by air, each
package must be designed and constructed to prevent
leakage caused by a change in altitude or temperature. Although most cargo areas are pressurized, a

TABLE 3. Transport groups for commonly used
radionuclides[a]

Radionuclide	Transport group
Radium-226	I
Strontium-90	II
Iodine-125, iodine-131	III
Carbon-14	IV
Tritium (^{3}H)	IV
Phosphorus-32	IV
Chromium-51	IV
Xenon-131m	V
Krypton-85	VI
Tritium gas (^{3}H)	VII

[a]Adapted from CFR Title 49, Part 173.390, "Transport
groups of radionuclides."

sudden change in pressure may cause liquids to vaporize. Stoppers, corks, or other friction devices must be held securely in place with wire or tape. Even screw-type closures must be secured or sealed to prevent the caps from loosening.

Labeling and marking

Special radioactive-hazard identification labels, which indicate the handling required, are available (Fig. 5). Each label indicates the maximum level of external radiation exposure at the surface and at 3 ft (ca. 0.9 m) from the container. If the exposure at the surface is less than 0.5 milliroentgens (mR)/h, a "Radioactive White-I" label (Fig. 5A) is appropriate. As the external radiation exposure increases, a "Radioactive Yellow-II" or "Radioactive Yellow-III" label is required (Fig. 5B and C, respectively). The criteria for these labels are listed in Table 5. No label is required if a package contains a "small quantity" or less (as described in Table 4) and if the exposure rate at the surface is less than 0.5 mR/h; however, the inner container must be visibly labeled "Radioactive—No Label Required." Only packages that do not require labels may be sent in the mail. All other packages of radioactive material must be labeled on opposite sides of the package according to the exposure rate.

Lower portions of the labels must be filled in with the appropriate information, as shown in Fig. 5, for the various radionuclides. The space marked "Contents" must contain the name of the radionuclide or the approved symbol, for example, ^{125}I. If a package contains a mixture of radionulcides, the most radiotoxic must be listed on the label. The next space, "Activity," must contain the quantity of the

material in the units of curie (Ci), millicurie (mCi), or microcurie (μCi). The square near the bottom of the label must indicate the value of the transport index; this unit gives the exposure rate 3 ft from the surface of the container and is used to determine how packages are placed in a carrier. An exposure rate of 2.5 mR/h equals a transport index of 2.5. The maximum transport index for a package shipped by common carrier is 10.

The outside of the container must be marked with the proper shipping name, "Radioactive Material, n.o.s." or "Radioactive material, L.S.A., n.o.s." Material shipped in dry ice, an ORM-A (Other Regulated Material-A) material, must have "ORM-A, Dry Ice" marked on the top or side. If the positioning of the box is essential to prevent leakage, a "This Side Up" marking must be clearly visible. The package must be sealed by a "security seal." The seal must be designed so as not to be readily broken from normal handling, and while it is intact, it should be evident that the package has not been opened. If a printed security seal is not available, a secure piece of wrapping tape with the handwritten words "Security Seal" will suffice.

USPS requirements

Some packages of radioactive material may be sent through the postal system. Only packages classified as small or exempt quantities and with an exposure rate at the surface of less than 0.5 mR/h may be mailed. The material must be securely contained in strong, tight packages to prevent the contents from leaking during normal postal handling, according to USPS Publication 6. No significant radioactive contamination may be present on the package. It is

TABLE 4. Quantity limits by transport groups[a]

Transport group	Exempt quantity (mCi)	Small quantity (mCi)	Type A quantity (mCi)
I	0.05	0.01	1.0
II	0.10	0.10	50.0
III	1.0	1.0	3,000.0
IV, V, VI	1.0	1.0	20,000.0
VII	25.0	25.0	1,000,000.0

[a]Adapted from CFR Title 49, Parts 173.389 and 173.391.

TABLE 5. Label requirements for radioactive material packages[a]

Label	Exposure rate limits (mR/h) at:		Transport index
	Surface	3 ft (ca. 0.9 m)	
Radioactive White-I	0.5	0	0
Radioactive Yellow-II	50.0	1.0	1
Radioactive Yellow-III	200.0	10.0	10

[a]Adapted from CFR Title 49, Part 172.403, 436–440.

extremely important that these packages contain no exterior warning label, but the inner container must be marked "Radioactive Material—No Label Required." The Post Office will not handle any package bearing a DOT warning label as shown in Fig. 5. If the package is sent with a label, the shipper will have to retrieve it from the Post Office in person. The best advice is not to send radioactive material in the mail.

Radiocontamination

The external surface of the package must be free of significant removable contamination. For beta- and gamma-emitting radionuclides used in the laboratory, the limit is 220 dpm/cm^2 when surveyed over an area of at least 300 cm^2. Measurements must be taken in the most appropriate locations (for example, the top and front of the container) by wiping the surface with an absorbent material such as a piece of filter paper. The wipe should then be analyzed in a counter appropriate for the radionuclide packaged. If the contamination on the wipe is less than 10% of the prescribed limits, or 22 dpm/cm^2 above background in this example, it may be assumed that the package does not exceed the limits and is free of significant contamination. The results should be recorded and maintained for reference should a question arise about the condition of a package when it was shipped. If the package is contaminated, the inner container should be checked, and if necessary, both containers should be replaced.

Shipping documents

Each package of radioactive material must have properly completed shipping papers. The proper shipping name and the name of each radionuclide (or an abbreviation) must be on the document. A description of the physical and chemical form and activity must be included. The category of the warning label, for example, Radioactive White-I, and the transport index must be on the shipping papers. In addition, the legally binding certification statement, as follows, signed by the shipper, must be included: "This is to certify that the above-named materials are properly classified, described, packaged, marked, and in proper condition for transportation according to the applicable regulation of the Department of Transportation."

If the material is to be sent by air, a Shipper's Certification for Radioactive Material form, as required by ATA, must accompany the package. Only material intended for use in or incident to research, medical diagnosis, or treatment may be shipped on a passenger-carrying aircraft. A signed statement certifying this fact must be on the shipping papers.

Regulations for the international shipment of radioactive material are administered by the International Atomic Energy Agency. Recent changes

in U.S. shipping regulations correspond to those required by this agency. However, it is best to obtain information on the shipping and, especially, the labeling requirements in a foreign country before the package is sent. Having a package sitting in a customs office, incorrectly labeled according to a local regulation and with no means of retrieval, is extemely frustrating. A *written* statement should be obtained from the government of the foreign country requesting the material. This places on that country the responsibility of escorting the material through customs.

Common Packaging Problems

Two infractions in packaging radioactive material occur most frequently and lead to the greatest safety problems. A few minutes of careful thought could eliminate these problems.

(i) **Excessive radiation.** Radiation levels at the exterior of the package are excessive for the warning label attached. This may be caused by too much material in the package, too small a container, or insufficient shielding. Since radiation intensity decreases as the distance from the source increases, a larger container may reduce the exposure rate without the need for expensive and heavy shielding. However, a warning indicating that the material is unshielded and that necessary precautions must be taken during opening procedures should be placed on the inner container.

(ii) **Improper packaging.** Improper packaging in unapproved cartons or weak packages can lead to broken shielding or a leaking package which may endanger anyone near the material. One of the requirements stated in the DOT regulations (49 CFR, 173.22) is, ". . . the shipper shall be responsible to determine that shipments of hazardous materials are made in containers which . . . have been made, assembled with all their parts or fittings in their proper place, and marked in accordance with applicable specifications." In other words, the individual making the shipment is responsible for any damage, such as contamination of an airplane cargo space and the baggage, that occurs if a package of radioactive material breaks open. Therefore, before shipping the material, be sure that it is in a strong container, is properly labeled, and has the correct shipping papers.

Some Packaging Examples

Consider a situation in which an individual wants to send 75 μCi of liquid iodine-125 in a 1-ml vial to a small clinical laboratory. An outline of the proper packaging procedures follows.

(i) Determine whether the laboratory has a

license to receive iodine-125. Call the investigator and get the number of his Nuclear Regulatory Commission or state license.

(ii) Determine the transport group from Table 3 (CFR 49, Part 173.390). Iodine-125 is in transport group III.

(iii) Determine the container required for 75 μCi of group III material from Table 4 (173.389). This is a small quantity requiring a strong container. Since the material is a liquid, a leakproof inner container with absorption material sufficient to contain twice the volume of liquid must be used.

(iv) Measure the external contamination of the package to ensure that it is below the limits of 220 dpm/cm^2 (173.397). Careful preparation should eliminate any contamination problem.

(v) Measure the external exposure rate (172.403) and, if applicable, determine the correct warning label from Table 5. The exposure rate at the surface of the box, measuring 15 cm per side, containing 75 μCi of iodine-125, is approximately 1.8 mR/h, with no measurable exposure at 3 ft. Thus, even though the quantity is small, a Radioactive Yellow-II label is required, since radiation can be measured, and the package cannot be sent through the mail. For this package, the Transport Index is 0.

(vi) Prepare the proper shipping papers with the following information:

Shipping name: Radioactive material
                       L.S.A., n.o.s.
                       Transport group II

Contents: Iodine-125, liquid

Chemical form: Sodium iodine (NaI)

Activity: 75 μCi

Label requirements: Radioactive
                       Yellow-II

Transport index: 0

Shipper Certification: (Listed in text)

(vii) Complete all information on a radioactive Yellow-II warning label, as in Fig. 5B. All of the spaces on the warning label must be filled (172.403).

Contents: ^{125}I

Activity: 75 μCi

Transport index: 0

CHEMICAL MATERIALS

Material Classification

The shipment of chemicals is governed by many of the previously discussed regulations for shipment of radioactive materials. Chemicals are classified by their physical characteristics and reactive properties; flammable liquids, corrosives, and poisons are a few of the more commonly known groups. The Hazardous Materials Table (Table 1) lists many of the hazardous chemicals. This table is much more useful for locating regulations for chemicals than those for radioactive or biological materials. Should a chemical not be listed in this table, refer to a technical manual, such as the "Hazardous Materials Data Handbook" produced by the National Fire Protection Association (6). This will give the physical properties and description of the material. For example, coprylyl chloride is a corrosive liquid not listed in the table, but can be found in the handbook. The shipping name and description for this material would be "Corrosive Material, n.o.s."

The regulations and exceptions for this chemical will be covered under the references listed for the shipping name "Corrosive Liquid, n.o.s." in the Hazardous Materials Table. A second example in Table 1 would be ethyl alcohol. It is not specifically listed in the table, but the information can be found under "Alcohol, n.o.s.," as either a flammable liquid or a combustible liquid, depending on the flash point. Ethyl alcohol (10%) is a combustible liquid with a flash point of 120°F, but 96% ethyl alcohol, with a flash point of 55°F, is a flammable liquid. To package the material properly, the shipper must have correct information.

Packaging Requirements and Instructions

Containers

The standard packaging requirements for chemicals are basically simple. The container must be constructed so that during normal transportation no significant amount of material will be released into the environment. Normal handling should not cause a change that would make the container less effective. Normal handling, for example, would not cause an internal sharp object to pierce the package or a mixture of vapors or gas inside the package to explode and damage the box.

New cardboard or fiberboard boxes and reinforced tape should be used to assemble a strong outer container. If the material can create vapor or pressure when confined, the container must be designed to withstand these conditions or must have a venting system that will capture any release.

Since normal handling during transportation may be rough, inner containers should be cushioned with nonreactive material. This is especially true when the inner container is made of earthenware, glass, or brittle plastic.

Liquids must be packaged in leak- and corrosion-resistant inner container with sufficient space at the top to allow for an increase in volume caused by temperatures above 130°F (ca. 55°C). In the summer months, during truck transportation, such temperatures often occur. The inner container must contain

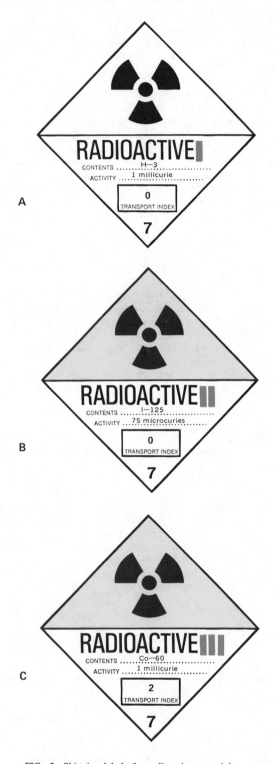

FIG. 5. Shipping labels for radioactive material.

FIG. 6. Shipping labels for hazardous chemical materials.

enough nonreactive absorbent material to collect at least twice the volume of the liquid.

All packages shipped by aircraft must be designed and constructed to prevent leakage caused by a change in altitude or temperature. Although most cargo areas are pressurized, liquids may become vaporized by a sudden change in pressure. Stoppers, cork, or other friction devices must be held in place by wire or tape. Screw-type closures must be secured or sealed to prevent loosening.

As a rule, the volume of chemicals shipped by medical or research institutions is less than 50 ml. The hazards associated with this amount of material are usually minimal, and therefore the chemicals may be exempt from many packaging requirements by the "limited-quantity" exception. The regulation covering the exception for a chemical is listed in column 5a of the Hazardous Materials Table. In most instances, the material must be in a specific inner container, for example, a metal can, earthenware jar, or fiberboard box, packed in a strong outside container. No label is required for a limited-quantity amount, unless it is shipped by aircraft; criteria and packaging requirements for these levels are found in the regulation cited in column 5a of the Hazardous Materials Table. Some chemicals, however, are extremely hazardous, even in small quantities. These, of course, are not exempted.

Labeling and marking

Special warning labels have been designed to describe visually the hazard class of the material (Fig. 6). Packages shipped under exceptions listed in column 5a of the Hazardous Materials Table do not need a label, unless they are shipped by air, in which case the correct labels are required. Samples of material submitted for laboratory analysis for which a reasonable doubt exists as to its class and labeling requirement should be labeled according to the shipper's knowledge of the material.

Remove all unnecessary warning labels from containers before reusing the containers for shipments. If an empty container is being shipped, place an "EMPTY" label on the outside container and remove all other labels.

Containers must be properly marked on the outside. The shipping name and consignee's or consignor's name and address must be clearly visible on the outside container in durable, English lettering, which should be printed or typed. Locate the information away from any advertising or other writing. Containers holding flammable liquids must be marked with an orientation marking, "This Side Up." If a container constructed under certain DOT specifications (that is, a cardboard box) is used, that specification designation (that is, Specification 7A) must be visible on the container. Material classified as ORM must be marked as such, following or below the shipping name on one end or side of the con-

tainer. For example, frozen samples sent in dry ice should be marked "Dry Ice, ORM-A." This marking certifies that the package is properly described, classified, and packaged as ORM-A material.

Shipping documents

Each package of chemicals submitted for transportation must have properly completed shipping papers. If hazardous and nonhazardous materials are shipped in the same container, both must be described on the shipping paper. The description of the hazardous material must be written in a color of ink that contrasts with the description of the nonhazardous material. The shipping name and classification, as listed in the Hazardous Materials Table, should be included in the description. The quantity of the material and any exemption pertaining to the packaging must be listed on the paper. In addition, the notation "DOT-E" and the regulation giving the exemption must follow the shipping name. Materials shipped as limited quantities must have the words "Limited Quantity" or "Ltd Qty" after the description.

As with radioactive materials, the following certification statement must be included: "This is to certify that the above-named materials are properly classified, described, packaged, marked, and in proper condition for transportation according to the applicable regulations of the Department of Transportation."

If the material is to be sent by air, a Shipper's Certification for Restricted Articles must be included.

Packaging Example

Consider a situation in which an individual wants to send 1 pint of a special acetone mixture to a small clinical laboratory. Here is an outline of the packaging procedures.

(i) Determine whether acetone is a hazardous material by reviewing the Hazardous Materials Table (172.101). For this example, acetone is listed in the table, but if the material is not listed in the table, check another reference such as the *Hazardous Materials Data Handbook*, National Fire Protection Association Publication 49 (6).

(ii) Determine the hazard class from column 2 of the Hazardous Materials Table. Acetone is classified as a flammable liquid.

(iii) Determine whether there are any exceptions to packaging acetone. Check column 5a. The reference 173.118 lists the limited-quantity exemption. Under this exemption, a 1-pint volume can be packaged in any suitable inside container and strong outside container. No label is required unless the material is to be shipped by aircraft. (If the amount of acetone had been more than the amount exempted in 173.118, column 5a, the reference listed in column 5b, 173.119, would give the regulation for properly packaging the larger amount.)

(iv) Prepare the proper shipping papers; the papers must contain the following information:

Shipping name: Acetone
Hazard class: Flammable liquid
Contents: 1 pint
Shipper certification: (Listed in text)
Exception: Limited quantity

(v) Mark and label the outside container. Since this is a limited quantity, no label or marking is required, unless the package is sent by aircraft, in which case the "Flammable Liquid" label (Fig. 6A) must be attached.

LITERATURE CITED

1. **Air Transport Association.** 1984. Restricted Articles Tariff 6-D. Airline Tariff Publication Co., Dulles International Airport, Washington, D.C.

2. **Block, S. S.** 1977. Disinfection, sterilization, and preservation. Lea & Febiger, Philadelphia.

3. **Chemical Rubber Company.** 1979. Handbook of chemistry and physics. The Chemical Rubber Co., Cleveland.

4. **International Air Transport Association.** 1980. Restricted Articles Regulations. International Air Transport Association, Montreal.

5. **National Committee for Clinical Laboratory Standards.** 1979. Standard procedures for handling and transporting of domestic diagnostic specimens and etiologic agents. NCCLS Approved Standard ASH-5. National Committee for Clinical Laboratory Standards, Villanova, Pa.

6. **National Fire Protection Association.** 1978. Hazardous materials data handbook, NFPA no. 49. National Fire Protection Association, Boston.

7. **Pike, R. M.** 1976. Laboratory-associated infections: summary and analysis of 3,921 cases. Health Lab. Sci. **13:**105–114.

9. **Pike, R. M.** 1979. Laboratory-associated infections: incidence, fatalities, causes, and prevention. Annu. Rev. Microbiol. **33:**41–66.

10. **U.S. Department of Agriculture.** 1980. Code of Federal Regulations Title 9, Parts 71–123:118–456.

11. **U.S. Department of Health and Human Services.** 1984. Biosafety in microbiological and biomedical laboratories. HHS Publication no. (CDC) 84–8395. U.S. Government Printing Office, Washington, D.C.

12. **U.S. Department of Transportation.** 1979. Code of Federal Regulations Title 49, Parts 100–199. U.S. Government Printing Office, Washington, D.C.

13. **U.S. Nuclear Regulatory Commission.** 1980. Code of Federal Regulations Title 10, Parts 20, 31, 33, and 71. U.S. Government Printing Office, Washington, D.C.

14. **U.S. Postal Service.** 1980. International mail—hazardous material. Postal bulletin **21246:**11–12.

15. **U.S. Postal Service.** 1984. Postal mail manual. U.S. Government Printing Office, Washington, D.C.

16. **U.S. Postal Service.** 1984. Postal operations manual. U.S. Government Printing Office, Washington, D.C.

17. **U.S. Postal Service.** 1984. Postal service manual. U.S. Government Printing Office, Washington, D.C.

18. **U.S. Postal Service.** 1984. International postage rates and fees. Post Office Publication no. 51. U.S. Government Printing Office, Washington D.C.

19. **U.S. Public Health Service.** 1970. Radiological health handbook. U.S. Government Printing Office, Washington, D.C.

20. **U.S. Public Health Service.** 1974. Laboratory safety at the Center for Disease Control. DHEW Publication no. CDC 79–8118. Center for Disease Control, Atlanta, Ga.

21. **U.S. Public Health Service.** 1974. Collection, handling and shipment of microbiological specimens. DHEW Publication no. (CDC) 80–8263. Center for Disease Control, Atlanta, Ga.

22. **U.S. Public Health Service.** 1978. NIH laboratory safety monograph. Office of Research Safety, National Cancer Institute, Bethesda, Md.

23. **U.S. Public Health Service.** 1980. Code of Federal Regulations Title 42, Part 72, Interstate shipment of etiologic agents. U.S. Government Printing Office, Washington, D.C.

24. **U.S. Public Health Service.** 1980. Reference diagnostic specimens. HHS Publication no. (CDC) 81–8350. Centers for Disease Control, Atlanta, Ga.

25. **U.S. Public Health Service.** 1980. Radiation safety guide. Office of Safety, National Institutes of Health, Bethesda, Md.

Chapter 26

Techniques for Toxic and Carcinogenic Chemicals

JAMES N. KEITH

In the past, it has been possible to consider the hazards of occupational exposure to chemical carcinogens as confined to certain industrial settings and laboratories involved in the synthesis or bioassay of such chemicals. It has become apparent in recent years, however, that although only a limited number of chemicals are carcinogens, they are found in nearly all chemical classes, and they seem to be ubiquitous. Furthermore, with the growing concern over the presence of carcinogens in the environment and in numerous consumer products, a great variety of laboratory activities currently involve testing of chemicals for carcinogenic or mutagenic activity or searching for carcinogens in other products. Many types of biological and chemical laboratories are involved, from large-scale synthesis or bioassay facilities to those devoted to analyzing trace levels of carcinogens in food, air, or drinking water.

In contrast with the industrial situation, where the expected exposures involve only a limited number of chemicals, the research laboratory worker may be exposed to numerous hazardous chemicals and must often handle compounds for which little or no toxicological information exists. Additional safety margins must be included in planning work with chemicals that have delayed health effects, since considerable uncertainty exists concerning the existence of thresholds, the extrapolation of animal data, and synergistic effects. The fact that an exposure and its results may be separated in time by months or even years makes it possible for faulty control procedures to go undetected for long periods, resulting in the exposure of large numbers of workers.

The complexity of the research activities involved makes it impossible to propose a single set of safety rules for working with carcinogens under all laboratory situations. Safety procedures for specific types of laboratories are available and should be consulted by those planning new facilities (4, 7, 8, 12, 19). In this chapter the emphasis will be on the relationship of the control procedures to the properties of carcinogens, especially those properties which distinguish them from other hazardous materials such as pathogens. No attempt is made to describe a specific type of laboratory in detail, but hazards and procedures appropriate to most laboratories are included.

It should also be evident that mutagens and teratogens should be treated in the same manner as carcinogens; in fact, it is rather arbitrary to regard them as distinct classes, since many carcinogens are also mutagenic, teratogenic, or both. Highly toxic compounds, such as 2,3,7,8-tetrachlorodibenzo-p-dioxin, or potent cholinesterase inhibitors are generally handled using the same procedures that are used for carcinogens. In production facilities a considerably higher degree of containment and more elaborate precautions are necessary, because of the much greater risk of a fatal exposure (12).

The basic philosophy espoused is essentially the following: (i) that carcinogens can be controlled by procedures already available; (ii) that the emphasis should be on engineering and work practice controls; and (iii) that carcinogens must be viewed individually and the biological, chemical, and physical properties of each compound must be considered. It should be emphasized here that the problems involved in developing adequate control procedures for such work can be complex, and it is necessary to make the best use of the scientific resources of the organization. Safety committees should include staff members experienced in toxicology, chemistry, laboratory safety, engineering, and medicine and should be chosen for their commitment to safety.

HAZARDS OF CHEMICAL CARCINOGENS

The principal routes of exposure in carcinogen handling are dermal contact and inhalation of contaminated aerosols. The physical properties of aerosolized carcinogens are not significantly different from those of aerosols encountered in microbiological work. However, biohazard control methods must be applied to chemicals with caution, since chemicals often possess properties which dictate additional controls (10). A few of these are summarized here.

(i) Carcinogens are found in nearly all chemical classes, and the more conventional hazardous properties may be displayed; thus, some carcinogens are highly flammable, explosive, or corrosive. They may react readily with water or other materials to produce hazardous products.

(ii) Many carcinogens are volatile at ambient temperatures. This means that loss of sample

228

can occur during an experiment. Filter-type respirators are ineffective, and cross-contamination can occur when, for example, culture dishes are stored together in an incubator.

(iii) Chemicals are usually not destroyed by thermal treatment except at high temperatures. Their stability and reactivity range from highly reactive to intractable, so that decontamination and disposal require detailed knowledge of the chemistry of the specific compound.

(iv) Both carcinogens and the solvents in which they are handled may readily penetrate the intact human skin. Frequent use of organic solvent can result in increased sensitivity of the skin to chemicals and susceptibility to bacterial infection. Dimethyl sulfoxide is particularly hazardous since it penetrates the skin very readily and can carry solutes directly into the bloodstream.

(v) Chemicals, especially when dissolved in certain organic solvents, are capable of permeating laboratory glove materials (11). They can sometimes be absorbed by gloves and slowly released with continued use.

Fortunately, carcinogens are not viable. However, the existence of thresholds for carcinogens is still in doubt. Synergistic effects also occur, and it is probable that all of us are exposed to small amounts of promotors and cocarcinogens throughout our lives. Very little is known of the range of human susceptibility to carcinogens.

Safety protocols should take into account all of the above factors and should include specific information on the relevant properties of the compounds to be used, usually in the form of safety data sheets (see Chapter 14).

SAFETY PLANNING

NIH Guidelines

The Division of Safety, National Institutes of Health (NIH), has prepared "Guidelines for the Laboratory Use of Chemical Carcinogens" for use by its operating agencies. The guidelines have been reviewed by the agencies within NIH, and public comment has been taken into account. Copies are sent to new NIH contractors and grantees and are available to the public from the Division of Safety or the U.S. Government Printing Office. The recommendations in this chapter have, in part, been based on the NIH Guidelines.

Eighteen carcinogens are specifically regulated by the Occupational Safety and Health Administration (OSHA) (16). Although laboratories are required to comply with them, these standards are designed for the industrial situation and do not contain provisions for the controls required in laboratory use of carcinogens. (Laboratory provisions formerly included in some of the standards were vacated by the Supreme Court.) Since the NIH Guidelines are consistent with the OSHA standards and do deal specifically with the laboratory situation, the best strategy is to use the Guidelines and refer to OSHA standards for specific provisions for regulated compounds, such as permissible limits, monitoring requirements, and respirator use.

Safety protocols

Although the organization will generally have prepared an overall safety plan, establishing policy and procedures for safety management and incorporating established procedures for safe operation of restricted facilities, it is a good practice to require a safety protocol for individual research programs (14). This is best done by the principal investigator at the beginning of the program, with provision for annual review.

The safety protocol for a research program should indicate the hazardous chemicals to be used, the type of experimental work to be performed, and the nature of the exposure risks. Safety cabinets and personal protective clothing and equipment should be identified, as well as the operations which will require their use. Plans for decontamination and other emergency procedures and for disposal of contaminated wastes should be described. If the work will require the involvement of service groups, such as for chemical analysis of diet or tissue samples, the protocol should describe procedures to be used by this group also, as well as for transfer of samples to other laboratories, etc. A copy of the safety protocol should be made available to everyone involved in the program.

Safety protocols should be reviewed by the safety committee, who will consider the appropriateness of the specific control equipment and procedures to be used in the program and ask for clarification or revision where needed. They should make recommendations for medical surveillance of laboratory personnel where the nature of the exposure risk indicates its advisability. The safety officer should inspect the facilities and equipment to be used and should require evidence of the effectiveness of alternative control procedures proposed by the principal investigator.

Medical surveillance

The OSHA standards require the establishment of a medical surveillance program for workers exposed to regulated carcinogens, and the NIH Guidelines recommend surveillance of persons assigned duties in work areas where carcinogens are used on a regular basis. A base-line preassignment physical should be followed by periodic examinations at a frequency determined by a qualified physician. Specific attention should be given to target organs indicated by available toxicity data.

The subject of medical surveillance of laboratory workers is rather controversial at the present time. There is considerable divergence of opinion as to the value and goals of such programs, as well as the content of a satisfactory medical examination for such personnel. Although it may be many years before sufficient epidemiologic data are available to fully evaluate the effect of medical surveillance in such cases, interesting results have been reported from an 11-year study of a group of 10,674 industrial workers (2). Death rates from "potentially postponable causes" were lower among those enrolled in a medical surveillance program, and a negative correlation was seen with the number of medical examinations undergone.

It should be emphasized that the goal of a medical surveillance program is not the prevention of occupational cancer. One of the most important functions may be in employee screening. The examination should include an evaluation of personal risk factors, occupation, medical, genetic, and environmental, including a counseling session in which these factors are explained by the physician and discussed in relationship to the employee's work assignment. Specific organic deficiencies may be detected, such as decreased liver function, which would indicate certain classes of chemicals to be avoided. Heavy smokers should be warned of possible synergistic effects resulting in increased risk. Methods are being developed for use in monitoring exposure to carcinogens and other chemicals (19). The interpretation of results of, for example, chromosome studies, however, is still open to question (1).

The medical officer should be kept informed of any work in progress or planned which might result in exposure to carcinogens or other toxic chemicals. He or she should be supplied with information resulting from any personnel or area monitoring that has been done and should require that all exposures be made a part of the employee's health record. A member of the medical staff should participate in the meetings of the safety committee.

THE LABORATORY

In preparing to initiate a research program involving regular manipulation of chemical carcinogens, it is important that the laboratory and equipment be carefully selected and the appropriate facilities be provided at the beginning of the program, rather than risk interruption of the research at a later time when the need for modification becomes apparent. Planning is particularly important when an existing laboratory is being renovated for a new program, since there is a tendency to "make do" with existing equipment and facilities which later prove to be inadequate. Several important principles are discussed in this section which should be incorporated in new facilities. Limitations may be imposed by the existing facility when a laboratory is renovated, but

compromises are sometimes possible, and operating procedures may be developed to compensate for inadequacies in design.

Isolation

If only one or a few laboratories are to be used for carcinogen work, rooms should be selected which are conveniently isolated from the rest of the facility. It is usually preferable that they be contiguous, so that service facilities can be easily shared. They may be on a single floor of the building, at the end of a corridor, etc. Corridors connecting nonrestricted areas should not pass through the facility. When several rooms are used, they may be isolated by a limited-access corridor with free access among the rooms. A locker and change room may separate the carcinogen laboratory area from a common corridor. Doors to such facilities should be kept closed at all times, and a sign posted at the entrance should identify the area as a restricted-access facility. When a multi-use laboratory works with carcinogens only intermittently, the door should be closed and a sign should be posted when carcinogens are being used. (See Chapter 17 for details on laboratory design.)

Access control

The number of persons who have access to laboratory areas where carcinogens are used should be restricted to the technical staff assigned to the research progam and the necessary support personnel. The principal investigator should provide a list of authorized personnel to the safety officer, who should recommend medical surveillance where it is indicated.

Casual visitors should be strictly forbidden. Occasional visitors for the purpose of maintenance inspection, etc., should require approval by the principal investigator and should be escorted. All visitors should have explained to them, and be required to comply with, the safety rules which apply to the restricted area.

The door to the restricted area should be kept closed at all times. For a multi-use laboratory, the door should be closed and a sign should be posted when carcinogens are used. The doors to the laboratory should generally be locked outside working hours, and emergency instructions should be given to security personnel. The laboratory should not be locked, however, when it is in use, since this would impede rescue in an emergency. A policy should be established by the safety officer on working after hours. This is a common problem at academic institutions, where graduate students often find evening or weekend work essential. All laboratory activities involving carcinogens or other hazardous materials should require the permission of the principal investigator, who should arrange for monitoring by himself or another staff member who will be available for emergencies. Solitary work should be forbidden.

Exhaust ventilation

Mechanical exhaust ventilation should be provided to ensure continuous movement of air from clean areas to potentially contaminated areas (as from the locker room into the laboratory). Air from the carcinogen laboratories should not be recirculated to other building areas, and it should be discharged well away from the building air intake. General ventilation exhaust air need not be treated. For maximum containment, the room should be maintained by an automatic control system at a lower pressure than surrounding areas, and an alarm should sound when this differential is lost. In such facilities, the exhaust system should be independent of exhaust systems for other areas.

Exhaust air from glove boxes and chemical fume hoods should be treated to remove contaminated aerosols, generally by high-efficiency particulate air (HEPA) filtration. The need for treatment of exhaust from biological safety cabinets should be determined by the safety officer and the laboratory supervisor. The pressure drop across the filters should be monitored regularly to determine the need for changing. The installation should be designed to permit changing of the filter elements without risk of exposure to maintenance personnel. Activated charcoal may be used to remove volatile carcinogens from the exhaust air, but reliable monitoring procedures have not been developed for this type of use, and the performance of the system in a laboratory handling organic solvents is rather uncertain. Charcoal has a low capacity for highly polar compounds such as nitrosamines, for example. It is a poor practice to manifold the exhaust ducts from two or more fume hoods, since exhaust system failure can, under some circumstances, result in transfer of contaminated air to another laboratory. If hoods are manifolded in an existing facility, special precautions should be taken, such as automatic dampers and exhaust failure alarms, to avoid exposure.

Surfaces

There should be a minimum of dust-collecting surfaces. Soffits and pilasters can be used to enclose ducts and plumbing services, reagent shelves can be replaced by enclosed cabinets, and all unnecessary furniture or equipment should be removed.

All surfaces should be easily cleanable. Floor coverings should be nonporous and monolithic, so that there are no crevices in which contamination can accumulate. Several materials are available. Polyurethane and epoxy floors are installed as liquid coatings and can be textured to give a nonskid surface. Polyvinyl chloride floors are installed in large sheets with welded seams. Vinyl is less suitable

for synthetic chemical laboratories, because it is less resistant to softening and swelling by solvents, but all three types of floor are suitable for many laboratory uses. The junction of floor and walls should be coved for easy cleaning, and the floor covering should be extended up the wall several inches.

The walls of the laboratory should be sealed with an epoxy paint and caulked carefully around door jambs, window frames, electrical and plumbing services, etc. It is preferable that there be no exterior windows, but if any are used, they should be sashless and double-glazed.

Facilities

A hand-washing facility should be available in the laboratory; where possible, foot- or elbow-operated faucets should be used. Liquid soap is recommended.

A shower in addition to the deluge shower should be available for emergency decontamination and for routine shower on leaving the facility when the procedures require it. This shower need not be adjacent to the laboratory, but should be readily accessible for daily use.

Each laboratory should have an eyewash station which is capable of washing both eyes continuously with potable water.

Vacuum services should be protected with an absorbent trap and a HEPA filter to prevent accidental contamination of the system. A separate vacuum pump, vented into a laboratory fume hood should be used for volatile carcinogens.

Containment equipment

Work with carcinogens should be performed in a suitable safety cabinet or other containment equipment, depending on the nature of the experiment.

Operations involving volatile carcinogens and aerosol-generating operations should be conducted in a suitable containment device. A glove box, Class I biological safety cabinet, or chemical fume hood should be used for handling pure carcinogens, including preparation of stock solutions for in vitro procedures, or for work with concentrated solutions containing carcinogens. Work with organic solvents and with toxic or corrosive chemicals should be done in a fume hood, including all neutralization procedures.

Laminar-flow biological safety cabinets, Class II B, should be used for in vitro procedures involving low concentrations of carcinogens. The essential feature of this cabinet design is that there is no contaminated positive air plenum. A Class II A cabinet may be used if provision is made for total exhaust.

Weighing of carcinogens should be done in a hood, glove box, or Class I biological safety cabinet. An alternative procedure is to dispense the carcinogen in the safety cabinet into a preweighed vial, which is sealed and placed in a tared weighing bottle for reweighing on an analytical balance. The weighing bottle must be treated as contaminated equipment when the operation is completed. Under these conditions, the weighing procedure should be planned in detail to avoid contamination of the analytical balance or the outside of the weighing bottle.

Poor placement and usage of safety cabinets can seriously affect their efficiency. Cross drafts, as from doorways and air intakes, can disturb the airflow pattern sufficiently that aerosols can actually be pulled out into the room, as can too rapid movement in front of the cabinet. For the same reason, safety cabinets should never be used for storage of chemicals or equipment. The proper functioning of safety cabinets should be tested annually by the maintenance staff or an outside contractor.

It is generally ineffective to place bulky analytical equipment, such as a gas chromatograph, in a fume hood, since the airflow pattern is seriously disturbed and effective capture of aerosols from the injection port is not possible. For such situations, local exhaust ventilation is more effective and economical. The design principles of ventilation systems are outlined by McDermott (6).

STORAGE AND STORABILITY

To exercise the proper inventory control of carcinogens required by the OSHA standards and to facilitate establishment of safe handling procedures for these compounds in the laboratory, they should be stored separately from other laboratory chemicals. When a large supply of carcinogens is kept, as for a bioassay program, a separate, properly ventilated storage room is needed. Some laboratories use a walk-in freezer; some have a separate reagent shelf separated by a partition from the laboratory; many use a separate storage cabinet. If only a few samples are kept, a single designated shelf in the reagent cabinet may be adequate. In any case, the storage area should be clearly posted as a carcinogen storage area, and the name of the person responsible for the inventory control should be listed.

Labeling

In most laboratories acronyms and abbreviations are used to simplify logbook entries and conversation. It is important to avoid any ambiguity whenever samples are removed from the facility and handled by nontechnical personnel, as in shipping of research samples and transport of contaminated waste. The labels on stock bottles should always bear the full chemical name or a widely recognized substitute, such as uracil mustard, methoxychlor, etc. All stock bottles should bear a warning label, "Potential Cancer Hazard" (NIH Guidelines), "Cancer Suspect Agent" (OSHA Standards), or "Chemical Carcinogen," which is clearly visible in letters larger than the other lettering on the label. If special storage conditions are required, these directions should also be included on the label.

Housekeeping

The preservation of the purity of chemical stocks requires care in packaging and storage. Labels can be preserved by the use of clear plastic tape or a plastic spray coating and should be replaced when they begin to show signs of deterioration. Spilled chemicals should be cleaned up immediately, and the other storage containers should then be inspected for contamination. All samples should be readily accessible; if a large number of small vials are used, a compartmented box may be used to keep them in order and to prevent accidental spillage. A simple method is to use paint cans, which also provide protection from light and a secondary barrier against contamination.

Compatibility

Consideration should be given to compatibility and stability in storage. Highly reactive chemicals should be segregated, so that oxidizers, such as N-methyl-N-nitroso-N'-nitroguanidine (MNNG), are not stored adjacent to antioxidants such as benzidine and β-naphthylamine, sodium azide is not stored with acids or heavy metal salts, etc. This is especially important in the case of bioassay studies, where rather large quantities may be stored for periods of several years.

In a laboratory where flammable solvents and corrosive liquids are used, the carcinogen storage should be well separated from these materials. Besides the risk of dissemination, a fire could result in the destruction of costly research materials. The resulting dislocation in the research program could be even more costly.

Chemicals stored for long periods can be degraded by thermal instability, atomspheric oxidation, photolysis, or hydrolysis. All carcinogen stocks should be stored in sealed containers, using chemically resistant cap liners and conical polyethylene inserts to reduce loss of volatile liquids. In cases of oxygen sensitivity, it may be necessary to seal the sample in an ampoule under nitrogen. Photolysis can be reduced by amber bottles, and most manufacturers supply chemicals in such containers.

Refrigerated storage

It is necessary to provide refrigerated storage space for unstable carcinogens and biological materials, but the use of refrigeration should be confined to those materials which actually require it. In the confined space of a refrigerator, appreciable vapor concentrations of volatile carcinogens and solvents can accumulate, producing several problems: (i) an explosion hazard from flammable solvents (the flash point of diethyl ether, for example, is $-39°C$); (ii) an inhalation exposure hazard, especially in walk-in freezers; (iii) contamination of the freezer ice, which will release volatile carcinogens on defrosting; (iv) cross-contamination.

Special precautions can and should be taken to reduce these hazards, such as the use of safety refrigerators, ventilation of walk-in freezers, separate storage of chemicals and biological materials, careful housekeeping, and redundant packaging. The best practice is to store stable volatile carcinogens in a ventilated storage cabinet. Flammable solvents should be stored only in a properly designed flammable safety cabinet.

Care must be taken to allow refrigerated samples to thaw completely, since moisture will condense on the sample and can accumulate in frequently used samples. Even if the sample is not hydrolyzed, the purity and therefore any quantitative results will be affected. It may be convenient to divide the stock into aliquots which may be opened weekly or monthly, so the entire stock need not be exposed each time.

Working quantities

The amount of carcinogen kept in the laboratory (i.e., outside the designated storage for stocks) should be kept to a minimum. The amount which will be considered a "working quantity" will vary considerably from one procedure to another, but the NIH Guidelines suggest a week's supply. This does not refer to analytical standards or similar diluted solutions, although these can also be dispensed in weekly batches.

TRANSPORT AND SHIPPING

Whenever it is necessary to transport stock containers of carcinogens from one laboratory area to another (e.g., removal from storage), an unbreakable secondary container should be used. For transport to laboratories outside the controlled area, the secondary container should be sealed and a label should be affixed identifying the contents and containing the warning, "Caution—Chemical Carcinogen." Analytical samples may be prepared in the facility and packaged in septum vials for chromatography, or the nuclear magnetic resonance tube, infrared cell, or UV cuvette may be filled before it is transported to the analytical laboratory. Laboratory carts with raised sides should be used for transporting samples between laboratories.

Samples of carcinogens to be sent to other facilities should be carefully packaged to prevent accidental damage or leakage during shipping. No shipping regulations exist for carcinogens as a class, so that many of them are not regulated by the Department of Transportation. Some carcinogens, of course, are regulated by name or by class because of other hazardous properties such as flammability, acute toxicity, etc. In such cases the packaging, labeling, and shipping requirements are specified.

Whether regulated or not, however, all carcinogens should be packaged so as to prevent exposure to handling personnel in the case of a transportation accident or to the recipient in the case of a leaking vial. The following is a convenient method for small samples. A sealed vial or ampoule serves as the primary container, which is sealed in a plastic bag to contain any leakage; this is surrounded by absorbent packing and sealed in a paint can, which is labeled to identify the contents. The can is also labeled "Caution—Chemical Carcinogen." Additional shock-proof packaging may be required between the paint can and the shipping container when sending large samples.

Although a Shipper's Certification and Department of Transportation labeling will be required for all restricted materials, it is probably not helpful to identify the outer package with the "Chemical Carcinogen" warning. This is not required by any shipping regulations and may slow down the shipment.

PROTECTION OF WORK SURFACES

Even the most careful laboratory techniques produce aerosols which, besides presenting a risk of inhalation exposure, result in the accumulation of low levels of contamination on working surfaces. Much greater contamination results from accidental spills, and although they are immediately apparent and are normally cleaned up immediately, complete decontamination is not always possible with ordinary laboratory surfaces. Resuspension of particulate matter and physical contact with contaminated surfaces can result in exposure to a carcinogen after the experimental work has been completed. For these reasons, it is advisable to protect work surfaces from contamination, especially during operations most likely to generate aerosols or result in spillage.

Even with "impermeable" work surfaces, routine housekeeping and emergency spill control can be simplified by the use of protective coverings. The most common type is the absorbent, plastic-backed bench paper which is simply rolled up and disposed of when contaminated. In some cases it is convenient to use a strippable paint, which can be used to protect the entire interior of a hood. On completion of the operation, the paint is carefully peeled off, leaving the hood surfaces uncontaminated.

In conducting any work involving a risk of spillage of a liquid carcinogen, a metal, glass, or enameled tray or pan may be used. It should be large enough to contain the liquid spilled, but should not interfere with the transfer operation so as to increase the probability of a spill. The tray may be lined with bench paper to absorb small spills. Many laboratories use a disposable diaper in such situations.

PERSONAL PROTECTION

The choice of personal protective clothing and equipment is not independent of laboratory design, and there are various approaches to this problem. Obviously many factors enter into this decision, including the properties of the carcinogens being handled, the concentrations and amounts involved in the experiment, the safety cabinets used, and the presence of other hazards, such as in the chemical synthetic laboratory, the animal treatment room, etc.

The choice of protective clothing is the responsibility of the principal investigator, but the safety officer should advise on the type of clothing available and its effectiveness.

Laboratory clothing

When one is working exclusively in a glove box or a similar isolated system, a fully fastened lab coat or smock may be adequate. Many laboratories use cloth lab coats over street clothes. This, of course, is the minimum protection, and this laboratory clothing should not be permitted in nonlaboratory areas, such as offices and lunch rooms. This practice can be conveniently controlled by using distinctive clothing, either color-coded, monogammed, or of a different design from that used elsewhere. Cloth lab coats should be laundered separately from those of neighboring laboratories and should not be sent to an outside uniform service. A simpler solution is to permit the use only of disposable lab coats in these laboratories.

For research with high risk of exposure to aerosols, as in diet preparation, cage cleaning, etc., various gradations in increased protection are available, such as coveralls, lab coats worn over uniform shirts and trousers which are changed before leaving the facility, jump suits, and various special cleanroom clothing. The use of head covering and shoe covers will depend on the type of research; such covering should be mandatory in, for example, animal rooms and diet preparation areas, but is usually unnecessary in the analytical laboratory.

A variety of disposable garments are available that are made of Tyvek, Du Pont's spun-bonded polyolefin fabric, which has excellent chemical resistance but is porous. It is also available with a polyethylene backing for those uses where a less permeable garment is needed; however, the more impermeable materials are uncomfortable to wear for extended periods. Another option is the use of plastic disposable aprons for handling acids, certain solvents or solutions, etc.

Clothing samples, which will be willingly furnished by most suppliers, should be examined for proper fit, reliable closure of buttoned garments, coverage of the forearms and neck, and the permeability of the fabric. If organic or corrosive solvents are used, samples of the fabric should be tested for chemical resistance and permeability.

Gloves

The item most often misused is the laboratory glove. Common practice in chemical laboratories is to use latex or neoprene gloves for handling corrosive materials, strong oxidizers, and toxic chemicals. Too often, the gloves are reused for many months if no obvious damage is detected.

Recent work on permeability of glove materials has demonstrated rather clearly that there is no such thing as a universal laboratory glove, and that even good ones should be viewed with suspicion after they have been used for a while (11). Although the gloves may be relatively resistant to penetration by the pure carcinogen or a methanol solution, permeability is also a function of the solvent; chlorinated solvents, such as methylene chloride, often used with nitrosamines, are notorious for their ability to penetrate most gloves.

The best practice is to buy good-quality, tear-resistant disposable gloves and discard them frequently: between handling operations, on leaving the work area, immediately after a spill or noticeable contamination. For handling liquids and solutions, many laboratories use two pairs of gloves. A pair of latex gloves is worn at all times in the laboratory. When handling carcinogens in the hood, a second pair of gloves is added and discarded on completion of the task. The outer pair may be of a different material. This practice can thus increase the range of protection if a variety of organic compounds are handled. Care should be taken to select a glove which will completely cover the wrist; latex gloves are available which extend past the wrist. It may be necessary to tape the wrist of the glove to the sleeve of the lab coat to prevent creeping.

Glove box gloves are a special problem, since they are obviously not replaced frequently despite the fact that they are exposed for long periods to carcinogens and solvent vapors in a confined space. Glove box gloves should therefore be regarded as permanently contaminated. For this reason, it is good practice to wear disposable gloves even when working in the glove box, or at minimum to require washing after every use of the glove box. The physical integrity of the gloves should be tested regularly. A simple pressure test requires only a few minutes per glove.

Respiratory protection

Except in operations where the risk of exposure to aerosols is high, such as diet preparation, cage cleaning, etc., routine use of respirators should be uncommon in research laboratories. In any laboratory where more than minimal amounts of carcinogens or other toxic chemicals are handled, appropriate respirators should be available for emergency use, and laboratory personnel should be trained in their selection and use. Only National Institute of Occupational

Safety and Health (NIOSH)-MSHA-approved respirators should be permitted.

The OSHA Cancer Policy emphasizes reliance on engineering controls for primary protection and the use of respirators only as a supplement in special circumstances (17). The types of respirators which may be used at specified concentrations are given in each standard (see, for example, the acrylonitrile and dibromochloropropane standards [16]). Specific recommendations for respirator selection are also included in NIOSH Criteria Documents.

In operations requiring respiratory protection, consideration must be given to the presence of volatile carcinogens and the vapors of solvents, against which filter-type respirators do not provide protection. Cartridge respirators are available for many chemicals, principally for organic vapors, acids, and amines. Training should include familiarization with the respirators available and with their proper use and limitatons. Respirator cartridges have a limited effective use period, which is normally determined by detection of the odor of the hazardous vapor. Obviously, in the case of volatile carcinogens this warning is inadequate. The best practice, though less comfortable, is the use of air-supplied respirators; if cartridge respirators are used, the cartridges should be replaced at least at the end of each day.

Respirator fit is much more important than most people who use the equipment infrequently seem to recognize. Respirators are usually designed for a common type of facial structure, and people with narrow faces, small features, etc., may have trouble getting a proper fit. Everyone using a respirator should be given a respirator fit test before use and should be instructed in the proper use of the equipment. Respirators should be individually assigned, and each person should be responsible for the care of his or her own equipment, cleaning it after every use. Respirators should be stored in a designated area (the locker or change room if one is available) where they will be available in an emergency and will be unlikely to be contaminated in case of an accident in the laboratory. In some operations, laboratory workers may be required to carry or wear respirators during the entire work shift.

In any facility where the use of respirators is necessary, the establishment of a respiratory program is required by OSHA standards (15).

Eye protection

Eye protective equipment should be used in all laboratory work. The type of protection will depend on the nature of the work performed, ranging from safety glasses to chemical splash goggles or full-faced shields, which may be required for some synthetic work. The appropriate eye protection should be worn at all times in the laboratory, and approved visitors should be required to comply. Contact lenses should not be permitted in the laboratory.

MISCELLANEOUS PRACTICES

Eating, drinking, gum chewing, the use of tobacco products, and the application of cosmetics should not be allowed in laboratories using carcinogens. These articles and utensils should be stored in office or locker room areas and used only outside the restricted area or in rest areas designated for this use.

Mouth pipetting of solutions containing carcinogens should not be permitted. Numerous mechanical pipetting aids are currently available for every conceivable purpose. The appropriate mechanical device should be selected for each procedure, and all laboratory staff should be required to become proficient in its use.

Laboratory personnel should wash their hands, in a sink provided within the laboratory, upon completion of handling operations and on each exit from the laboratory. A personal shower should be taken after overt exposure. In some procedures, such as powdered-diet preparation, contaminated-cage cleaning, etc., where the risk of exposure to aerosols is high, a shower may be required upon each exit from the laboratory. The principal investigator should incorporate this requirement in the safety protocol when it is appropriate.

Among the less noted actions which can result in contamination of the body are various nervous habits, such as pencil chewing, touching the face, adjusting eyeglasses, scratching the head, etc. Laboratory personnel should be made aware of the contamination potential of such actions and should be encouraged to avoid them in the laboratory.

HOUSEKEEPING, DECONTAMINATION, AND DISPOSAL

Routine laboratory operations produce small quantities of aerosols, even under the best conditions, which may result in some contamination of work surfaces and other areas (9). To minimize the possibility of resuspension, contact with contaminated surfaces, or dissemination to other areas, careful attention to housekeeping is needed in work with carcinogens.

Routine housekeeping

Personnel involved in housekeeping in carcinogen laboratories should understand the need to minimize the generation of aerosols. Dry sweeping should not be allowed; floors may be cleaned with a HEPA-filtered vacuum cleaner or by wet mopping. Protection of work surfaces with bench paper will help to keep these areas clean, but the covering should be changed regularly and disposed of immediately after a spill. Cleaning utensils should be kept in the controlled area and not used in other laboratories. Although custodial staff may be responsible for housekeeping in many facilities, it is advisable to have these tasks performed by the technical staff in any facility handling carcinogens regularly. Glassware should be decontaminated before being delivered for washing. Specific housekeeping and spill control procedures should be included in the safety plan. (See Chapter 23 for housekeeping details.)

Spill control

Fortunately, the levels of contamination from routine laboratory procedures can be minimized by the use of safety cabinets and by the precautions described above. Accidents, however, such as fires, explosions, or broken stock bottles, may produce dangerous aerosols of concentrated carcinogen and can heavily contaminate equipment and work areas.

In any case of accidental spillage, aerosols produced on impact may be carried by air currents for considerable distances. Unless the amount spilled is quite small or it is well contained by a safety cabinet, the room should be evacuated immediately. Personnel reentering the laboratory for decontamination should wear adequate protective clothing including a clean disposable coverall, jump suit, or other garment, shoe covers, gloves, hair covering, and respirator. The decontamination procedure should be reviewed with those assigned to the clean-up task, including the reagents to be used, the specific procedures for recovering the bulk of the sample, rinsing, neutralization of the washings, chemical neutralization of the residue on the work surface, etc. Shifts should be arranged if the cleanup may take more than 60 min, and fresh disposable protective clothing should be made available for each shift. The persons assigned to decontamination should be monitored visually by someone stationed outside the room, prepared for emergency rescue.

The first step is to remove the bulk of the spilled material. No attempt should be made to neutralize a spilled carcinogen until most of it has been removed by mechanical means. Any "floor chemistry" attempted with a large spill is likely to generate a substantial aerosol or hazardous vapor.

Spill-control systems using granular absorbents are quite effective for conventional corrosive reagent spills, but with carcinogens, the absorbent itself represents a potential source of contaminated aerosol during subsequent cleanup. Nonparticulate absorbents, such as paper toweling, rags, or sponges should be used. Spill control bags are available containing amorphous silicate, which is effective for all liquid spills except hydrofluoric acid. The bag is simply placed in the liquid spill before leaving the laboratory, picked up on returning, and placed in a leak-proof container in a fume hood to await appropriate treatment.

Great care must be taken to minimize generation of aerosols during the cleanup of powdered chemicals. All unnecessary movement should be avoided, and steps should be taken immediately to contain the sample. Small spills may be covered carefully with paper towels moistened with water or a solvent. With larger spills a HEPA-filtered vacuum cleaner can be used. Commercial equipment is available, and a device has been described which is equipped with a cyclone separator for recovering powders for possible reuse (20). If this equipment is not available, a simple remedy is to cover the spilled powder with foamed shaving cream. The mixture can then be carefully scooped up and placed in a jar for disposal.

If the chemical is sufficiently soluble in water, the residue should be removed as completely as possible by rinsing and collected in a chemical fume hood. In many cases a decontamination procedure is available which may convert the carcinogen to a less hazardous product. The reagent may sometimes be applied to the floor or work bench, but only after the bulk of the spill has been removed. Upon completion of decontamination, all surfaces should be washed with soap and water and then rinsed with clean water. The affected surfaces should be checked at this point, and the procedures should be repeated if necessary. If rinse water is to be treated, it should be collected in a large beaker in a fume hood and reagents should be added slowly, to avoid sudden violent reaction and the generation of aerosols.

In some cases, of course, no effective neutralization procedure is available. If this is the case, the only disposal methods available may be incineration in an organic solvent or adsorption from the aqueous solution onto charcoal, followed by incineration of the charcoal. This latter method is often used by pesticide manufacturers for removal of traces of residues from wastewater.

Chemical treatment

The choice of a method for chemical treatment before ultimate disposal must be made with care. Too often, in the attempt to accomplish the destruction of a toxic chemical quickly, powerful reagents are used, resulting in violently exothermic reactions which may be more hazardous than the material being treated. Several general principles should be kept in mind when planning treatment procedures, as follows.

1. Be sure that the products of the reaction are actually less toxic than the original waste.
2. Avoid unnecessarily powerful reagents; do not use permanganate where hypochlorite would be effective.
3. Avoid high concentrations of waste material or reagents. Dilute the waste before adding reagents whenever a strongly exothermic reaction is anticipated.
4. Avoid generation of toxic gaseous by-products, or provide for treatment of the exhaust gases.
5. Reductions with metals or hydrides will often produce hydrogen. Provide good ventilation

and remove all ignition sources.

6. Use a chemical fume hood for all treatment procedures.

7. Take into account the effect of limited solubility. It may be necessary to use a mixed solvent or vigorous mixing to accelerate the reaction.

8. Allow ample time for completion of the reaction before disposing of the products.

A few chemical neutralization procedures are available, but documentation of their effectiveness is sometimes lacking, or they may have been checked with only one or two members of a chemical class. It is important that the procedure is checked with the specific compound, under conditions closely approximating the expected use. Failure to take into account differences in kinetics could result in an excessively exothermic reaction or in incomplete neutralization.

If a specific neutralization procedure is included in the safety protocol, a detailed procedure should be prepared, including a summary of the chemical basis of the procedure, specific reagents and concentrations, limitations on the use of the procedure, cautionary statements regarding secondary hazards, and disposition of the reaction products.

In 1978, collaborative studies were initiated to develop and evaluate a series of procedures for the destruction of chemical carcinogens in laboratory wastes. Under joint sponsorship of the Division of Safety, NIH, and the International Agency for Research on Cancer, five monographs have been published, and three more are in preparation (3). For each chemical type, several methods are presented, with recommendations for use with a variety of laboratory wastes (aqueous, culture media, animal bedding, organic solutions, etc.). Detailed procedures are outlined, including concentrations, reaction times, safety precautions, and methods of analytical validation. Each procedure has been checked in a collaborating laboratory, and its effectiveness has been determined with representative members of the class of carcinogens to be treated.

Several of the important problems which must be considered in selecting a treatment procedure are illustrated in the brief descriptions of the methods given below.

Primary aromatic amines. Weeks and Dean (20) have described the use of a 6 N hydrochloric acid-methanol mixture for cleanup of spills of aromatic amines and monitoring by spot tests using Ehrlich's reagent and two fluorogenic reagents. The amine is not neutralized by the procedure but simply converted to the more soluble hydrochloride. These amines are susceptible to oxidation, of course, but the products of mild oxidation will be nitro aromatics, which may also be carcinogenic. The washings should be neutralized in a fume hood and disposed of by incineration, adsorption on activated charcoal, or other suitable method.

Aflatoxins. A common reagent for destruction of aflatoxins in the laboratory is commercial bleach; the process is monitored by UV fluorescence. Recent work at the International Agency for Research on Cancer has shown that this procedure is not as simple as described, since free chlorine present in commercial bleach chlorinates aflatoxin B_1 to the mutagenic 2,3-dichloro derivative. Fortunately, this product can be destroyed by the addition of acetone, after the mixture is diluted to avoid a violent haloform reaction. The final product is a nonmutagenic dihydroxy derivative.

Thioureas. In alkaline solution, thioureas are oxidized by hypochlorite to the corresponding ureas, which are generally noncarcinogenic. This method has been recommended for ethylene thiourea. However, ethylene urea has been reported to be carcinogenic.

Nitrosamines. Several methods have been proposed for the neutralization of nitrosamines. The ideal method, of course, would be one which removes the nitroso function irreversibly, but not all methods do this. The nitroso group can be reduced by aluminum in alkaline solution, or by zinc in hydrochloric acid. A recent report has shown that the principal product of Al/KOH reduction of dialkyl nitrosamines is the corresponding hydrazine, often in better than 90% yield. An improvement of this method, using a 50% aluminum-nickel alloy, eliminates hydrazine formation completely, and the same alloy can be used for destruction of hydrazine-contaminated wastes (3).

Hydrogen bromide in glacial acetic acid is often recommended for destruction of nitrosamines in organic solvents. This method requires nearly anhydrous conditions for rapid reaction, and the reagent is a rather hazardous material to work with. Provision must also be made for the removal and destruction of the gaseous product, nitrosyl bromide. This is essential, since the reaction is reversible. This may be accomplished by bubbling nitrogen through the mixture and then through a solution of sulfamic acid.

Nitrosamides. Since nitrosamides such as nitrosomethyl urea and MNNG are unstable in alkaline media, it is tempting to use alkaline hydrolysis to accomplish their destruction. Unfortunately, the product is diazomethane, a very hazardous gas which is highly toxic, carcinogenic, and dangerously explosive. The safest methods of destruction of the nitrosamides are those which can be accomplished under acidic conditions. Several are given in the International Agency for Research on Cancer monographs, including treatment with sulfamic acid or hydrogen bromide-glacial acetic acid. It should be kept in mind that these reagents are also hazardous, and the procedures given should be followed closely.

An alternative method reported for MNNG, which also seems quite safe, is treatment with aqueous sodium thiosulfate (2%, buffered to pH 8 to 9 with phosphate). This procedure is claimed to require about 1 h for complete destruction of MNNG (5).

This method should be effective for nitrosomethyl urea and other nitrosamides. Care should be exercised in adapting such a method, of course, taking into account the possibly more vigorous reaction with nitrosomethyl urea.

MONITORING

Evaluation of the operating procedures used in the laboratory should include some measurement of their effectiveness in reducing employee exposure and limiting surface contamination. Personnel monitoring is best performed by an industrial hygienist. Standard procedures are available for many carcinogens, although much work needs to be done in this area. There is a great need for the development and evaluation of simple, inexpensive monitors, such as the passive vapor dosimeters currently appearing on the market, and of general methods sensitive to a large number of compounds. An interesting dosimeter has been described recently for monitoring exposure to direct-acting alkylating agents (19). There are monitoring requirements for only a few OSHA-regulated carcinogens at present, but such requirements will be included in future standards. In most cases initial monitoring will be required, with periodic monitoring when the initial monitoring shows exposure above the action level.

Although the results of monitoring of surface contamination are more difficult to interpret, it is possible to use a number of rapid, simple procedures to monitor the effectiveness of housekeeping and decontamination. Laboratories using polycyclic aromatic hydrocarbons and other fluorescent compounds routinely use a hand-held UV light which is passed over the work area upon completion of an operation to detect spillage or contaminated equipment. Simple spot tests can be used to detect surface contamination when handling many aromatic amines (19,20). These tests have the advantage that they are inexpensive and rapid and do not require special analytical techniques or equipment. They are also nonspecific, so that several potential contaminants could be monitored with the same test. Chromatographic analysis of wipe samples can be used to supplement the simple tests if it seems warranted.

At the beginning of a new procedure, it is sometimes useful to use modeling experiments to determine the effectiveness of control methods (9). An easily detected compound such as a highly fluorescent dye is used to simulate the carcinogen, and air sampling or wipe tests are performed during the operation and analyzed. In this way, more effective controls can be installed before work begins and the laboratory becomes contaminated.

A rather interesting monitoring procedure which in vitro laboratories might find convenient is the use of the Ames test to detect mutagenic activity in wipe samples (19). Although this type of test has a high specificity for mutagens and carcinogens, several days are required for interpretation, and the test is beyond the capability of the average laboratory.

WASTE DISPOSAL

Although research laboratories are among the smaller generators of hazardous wastes, the disposal of several hundred pounds of hazardous waste per month can become expensive. It is best, of course to destroy carcinogenic waste as close to the point of generation as possible. If a suitable neutralization procedure is available and has been proven effective with actual wastes, it should be used for pretreatment. Properly treated, aqueous wastes may be discharged to the sewer. Organic solutions must be handled as waste solvents.

Although several intriguing possibilities for hazardous waste disposal have been proposed in recent years (19), there are only two methods which are considered practical at the present time: incineration and landfill. Although the bulk of the hazardous waste from laboratories has probably been disposed of by landfill, recent environmental regulations (18) are beginning to make this option less convenient. As "gypsy haulers" and unauthorized landfill sites disappear, the cost of disposal by this method will increase dramatically.

Many laboratories are considering the purchase of incineration equipment for handling their hazardous wastes as well as their general combustible wastes. Several designs are available for high-temperature incineration which would be suitable for a research facility. The evaluation of several such incinerators has been described (19).

Flammable organics may be incinerated by spraying directly into the firebox through a nozzle. Careful maintenance and control of the type of wastes burned is necessary to avoid damage to the system or exposure of maintenance personnel. For small amounts (pints to gallons) of flammable waste the solution should be poured over vermiculite or, preferably, sawdust in a plastic bottle before delivery to the incinerator. Bulk carcinogens should be diluted with combustible material before incineration, especially if the carcinogen is not highly flammable. Large amounts of chlorine-, nitrogen-, or sulfur-containing compounds should not be incinerated unless the exhaust is scrubbed.

An alternative procedure for treatment of wastewater is adsorption on activated charcoal or resin filters, which can then be incinerated. Provision should be made for routine sampling of the effluent to determine the lifetime of the adsorbent. This method has been used successfully in animal treatment facilities.

It is difficult to provide adequate containment for cleaning of animal cages and disposal of bedding.

Canopy hoods can be used to limit dissemination of aerosols during cage dumping, and the bedding should be bagged immediately. Respiratory protection should be provided for this operation, showering requirements should be enforced, and technicians should be warned of the possible presence of carcinogens or their metabolites in the bedding and excreta. Contaminated bedding should not be allowed to accumulate, but should be disposed of as soon as possible after cage cleaning.

In the transport of contaminated waste to the disposal site, the material will usually be handled by nontechnical personnel. All contaminated waste should be placed in sealed containers before it leaves the laboratory and should be labeled "Caution—Chemical Carcinogen." Waste from an operation involving relatively large amounts of one or a few specific carcinogens should also be labeled with the names of the carcinogens. If the waste is handled by a disposal contractor, he will have specific requirements for packaging hazardous wastes, usually consisting of sealed steel drums. Plastic bags are usually used if the wastes are to be incinerated on site.

EXPOSURE TREATMENT

Even with careful training, modern laboratories and equipment, and careful supervision, accidental exposures to carcinogens will occasionally occur, through spillage, minor explosions, faulty laboratory gloves, and a plethora of other mishaps familiar to everyone who has worked in a laboratory. These exposures should be expected and planned for, so that equipment and supplies are on hand in such emergencies and laboratory workers know in advance what needs to be done. The procedures to be followed for exposure should be made part of the institution's safety plan and discussed with the medical advisor and safety committee.

In case of direct skin contact with a carcinogen, the affected area should be immediately flushed with plenty of water. Generally a good lather of soap should be used to aid in physical removal, except for the aromatic amines, which form insoluble products with soaps. Washing should continue for at least 15 min. Cool water should be used, and vigorous rubbing should be avoided.

When a reactive chemical has been spilled on the skin, it is tempting to use another chemical to neutralize it in situ. Although the application of topical ointments may accomplish this purpose after complete cleansing, it is a poor idea to use the skin as a substrate for a chemical reaction. In addition to the possible skin irritation by such reagents, the chemical reaction in close contact may produce some heating, accelerating any reaction with the skin and opening the pores as well. Cold water, on the other hand, will close the pores, lower the rate of chemical reaction with the skin, and, even with water-insoluble materi-

als, provide mechanical removal of the chemical, which can be aided by the detergent action of the soap. The use of organic solvents to remove skin contamination may be extremely risky, since in most cases the penetration of the chemical will be increased and the exposed area will be enlarged (13).

There is a growing concern about the use of eyewash fountains in the laboratory, and certainly they should be available in any laboratory where the danger of splashing chemicals in the eyes exists. Good industrial-type eyewash stations are expensive, however, and numerous inexpensive substitutes have appeared on the market recently, some of extremely dubious value. The important thing to remember when buying one of these is that the usual recommended treatment for chemicals splashed in the eyes is a minimum of 15 min of washing with plenty of water. The NIH Guidelines require that the device be able to wash both eyes simultaneously with a continuous stream of potable water.

In the case of exposure of a large area of the body, a complete shower should be taken. Cool water should be used, and the exposed areas should be washed for at least 15 min. Clean clothing should be supplied; contaminated clothing should not be reworn. Clothing should not be combined with other laundry for decontamination. Leather goods can be a serious problem, because if they become soaked with an organic liquid it is very difficult to clean them completely, and reuse can cause serious chemical burns or dermatitis (13).

All of the above, of course, presupposes that there is no need for immediate emergency medical attention. Unless such emergencies exist, the above procedures can be applied by the person affected or, in case of minor incapacitation, by other workers while awaiting the arrival of the medical advisor. Full details of the chemical exposure should of course be given to the medical advisor.

SOURCES OF INFORMATION

Recently several states have enacted so-called "right-to-know" legislation, requiring employers to provide employees with Material Safety Data Sheets whenever they are required to work with toxic or hazardous chemicals, and to provide safety training. Such laws and other legislation on hazardous chemicals have resulted in a greatly increased availability of safety data from the chemical suppliers. Although only limited data are available for many research chemicals, the suppliers will provide Material Safety Data Sheets for many compounds. Some safety data are included by some suppliers in their catalogs, and several suppliers have begun to include more useful information on their labels.

Although it is sometimes difficult to locate specific safety data for individual carcinogens, there is a large and growing body of literature on chemical hazards and safety procedures. Several pertinent references are listed at the end of this chapter, but a truly comprehensive bibliography would be a monumental task. The bibliographies given in several of the references are rather extensive and should be a useful source of new material.

Safety-related information is also available from a number of government agencies. Many of the pertinent reports are available directly from the agencies. In other cases, they can be purchased from the U.S. Government Printing Office or from the National Technical Information Service.

Two agencies whose reports are a valuable source of information are the Environmental Protection Agency and NIOSH. Catalogs of publications are available from the following addresses:

Publications Disseminations, DTS
National Institute for Occupational Safety and Health
4676 Columbia Parkway
Cincinnati, OH 45226

U.S. Environmental Protection Agency
Office of Research and Development
Environmental Research Information Center
Cincinnati, OH 45268

National Technical Information Service
U.S. Department of Commerce
Springfield, VA 22151

In the industrial hygiene field, the NIOSH publications are particularly useful. The Registry of Toxic Effects of Chemical Substances is a bibliography of toxicity information, giving data and references on over 59,000 chemicals. Although the data are not evaluated but simply tabulated uncritically, this source can be an extremely useful starting point. The principal function of NIOSH is to gather data by literature survey or research on hazards and to prepare comprehensive evaluations to provide the basis for proposed OSHA standards. Newly reported hazards are frequently reported in a brief (usually 5 to 6 pages) report, *The Current Intelligence Bulletin,* whose purpose is to alert workers in the field to an existing or potential hazard. At a later time, a more intensive study may produce a "Criteria for a Proposed Standard," which is a comprehensive survey on the hazard potential of a specific chemical or class of chemicals, with recommendations for control methods, permissible exposures, etc. Most of the information in the Criteria Document may appear in the OSHA standard. The Criteria Documents are excellent sources of information on specific compounds, including physical and chemical properties, toxicological and pharmacological information, recommendations for protective clothing and equipment, etc. NIOSH also operates a certification program for protective equipment, and a list of NIOSH-certified equipment may be obtained.

A number of computer searchable data bases are available, including the well-known MEDLINE, TOXLINE, CANCERLIT, etc. A useful description of such systems and of the operation of a computerized information system at Oak Ridge National Laboratory is given by J. S. Wassom (19).

On 22 January 1980 the OSHA Cancer Policy was published, outlining the policy established by the Department of Labor for the identification and regulation of chemicals posing a potential occupational carcinogenic risk (17). The rather lengthy preamble provides a summary on the historical background, an analysis of the technical issues involved, and an explanation of the rationale for the final policy. A detailed explanation of the Model Standard is given, and numerous references are listed. A copy of the OSHA Cancer Policy is available from the OSHA Office of Public Affairs or any regional or area office.

As indicated above, the NIH Guidelines are available to the public, and a copy is sent to principal investigators of new contracts and grants. To supplement the guidelines, NIH has prepared a set of safety data sheets in which specific information is given on the chemical, physical, and toxicological properties of over 65 compounds and recommendations are made for personal protection, decontamination, and emergency response.

The topic of carcinogen laboratory safety has been discussed at length in countless technical meetings during the past several years. One of the most useful sessions was the symposium conducted at the American Chemical Society/Chemical Society of Japan Chemical Congress, Honolulu, Hawaii, in April 1979 (19). Many useful new results were reported, and the descriptions of facilities and procedures for safety management will be very helpful for those organizations who are planning new facilities.

Much of the information included in this chapter was assembled for presentation in the short training course, "The Safe Handling of Chemical Carcinogens in the Research Laboratory." Funding for the development of the course was provided by the Office of Research Safety, National Cancer Institute, and the Division of Safety, NIH, under contracts no. N01-C0-65278 and N01-C0-95425, and its presentation to over 30 classes from 1977 to 1982 was supported by the above contracts and by EPA contract no. WA80-C159.

LITERATURE CITED

1. **Cralley, L. J., and L. V. Cralley (ed.).** 1979. Patty's industrial hygiene and toxicology, vol. III. John Wiley and Sons, New York.
2. **Dales, L. G., G. D. Friedman, and M. F. Cullen.** 1979. Evaluating periodic multiphasic health checkups: a controlled trial. J. Chron. Dis. **32:**385–404.
3. **International Agency for Research on Cancer.**

1980–1983. Laboratory decontamination and destruction of carcinogens in laboratory wastes. IARC Scientific Publications no. 37, 43, 49, 54, and 55. WHO Publications Center USA, Albany, N.Y.

4. **Jurinski, N. B.** 1974. Carcinogenesis bioassay health and safety plan. Tracor Jitco, Inc., Rockville, Md.

5. **Kilbey, B. J., M. Legator, W. Nichols, and C. Ramel (ed.).** 1977. Handbook of mutagenicity test procedures. Elsevier Scientific Publishing Co., Amsterdam.

6. **McDermott, H. J.** 1976. Handbook of ventilation for contaminant control. Ann Arbor Science, Ann Arbor, Mich.

7. **National Cancer Institute.** 1977. Safety and operation manual, high containment research facility. Office of Biohazard Safety, National Cancer Institute, Bethesda, Md.

8. **National Cancer Institute.** 1979. Chemical carcinogen hazards in animal research facilities. Office of Biohazard Safety, National Cancer Institute, Bethesda, Md.

9. **Sansone, E. B., A. M. Losikoff, and R. A. Pendleton.** 1977. Sources and dissemination of contamination in material handling operations. Am. Ind. Hyg. Assoc. J. **38:**433–442.

10. **Sansone, E. B., and M. W. Slein.** 1976. Application of the microbiological safety experience to work with chemical carcinogens. Am. Ind. Hyg. Assoc. J. **37:**711–720.

11. **Sansone, E. B., and Y. B. Tewari.** 1978. The permeability of laboratory gloves to selected solvents. Am. Ind. Hyg. Assoc. J. **39:**169–173.

12. **Scott, R. A., Jr. (ed.).** 1979. Toxic chemical and explosive facilities. Safety and engineering design. ACS Symposium Series 96. American Chemical Society, Washington, D.C.

13. **Steere, N.** 1971. Handbook of laboratory safety, 2nd ed. Chemical Rubber Co., Cleveland.

14. **U.S. Department of Health and Human Services.** 1981. NIH guidelines for the laboratory use of chemical carcinogens. NIH Publication no. 81-2385, May, 1981. U.S. Government Printing Office, Washington, D.C.

15. **U.S. Department of Labor.** 1980. Personal protection. 29 CFR Parts 1910.132–1910.140. U.S. Government Printing Office, Washington, D.C.

16. **U.S. Department of Labor.** 1980. Toxic and hazardous substances. 29 CFR Parts 1910.1001–1910.1045. U.S. Government Printing Office, Washington, D.C.

17. **U.S. Department of Labor.** 1980. Identification, classification and regulation of potential occupational carcinogens FR 45 pp 5001–5296, January 22, 1980. U.S. Government Printing Office, Washington, D.C.

18. **U.S. Environmental Protection Agency.** 1980. Hazardous waste management system. 40 CFR Parts 260–265. U.S. Government Printing Office, Washington, D.C.

19. **Walters, D. B. (ed.).** 1980. Safe handling of chemical carcinogens, mutagens, teratogens and highly toxic chemicals. Ann Arbor Science, Ann Arbor, Mich.

20. **Weeks, R. W., Jr., and B. J. Dean.** 1978. Decontamination of aromatic amines cancer-suspect agents on concrete, metal or painted surfaces. Am. Ind. Hyg. Assoc. J. **39:**758–762.

Chapter 27

Techniques for Safe Handling of Radioactive Materials

JEROME W. STAIGER

The purpose of this chapter is to provide basic guidelines and information concerning the use of sources of ionizing radiation and the control of radiation exposure to persons who handle such sources. The materials presented will be limited primarily to sources of ionizing radiation encountered in the laboratory environment.

RADIOACTIVE MATERIALS: RADIATION AND RADIOACTIVITY

All matter is made up of atoms which contain protons and neutrons in their nuclei and have orbital electrons which rotate in specific energy levels (orbits) around the nucleus. The number of protons and neutrons in the nucleus of an atom of a particular material determines the elemental nature of the material (1 proton = hydrogen, 2 protons = helium, etc.). The number of neutrons in the nucleus does not change the element, but changes the atomic mass of the nucleus and therefore affects the stability of the nucleus. Materials which have the same number of protons (same element) but a different number of neutrons in the nucleus are called "isotopes" of the element (^1_1H = hydrogen, ^2_1H = deuterium, and ^3_1H = tritium are the three isotopes of the element hydrogen). A very useful listing of all known proton and neutron combinations (nuclides) is available in the form of a chart called "The Chart of the Nu-

clides," available from the Superintendent of Documents, U.S. Government Printing Office. The chart is arranged according to element and lists each isotope, both stable and radioactive (unstable), for each element.

Each element has one or more isotopes which are unstable. Such isotopes are called "radioisotopes" or "radionuclides" because their unstable nuclei give off energy in the form of ionizing radiation through a process called nuclear decay. The energy given off in this process is in the form of one or more of three types of ionizing radiation: alpha particles, beta particles, and gamma rays.

Table 1 lists the characteristics of these and other types of ionizing radiation. Sources of neutron and X-ray radiation are not likely to be encountered in the laboratory and will not be discussed in this chapter. Gamma and X rays are both electromagnetic radiations and differ only in the way in which they are produced. Gamma rays result from decay transitions within radioactive nuclei, and X rays result from high-energy particle interactions near the nucleus of atoms of an absorbing material or from electron transitions in ionized atoms. The range of alpha particles, beta particles, and gamma rays varies depending on the mass, charge, and energy of the radiation. Alpha particles have a very limited range because of their large mass and charge (5-MeV alpha range is $\simeq 5$ cm in air and <0.07 mm in tissue). Beta particles have a smaller mass and charge and, there-

TABLE 1. Types of ionizing radiation

Type of radiation	Symbol	Mass (amu)[a]	Charge	Common source
Alpha particles	$\alpha(^4_2He)$	~ 4	$+2$	Radioisotope decay
Beta particles				
Negatron (electron)	$\beta^-(-^0_1e)$	0.00055	-1	Radioisotope decay
Positron	$\beta^+(+^0_1e)$	11	$+1$	Radioisotope decay
Gamma rays	γ	0	0	Radioisotope decay
X rays	x	0	0	X-ray machines (medical, research); X-ray defraction, electron microscope, characteristic X rays
Neutrons	$n(^1_0n)$	~ 1	0	Nuclear reactors (fission), high-energy accelerators, nuclear reactions (α, n)

[a]amu, Atomic mass unit.

fore, have a somewhat greater range than alpha particles. Gamma rays have no mass or charge and consequently have a considerably greater penetrating ability than alpha or beta radiations of comparable energy (5).

The radioactivity of a particular radioisotope is a measure of the rate at which the nuclei of the radioisotope decay or disintegrate. The commonly used unit of radioactivity is the curie (Ci) and is, by definition, that amount of a particular radioisotope which has a nuclear decay rate of 3.7×10^{10} disintegrations per s (dps). This unit can also be expressed in terms of disintegrations per min (dpm; 3.7×10^{10} dps $\times$ 60 s/min $= 2.22 \times 10^{12}$ dpm). Other useful units of radioactivity are the millicurie (mCi), which is 1/1,000 of a curie or 10^{-3} Ci, and the microcurie (μCi), which is 1/1,000,000 of a curie or 10^{-6} Ci. Recently there has been an effort to introduce a new unit of radioactivity under the International System of Units (SI units), the becquerel (Bq), defined as equal to 1 dps. Table 2 shows the relationship between curie and becquerel units (5).

RADIOISOTOPES: CHARACTERISTICS AND CATEGORIZATION

All radioisotopes have certain characteristics which are constant and which may be used in identifying and categorizing them. Two defining characteristics that are very helpful in assessing the radiation protection implications of a radioisotope are the half-life and the radiation decay scheme.

The half-life ($t_{1/2}$) is a measure of the time it takes for one half of the nuclei of a particular radioisotope to decay to a more stable state, and is a constant for the particular radioisotope. During each successive half-life, half of the remaining radioactive nuclei will have decayed and converted to a different substance (in many cases a stable or nonradioactive material). The fraction of the original activity (A_0) remaining after a particular number of half-lives (n) can be calculated using the relationship shown in Table 3.

The rate at which a radioisotope decays, or its disintegration constant (λ), is inversely proportional to the half-life of the radioisotope; $\lambda = 0.693/t_{1/2}$. The activity of a particular radioisotope can be calculated for any time t after the assay or determination of the original activity (A_0) by using the following formula (8):

$$A_t = A_0 e^{\dfrac{-0.693t}{t_{1/2}}}$$

where A_t = activity after time t; A_0 = initial activity at $t = 0$; t = time since determination of A_0; and $t_{1/2}$ = half-life of radioisotope in the same time units as t.

The decay scheme of a radioisotope is the type and energy of ionizing radiations emitted by the nuclei of a radioactive material when they decay. Most radioisotopes emit beta or gamma radiations or both.

TABLE 2. Relationship of units[a]

Radiological quantity	Old unit	SI Unit	Relationship between units		
Activity of a radioactive material	The curie 1 Ci $= 3.7 \times 10^{10}$ dis/s	The becquerel 1 Bq $= 1$ dis/s 10^3 Bq $= 1$ kilobecquerel (kBq) 10^6 Bq $= 1$ megabecquerel (MBq) 10^9 Bq $= 1$ gigabecquerel (GBq) 10^{12} Bq $= 1$ terabecquerel (TBq) 10^{15} Bq $= 1$ petabecquerel (PBq) 10^{18} Bq $= 1$ exabecquerel (EBq)	1 Bq $= 2.7 \times 10^{-11}$ Ci 1 kBq $= 2.7 \times 10^{-8}$ Ci 1 MBq $= 2.7 \times 10^{-5}$ Ci $= 27\ \mu$Ci 1 GBq $= 27$ mCi 1 TBq $= 27$ Ci 1 PBq $= 27$ kCi 1 EBq $= 27$ MCi	$1\ \mu$Ci $= 37$ kBq 1 mCi $= 37$ MBq 1 Ci $= 37$ GBq 10^3 Ci $= 37$ TBq 10^6 Ci $= 37$ PBq 10^9 Ci $= 37$ EBq	
Exposure	The roentgen 1 R $\equiv$ the production of ions (of one sign) carrying a charge of 2.58×10^{-4} C/kg of air	No special named unit for exposure. The unit for ionization is C/kg, and this can be used to express the results of ionization chamber measurements as an intermediate step in the determination of absorbed dose	—	—	
Absorbed dose	The rad 1 rad $= 0.01$ J/kg	The gray 1 Gy $= 1$ J/kg 1 Gy $= 10^3$ mGy $= 10^6\ \mu$Gy	$1\ \mu$Gy $= 0.1$ mrad 1 mGy $= 100$ mrad 1 Gy $= 100$ rad	1 mrad $= 10\ \mu$Gy 1 rad $= 10$ mGy 100 rad $= 1$ Gy	
Dose equivalent	The rem 1 rem $= 1$ rad $\times Q$ Q is the quality factor	The sievert 1 Sv $= 1$ Gy $\times Q \times N$ 1 Sv $= 10^3$ mSv $= 10^6\ \mu$Sv N is the product of all other modifying factors (currently taken as 1 by ICRP)	$1\ \mu$Sv $= 0.1$ mrem 1 mSv $= 100$ mrem 1 Sv $= 100$ rem	1 mrem $= 10\ \mu$Sv 1 rem $= 10$ mSv 100 rem $= 1$ Sv	

[a]From reference 5 with permission.

TABLE 3. Fraction of original radioactivity related to half-lives[a]

No. of half-lives (n)	A_n	Fraction of A_0 remaining
1	$1/2 \, A_0$	0.5
2	$1/4 \, A_0$	0.25
3	$1/8 \, A_0$	0.125
4	$1/16 \, A_0$	0.0625
5	$1/32 \, A_0$	0.03125
6	$1/64 \, A_0$	0.0156
7	$1/128 \, A_0$	0.0078

[a] $A_n = A_0(1/2)^n$.

Some radioisotopes, such as ^{3}H, ^{14}C, ^{32}P, and ^{35}S, decay by beta-particle emission only. The maximum beta energy (β_{max}) emitted by each of these radioisotopes is a constant. Radioisotopes such as ^{51}Cr and ^{125}I are gamma-emitting radioisotopes, and ^{131}I is a beta- and gamma-emitting radioisotope. Only high-atomic-number radioisotopes (heavy elements) emit alpha-particle radiations. Information on the half-lives and decay schemes of some of the radioisotopes commonly used in clinical and research laboratories is listed in Table 4. Additional information pertinent to the assessment of protective precautions in the use of the radioisotopes is also listed.

RADIATION DOSE: UNITS AND PROTECTION LIMITS

Three terms or units are used when expressing radiation exposure and dose from sources of ionizing radiation: the roentgen, the rad, and the rem. The term used for radiation exposure is the roentgen (R); it is by definition a specific amount of ionization in air (2.58×10^{-4} C/kg of air). The unit of absorbed radiation dose is the rad (0.01 J/kg), a measure of the energy deposited in an absorbing material. The rem is the unit of biological dose or dose equivalence, used to express the dose from various types of ionizing radiation in common units. The rem dose is calculated by the equation: rem = rad $\times$ Q, where Q is the quality factor of the particular type of ionizing radiation. For the most commonly encountered types of ionizing radiation (beta, gamma, and X-ray radiations) the quality factor is approximately equal to 1. In such cases, the rem dose is equal to the rad dose. Also, for radiation protection purposes, the rad dose in tissue is approximately equal to the air exposure in roentgens for these same types of ionizing radiations. Therefore, the approximate rem dose can be determined or projected based on instruments which measure radiation exposure (R) in air. Table 2 lists the SI units, the gray (Gy) and the sievert (Sv), which may be substituted for the rad and rem, respectively, using the conversion factors indicated in the table (5).

Radiation exposure is often expressed in terms of milliroentgens (mR) and is commonly measured as an exposure rate per unit of time (e.g., mR/min, mR/h). Units of absorbed dose (rads) and dose equivalence (rems) are also expressed in terms of rads per unit time and rems per unit time, respectively.

All individuals are exposed to varying levels of ionizing radiation from natural background radiation sources present here on the earth. Natural background radiation results from three primary sources, (i) cosmic or space radiations which penetrate the earth's atmosphere, (ii) terrestrial radiations emitted by naturally occurring radioisotopes in our environment (e.g., soils, air, building materials), and (iii) internal radiations from naturally occurring radioisotopes present in our own bodies (primarily ^{40}K). Natural background radiation exposure varies depending on where a person lives. In the United States the average dose received by the population is ca. 120 mrem/year.

It is generally accepted that exposure to ionizing radiation presents a risk of potential biological damage and that the degree of the risk increases with increasing amounts of ionizing radiation dose received by an individual. The probability of harmful effects from ionizing radiation, such as leukemia and other cancers and genetic effects on future generations, is dependent on the total dose of radiation. The biological effects of ionizing radiation exposure have been thoroughly studied by scientific review groups such as the Committee of Biological Effects of Ionizing Radiation (BEIR III Committee) of the National Academy of Science. The BEIR III Committee report presents a range of risk probabilities for various radiation-induced effects, based on different dose-response relationships (3). The report indicates that probabilities of cancer incidence are small for low doses of ionizing radiation (2 to 3 in 10,000 persons for a 1-rem dose to each person), and that former dose-response models may have overestimated the risk probabilities at low dose. Some perspective on the relative risks of ionizing radiation compared to other health risks is given in Table 5.

To protect persons exposed to ionizing radiation, federal regulatory agencies such as the Nuclear Regulatory Commission (NRC) and the Occupational Safety and Health Administration have adopted radiation dose limits and guides recommended by international and national standards-setting groups such as the National Council on Radiation Protection. Table 6 lists the recommended protection limits for various population groups. The dose limit for occupationally exposed persons is based on the formula 5(N-18) rem, where N = age in years. Under most situations this translates into a maximum permissible dose of 5 rem/year or 1.25 rem per calendar quarter for whole-body and gonadal exposure. The occupational dose limit for the hands and feet is considerably higher than that for the whole body. The limit for hands and feet is 75 rem/year or 18.75

TABLE 4. Half-lives and decay schemes of some commonly used radioisotopes

Radiotoxicity group, isotope	Half-life		Radiation emitted		Specific gamma constant (Γ)[a,b]	Maximum permissible body burden[c] (μCi)	MPC[d] (μCi/ml)			
	Physical[a]	Biological[c]	Type	Energy[e] (MeV)			Air, 40 h, occupational	Air, 168 h, environmental	Water, 40 h, occupational	Water, 168 h, environmental
Low radio-toxicity group										
³H	12.3 yr	12 days	β⁻	0.0186 (max)		1,000	5 × 10⁻⁶	2 × 10⁻⁷	1 × 10⁻¹	3 × 10⁻³
¹⁴C	5,730 yr	10 days	β⁻	0.156 (max)		300	4 × 10⁻⁶	1 × 10⁻⁷	2 × 10⁻²	8 × 10⁻⁴
Intermediate radiotoxicity										
³²P	14.3 days	19 days	β⁻	1.71 (max)		6 (bone)	7 × 10⁻⁸	2 × 10⁻⁹	5 × 10⁻⁴	2 × 10⁻⁵
³⁵S	87.4 days	90 days	β⁻	0.167 (max)		400	3 × 10⁻⁷	9 × 10⁻⁹	2 × 10⁻³	6 × 10⁻⁵
⁵¹Cr	27.7 days	616 days	γ	0.320 (9.8%)	0.16	800	1 × 10⁻⁵	4 × 10⁻⁷	5 × 10⁻²	2 × 10⁻³
			x⁻	0.005 (22%)						
High radio-toxicity										
¹³¹I	8.04 days	138 days	β⁻	0.606 (22%)	2.2	0.7 (whole body)	9 × 10⁻⁹	1 × 10⁻¹⁰	6 × 10⁻⁵	3 × 10⁻⁷
			γ	0.364 (81.2%)						
				0.637 (7.3%)						
				0.284 (6.1%)						
¹²⁵I	60 days	138 days	γ	0.035 (6.5%)	0.7	0.14 (thyroid)	5 × 10⁻⁹	8 × 10⁻¹¹	5 × 10⁻⁵	2 × 10⁻⁷
			KαX	0.027 (113%)		0.65 (thyroid)				
			KαX	0.031 (25%)						
⁴⁵Ca	163 days	1.8 × 10⁴ days	β	0.257 (max)		30	3 × 10⁻⁸	1 × 10⁻⁹	3 × 10⁻⁴	9 × 10⁻⁶

[a] *Radiological Health Handbook* (2).
[b] The specific gamma constant, Γ, is in units of roentgens per hour per millicurie at 1 cm, or Γ/10 = roentgens per hour per curie at 1 m.
[c] ICRP Committee II report (4).
[d] 10 CFR 20.
[e] Parentheses indicate percent of time that a disintegration results in this type of radiation emission. max, Maximum energy of the beta particle emitted by the radioisotope. Kα, X-ray emission corresponding in energy to the Kα electron orbit of the radioisotope.

TABLE 5. Estimated loss of life expectancy from health risks[a]

Health risk	Estimated avg days of life expectancy lost
Smoking (20 cigarettes per day)	2,370 (6.5 years)
Overweight (by 20%)	985 (2.7 years)
All accidents combined	435 (1.2 years)
Auto accidents	200
Alcohol consumption (U.S. avg)	130
Home accidents	95
Drowning	41
Safest jobs (such as teaching)	30
Natural background radiation, calculated	8
Medical X rays (U.S. avg), calculated	6
All catastrophes (earthquake, etc.)	3.5
1 rem occupational radiation dose, calculated[b]	1
1 rem/year for 30 years, calculated	30
5 rems/year for 30 years, calculated	150

[a]From Draft NRC Regulatory Guide, Division 8, Task 0H902-1, Office of Standards and Development, USNRC, 1980.
[b]Industry average is 0.34 rem/year.

TABLE 6. Radiation dose limits and recommendations

Persons exposed	Maximum permissible dose[a]
Occupationally exposed (restricted areas)[b]	
Whole body, head and trunk, active blood-forming organs, gonads, lens of eyes	5 rem/year, 1.25 rem/quarter
Skin of whole body	30 rem/year, 7.5 rem/quarter
Hands and forearms, feet and ankles	75 rem/year, 18.75 rem/quarter
Thyroid	30 rem/year, 7.5 rem/quarter
Other body organs (not listed above)	15 rem/year, 3.75 rem/quarter
General public (unrestricted area)[b,c]	0.5 rem/year
Special exposure considerations[d]	
Occupationally exposed pregnant women[e]	0.5 rem during gestation period
Students[f]	0.1 rem/year

[a]Dose of 3 rem/quarter maximum allowed (12 rem/year) only if a cumulative lifetime exposure history is maintained and the individual does not exceed $5(N - 8)$ rem, where N = age in years.
[b]10 CFR 20.
[c]One-tenth the occupational maximum permissible dose.
[d]NCRP Report no. 39, Basic Radiation Protection Criteria (NCRP Publications, Washington, D.C., 1971).
[e]Recommended dose guide to protect the fetus.
[f]From course-related exposure. Recommended guide, not in addition to the 0.5 rem/year public maximum permissible dose.

rem per quarter. Persons under 18 years of age are not permitted in jobs where occupational exposure limits apply.

The dose limit designated for individuals in the general public who live or work near sources of ionizing radiation is 1/10 of the occupational limit or 0.5 rem/year. For protection of the unborn child, it is recommended that an occupationally exposed woman who is pregnant not receive more than 0.5 rem during the 9-month gestation period (NRC Regulatory Guide 8.13 [9]). The radiation dose should be limited especially during the first trimester of pregnancy, because of the greater sensitivity of the fetus to ionizing radiation.

The protection limits and guides specified in Table 6 are intended as maximum levels to assure the protection of personnel and the public. In addition to these limits, the NRC regulations (Code of Federal Regulations, Title 10, Part 20 [10 CFR 20]) require that radiation exposures be maintained "as low as reasonably achievable" (ALARA) within these limits. This ALARA principle is enforced by the NRC, and employers are expected to have an operational ALARA program.

RADIATION EXPOSURE CONTROL

Control of ionizing radiation exposure is commonly divided into two categories: (i) control of exposure from external sources and (ii) control of exposure from internal sources. External radiation sources are sources that are located outside a person's body, and internal sources are sources that are located inside (taken into) a person's body. Alpha, beta, and gamma radiations differ in potential of causing a radiation dose when they are emitted by a source that is external or internal to the body. Of these three radiations, gamma rays present the greatest external exposure potential because of their ability to penetrate the human body. Beta particles present the next greatest external exposure potential, and alpha particles present essentially no external exposure potential because their range in tissue is less than the thickness of the dead layer of skin cells (0.07 mm) which covers the body. If these radiations are emitted inside the body, alpha particles present the greatest exposure potential because they deposit a large amount of energy in a small volume. Beta particles would present the next greatest internal exposure potential, and gamma rays would present the smallest exposure potential because of the relatively small amount of energy which they deposit per unit mass as compared to alpha and beta radiations of equal energy.

Control of external exposure

The exposure from external radiation sources can be controlled through the application of four principles: (i) time, (ii) distance, (iii) shielding, and (iv) substitution.

(i) Time. Any radiation dose received by a person is a product of the dose rate (dose per unit time) and the time the person is present in the dose field (dose = dose/unit time × time). Therefore, the dose received will be reduced in direct proportion to the reduction of the time of exposure. Practice or dry runs (without radioactive material) often prove helpful in perfecting and modifying procedures to reduce the time necessary to perform a particular manipulation, thereby reducing radiation dose to personnel.

(ii) Distance. Increasing the distance from a radiation source is a very effective way to reduce external radiation exposure levels. Most sources of ionizing radiation are contained in a relatively small volume and approximate a point source. The radiation intensity from a point source decreases very rapidly with distance. This decrease in intensity is inversely proportional to the square of the distance from the source ($I \propto 1/d^2$). The exposure rate from a point source can be approximated by using the inverse-square relationship represented by the formula: $I_1/I_2 = (d_2)^2/(d_1)^2$.

The exposure rate (I_2) at any distance (d_2) from a point radiation source can be calculated from a single measurement of exposure rate (I_1) at a known distance (d_1): $I_2 = I_1 (d_1)^2/(d_2)^2$. For example, if the exposure rate at 1 cm from a source is 100 mR/h, the exposure rate at 10 cm from the source will be: $I_2 = 100$ mR/h $\times (1)^2/(10)^2 = 1$ mR/h (factor of 100 reduction).

The specific gamma constant, Γ, is a unit which is useful when determining the radiation exposure rate at various distances from gamma-emitting radioisotopes. The specific gamma constant has been calculated for a number of different radioisotopes that emit gamma radiation and is an expression of the exposure rate in roentgens per hour at a distance of 1 cm from a 1-mCi point source of the particular radioisotope (Γ = R/h per mCi at 1 cm, or $\Gamma/10$ = R/h per Ci at 1 m). Table 4 lists values of Γ for some commonly used radioisotopes. Values of Γ for other gamma-emitting radioisotopes are available in publications such as the *Radiological Health Handbook* (2). In addition to tables of Γ, the following formula can be used to approximate the exposure rate from a gamma-emitting radioisotope: DR $\simeq 6AEn$, where A = activity in millicuries; En = mean gamma energy per disintegration in mega-electron volts (gamma energy × fraction of time emitted); and DR is the exposure rate in milliroentgens per hour at 1 ft for a point source (6).

The radiation intensity from a source may decrease more rapidly than projected by inverse-square calculations. This occurs when there is significant attenuation of the radiation (especially true for beta radiations) before it reaches the distance at which the exposure rate is calculated. Beta radiations have a maximum range in absorbing materials, and beyond this range the exposure rate from the beta particles will be essentially zero. Also, in cases where the source does not approximate a point source, the radiation exposure rate will not decrease as rapidly as projected by inverse-square calculations. This is true when the physical dimensions of the source are a significant percentage of the distance at which the exposure rate is calculated. For example, the radiation exposure rate from a line radiation source varies inversely proportionally to the distance ($I \propto 1/d$), not to the distance squared as in the case of a point source, because the activity is distributed uniformly along a line rather than at one point (1a).

Items such as long-handle tongs, forceps, clamps, and remote pipettors and tools are commonly used to increase distance from a source and are very effective in greatly reducing the radiation dose received by personnel.

(iii) Shielding. The positioning of radiation-absorbing or -attenuating materials between a radiation source and the area occupied by personnel is a very good method for control of external radiation exposure. The type and thickness of shielding material used depends on the initial intensity of the radiation (activity of the radioactive material) and the type and energy of the ionizing radiations emitted. In the case of alpha particles emitted by radioisotopes, the energy ($\simeq 4$ to 6 MeV) is such that their range in tissue is less than the thickness of the dead cell layer which covers the human body. Therefore, for alpha radiation no shielding is necessary for personnel protection.

Beta particles, like alpha particles, have a definite range in absorbing materials. The thickness of shielding necessary for beta sources depends on the energy of the beta particles emitted. For low-energy beta-emitting radioisotopes such as ^3H and ^{14}C, essentially no shielding is required because of the very limited range of such radiations in air and other absorbing materials. High-energy beta-emitting radioisotopes such as ^{32}P have a much greater range and can result in a significant skin dose to personnel. The range in various absorbing materials of beta particles having various energy levels can be determined by using Fig. 1 (*Radiological Health Handbook* [2]). For example, the range of the maximum beta particle emitted by ^{32}P ($\beta_{max} = 1.71$ MeV) is approximately 0.3 cm in lucite. Therefore, if a thickness of 0.3 cm of lucite is used to shield ^{32}P, no beta radiation will penetrate through the shield. However, a note of caution is necessary when selecting a shielding material for a beta-emitting radioisotope. When beta radiations are absorbed by a shielding material they interact with the atoms of the shielding material in such a way that "bremsstrahlung" (braking radiation) is generated. Bremsstrahlung is X-ray radiation which is emitted as a result of loss of energy by beta particles when they interact with the positively charged nuclei of the atoms of the shielding material. The amount of bremsstrahlung X radiation produced is directly proportional to the atomic number (Z) of the shielding material. The fraction of the beta-particle energy converted to bremsstrahlung is approximately equal to the average beta energy ($E_{avg} \simeq 1/3E_{max}$) $\times$ 0.001 $\times$ Z. Because of this dependency on Z of the shielding material, it is preferable to use low-Z materials (lightweight elements) such as lucite or wood to shield intermediate- and high-energy beta emitters. Lucite or plexiglass has been found to be an excellent beta-shielding material because it is lightweight, readily available, and transparent (to allow for visual observation of material transfer). If large quantities (>10 mCi) of a high-energy beta emitter are used, it may be necessary to add lead shielding to the exterior of the lucite shield to reduce the bremsstrahlung exposure level to well within protection limits.

The thickness of shielding necessary to adequately attenuate gamma and X-ray radiations depends on their energy and the type of material used for shielding. Gamma- and X-ray shielding differs from beta shielding in that these electromagnetic radiations have no definite range in an attenuating material. Such electromagnetic radiations are absorbed in an exponential manner. The following formula may be used to calculate this exponential attenuation (8): $I = I_0 e^{-\mu x}$, where I = intensity after attenuation; I_0 = original intensity; μ = attenuation coefficient in units of centimeters^{-1} for the particular energy and shielding material; and x = shielding material thickness in centimeters.

A very useful factor to know when evaluating gamma shielding is the half-value layer (HVL) of various shielding materials for a certain gamma-ray energy. The HVL ($x_{1/2}$) is the thickness of shielding material that will reduce the intensity from a particular gamma energy by one half. The HVL is related to the attenuation coefficient by the following relationship: HVL or $x_{1/2} = 0.693/\mu$.

Table 7 lists the HVLs for several different gamma energies in various attenuating materials. Also, as in the case of half-life, the fraction to which the intensity is reduced after n HVLs can be determined by using the formula $I = I_0(1/2)^n$. Table 8 lists values for $(1/2)^n$ for values of $n = 0$ to $n = 6.6$. Using this HVL information, the reduction in intensity provided by a known shielding thickness, or the thickness of a particular shielding material necessary to reduce the gamma intensity by a known factor, can be easily calculated. For example, the thickness of lead required to reduce the intensity of 1 MeV gamma radiation to 0.01 (1/100) of the unshielded intensity can be calculated as follows. From Table 8 the number of HVLs for a reduction to 0.01 is 6.6. From Table 7, the HVL for 1.0 MeV gamma rays in lead is 0.90 cm. Therefore, the thickness required to reduce the original intensity to 0.01 is 6.6 $\times$ 0.9 cm = 5.94 cm of lead (6).

In situations where scatter radiation contributes significantly to the intensity of radiation transmitted through a shield (true for thick shields), it is necessary to take into account the buildup due to the scatter. The formula used to calculate the contribution from buildup of scatter radiation is: $I = BI_0 e^{-\mu x}$, where B is the buildup factor. Tables listing values for μ, μx, and B are available in the National Bureau of Standards *Handbook 92* (8), the *Radiological Health Handbook* (2), and other texts on radiation shielding.

The most commonly used shielding materials for gamma radiations are lead, concrete, and steel. Lead is more commonly used for localized shielding, and concrete and steel are used for structural shielding of large barriers such as those used in hot cells and irradiation facilities.

(iv) Substitution. Before radioactive materials are used, a determination should be made as to whether there is a comparable method which does not require the use of radioactive materials. If radioactive materials are required to obtain the desired results, the activity used should be no more than necessary, and if possible, a low-radiotoxicity radioisotope should be used. These considerations will reduce both the external and internal exposure potential.

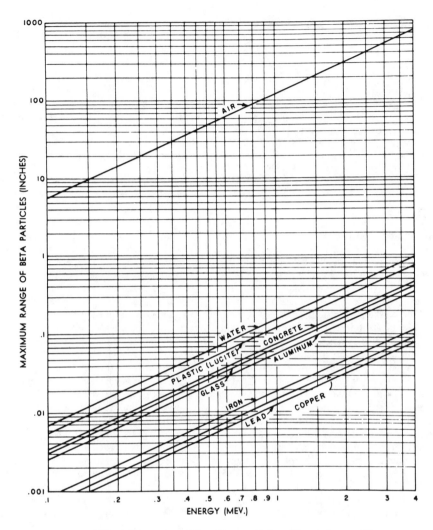

FIG. 1. Penetration ability of beta radiation. (From reference 2.)

TABLE 7. HVLs for various materials against gamma rays[a]

Gamma energy (MeV)	HVL								
	Aluminum (cm)	Iron (cm)	Copper (cm)	Lead (cm)	Lead (X ray)[b] (cm)	Water (cm)	Air[c] (m)	Concrete[d] (cm)	Concrete (X ray)[b] (cm)
0.1	1.60	0.26	0.18	0.012	0.024	4.14	35.5	1.75	1.75
0.2	2.14	0.64	0.53	0.068	0.050	5.10	43.6	2.38	2.54
0.5	3.05	1.07	0.95	0.42	0.31	7.17	61.9	3.40	3.26
1.0	4.17	1.49	1.33	0.90	0.80	9.82	84.5	4.65	4.51
2.0	5.92	2.09	1.86	1.34	1.20	14.05	120.5	6.60	6.12
5.0	9.11	2.84	2.47	1.44		23.02	195.8	10.28	

[a]From H. F. Henry, *Fundamentals of Radiation Protection*, Wiley-Interscience, New York, 1969, with permission of the publisher.
[b]HVLs for heavily filtered constant-potential X rays.
[c]Inverse-square distance effects are neglected.
[d]Average concrete density, 2.35 g/cm^3.

TABLE 8. Powers of one-half for attenuation and decay calculations $[(1/2)^n]^a$

n	.0	.1	.2	.3	.4	.5	.6	.7	.8	.9
0	1.000	0.933	0.871	0.812	0.758	0.707	0.660	0.616	0.578	0.536
1	0.500	.467	.435	.406	.379	.354	.330	.308	.287	.268
2	.250	.233	.217	.203	.190	.177	.165	.154	.144	.134
3	.125	.117	.109	.102	.095	.088	.083	.077	.072	.067
4	.063	.058	.054	.051	.047	.044	.041	.039	.036	.034
5	.031	.029	.027	.025	.024	.022	.021	.019	.018	.017
6	.016	.015	.014	.013	.012	.011	.010			

[a]From reference 6 with permission of the publisher. *Note:* n is the number of HVLs or half-lives. In 2.6 half-lives, the activity or number of atoms will be reduced to 0.165 of the original value. In 2.6 HVLs, the number of photons will be reduced to 0.165 of the number entering the shield.

Internal exposure control

Once a radioactive material has been taken into the body there is little that can be done to accelerate its removal. For this reason, the control of internal radiation exposure necessarily emphasizes the prevention of the uptake of radioactive materials via the routes of (i) inhalation, (ii) ingestion, (iii) skin absorption, and (iv) injection or puncture.

(i) Control of inhalation uptake

- Use a properly designed radioisotope hood with an airflow rate of 100 to 150 linear ft/min (3.0 to 4.6 m/min) with the hood sash opened to control the release of contaminants into the breathing zone. The hood may require filtration of the air stream before discharge to the environment to assure that environmental releases are well within permissible release concentrations. For particulate filtration a high-efficiency particulate air (HEPA) filter (99.97% efficient at 0.3 μm) is recommended. For gaseous contaminants such as radioiodine an activated charcoal filter is recommended.
- A glove box or isolated room with remote-source manipulators may be required for large quantities or highly radiotoxic radioactive materials. Again, the air discharged from such containment facilities must be properly filtered and monitored to protect individuals in the environment surrounding the facility.
- General laboratory room air ventilation must be changed frequently enough to minimize the buildup of possible radioisotope contaminants. Room ventilation should be designed to maintain a negative room air pressure with respect to surrounding corridors and unrestricted areas. If possible, room air should not be recycled, but exhausted through the radioisotope hoods located in the laboratory, which should be single stacked to the roof with the duct under negative pressure.
- Personal respiratory protection in the form of particle or gaseous filter respirators or supplied air respirators may be necessary in some situations where adequate ventilation systems are not available for contamination control (accidental releases, spills, etc.). Care must be taken when establishing a program for use of respirators to assure that personnel have been properly trained in their use and that the appropriate health evaluations have been made of the personnel who will wear the respirators (NRC Regulatory Guide 8.15).

(ii) Control of ingestion uptake

- Disposable gloves should be used which are removed and properly disposed of before the worker leaves the radioisotope work station.
- Protective clothing such as a laboratory coat should be worn. Laboratory coats should not be worn outside of the laboratory and should be checked after each use. In some laboratory situations, protective covering for shoes may be necessary.
- Eating, smoking, drinking, or food and beverage storage is not allowed in the laboratory.
- Mouth pipetting is not allowed, even when there is a method to prevent liquids from entering the mouth. Volatile radioisotopes may be contained in the air drawn into the mouth during mouth pipetting.
- Frequent hand washing should be required, especially before leaving the laboratory for lunch breaks or to return home. Hands should be surveyed with an appropriate radiation monitor after each radioisotope use.
- Trays, plastic-backed absorbent pads, and remote-handling tools should be used to confine spills and contaminants and to minimize hand contact with potentially contaminated objects. All radioisotope containers and contaminated items should be properly labeled.
- General laboratory design and use should minimize contamination spread, which includes locating the area of highest radioisotope activity and usage in a low-traffic area (back of laboratory) and constructing hoods and work surfaces of stainless steel or some other nonporous, easily cleaned material. Floors, walls, and other surfaces should be covered with a nonporous, nonpermeable material that is easily cleaned.
- Contamination surveys of all use, storage, and waste collection areas in the laboratory should be undertaken frequently to assure that contamination is detected early and spread is minimized.

(iii) Control of absorption uptake

- Cover skin surface exposed to contaminants with a nonpermeable barrier. Hands and forearms are the most likely area for skin absorption. A long latex glove pulled over the laboratory coat sleeve usually provides adequate protection for hands and forearms.
- Double gloving, with frequent surveys and changes of the outer glove, proves very effective in preventing penetration of the contaminant to

the skin surface. A latex surgery glove for the inner glove and a lightweight, disposable poly or latex glove for the outer glove are satisfactory for most contaminants.

- Restrict the use of volatile contaminants readily absorbed through the skin, such as ^{3}H-labeled water and ^{125}I-labeled sodium iodide, to a properly ventilated hood.
- If volatile radioactive materials which can be absorbed through the skin are spilled in areas where there is inadequate ventilation, cleanup personnel will require nonpermeable protective clothing, gloves, and appropriate respiratory protection. A nonpermeable, supplied-air decontamination suit should be available if use procedures could result in the spillage in the general laboratory of significant activities (greater than the low levels in Table 9) of these radioisotopes.

(iv) Control of uptake via injection or puncture

- Use remote syringes for withdrawal and injection of radioactive materials, and use remote-holding devices for radioactive material stock containers.
- Personnel should be properly trained in the techniques for handling syringes, needles, and other sharps. Practice sessions before the actual use of radioactive materials can minimize the likelihood of error.
- Reduce the potential for puncture wounds by covering needles and other sharps with a protective cap or shield when not in use. Used sharps should be disposed without recapping in a proper sharps disposal box.
- Dispose of used needles and other sharps in a specially designated and labeled protective container. Instruct waste-handling personnel not to compress waste bags and containers by using their hands.

The exposure potential or degree of radiotoxicity of a radioisotope which is internally deposited in the body is dependent on a number of factors including: the type and energy of radiation emitted, the half-life of the radioisotope, its chemical and physical form, the distribution and localization (critical organ) of the radioisotope within the body, the mass of the critical organ, and the rate of biological removal from the body (biological half-life). These factors are used to derive the activity of a radioisotope which if sustained in the critical organ would result in a radiation dose to the organ equal to the maximum permissible dose. This activity is known as the maximum permissible body burden for the particular radioisotope. Based on these values, it is possible to derive concentration values in air and water for the working environment which are known as maximum permissible air and water concentrations (MPC_a and MPC_w). The MPC_a (if breathed) or the MPC_w (if ingested) for 40 h/week and 50 weeks/year would result in the body accumulating and sustaining the maximum permissible body burden. Table 4 lists MPC_a and MPC_w values for several radioisotopes for both occupational and general public exposure groups. A more complete listing of MPC values is available in the NRC regulations (10 CFR 20, Appendix B). If more than one radioisotope is released in a particular environment, it is necessary to determine that the sum of the fractional percentages of the MPC for each radioisotope is less than or equal to unity ($[C_1/MPC_1] + [C_2/MPC_2] \ldots + [C_n/MPC_n] \le 1$) (4).

Table 9 is a listing of several radioisotopes grouped according to relative internal radiation hazard (radiotoxicity). The table includes low, medium, and high activity ranges for each radiotoxicity grouping. The information available from such listings is helpful in determining the precautions and controls necessary to assure personnel protection in the use of various amounts of different radioisotopes. In general, low-level activities can be safely used in a well-designed chemistry laboratory with a chemical hood. Medium-level activities should be used in a well-designed radioisotope laboratory with a properly operating and filtered radioisotope hood. High-level activities will require further control measures to assure personnel protection, such as filtered glove boxes and isolated hot laboratories equipped with remote-handling devices and special filtered ventilation. When the radioisotope used is potentially volatile, the lower limit of the range categories should be used to determine the precautions and control measures necessary for safe use (8).

Laboratories where radioisotopes are to be used should be designed with consideration for ease of contamination control and decontamination. The laboratory layout should be such that areas for handling radioisotopes of high concentration or activity are in remote, low-traffic areas away from doors. Work surfaces and hoods should be constructed of a nonporous, nonpermeable material such as polished stainless steel. Floor areas should be covered with a continuous sheet covering which is nonporous and sealed with a protective covering. The walls and ceiling should be covered with a smooth, nonporous covering or a strippable surface covering. Light fixtures should be recessed with a flush cover plate. Nonporous, nonpermeable trays lined with poly-backed absorbent pads should be used in the hood and at work stations to restrict and confine radioisotope use areas.

PERSONNEL RADIATION MONITORING

NCR regulations require that occupationally exposed personnel be monitored for external or internal radiation dose, or both, when there is a potential for them to receive a dose greater than $\frac{1}{4}$ of the quarterly dose limits specified in Table 6. A permanent record of radiation dose received by personnel is required by

TABLE 9. Hazard from absorption into the body[a]

Group 1. Very High Hazard.

10μc	100μc	1mc	10mc	100mc	1c	10c
Low Level	Medium Level		High Level			

*Pb²¹⁰, Po²¹⁰, *Ra²²⁶, *RA²²⁸, Ac²²⁷, Th²²⁸, Th²³⁰, Np²³⁷, Pu²³⁸, Pu²³⁹, Pu²⁴⁰, Pu²⁴¹, Pu²⁴², *Am²⁴¹, Cm²¹²

Group 2. High Hazard.

10μc	100μc	1mc	10mc	100mc	1c	10c
Low Level		Medium Level		High Level		

*Na²², Ca⁴⁵, *Sc⁴⁶, *Co⁶⁰, Sr⁹⁰, *Ru¹⁰⁶, I¹²⁹, *I¹³¹, *Cs¹³⁷, *Ce¹⁴⁴, *Eu¹⁵⁴, *Ta¹⁸², Bi²¹⁰, At²¹¹, Ra²²⁴, U²³³

Group 3. Medium Hazard.

10μc	100μc	1mc	10mc	100mc	1c	10c
Low Level			Medium Level		High Level	

C¹⁴, *Na²⁴, Si³¹, P³², S³⁵, Cl³⁶, *K⁴², Sc⁴⁷, *V⁴⁸, *Cr⁵¹, *Mn⁵⁴, *Mn⁵⁶, Fe⁵⁵, *Fe⁵⁹, *Cu⁶⁴, *Zn⁶⁵, *Ga⁷², *As⁷⁶, *Rb⁸⁶, Sr⁸⁹, Y⁹⁰, Y⁹¹, *Zr⁹⁵, *Nb⁹⁵, *Mo⁹⁹, *Ru¹⁰³, *Rh¹⁰⁵, Pd¹⁰³, Ag¹⁰⁵, Ag¹¹¹, *Cd¹⁰⁹, *Sn¹¹³, *Te¹²⁷, *Te¹²⁹M, *Ba¹⁴⁰, *La¹⁴⁰, Pr¹⁴³, Pm¹⁴⁷, Sm¹⁵¹, *Ho¹⁶⁶, *Tm¹⁷⁰, *Lu¹⁷⁷, *Re¹⁸³, *Ir¹⁹⁰, Ir¹⁹², *Pt¹⁹¹, *Pt¹⁹³, *Au¹⁹⁶, *Au¹⁹⁸, *Au¹⁹⁹, *Tl²⁰⁰, *Tl²⁰¹, Tl²⁰², Tl²⁰⁴, *Pb²⁰³, Rn²²⁰, *Rn²²², U²³⁵

Group 4. Low Hazard.

10μc	100μc	1mc	10mc	100mc	1c	10c
Low Level				Medium Level		High Level

H³, *Be⁷, C¹⁴, F¹⁸, Ni⁵⁹, Zn⁶⁹, Ge⁷¹, U²³⁸, Natural Thorium, Natural Uranium, Noble Gases.

[a]From reference 8. μc, μCi; mc, mCi; c, Ci. *, Emits gamma radiation in significant amounts.

the regulations. Personnel dosimetry reports are a legal record and document whether the radiation dose received by an individual has been maintained within protection limits. The routine dosimetry reports (usually monthly) are helpful in assessing the adequacy of personnel protection, and in the event of an elevated dose they will serve to alert personnel of a need for an improvement or change in procedures.

External dose monitoring

There are three commonly used devices for monitoring external radiation dose: (i) film badge dosimeters, (ii) thermoluminescent dosimeters (TLDs), and (iii) pocket ionization dosimeters (7). All of these dosimeters are available from commercial suppliers.

(i) Film badge dosimeters. The film badge is the most widely used method of measuring external radiation dose. It has the advantages of being quite inexpensive and readily available, is capable of integrating doses over a wide range ($\approx$10 mR to 500 R), and is able to measure different radiation energies and types with reasonable accuracy. The developed film also provides a permanent record which can be reread at a later date to recheck a reported radiation dose. Some of the disadvantages of the dosimeter are that it has an energy-dependent response (especially for low-energy radiations), response varies with angle relative to incident radiation, it is sensitive to heat and humidity, and there is latent image fading of the film which limits the wear period (1 month is recommended).

(ii) TLDs. TLDs are also widely used for external dose monitoring. Often they may be used in combination with film badges, e.g., a film badge for whole-body dose monitoring and a TLD ring for hand dose monitoring. The TLD has the advantages of a more energy-independent response, a wide dose integration range, and no appreciable latent image fading or angular response variation, and it is not as sensitive to the effects of heat and humidity. Disadvantages include the facts that TLDs cost more; the dose information is destroyed at readout, which eliminates the possibility of a recheck of the dose at a later date; and the TLD crystals or powder can give erroneous readings if damaged or if they become dirty through improper handling.

(iii) Pocket ionization dosimeters. Pocket ionization dosimeters are not commonly used. Their primary use is for monitoring X or gamma radiation in situations where immediate dose readout capability is needed because of a potential for high radiation doses to personnel. Most pocket ionization dosimeters are self-reading (calibrated scale or digital readout) so that the dose can be assessed at any phase of an operation involving potentially high doses. Such dosimeters are usually worn in conjunction with a film badge or TLD, which is used to provide the permanent record of radiation dose to the individual. Some of the disadvantages are that they cost more, are insensitive to low-energy radiations, can lose dose due to charge leakage, and are easily damaged.

General guidelines for the use of these dosimeters are as follows:

- Store the dosimeters in a cool, dry area away from radiation storage and use areas.
- Do not wear the dosimeter home, but return it to the storage area after each use.
- Wear dosimeters at waist or chest height for whole-body monitoring and on the finger or wrist for hand and forearm monitoring. For hand

dose, be sure the dosimeter is turned toward the area of highest potential dose. Wear film badges with label (open window) facing away from the body.

- No dosimeter is necessary for personnel who use low-energy beta emitters (^{3}H, ^{14}C, ^{35}S).
- Report lost or damaged dosimeters immediately and obtain a replacement.

Internal dose monitoring

The method used to monitor personnel dose for internally deposited radioisotopes depends on the way in which the radioisotope is distributed in and eliminated from the body. Because many radioisotopes, e.g., ^{3}H and ^{32}P, are eliminated primarily by urinary excretion, urine analysis is a reliable method to monitor the amount present in the body. The activity of some other radioisotopes in the body (primarily gamma-emitting radioisotopes) can be determined by a direct count of a particular organ or the whole body. In the case of radioiodines, the thyroid gland can be counted with an NaI scintillation detector and analyzer to measure the amount of a particular radioiodine (^{125}I, ^{131}I) present. Whole-body counting is a less prevalent method because of the cost to purchase and maintain such a system. Some whole-body counters are capable of measuring radioisotope content in a number of different body locations or in the body as a whole and can be used to determine the amount of several different radioisotopes present in the body. Once the activity of a radioisotope in an organ or in the whole body has been determined, it is necessary to apply dose conversion factors to determine the radiation dose received by the individual. Such factors are available from various ICRP (International Commission on Radiation Protection) publications (no. 2, 10, and 10A), and from Medical Internal Radiation Dose Reports from the Society of Nuclear Medicine. Commercial services are available for assessment of internal radiation dose. Also, many institutions have radiation protection programs and professional staff who can assist in internal dose assessment (7).

RADIATION MONITORING INSTRUMENTATION AND CONTAMINATION SURVEYS

Instrumentation

Radiation monitoring instrumentation is generally divided into two categories, portable survey instruments and fixed analytical instruments.

Portable instruments. Common types of portable radiation survey instruments include Geiger-Mueller (G-M), gamma scintillation, and ionization survey instruments.

(i) G-M survey instruments. The G-M is by far the most commonly used and versatile of the portable survey instruments. It is capable of detecting medium- and high-energy beta radiation (low-energy beta radiation from ^{3}H will not penetrate the detector window) and gamma radiation and is available with a number of different types of detection probes. The preferred probe for laboratory use is a thin-window (≈ 1 mg/cm^2) pancake G-M probe. The pancake probe is excellent for contamination surveys of hands, feet, clothing, and work surfaces because of the large surface area of the window. The G-M can also be used to monitor direct radiation levels; however, because it has an energy-dependent response, it must be calibrated to the energy of the radiation being monitored to accurately measure exposure rates. It is relatively inexpensive ($300 to $500), quite rugged, easily maintained, and very sensitive to most beta radiations.

(ii) Gamma scintillation instruments. When a greater sensitivity (higher efficiency) is needed for X or gamma radiations, a gamma scintillation detector should be used. These detectors are available in different sizes and thicknesses (thin crystal for low-energy gamma, thick for high energy). Their efficiency for detecting gamma radiations is about 10-fold higher than that of the G-M. In many cases the same instrument meter that is used for the G-M probe can be used for the gamma scintillation probe. The cost of the portable scintillation probe may vary from $200 to $500, depending on crystal dimensions. Users of the probe must be careful not to crack the crystal or break the protective covering.

(iii) Ionization survey instruments. Portable ionization survey instruments (''cutie-pies'') are used primarily for the measurement of external radiation exposure levels. They are not an appropriate instrument for contamination surveys because of their relatively low sensitivity. The instrument response is energy independent over a wide range of gamma energies (~ 0.03 to 2 MeV). The exposure rate readout is usually in milliroentgens per hour and is displayed on a meter or digital display. Some instruments have an exposure integrate mode (in milliroentgens) which can measure the accumulation of radiation during a known exposure time. The instrument will require periodic calibration (once every 6 months or a year) to assure that it is adjusted to accurately indicate exposure rates when checked against known exposure rates. Commercial calibration services are available.

Fixed analytical instruments. The most commonly used analytical instruments for counting ionizing radiation in the radioisotope laboratory are the liquid scintillation counter and the automated gamma scintillation counter. These instruments are used primarily for quantitative analysis of experimental samples but can also be used for routine contamination monitoring of the laboratory.

(i) Liquid scintillation counter. The liquid scintillation counter is the preferred method for counting beta-emitting radioisotopes, especially low-energy beta emitters such as ^{3}H. A critical aspect of liquid scintillation counting is the preparation of the scintillation counting solution which is placed into the counting vials with the sample to be analyzed. The correct counting solution (organic solvent and scintillation fluor) must be chosen for the particular type of sample to be analyzed. Various types of counting solution mixtures for specific counting needs are available from several commercial suppliers. Because the solutions contain flammable and toxic organic solvents, they should be stored and transferred to counting vials in a well-ventilated hood. Used counting vials containing these solutions (especially polyethylene vials, which are permeable to organic solvents) should also be stored in a properly ventilated area.

(ii) **Auto-gamma scintillation counter.** The automated gamma scintillation counter is often used in radioisotope laboratories where gamma emitters such as ^{125}I and ^{51}Cr are analyzed, because the system has a relatively high counting efficiency for gamma radiations. The detector is usually a well-type scintillation crystal (NaI), and the sample to be counted is lowered or raised into the center of the detector well for maximum counting efficiency. The advantages of the auto-gamma counter are that it involves much less sample preparation than the liquid scintillation counter and has a higher efficiency for gamma radiation. The cost of such a system is usually well over $15,000.

Contamination surveys

Contamination smear surveys should be performed in all radioisotope storage and use areas to assure that removable activity levels do not present an appreciable radiation exposure potential. The recommended limits for removable contamination in the laboratory use areas are 250 dpm/100 cm^2 for intermediate- and high-energy beta and for gamma emitters, and 500 dpm/100 cm^2 for low-energy beta emitters. With respect to personnel skin contamination, the limit is 100 dpm/100 cm^2. The usual method for determining removable contamination is a wipe test (filter paper smear) of a defined area of 100 cm^2 (approximately equal to the area covered by a 1-in. [ca. 2.5-cm]-diameter filter paper moved 20 in. [ca. 50 cm] along a surface). Care should be taken to prevent hand contamination and cross-contamination of the individual wipes. The wipes should be numbered or coded before the survey, and the person performing the survey should wear disposable gloves. Wipes should be stored separate from each other, and a forceps should be used to place them into individual counting vials or tubes. Survey results should be recorded on a standard form which identifies laboratory areas surveyed (numbered areas on sketch of lab) and lists the results (dpm/100 cm^2) for each area surveyed. These records are to be kept for inspection

purposes. To determine the activity on a wipe in dpm per 100 cm^2 it is necessary to determine the fractional efficiency (E) of the counting system for the particular radioisotope. If the wipe is to be counted on a liquid scintillation, auto-gamma, or other system, the efficiency of the system is determined by placing a known amount (standard) of the radioisotope on a wipe and counting it under the same conditions as the contamination wipes. The background count rate is obtained by counting an unused wipe under the same conditions (net counts per minute [cpm] = gross cpm − background cpm). The fractional efficiency (E) = net cpm of standard/dpm of standard. The dpm for the standard = activity of the standard (μCi) × 2.22 × 10^6 dpm/μCi. From the fractional efficiency, the amount of activity on a wipe can be calculated as follows: dpm/100 cm^2 = (net cpm/100 cm^2)/E.

In the case of contamination surveillance of radioisotopes in the laboratory air or in effluent air streams released to the environment, an appropriate particulate air filter or gaseous trap (activated charcoal for radioiodine) is used for sampling. A known volume of air (V) (cm^3) is drawn through the filter or trap, and the filter or trap is counted on an appropriate system under the same conditions indicated for contamination wipes. The concentrations in air can then be calculated by the following formula:

Concentration (μCi/cm^3)
$$= \frac{\text{Net cpm}}{E} \times \frac{1}{V} \times \frac{1}{2.22 \times 10^6 \text{ dpm/}\mu\text{Ci}}$$

where E is the fractional efficiency of the counting system.

This calculation assumes 100% collection efficiency for the filter or trap used. If the true collection efficiency varies appreciably from 100%, an appropriate correction factor will need to be applied to correct for the fact that not all of the contaminant was collected (7).

STATISTICAL CONSIDERATIONS OF SAMPLE COUNTING

Radioactive decay of any radioisotope is a random process. Therefore, the count of a sample which contains a radioisotope requires, from a statistical standpoint, that the results be expressed in terms of the degree of confidence of the count. The statistical distribution of successive counts of a sample closely approximates a normal error curve (normal distribution). This curve can be drawn from just two parameters, the average value of all measurements, $\bar{x}$, and the standard deviation of the measurements, SD. For the normal distribution, confidence limits can be derived from the standard deviation: 68% of all measurments will fall within ±1 standard deviation from the average, 95% will fall within ±2 standard deviations, and 99% will fall within ±2.58 standard deviations. For a single count or measurement, N, of a sample, the standard deviation and confidence limits can be determined (assuming a normal distribution). The standard deviation, $SD = \sqrt{N}$, and the confidence limits are expressed as follows: 68% confidence interval $= N \pm \sqrt{N}$; 95% confidence interval $= N \pm 1.96\sqrt{N}$; 99% confidence interval $= N \pm 2.58\sqrt{N}$.

Table 10 lists the confidence intervals (68 and 95%) expressed as counts or fractions of measured value, N.

When a measurement includes a significant background count (low count rate samples), the background must be subtracted to obtain the net count of the sample being counted (gross count − background count = net sample count). If the counting times for the gross count and the background count are equal, then the best estimate of the standard deviation for the sample count (net count) is given by the square root of the sum of the two measurements:

$$SD_{net\ count} = \sqrt{gross\ count + background\ count}$$

When the counting times for sample and background are different, the net count rate is obtained as follows: $R = S/t_s - B/t_B$, where R = net count rate; S = gross count; t_s = time of gross count; B = background; and t_B = time of background count. The standard deviation is then estimated by:

$$SD_{net\ count} = \sqrt{\left(\frac{S}{t_s^2} + \frac{B}{t_B^2}\right)}$$

When a finite time is available for counting both the sample and the background, the most efficient partitioning of the time between sample counting time and background counting time is obtained as follows:

$$t_s/t_B = \sqrt{R_s/R_B}$$

where R_s = sample count rate and R_B = background count rate.

TABLE 10. Limits of confidence intervals (CI)[a]

| Accumulated counts (N) | Limits of CI expressed as counts and fraction of measured value | | | |
| | 68% CI | | 95% CI | |
	$\sqrt{N}$	$\sqrt{N}/N$	$1.96\sqrt{N}$	$196\sqrt{N}/N$
20	4.47	0.22	8.75	0.432
60	7.74	0.13	15.2	0.255
100	10.0	0.10	19.6	0.196
600	24.5	0.041	47.9	0.080
1,000	31.6	0.032	61.9	0.062
6,000	77.4	0.013	152.0	0.026
10,000	100.0	0.010	196.0	0.020

[a]From reference 6 with permission of the publisher.

In many cases it is desirable to calculate the minimum significant difference between gross sample and background counts to determine the minimum count above a given background that can be considered significant. The minimum detectable net count is obtained as follows:

$$(S - B)_{min} = 1/2(K_\alpha^2 + \sqrt{K_\alpha^4 + 8K_\alpha^2 B})$$

For a 95% confidence level ($K_\alpha = 1.645$), the net count, which is significantly different from background, is given by the equation (6):

$$(S - B)_{min} = 1/2[(1.645)^2 + \sqrt{(1.654)^4 + 8(1.645)^2 B}]$$

USER GUIDELINES

General guidelines

The following is a listing of some generally accepted precautions and guidelines for the use and storage of radioisotope materials.

• Label all use and storage areas with required caution signs: "Caution Radioactive Materials" sign on doors and storage areas of all restricted use areas; "Caution Radiation Area" signs in areas where the radiation exposure level to personnel is between 5 and 100 mR/h. Above 100 mR/h, post "Caution High Radiation Area" sign.

• Inspect and survey radioisotope shipments upon receipt. Wear disposable gloves when handling the shipping container, and open the shipment in a hood if the radioisotope is potentially volatile. Properly dispose of shipping carton by removing labels and blanking out radioactive-material markings (if contaminated, dispose in radioactive-waste container).

• Wear appropriate protective clothing and equipment, i.e., a lab coat, disposable gloves, safety glasses or goggles, shoe covers, etc. Do not wear lab coat, disposable gloves, etc., outside the laboratory. Survey hands, feet, and clothing before leaving the use area. Wash hands after each use of the material and before leaving the laboratory.

• Obtain and wear appropriate personnel radiation dosimeters (body and hand dosimeters when there is the potential of exposure to penetrating beta and gamma emitters).

• Obtain and use appropriate shielding: lucite for intermediate- to high-energy beta emitters; lead for gamma emitters (thickness required depends on the energy of radiation).

• Use remote-handling tools (forceps, tongs, clamps, etc.) to prevent high hand dose and possible hand contamination. Use rubber tubing on forceps, tongs, and other tools to provide a better grip on containers.

• Do not eat, smoke, drink, chew gum, or apply cosmetics in the laboratory. Do not store food or beverages in the laboratory.

• Do not allow mouth pipetting. Provide appropriate pipetting aids.

• Use trays lined with absorbent plastic-backed pads to confine radioisotope use areas and to contain contamination in the event of spills.

• Obtain and use appropriate radiation monitoring instrumentation. Use a portable instrument to survey hands, clothing, and shoes before leaving the work station. Perform routine (recommended, once a week) contamination wipe surveys of all use, storage, and laboratory access areas.

• Label all radioisotope containers with a "Caution Radioactive Material" tag or tape indicating radioisotope activity and date of assay.

• Restrict all use of potentially volatile or airborne radioisotopes to a properly ventilated hood or glove box.

• Conduct dry runs involving no radioactive materials for new protocols and procedures which involve significant exposure potential.

• Store radioactive wastes in appropriately labeled and sealed waste containers.

• Maintain appropriate records on all radioisotope receipts, personnel dosimetry reports, routine surveys, and waste disposal.

Special procedures for ^{32}P

• Use lucite stock container shields (lead may be needed around lucite) and lucite barriers for body and head (lens of eyes) shielding. Lucite should be ⅜ in. (ca. 9.5 mm) thick or greater.

• Direct handling of stock vial or other high-concentration sources (very high dose rate, several rems per hour at contact with vial) should not be permitted. Use remote-handling tools and remote pipettes or syringes to transfer concentrated material. Conduct dry runs for all new personnel and procedures.

• Wear double gloves and frequently monitor and change outer glove. Wash and survey hands after use of materials.

• Wear a long-sleeved lab coat with gloves pulled over sleeves. Wear safety goggles.

• Restrict stock solution transfer procedures to a well-ventilated hood.

• Wear body badge and ring dosimeter.

• Dispose of highly contaminated items such as pipette tips, stock vials, etc., in a shielded waste storage receptacle that is properly labeled. High-activity wastes should be held in a remote shielded area for decay.

• Use a G-M survey instrument to survey hands, clothing, feet, and use area during and after each use.

• Use special care when centrifuging samples. Place sample in double containment, preferably a sealed, unbreakable outer container.

Special procedures for volatile radioiodine

• Restrict all usage to a well-ventilated hood. If there is the potential for significant radioiodine re-

lease to the environment, the hood should have an activated charcoal filtration system to remove radioiodine.

• Conduct dry runs for all new personnel and procedures to reduce potential release and exposure.

• Monitor concentration in air in breathing zone and effluent discharge (use activated-charcoal sampling trap).

• Wear body badge and ring dosimeter.

• Wear long-sleeved lab coat and double gloves (inner latex surgical glove and outer disposable glove). Monitor and dispose of outer glove frequently. Pull surgical glove over lab coat sleeve.

• Direct handling of stock vial or other high-concentration sources should not be permitted. Use remote handling tools, pipettes, and syringes.

• Use sheet-lead and lead-glass work station shields for body and eye shielding.

• Use techniques to minimize volatile radioiodine release: (i) maintain high pH; (ii) do not open stock vial to air, but add and withdraw material through a septum with a syringe; (iii) if it is necessary to open the stock vial, first purge the air space above the solution and force the air through a charcoal trap (syringe barrel filled with activated charcoal); (iv) use a septum on the separatory column and use a syringe to add radioiodine solution and other solutions to the column; (v) use a solution of 0.1 M NaI–0.1 M NaOH–0.1 M Na$_2$S$_2$O$_3$ to stabilize any radioiodine-contaminated items, waste solutions, columns, pipettes, or needles; and (vi) place all contaminated items in a properly sealed disposal container (double bagged and taped, or in a sealed plastic jar).

• Do not remove any items or wastes from the hood until properly decontaminated or treated with Na$_2$S$_2$O$_3$ solution and properly packaged and sealed.

• Monitor hands, clothing, and use area during and after use, using a thin NaI gamma scintillation probe for ^{125}I or a pancake G-M probe for ^{131}I. If contamination is detected, immediately decontaminate the area with Na$_2$S$_2$O$_3$ solution.

• Perform a thyroid count of personnel involved in radioiodine handling within 1 week of use. Use a properly calibrated NaI scintillation probe and analyzer for thyroid counts.

• In the case of an accident in which radioiodine is released into the general laboratory air, thyroid uptake by personnel can be blocked (greater than 90%) by administering potassium iodide (KI). However, KI should be administered only under a physician's surveillance.

Special procedures in the event of accident or spill

• In the event of an radiation accident involving injury, restrict access to area, provide necessary first aid or life-saving procedures, and notify appropriate emergency personnel.

• If the person is contaminated and must be transported out of the area for medical attention: (i) remove contamination if possible (remove contaminated clothing, lightly wash skin surface); (ii) if decontamination is not possible, wrap person in blanket or sheet to minimize contamination spread; (iii) instruct medical personnel concerning need for contamination control and radiation surveillance at medical facility.

• If the injury does not require immediate medical care, have the person remove contaminated clothing and wash contaminated skin surfaces with water first and then mild soap and water. Resurvey contaminated skin surfaces after each washing until contamination is reduced to less than 100 dpm/100 cm^2. More extensive skin and area decontamination procedures are outlined in publications such as the National Bureau of Standards *Handbook 92* (8) and the *Radiological Health Handbook* (2).

• Notify appropriate authorities (radiation protection staff, building security, etc.).

• Assess the radiation dose, both external and internal (urine assay, thyroid count), received by personnel involved in the accident, and maintain appropriate records.

• In the event of a spill of radioactive materials, use the following procedures. (i) Restrict access to the spill area and instruct those who may have been contaminated not to leave the lab but to survey and remove contaminated clothing and to wash contaminated skin surfaces. (ii) Put on appropriate protective clothing (lab coat, disposable gloves, and shoe covers; respirator or self-contained breathing apparatus in the event of airborne contaminants) and confine the spill with absorbent material or pads. (iii) Notify the appropriate authorities (radiation protection staff, etc.) and request assistance. (iv) Monitor personnel for contamination and decontaminate them before they leave the laboratory. (v) Proceed with spill cleanup under appropriate supervision and using necessary protective equipment. (vi) Begin decontamination efforts from the perimeter (low activity) of the spill and proceed toward the center of the spill (higher activity). (vii) Use water, soap, cleansers, and absorbent cleaning pads to decontaminate the area. (viii) Resurvey the area and continue decontamination until levels are reduced to acceptable limits (250 dpm/100 cm^2 for most beta and gamma emitters). (ix) Place decontamination materials, gloves, shoe covers, etc., in an appropriate waste container. (x) Survey all personnel before they leave the laboratory. (xi) Conduct necessary external and internal dose assessments and maintain records.

Special procedures for radioisotope storage: considerations

All radiochemicals undergo decomposition with time due to a number of factors (ionization caused by radiation energy absorbed in radiochemical solutions; temperature and environmental instabilities). Such decomposition affects the useful life of the radiochemical. The following are some guidelines for storage which will aid in controlling decomposition.

• Store radiochemicals at the lowest possible molar specific activity.

• Disperse solids as much as possible and store in a dry, inert atmosphere.

• Store radiochemicals in a purified aromatic solvent when possible. If benzene is used, do not freeze, but store just above freezing.

• If radiochemicals must be stored in an aqueous solution, add 2 to 10% ethanol to scavenge free radicals.

• To prevent light (UV) decomposition, store materials in a dark area (dark-colored vial; sealed container or box).

• To prevent microbial growth and decomposition, use special care in withdrawal of stock (septum transfers with sterile needle if possible) and add a bacteriostat when necessary.

• Optimum storage temperature for aqueous solutions depends on the radiochemical; generally, ^{14}C, ^{32}P, and ^{35}S materials should be stored at as low a temperature as possible (^{14}C-amino acids of high molar specific activity should be stored at 2°C with ethanol). ^{3}H-labeled materials should be stored at 2 or −140°C, or lower (freezing above −140°C may cause molecular clustering which accelerates decomposition).

• Typical decomposition rates under optimum storage conditions are: ^{14}C, ≃1 to 3% per year; ^{3}H, ≃1 to 3% per year; ^{35}S, ≃2 to 3% per year; ^{32}P, ≃1 to 2% per week; ^{125}I, ≃5% per month.

• ^{3}H-radiochemicals and radioiodines should be stored in gas-tight and nonpermeable containers to prevent leakage and diffusion from the storage container. Glass vials and covers with metal foil seals help to minimize release. Double containment is recommended.

• Routinely survey storage area for possible contamination, especially ^{3}H storage refrigerators and freezers.

• Reanalyze compounds immediately before use to be certain the radiochemical is still usable.

Much of the information listed above is available from Amersham Corp. (1). Information on optimum storage conditions for specific radiochemicals may be obtained from the commercial supplier.

OTHER CONSIDERATIONS

Regulatory requirements

The following is a listing of regulations which may apply to users of radioactive materials.

NRC Regulations (10 CFR)
1. 10 CFR 19.12, Instructions to workers
2. 10 CFR 20.101, Radiation dose limits

3. 10 CFR 20.103, Radioisotope concentration (restricted area)
4. 10 CFR 20.106, Radioisotope concentration (unrestricted area)
5. 10 CFR 20.201, Survey requirements
6. 10 CFR 20.202, Personnel monitoring requirements
7. 10 CFR 20.203, Posting and labeling requirements
8. 10 CFR 20.205, Shipment receipt of surveys
9. 10 CFR 20.303, Waste disposal (sanitary sewer)
10. 10 CFR 20.305, Waste disposal (incineration)
11. 10 CFR 20.306, Waste disposal (^{3}H, ^{14}C wastes)
12. 10 CFR 20.401, Records requirements
13. 10 CFR 20, Appendix B, MPCs for air and water
14. 10 CFR 20, Appendix C, Exempt quantities
15. 10 CFR 30, By-product material licensing
16. 10 CFR 71, Packaging and transport

NRC Regulatory (Reg.) Guides
1. Reg. Guide 8.10, Maintaining occupational exposure as low as reasonably achievable (ALARA)
2. Reg. Guide 8.13, Prenatal radiation exposure instructions
3. Reg. Guide 8.15, Respiratory protection programs
4. Reg. Guide 8.18, ALARA at medical institutions
5. Reg. Guide 8.20, Bioassay for ^{125}I and ^{131}I
6. Reg. Guide 8.23, Radiation safety surveys at medical institutions

Department of Transportation (DOT) Regulations (Title 49 CFR)
1. 49 CFR 172.30, Marking requirements
2. 49 CFR 172.403, Labeling requirements
3. 49 CFR 172.507, Vehicle placarding
4. 49 CFR 173.403, Radioactive material definitions
5. 49 CFR 173.421, Limited quantities
6. 49 CFR 173.423, Table of activity limits
7. 49 CFR 173.425, Low specific activity
8. 49 CFR 173.433, Requirements for determination of A_1, and A_2
9. 49 CFR 173.443, Contamination control
10. 49 CFR 173.448, General transportation requirements

These and other DOT regulations pertain to packaging, labeling, and transport requirements for radioactive material shipments and radioactive waste shipments.

Radioactive waste considerations

The following is a list of general guidelines and precautions in the handling, storage, and disposal of low-level radioactive wastes.
• Short-half-life waste (^{32}P, ^{131}I) should be sepa-rated from longer half-life waste (^{35}S, ^{125}I). Short-half-life waste may be held for decay for a period sufficient to decrease the radioactive content to essentially background level. Such waste should be labeled with the activity of each radioisotope and dated when placed in storage for decay. After a decay period (calculated based on activity and half-life of radioisotope) sufficient to decrease the activity to essentially background, the waste may be released and handled as nonradioactive waste. However, the waste should be surveyed before release, using a radiation survey instrument that is sensitive to the type and energy of the radiation emitted by the radioisotope (G-M for intermediate- and high-energy beta, gamma scintillation for gamma). If the instrument indicates no activity above background, record the reading and maintain a record of the survey result and date of disposal. File this information with the initial record of the original content of the waste at the start of the storage period.

• Separate solid and liquid radioactive wastes in the laboratory. Short-half-life liquid wastes may be held for decay and then disposed of as nonradioactive liquids, after an appropriate survey to document sufficient decay.

• Separate ^{14}C and ^{3}H liquids, animal carcasses, and liquid scintillation vials from other wastes. Under 10 CFR 20.303 and 20.306, respectively, certain quantities of low-level ^{14}C and ^{3}H aqueous wastes may be discharged to the sanitary sewer, and ^{14}C- and ^{3}H-containing animal carcasses and scintillation vials may be classified as nonradioactive waste if the concentration is less than 0.05 μCi/g. Accurate records must be maintained to document that disposal is within designated limits and in accordance with the conditions of these sections. Small amounts of other radioisotopes contained in liquid aqueous wastes may also be disposed to the sanitary sewer provided the limits of 10 CFR 20.303 are not exceeded.

• Flammable organic solvent wastes should be collected in appropriate safety containers. These wastes should be stored and transferred in a properly ventilated area.

• Powdered and potentially volatile wastes should be properly sealed and packaged before being removed from the hood, to prevent release of airborne contaminants.

• Pathogenic and biohazardous agents combined with radioactive wastes should be inactivated and destroyed before the waste is discarded. Care must be taken to ensure that the treatment for biohazard deactiviation does not release airborne radioactive contaminants (chemical inactivation using sodium hypochlorite may release volatile radioiodine from waste; steam or heat sterilization may result in the release of gaseous ^{3}H, ^{14}C, or radioiodines).

• Hazardous chemicals contained in the waste must be properly handled. Neutralize strong acids and bases; do not combine acidic liquid waste with wastes containing cyanide (hydrogen cyanide will be generated); do not combine liquids containing biodegradable materials such as blood, tissue, etc., with acidic liquid waste (gaseous releases may result).

• Maintain accurate records on the content of each waste container. Record radioisotopes and activity of each radioisotope present in each container.

• Package and label all wastes in accordance with DOT (Title 49 CFR) regulations. Absolutely no liquids should be allowed in solid-waste containers; all liquids must be properly absorbed (twice the volume of absorbent necessary to absorb liquid). Noncompatible liquids must not be packaged in the same container. Use only DOT-approved shipping containers.

• Disposal of animal carcasses containing radioactive materials may pose special handling and packaging problems because of the biodegradation of the animal carcasses. If carcasses are to be shipped for disposal via shallow land burial, they must be packaged in accordance with DOT regulations and the requirements of the disposal facility. One accepted method is to package the animals in a DOT-approved double-wall drum (30-gallon steel within a 55-gallon steel drum). The 30-gallon drum is lined with a plastic bag. The animal carcasses must be surrounded by an approved lime-absorbent mixture within the 30-gallon drum. The space between the 55-gallon drum and the 30-gallon drum must be filled with an approved absorbent. A preferred alternative to this method would be incineration of the animal carcasses in a properly designed, NRC-licensed incinerator. This method eliminates the problem of biological degradation and greatly reduces the volume of the waste and thereby the cost for disposal at the burial facility. Under 10 CFR 20.306, animal carcasses which contain ^{14}C or ^{3}H in a concentration of less than 0.05 μCi/g are no longer required to be disposed of as radioactive waste but may be handled with the nonradioactive animal waste stream. In some instances it may be feasible to hold animal carcasses containing short-half-life radioisotopes for sufficient time to allow disposal as nonradioactive waste.

• Package wastes in a properly ventilated and controlled area which provides properly designed transfer facilities and protective clothing and equipment for personnel.

GLOSSARY

Alpha particle. (Symbol, α) Positively charged particle emitted by certain radioactive materials. It is composed of two neutrons and two protons bound together and hence is identical with the nucleus of a helium atom. It is the least penetrating of the three common types of radiation (alpha, beta, gamma) emitted by radioactive material. Alpha particles can be stopped by a sheet of paper. Alpha particles are not dangerous to plants, animals, or humans unless the alpha-emitting substance has entered the body.

Beta particle. (Electron, positron; symbol, β) An elementary particle emitted from a nucleus during radioactive decay, with a single electrical charge and a mass equal to 1/1,837 that of a proton. A negatively charged beta particle is identical to an electron. A positively charged beta particle is called a positron. Beta radiation may cause skin burns, and beta emitters are harmful if they enter the body. Most beta particles are easily stopped by a thin sheet of metal.

Contamination (radioactive). The presence of radioactive material on materials or places where it is undesirable.

Curie (Ci). Quantity of any radioactive material in which the number of disintegrations is 3.700 × 10^{10}/s. Divisions of this unit include millicurie (mCi; 3.7 × 10^{7} dps), microcurie (μCi; 3.7 × 10^{4} dps), and picocurie (pCi; 3.7 × 10^{-2} dps or 2.22 dpm).

Dose equivalent (DE). Quantity used in radiation protection, expressing all radiation on a common scale for calculating the effective absorbed dose. The unit of dose equivalent is the rem, which is numerically equal to the absorbed dose in rads multiplied by certain modifying factors such as the quality factor and the distribution factor.

Gamma ray. (Symbol, γ) High-energy, short-wavelength electromagnetic radiation. Gamma radiation frequently accompanies alpha and beta emissions and always accompanies fission. Gamma rays are very penetrating and are best stopped or shielded by dense materials such as lead or depleted uranium. Gamma rays are essentially similar to X rays but are usually more energetic and are nuclear in origin.

Half-life, biological (t_b). Time required for the body to eliminate one-half of an administered dose of any substance by the regular processes of elimination. This time is approximately the same for both stable and radioactive isotopes of a particular element.

Half-life, effective (t_{eff}). Time required for a radioactive nuclide in a system to be diminished 50% as a result of the combined action of radioactive decay and biological elimination: ($t_{eff} = t_b \times t_r/t_b + t_r$).

Half-life, physical (radiological, T_r). Time required for a radioactive substance to lose 50% of of its activity by decay. Each radionuclide has a unique half-life.

Half-value layer. The thickness of any specified material necessary to reduce the intensity of an X-ray or gamma-ray beam to one-half its original value.

Ionization. Process of adding one or more electrons to, or removing one or more electrons from, atoms or molecules, thereby creating ions. High temperatures, electrical discharges, or nuclear radiations can cause ionization.

Ionizing radiation. Any electromagnetic or particulate radiation capable of producing ions, directly or indirectly, by interaction with matter.

Rad. The unit of absorbed radiation dose equal to 100 ergs/g or 0.01 J/kg of absorbing material.

Radioactive decay (disintegration). The spontaneous transformation of one nuclide into a different nuclide or into a different energy state of the same nuclide. The process results in a decrease, with time, of the number of the original radioactive atoms in a sample. It involves the emission from the nucleus of alpha particles, beta particles (or electrons), or gamma rays; the nuclear capture or ejection of orbital electrons; or fission. Also called radioactive disintegration.

Radioactivity. The spontaneous decay or disintegration of an unstable atomic nucleus, usually accompanied by the emission of ionizing radiation. (Often shortened to "activity.")

Radioisotope (radioactive material). A radioactive isotope. An unstable isotope of an element that decays or disintegrates spontaneously, emitting radiation. More than 1,300 natural and artificial radioisotopes have been identified.

Radiotoxicity. Term referring to the potential of an isotope to cause damage to living tissue by absorption of energy from the disintegration of the radioactive material introduced into the body.

Rem. The special unit of dose equivalence (rem = rads × QF [quality factor]). One millirem (mrem) is equal to 1/1,000 rem.

Restricted area (controlled area). Any area to which access is controlled for the purpose of protection of individuals from exposure to sources of ionizing radiation.

Roentgen (R). A unit of exposure to ionizing radiation, that amount of gamma or X rays required to produce ions carrying one electrostatic unit of electrical charge (either positive or negative) in 1 cm^3 of dry air under standard conditions. Named after Wilhelm Roentgen, the German scientist who discovered X rays in 1895. One milliroentgen (mR) is equal to 1/1,000 R.

Unrestricted area (noncontrolled area). Any area to which access is not controlled for the purpose of protection of individuals from exposure to sources of ionizing radiation.

X rays. Penetrating electromagnetic radiations having wave lengths shorter than those of visible light. X rays are usually produced by bombarding a metallic target with fast electrons in a high vacuum. In nuclear reactions, it is customary to refer to photons originating in the nucleus as gamma rays and those originating in the extranuclear part of the atom as X rays. Sometimes called roentgen rays after their discoverer, Wilhelm Roentgen.

LITERATURE CITED

1. **Amersham Corp.** 1982. Amersham review 16, Self-decomposition of radiochemicals. Amersham Corp., Arlington Heights, Ill.

1a. **Attix, F. H., W. C. Roesch, and E. Tochilin (ed.).** 1969. Radiation dosimetry, 2nd ed., vol. 3. Academic Press, Inc., New York.

2. **Bureau of Radiological Health, U.S. Department of Health, Education and Welfare.** 1970. Radiological health handbook. U.S. Government Printing Office, Washington, D.C.

3. **Committee on Biological Effects of Ionizing Radiation (BIER III), National Research Council.** 1980. The effects on populations of exposure to low levels of ionizing radiation. National Academy Press, Washington, D.C.

4. **ICRP Committee II.** 1960. Report of ICRP Committee II on permissible doses for internal radiation. Health Physics J. **3**:1–380.

5. **Martin, A., and S. A. Harbison.** 1979. An introduction to radiation protection, 2nd ed. Halstead Press, Div. of John Wiley & Sons, New York.

6. **Shapiro, J.** 1981. Radiation protection: a guide for scientists and physicians, 2nd ed. Harvard University Press, Cambridge, Mass.

7. **Slobodien, M.** 1980. Radiation hazards in the laboratory. *In* Fircaldo, Erlick, and Hendman (ed.), Laboratory safety: theory and practice. Academic Press, Inc., New York.

8. **U.S. Department of Commerce.** 1964. National Bureau of Standards handbook no. 92, Safe handling of radioactive materials. U.S. Government Printing Office, Washington, D.C.

9. **U.S. Nuclear Regulatory Commission.** 1975. Regulatory guide 8.13. Instructions concerning prenatal radiation exposure. Revision 1, November 1975. Office of Standards Development, U.S. Nuclear Regulatory Commission.

ACCIDENTS AND MEDICAL EMERGENCIES

Chapter 28

Laboratory Accidents with Infectious Agents

DIETER H. M. GRÖSCHEL, KERMIT G. DWORK, RICHARD P. WENZEL, AND LEONARD W. SCHEIBEL

Accidental exposure to infectious microorganisms is an expected, although rare, event in hospital, research, and industrial laboratories. Earlier chapters in this volume describe medical programs for laboratory safety, hazard assessment, and epidemiologic considerations for work with various infectious agents. In this chapter, first aid to laboratory employees after accidental exposure to microorganisms is discussed.

After known exposure, no emergency procedures can be wholly relied upon to prevent infection. Indeed, the availability of emergency measures should never permit the worker an illusion of complete protection lest he be lured into lowering the necessary standards of safety.

The person in charge of first aid assistance should be familiar with the potentially hazardous microorganisms used in the laboratory and, in a hospital, be aware of the prevalent nosocomial and communicable diseases. First aid after exposure consists of three steps: (i) removal or dilution of infectious material and institution of first aid measures; (ii) assessment of the infection risk; (iii) referral to a physician for treatment and evaluation for prophylaxis.

Infectious material can be removed or diluted from the intact or broken skin and mucosa by washing with copious amounts of water and soap. Antiseptics such as alcohol, tincture of iodine, idophor, or chlorohexidine preparations should be applied. Contamination of the eyes requires immediate flushing with water or ophthalmic saline irrigation solution. Puncture or bite wounds should be left bleeding freely. Sucking of wounds may assist in removing inoculated materials and stimulate bleeding, but too vigorous manipulation of closed deep wounds may assist in the hematogenous and lymphatic spread of pathogens. If contact has occurred with larvae of strongyloides, hookworm, or *Ancylostoma brasiliense*, the involved area should be rubbed vigorously with a 10% aqueous suspension of thiabendazole (Mintezole; 12). Penetration of the skin by cercariae of avian or human schistosomes takes time and is said to be limited by vigorous rubbing with a towel (4,5) and povidone iodine solution.

If oral contamination has occurred, the mouth should be rinsed immediately with tap water. If dangerous microbiologic material is swallowed, several glasses of water should be drunk and then vomiting should be induced by stimulating the back of the throat with the tip of a finger. However, if ova of *Taenia solium* are involved, induced vomiting could possibly result in future cysticercosis.

Ectoparasites should be eliminated from the skin as soon as possible. Ticks can be removed carefully with forceps while applying pressure to the tick's head. A drop of glycerol, alcohol, or fingernail polish may help make the tick release its grip. The mouth parts of the tick should not remain in the skin since they may cause a granuloma. The use of an antiseptic may prevent secondary infection of the bite wound.

The assessment of the infection risk should include an evaluation of the type and amount of infectious material, the mode of transmission, the portal of entry, and the general and specific conditions of the host. For some infectious diseases the infectious dose has been established by voluntary or accidental inoculation. Wedum and collaborators (15,16) collected data from the literature for infectious doses for human volunteers; these are summarized in Table 1. For staphylococci, an accidental subcutaneous dose even of as many as 10^8 or 10^9 organisms will rarely result in the formation of an abscess (9), but in the presence of a foreign body (suture) 100 *Staphylococcus aureus* units will result in an infection (6). In research and industrial laboratories employees are often immunized against the microorganisms with which they are working. This information should be known to the person administering first aid. The employee's health status needs to be assessed, because immunosuppressive or other chronic diseases or pregnancy may influence the steps to be taken for definitive treatment of the accident victim.

Whether or not prophylactic measures are indicated needs to be decided by a physician or his delegate. Many laboratories have established guidelines or procedural policies regarding occupational exposure to infectious materials. The recommended prophylaxis or therapy is sometimes administered in

TABLE 1. Infectious dose for human volunteers (15, 16)

Disease or agent	Inoculation route	25 to 50% infectious dose[a] (no. of organisms)
Anthrax	Inhalation	$\geq 1,300$
Cholera	Ingestion	10^8
Malaria	Intravenous	10
Q fever	Inhalation	10
Scrub typhus	Intradermal	3
Shigella flexneri	Ingestion	180; 10^8
Syphilis	Intradermal	57
Tularemia	Inhalation	10; 10^8
	Intradermal	10
Typhoid fever	Ingestion	10^5

		$TCID_{50}$[a]
Measles virus	Intranasal spray	0.2
Rhinovirus	Nasal drops	≤ 1
Venezuelan encephalitis virus	Subcutaneous	1AIU[b]
West Nile fever virus	Intramuscular	1AIU
Rubella virus	Pharyngeal spray	≤ 10
	Subcutaneous	30
	Nasal drops	60
Influenza A2 virus	Nasopharyngeal	≤ 790

[a] $TCID_{50}$, 50% infectious tissue culture dose.
[b] AIU, Animal infectious unit.

the first aid station or the employee health service (1). The laboratory director should advise the contract services of a clinic or hospital or the employee's personal physician about the spectrum of organisms present in the laboratory and the specific prophylactic or therapeutic measures which are recommended.

Prophylactic measures should be applied only under the direction of a physician and may include local therapy such as instillation of antiseptic or antibiotic eye drops; administration of specific or nonspecific immunoglobulin; vaccination, e.g., for the prophylaxis of tetanus in a previously nonimmunized patient; and the use of specific antimicrobial substances. If the laboratory uses a limited number of potentially pathogenic microorganisms, their antimicrobial susceptibility profile should be established and communicated to the employee's medical care facility.

In cases of accidents with microorganisms for which serological tests are available or can be developed, a base-line serum sample should be obtained at the time of occurrence. Preferably, baseline sera from all laboratory personnel should have been collected and stored, and a second serum sample is obtained at the time of exposure or onset of symptoms.

All accidents should be reported to the laboratory safety officer and the laboratory director. A report should be filed and maintained for medical and epidemiological purposes.

Tables 2 through 5 list the sources and routes of entry of certain bacteria, fungi, viruses, and parasites. The organisms were selected from lists in the Centers for Disease Control/National Institutes of Health biosafety guidelines (14) and several unpublished addenda, and from information collected by Pike (11), Hanel and Kruse (7), and Wedum and Kruse (16). Table 5 lists also the infective stages of parasites pathogenic to humans, including the recently recognized cryptosporidia (10). Obviously, not all microorganisms found in microbiology laboratories can be included. For additional information, the reader is advised to review the previous chapters in this volume, textbooks (3,4,8), the American Public Health Association booklet on the control of communicable diseases (2), and review articles in the literature (7,11,13,16).

The tables provide information about proven or potential routes of laboratory-acquired infections. Such routes include the following.

(i) Intact mucosa. Infectious droplets may reach the mucosal surfaces of the eye, nose, or oral cavity. Organisms may be transmitted to the mucosa through devices such as pipettes, by spraying from syringes, by droplets and aerosols generated by laboratory procedures such as high-speed blending and centrifugation or centrifuge accidents, and by direct contact (finger) from the source or from contaminated fomites.

(ii) Broken skin. Infectious material may enter directly or via fomites through abrasions, small cuts, or larger accidental wounds or through inoculation with needles or other sharp items. Insect vectors may transmit disease through stings, bites, or contamination of a bite wound. Certain microorganisms may enter the body through the unbroken skin, e.g.,

TABLE 2. Accidental infections with bacteria

Organism	Route of infection[a]				
	Intact mucosa	Broken skin	Animal contact	Inhalation	Ingestion
Bacillus anthracis (V)[b]	X	X	X	X	
Bordetella pertussis (V)	X	?		X	
Borrelia recurrentis	X	?	X		
Brucella spp.	X	X	X	X	
Campylobacter spp.		X	X	?	X
Chlamydia spp.	X			X	
Clostridium botulinum (toxin) (V)	X	X		X	X
Clostridium spp. (V)		X			X
Clostridium tetani (V)		X			
Corynebacterium diphtheriae (V)	X	X		X	X
Erysipelothrix insidiosa		X	X		
Francisella tularensis (V)	X	X	X	X	X
Leptospira interrogans	X	X	X	?	X
Listeria monocytogenes	X		X	?	?
Mycobacterium avium-intracellulare, M. fortuitum, M. kansasii		X		X	
Mycobacterium tuberculosis, M. bovis (V)		X		X	X
Neisseria gonorrhoeae	X	X		?	
Neisseria meningitidis (V)	X	X		?	X
Pasteurella multocida	X	X			
Pseudomonas mallei	X	X		X	
Pseudomonas pseudomallei		?		X	
Rickettsial agents (V)	X	X	X	X	
Salmonella typhi (V)		X		?	X
Salmonella spp. (V)		X	X	?	X
Shigella spp.		X		?	X
Staphylococcus aureus	X	X	X	?	?
Streptobacillus moniliformis		X	X		X
Streptococcus agalactiae	X	X	X		
Streptococcus pyogenes	X	X			
Treponema pallidum	X	X		X	
Vibrio cholerae (V)		X		?	X
Vibrio spp.		X	X		X
Yersinia enterocolitica	X				X
Yersinia pestis (V)	X	X	X	X	X
Yersinia pseudotuberculosis	X				X
Enterobacteriaceae and other gram-negative bacteria		X			

[a]X, Proven or likely hazard; ?, doubtful or not determined. See the text for discussion of routes of infection.
[b](V), Vaccination available and advised.

TABLE 3. Accidental infections with fungi

Organism	Route of infection[a]				
	Intact mucosa	Broken skin	Animal contact	Inhalation	Ingestion
Blastomyces dermatitidis		X		?	
Coccidioides immitis		X		X	
Cryptococcus neoformans		X	X	?	
Histoplasma capsulatum		X		X	
Paracoccidioides brasiliensis		X			
Sporothrix schenckii	X	X	X		
Dermatophytes		?	X		

[a]See Table 2, footnote *a*.

TABLE 4. Accidental infections with viruses

Organism	Route of infection[a]				
	Intact mucosa	Broken skin	Animal contact	Inhalation	Ingestion
Adenoviruses	X			X	?
Arenaviruses					
Lymphocytic choriomeningitis	X	X	X	X	X
Lassa	X	X	X	X	X
Herpesvirus group					
Epstein-Barr	X			?	
Herpes simplex	X	X			
Cytomegalovirus	X	X			
Herpes simiae (B virus)	X	X	X	?	
Varicella	X	X		X	
Marburg virus, Ebola virus	X	X	X		
Myxo-paramyxoviruses					
Influenza (V)	X			X	
Measles (V)	X			X	
Mumps (V)	X			X	
Respiratory syncytial	X			X	
Picornaviruses					
Coxsackievirus	X			X	X
Echovirus					X
Poliomyelitis (V)	X				X
Rhinovirus	X				
Poxviruses					
Vaccinia (V)	X	X			
Monkey pox (V)	X	X	X	X	X
Papovaviruses	?	X			
Spongiform encephalopathy viruses					
Creutzfeldt-Jakob agent	?	X		?	?
Kuru agent	?	X		?	?
Retroviruses					
HTLV-III/LAV[b]	X	?			
Rhabdoviruses					
Rabies (V)		X	X	X	
Vesicular stomatitis	X	X	X	X	
Rotaviruses, Norwalk agent				?	X
Togaviruses					
Dengue	X	X			
Eastern eguine encephalitis (V)		X	X	X	
Rubella (V)	X				
Russian spring-summer encephalitis (V)		X	X		
Venezuelan equine encephalitis (V)		X	X	X	
Western equine encephalitis (V)		X	X	X	
Yellow fever (V)		X			
Other viruses					
Hepatitis (V)	X	X		?	X(A)[c]

[a]See Table 2, footnotes a and b.
[b]HTLV-III/LAV, Human T-cell lymphotropic virus type III/lymphadenopathy-associated virus.
[c]Hepatitis A.

spirochetes and schistosomes, or infect the skin tissue directly, such as dermatophytes.

(iii) **Animal contact.** Animal contact includes animal bites or scratches, transmission by vectors such as ticks or fleas, contamination of hands, and exposure to contaminated aerosols, dusts, body fluids, and excreta.

(iv) **Inhalation.** Entry into the body is achieved through droplets and aerosols such as those generated by sprays from syringes, centrifugation, and tissue mincing, or with dust (animal bedding).

(v) **Ingestion.** Microorganisms reach the gastrointestinal tract directly from an infected source (fecal-oral transmission) or through a contaminated carrier. Careless storage of food in a laboratory or eating and drinking at the laboratory bench are other causes of transmission.

In all cases of accidental exposure to infectious

TABLE 5. Accidental infections with parasites

Organism	Infective stage	Intact mucosa	Intact or broken skin	Animal contact	Inhalation	Ingestion
		Route of infection[a]				
Intestinal protozoa						
Entamoeba histolytica	Cysts in stool or culture					X
Giardia lamblia						X
Cryptosporidium spp.	Oocysts in feces			X	?	X
Central nervous system protozoa						
Naegleria fowleri	Trophozoites or flagellates in water	X			X	
Blood protozoa						
Plasmodium spp.	Mosquito-borne sporozoites; transfusion- or syringe-borne merozoites	X				
Babesia spp.	Developmental forms in ticks and in infected blood		X	X		
Trypanosoma cruzi	Metacyclic trypanosomes in reduviid bugs or in infected blood		X		X	
Trypanosoma brucei gambiense, *T. brucei rhodesiense*	Metacyclic trypanosomes in tsetse flies and in infected blood		?	?		
Leishmania spp.	Leishmaniae via phlebotomus bite or by direct contact with infected blood	X	X	X		
Other protozoa						
Toxoplasma gondii	Oocysts in cat feces; infected meat of various origins	X	X	X		X
Intestinal nematodes						
Enterobius vermicularis	Ova (infectious several hours after passage)				X	X
Ascaris lumbricoides	Ova containing developing larvae (infectious 3 weeks after passage)					X
Hookworms	Filariform larvae in soil		X			
Strongyloides stercoralis	Filariform larvae in soil		X			
Nonintestinal nematodes						
Ancylostoma brasiliense	Larvae in soil		X			
Toxocara spp.	Embryonated ova in soil					X
Cestodes						
Hymenolepis nana	Ova in feces of human, rat, or mouse			X		X
Echinococcus spp.	Ova in feces of dog, sheep, cattle					X
Taenia solium	Cysticercus cellulosae in pork; *T. solium* ova in humans					X
Trematodes						
Fasciola hepatica	Metacercaria on watercress-like plants		X			
Schistosoma spp.	Cercariae in fresh water in field or in laboratory tanks		X			

[a] See Table 2, footnote *a*.

265

agents, the exposed employee should be referred to a physician for evaluation and appropriate treatment or prophylaxis.

LITERATURE CITED

1. **Atuk, N. O., T. R. Townsend, E. H. Hunt, and R. P. Wenzel.** 1981. An employee health service for infection control, p. 211–250. *In* R. P. Wenzel (ed.), CRC handbook of hospital acquired infections. CRC Press, Inc., Boca Raton, Fla.

2. **Benenson, A. S. (ed.).** 1981. Control of communicable diseases in man, 13th ed. American Public Health Association, Washington, D.C.

3. **Braude, A. I. (ed.).** 1981. Medical microbiology and infectious diseases. W. B. Saunders Co., Philadelphia.

4. **Brown, H. W.** 1975. Basic clinical parasitology, 4th ed., p. 254. Appleton-Century-Crofts, New York.

5. **Canizares, O.** 1975. Clinical tropical dermatology, p. 233. Blackwell Scientific Publications, Oxford.

6. **Elek, S. D., and P. E. Conen.** 1957. The virulence of *Staphylococcus pyogenes* for man. A study of the problems of wound infection. Br. J. Exp. Pathol. **38:**573–586.

7. **Hanel, E., Jr., and R. H. Kruse.** 1967. Laboratory-acquired mycoses. Miscellaneous Publication no. 28. Industrial Health and Safety Office, Department of the Army, Fort Detrick, Md.

8. **Mandell, G. L., R. G. Douglas, Jr., and J. E. Bennet.** 1985. Principles and practice of infectious diseases, 2nd ed. J. Wiley & Sons, New York.

9. **Morse, S. I.** 1981. Staphylococci, p. 280. *In* A. I. Braude (ed.), Medical microbiology and infectious diseases. W. B. Saunders Co., Philadelphia.

10. **Navin, T. R., and D. D. Juranek.** 1984. Cryptosporidosis: clinical, epidemiologic, and parasitologic review. Rev. Infect. Dis. **6:**313–327.

11. **Pike, R. M.** 1979. Laboratory-associated infections: incidence, fatalities, causes and prevention. Annu. Rev. Microbiol. **33:**41–66.

12. **Scheibel, L. W.** 1973. Chemotherapy of parasitic diseases commonly seen in the United States. J. Florida Med. Assoc. **60:**17–24.

13. **Subcommittee on Arbovirus Laboratory Safety of the American Committee on Arthropod-Borne Viruses.** 1980. Laboratory safety for arboviruses and certain other viruses of vertebrates. Am. J. Trop. Med. Hyg. **29:**1359–1381.

14. **U.S. Department of Health and Human Services.** 1984. Biosafety in microbiological and biomedical laboratories. HHS Publication no. (CDC)84–8395. U.S Government Printing Office, Washington, D.C.

15. **Wedum, A. G., W. E. Barkley, and A. Hellman.** 1972. Handling of infectious agents. J. Am. Vet. Med. Assoc. **161:**1557–1567.

16. **Wedum, A. G., and R. H. Kruse.** 1966. Assessment of risk of human infection in the microbiological laboratory. Miscellaneous Publication no. 19. Industrial Health and Safety Office, Department of the Army, Fort Detrick, Md.

Chapter 29

Allergic Reactions and Poisoning

DANIEL A. SPYKER

What is a medical emergency?

For the purposes of this chapter a medical emergency is considered to be any circumstance where human injury may potentially occur. The chapter does not cover surgical problems such as cuts, fractures, and other trauma. Both intrinsic (such as low blood sugar) and allergic reactions will be discussed briefly, although the major emphasis will be on poisoning as a medical emergency.

Injury versus accident

It seems best to emphasize the concept ''injury'' rather than ''accident'' since the latter term connotes fate, luck, or unavoidability, with no indication of personal or property damage. Promotion of safety means reduction of morbidity and health care cost regardless of whether the injury was accidental or intentional. Thus the term ''injury'' will be used in preference to ''accident'' throughout this chapter.

Prevention: Haddon's ten countermeasures

Haddon (15) has proposed an organized approach to the prevention of injury. He asserts that injuries result from the transfer of energy in excessive amounts or rates (15). In the case of poisoning, injury results from the toxin's interference with the body's normal energy transfer. The first eight of these strategies involve primary prevention of the injury, and the last two represent minimizing the injury after the event, i.e., secondary prevention.

Haddon's strategies call for measures to:
1. Prevent accumulation of the agent
2. Reduce amount of the agent
3. Prevent inappropriate release of the agent
4. Modify release of the agent
5. Separate victim from agent in time or space
6. Erect physical barriers
7. Change surfaces and basic structures
8. Strengthen resistance of victim

After injury, effects are minimized by:
9. Rapid emergency response
10. Improved medical care and rehabilitation

Determining the cause of injury (differential diagnosis)

In approaching the injured patient the first responder to the emergency must begin to gather information (the history) at the same time as assuring support of vital functions (basic life support). Except when comatose, the patient provides the best source of information. Many people with a medical problem that might result in a loss of consciousness, such as diabetes mellitus or a serious allergy, will wear a MedicAlert identification bracelet (MedicAlert Foundation, Turlock, Calif.).

The first priority is to detect and treat potentially reversible problems. Support of respiration and cardiac functions and management of anaphylaxis are discussed below. Cerebrovascular accident (stroke) and myocardial infarction (heart attack) rank among the most common medical emergencies, but the first responder can do little more than provide basic life support and immediate transportation to medical care. Likewise, the first aid setting permits little specific treatment for seizures, infection, or neoplasm.

Intrinsic (metabolic) derangements of several kinds (low sodium, high calcium, high glucose, etc.) may cause stupor or coma. Hypoglycemia (low blood sugar) can cause virtually any central nervous system problem from dizziness to coma. The appropriate emergency treatment, sugar in orange juice or any glucose-rich food or drink, can easily be provided by the first responder. This treatment necessarily depends on recognition of the problem and the cooperation of the patient in ingesting the sugar.

The basic interaction between people and poisons can be of three types: allergic, idiosyncratic, or toxic (Table 1). An allergic or anaphylactic response depends on prior exposure to the offending agent and occurs independently of dose. The idiosyncratic reaction represents an unusual sensitivity of a particular person to a particular chemical and does depend on dose. The toxic reaction represents the expected effect of the poison on the human.

ALLERGIC REACTIONS

The term ''allergic'' applies to reactions mediated by the body's immunologic system, as contrasted to

TABLE 1. Classification of human reactions to chemicals

Reaction	Likelihood of effect (%)	Dose dependence	Dependence on prior exposure	Likelihood of effect with reexposure (%)
Allergic	<1	No	Yes	60
Idiosyncratic	<5	No	No	90
Toxic	100	Yes	No	100

reactions due to pharmacologic idiosyncrasy or direct toxicity. Such reactions may involve a mild or severe skin rash, fever, direct cytotoxic reactions to individual organs, serum sickness (fever, rash, adenopathy, arthritis), or anaphylaxis.

Anaphylaxis

Anaphylaxis is a clinical syndrome characterized by abrupt reaction to locally or systemically introduced agents. Severe anaphylaxis represents the gravest of emergencies in humans. It may include low blood pressure (hypotension), wheezing and troubled breathing (bronchospasm), hives (urticaria), laryngeal edema, cardiac arrhythmia, or seizures, either in combination or as isolated problems. Serious reactions most frequently occur after intravenous or intramuscular administration, although oral, skin (percutaneous), or respiratory exposures may produce a response in highly sensitive persons.

The more quickly the allergic response follows contact with the allergen, the more severe the reaction will likely be. Thus, immediate collapse after exposure to any chemical must alert observers to anaphylaxis. Noisy respirations (stridor or wheezing) frequently accompany severe anaphylaxis. A widespread skin reaction of large irregular wheals or hives (urticaria) almost uniformly accompanies severe anaphylaxis. Thus the diagnosis of anaphylaxis is not usually a problem, and definitive treatment must be instituted as soon as possible.

Reactions of this general type may occur without prior sensitization to the offending agent. Such reactions are termed anaphylactoid. They are clinically similar, but antibody involvement has not been demonstrated. These reactions probably result from direct release of histamine and other mediators by such substances as dextran, aspirin, and lidocaine. Anaphylactoid reactions are managed in the same fashion as true anaphylaxis.

Management of anaphylaxis. The central strategy in treating anaphylaxis or any other medical emergency is basic life support. This implies cardiac compression for the pulseless patient and mouth-to-mouth ventilation for the patient who is not breathing (apneic). Everyone working in a laboratory or hospital should be trained in basic life support as standardized by the American Heart Association (2).

The most effective treatment for anaphylaxis is the injection of epinephrine. For moderate reactions epinephrine may be injected subcutaneously, but for severe anaphylaxis, intravenous administration is preferred. Table 2 divides anaphylaxis into four therapeutic groups based on clinical presentation (14). As the patient becomes more severely reactive, he or she usually (though not always) evolves from the less severe (group 1) reaction to the more severe (group 4), thus permitting recognition and interruption of the process.

Causes of anaphylaxis. Agents which are known to trigger anaphylaxis include animal venoms, drugs, foods, hormones, sera, and contrast media. Between 100 and 500 people per year probably die from penicillin reaction, compared to about 40 deaths from hymenoptera stings. Table 3 lists some of the agents known to cause anaphylaxis (13). Personnel in industry and clinical laboratories may be exposed to large doses of antibiotic agents in powder form.

Prevention. Any systemic reaction (Table 2; grade II to IV), no matter how mild, is considered a warning of serious risk of fatal anaphylaxis. Such individuals must consider three potentially lifesaving steps.

(i) Prevention of exposure. Prevention of exposure represents a clearly superior solution. These patients should avoid exposure to the allergen as much as possible.

(ii) Emergency kit. Carry an emergency epinephrine supply in a preloaded injector. Several forms of injectable empinephrine are currently available, including:

Anakit (Hollister-Stier Laboratories, Spokane, WA 99207); kit contains 1 ml of 1:1,000 epinephrine in a preloaded syringe which administers 0.5 ml with each push, also four 2-mg tablets of chlorpheniramine; retail price about $9.66

EpiPen (Center Laboratories, Port Washington, NY 11050); a preloaded injector with 0.3 ml of 1:1,000 epinephrine; retail price about $19.86

Patients at risk should have at least one kit for the home, one for the car, and one to carry when outdoors. Personnel responsible for the safety of others engaged in outdoor activity, such as camp counselors and athletic instructors, should have kits available and be competent in their use (4).

TABLE 2. Clinical classification of anaphylactic reactions[a]

Grade	Findings	Treatment
I	Local reaction only (>15 cm)	Symptomatic: cold compresses, aspirin
II	Symptoms confined to skin (pruritis, urticaria)	Epinephrine (1:1,000), 0.3 ml SQ every 15 min; diphenhydramine (Benadryl) 50 mg PO or IM, TID for 1 to 2 days
III	Dyspnea, angioedema, nausea, vomiting	Epinephrine (1:1000), 0.3 ml SQ every 15 min
IV	Respiratory distress, asthma, dysphagia	Early intubation; epinephrine (1:1,000) SQ; aminophylline
	Cardiovascular distress, hypotension	Epinephrine (1:10,000), 1 to 2 ml IV if BP < 70; fluids, 0.5 to 2 liters of NS

[a] SQ, Subcutaneous injection; PO, by mouth (per os); IM, intramuscular; TID, three times a day; IV, intravenous; BP, blood pressure; NS, normal saline.

TABLE 3. Some chemical agents known to cause allergic reactions

Agent	Examples
Drugs	Penicillin, synthetic penicillins, cephalosporins, tetracyclines, nitrofurantoin, procaines, dextrans, aspirin, insulin, heparin
Diagnostic agents	Contrast media, halogenated compounds, skin testing agents
Animal venom	Honeybees, wasps, hornets, jellyfish
Foods	Seeds, crustaceans, fish, nuts, legumes (peanuts), rice, potatoes, tangerines, garbanzo beans, raw milk

(iii) Desensitization. Effective desensitization is now routinely available for insect sting allergy and should be seriously considered in any reactive person. Desensitization has been carried out for penicillin, but is not routinely practiced.

Occupational dermatoses

The skin serves as a principal route of absorption for many toxic substances and is the target organ for occupational dermatoses. Skin damage comes in a spectrum of injury from mild reddening (erythema) to lethal desquamation. These problems are divided into irritant and allergic reactions.

Irritant dermatitis. Most occupational dermatitis results from primary chemical irritant effects. Irritants can damage the skin by removal of the lipid film (solvents), denaturation of keratin, interference with the subcoreal barrier layer, protein precipitation (heavy metal salts), oxidation, or dehydration. An acid burn clinically resembles a thermal burn; strong alkalis combine with fats and dissolve protein. Less strong alkalis produce dehydration of the keratin, loss of cell cohesion, and skin cracking. Most cases of occupational dermatitis occur in new employees and are mild in character.

Good general hygiene and protective clothing represent the mainstay of preventing occupational dermatoses. The engineering approach, i.e., limiting the opportunity for exposure to the irritants, represents the most successful approach to reducing irritant exposure. Protective clothing (aprons, gloves, boots, face masks) is essential in some situations. When gloves cannot be worn, barrier creams can help to protect against irritants, although they will not prevent allergic reactions. They should be applied three or four times each shift and might also be of some value when applied to the face and neck to protect against some vapors and dust. The routine use of skin cleansers and general attention to personal and work area hygiene should not be overlooked in reducing irritant dermatitis (28).

Allergic dermatitis. Allergic dermatitis represents approximately 20% of occupational contact dermatitis. Any chemical can act as a sensitizer, but conspicuous causes include poison ivy, poison oak, epoxy monomers and their amine hardeners, potassium dichromate, nickel, and formaldehyde. Characteristically, sensitizers do not produce demonstrable injury on the first few contacts. After the 3- to 5-day sensitization period, the anamnestic response is evident. Most important of these is contact eczematous dermatitis. The principal management of this problem is simply prevention of contact.

Extensive tables of occupation-specific chemical risks and mechanisms of chemical dermatoses are available (19). Laboratory workers should be most concerned about exposure to acids, alkalis, chromates, synthetic detergents, organic chemicals, soaps, and solvents.

MANAGEMENT OF ACUTE POISONING

The diagnosis is usually apparent in the poisoned patient. Distinguishing between intrinsic causes of loss of consciousness and extrinsic factors (poison) as a cause of injury permits definitive therapy. However, the approach to the unconscious or the seriously compromised patient should be the same regardless of cause, that is, airway, breathing, and cardiac support—the ABCs of basic life support.

For the poisoned patient, the major addition to this strategy (the D of the ABCDs) is "decontaminate." Any possibility of an inhaled or skin contact of toxic substances requires removing the patient from the area, removing any contaminated clothing as appropriate, or both. In the event of skin exposure, the patient must be washed quickly and thoroughly with soap and water. Two washes for any toxic exposure to the skin and at least 20 min of continuous water lavage for eye exposures are minimum treatments (see Chapter 30).

Paracelsus quite correctly observed that "All substances are Poison; there is none which is not a Poison. The right dose differentiates a poison and a remedy." Although it is important to keep in mind that any substance can be ingested in such an amount as to cause injury, this chapter is principally concerned with substances which when ingested or otherwise absorbed in relatively small amounts cause injury by chemical action.

Fundamentals of treatment

The five fundamentals in managing any poison victim include basic life support, decrease in absorption, increase in elimination, identification of poison, and specific treatment (antidotes).

(i) Basic life support. Most deaths due to poisoning result from lack of respiration. The mnemonic ABC is a reminder to provide airway, breathing, and cardiac support. If the patient is not breathing, his airway should be opened; if he does not breathe spontaneously, immediate mouth-to-mouth resuscitation should be started; in the event that no pulse is palpable, closed-chest cardiac compression should be applied. Every member of the laboratory team should be trained and certified to provide cardiopulmonary resuscitation (2). This assistance must be continued as long as necessary, and the decision to discontinue must be made only by a physician informed as to the patient and the circumstances. The approach to the unconscious or the seriously compromised patient is the same regardless of cause.

(ii) Decrease absorption. The greatest impact in decreasing morbidity and mortality from acute poisoning can be made by measures to decrease absorption of the toxin. This means copious and continuous lavage for eye and skin exposures and removal from the source for inhalation exposure. Appropriate treatment of an ingested poison usually includes dilution with water, induction of emesis through ipecac (if soon after ingestion), adsorption with activated charcoal, and cathartics.

(iii) Increase elimination. Efforts aimed at enhancing elimination of a drug are, in general, applied only to hospitalized patients. Oral activated charcoal, however, does enhance elimination of some toxins.

(iv) Identification. Identification of the toxic substance can provide critical data in the treatment of the poisoned patient, but should not delay cardiopulmonary resuscitation or efforts to minimize absorption. Information that could identify the offending substance must be made available to the poison center or physician. Whenever possible, labels and samples of the poison should be secured.

(v) Specific treatment. Few poisonings require a specific "antidote" or pharmacologic antagonist. In general these are the province of the treating physician and should be avoided in the first-aid setting. The one exception to this is that high-flow oxygen should be used after exposure to carbon monoxide, cyanide, or other metabolic poisons. An exception to the blanket use of oxygen is that injury from the herbicide paraquat appears to be enhanced in the presence of elevated oxygen levels. Oxygen should be used with caution in patients with preexisting (chronic) lung disease.

Eye exposures

Virtually every solvent, cleaner, adhesive, or caustic can cause eye irritation. The initial management for all of these exposures is the same: copious irrigation. The importance of time so far outweighs other considerations that tap water or the first innocuous water source at hand should be used. The essential features of appropriate emergency treatment of eye exposures are the four Ls as follows.

(i) Large volume. An emergency shower, a large sink, or an eye wash device probably provides the best source of irrigating fluid for eye contamination. The use of small squeeze bottles of irrigation fluid for such purposes should be discouraged. Every laboratory worker should know the location of such emergency showers and eye washers, and this equipment should be checked periodically to ensure it is in operating condition.

(ii) Low pressure. The ideal irrigation is provided by a permanently installed eye wash station, by an intravenous bottle of normal saline, or by water poured from a pitcher held 12 to 14 in. above the eye. It is essential that the eyelid be retracted so that all damaged surfaces are exposed to the lavage.

(iii) Look. The eye must be examined, and contact lenses or other foreign bodies must be removed.

(iv) Long time. Lavage should continue for at least 20 min. If a caustic injury is suspected, then lavage must be continued throughout transport and emergency room evaluation. For minor exposures, lavage may be discontinued after 20 min, and the patient may then be evaluated.

If discomfort persists for more than 15 min after the discontinuation of lavage, and if there is any decrease in vision or any other symptom, the patient should be immediately evaluated by a physician, preferably an ophthalmologist.

Skin exposures

Skin exposure of any kind may be appropriately treated by immediate and copious irrigation. Caustics in particular should be flushed immediately with cool water for a minimum of 10 min. Emergency showers with large pull chains provide the best irrigation. Do not attempt to neutralize acids or bases, as the resulting heat of neutralization can cause greater injury to the skin (see Chapter 30).

Decontamination after a topical exposure to pesticides, hydrocarbons, etc., is also best accomplished with copious flushing. Two soap-and-water washes are recommended. For hydrocarbon-base liquids, tincture of green soap or alternate washes with rubbing alcohol (isopropranol) and soap and water may provide a better removal, but any soap may be used. All contaminated clothes should be removed and placed in plastic bags.

The patient should be evaluated after washing. Any redness, pain, or other symptom requires physician evaluation.

Inhalations

Inhalation of toxic vapors, mists, gases, or dust (particulates) can poison by absorption through the mucous membranes of the mouth and throat or the lungs. Serious injury may result from direct caustic effect on these tissues or through absorption into the lungs. The severity of injury will depend on the toxicity of the material, its solubility in tissue fluids, and the concentration and duration of exposure. The higher ventilation rate which accompanies physical exertion can increase the amount of toxin adsorbed four- to sixfold. Thus the risk is much greater for persons who are anxious or active.

The treatment in any event is rapid removal from the noxious environment. Cardiopulmonary resuscitation as described above may be necessary. Frequent deaths among the rescuers of poisoned victims underscore the importance of the warning: DO NOT APPROACH THE VICTIM WITHOUT ADEQUATE RESPIRATORY PROTECTION. Any patient with symptoms should be referred to an emergency room for medical treatment.

Ingestions

As with poisoning by other routes, support of vital functions is the principal objective in ingestion poisoning. In addition, the poison injury can be reduced through use of dilution, emesis, adsorption with activated charcoal, and cathartics.

Dilution. Dilution with water, while not antidotal, is a central part of the decontamination measures. Some other fluids might be desirable as a diluent for certain substances, such as milk for an alkaline ingestion, but time still outweighs other considerations, and water is the liquid of choice for dilutions. The use of neutralizing substances, such as vinegar for alkaline ingestions, causes more harm than good due to the heat of neutralization. Thus two large glasses (500 ml) of water should be the first therapeutic measure.

Emesis. There are only two contraindications to inducing emesis in an adult. Patients who cannot protect their airway (due to seizures or coma) and patients who have ingested a caustic should not receive ipecac. Syrup of ipecac is the agent of choice for inducing emesis in the absence of these situations. Adults should receive 30 ml (2 tablespoons) followed by more fluids. Fluids are essential but may be given before or after ipecac with equal efficacy. This dose of ipecac may be repeated once if vomiting does not occur within 30 min. Contraindications to ipecac or the presence of any symptoms is an indication for referral to the nearest emergency department.

Adsorbents. Activated charcoal is an inert nontoxic substance which binds (adsorbs) most organic compounds. No contraindications exist to the use of activated charcoal. Although the decrease in absorption of orally ingested drugs has been well documented, we now have data on the enhanced elimination of intravenously administered drugs. Prospective studies have shown charcoal to be superior to ipecac and, following European strategies, many hospitals are shifting to charcoal as the first line of treatment (35). Ten times the weight of the ingested substance or 1 g of charcoal per kg of body weight is usually recommended. Charcoal is best given with a cathartic or in a water slurry.

Cathartics. Cathartics enhance the elimination of bowel contents through osmotically pulling water into the bowel lumen. Oil-based cathartics such as castor oil should be avoided due to (i) enhanced absorption of some lipid-soluble toxins and (ii) the risk of aspiration pneumonia. Sodium sulfate or magnesium sulfate (Epsom salts) provides useful catharsis. The recommended dose of sodium sulfate or magnesium sulfate is 250 mg/kg. Two tablespoons of Epsom salts in two glasses of water provides effective catharsis in the home.

The saline cathartics do not interfere with charcoal absorption, and the combination of 80 g of charcoal and 20 g of sodium sulfate may be premixed and kept available in a large bottle ready for use. Charcoal may also be premixed as a 20% volume suspension in a 70% sorbitol solution. In comparative studies, this mixture appears to have superior cathartic properties, with a mean charcoal transit time of about 1 h. Thus, charcoal-sorbitol is generally considered the cathartic of choice for poisonings.

ACUTE POISONING: SPECIFIC AGENTS

The laboratory provides a wealth of chemical toxins and allergens. Appreciation of this fact, leading to routine basic precautions, represents the most imporant step in minimizing the hazard of the laboratory (18). The individual laboratory director must be cognizant of the toxicity of all substances contained in his or her laboratory. This section represents an introduction to some of the specific substances, but is not intended to be an exhaustive treatment. Regional poison control centers represent a valuable resource with which the laboratory director should establish a working relationship. A listing of Poison Control Centers by area code is provided in Appendix 3 of this volume.

Outbreaks of human intoxication have accompanied every major advance in technology. The toxicity of older chemicals is sometimes well documented, but the introduction of 500 new compounds per year and the absence of any structure-toxicity relationship portend many new cases of neurotoxicity (1). Both additive and synergistic neurotoxicity of solvents have been demonstrated, and the increasing use of complex (and often "trade secret") mixtures does little to assist detection and prevention.

Solvents and vapors

Virtually all solvents and vapors exert a general anesthetic effect after inhalation. Although many gases and vapors can cause problems at levels below the threshold of olfactory detection, a good general rule is that a vapor detected by smell probably exceeds the allowable standards of industrial exposure. Below are some of the more common solvents and vapors discussed with regard to their neurotoxicity and other toxic effects.

The Great Hydrocarbon Debate. The issue of whether to induce emesis in a patient ingesting hydrocarbons stands at the center of the "Great Hydrocarbon Debate." The enlightened approach to the treatment of hydrocarbon ingestion seems to be first to subdivide hydrocarbons according to toxicity. Hydrocarbon solvents with a viscosity greater than no. 2 diesel fuel represent little threat regardless of dose and may be managed with reassurance and follow-up. Ingestion of even a swallow of aromatic hydrocarbons, especially benzene, or chlorinated hydrocarbons represents potentially toxic exposure, and any ingestion of more than a lick should be treated with emesis. The ingestion of gasoline, kerosene, turpentine, and other common middle-viscosity solvents should be managed, according to dose, with ipecac-induced emesis for ingestions exceeding 1 ml/kg (23). As a practical guide, the average swallow represents 0.3 to 0.25 ml (19).

Aromatic hydrocarbons. (i) Benzene. Benzene, the parent compound of this group, is myelotoxic and implicated as a cause of leukemia, but has not been clearly linked to neurotoxicity (30). Appreciation of these risks has virtually eliminated benzene from the workplace. When used, it deserves the greatest care to minimize exposure.

(ii) Toluene. Toluene, the archetypal solvent of abuse, is present in many glues and commercial solvents. Toluene was intialy used as a substitute for benzene due to its much lower hematologic toxicity. Most cases of neurotoxicity relating to toluene probably result from hexane or other contaminants (9). Severe neurologic damage has resulted when manufacturers changed formulation to a more toxic compound wihtout notification to dedicated "huffers" (31).

In addition to the neurologic and psychiatric syndromes, a muscular weakness and gastrointestinal syndrome has been recognized in toluene-sniffing adults. Those patients presenting with primarily muscular weakness tend to have low serum potassium, phosphate, and calcium. Rhabdomyolysis is also common in this group of patients. Renal toxicity is usually clinically minor and may include proteinuria, microscopic hematuria, and pyuria. Renal tubular acidosis occasionally occurs with heavy abuse. Hepatic toxicity is rare and probably relates to exposure to other solvents. Toluene has not been shown to be cardiotoxic. Hematologic toxicity is uncommon (36). One patient abusing toluene for 14 years developed permanent encephalopathy and cerebral atrophy (20), but solid data supporting neurotoxicity are scanty for a substance in such wide use.

(iii) Xylene, styrene, naphthalene. Xylene, styrene, and naphthalene cause systemic toxicity and cerebral depression, but specific neurotoxicity has not been recognized.

Aliphatic hydrocarbons. (i) n-Hexane. n-Hexane, a principal component in many adhesives, has been clearly implicated in severe sensory and motor neuropathy, with the sensory defects prevailing. This neuropathy was reported in cabinet finishers (17) and then in several Italian shoe factories where n-hexane was used as the solvent in the glue (9).

(ii) n-Heptane. n-Heptane has a similar metabolic pathway and has been implicated in neurotoxicities similar to hexane.

(iii) Others. Ethane, acetylene, propane, propylene, isobutane, butane, and isopentane exert narcosis and cardiotoxicity, but neurotoxicity has not been documented.

Alcohols, ketones, and esters. The four alcohols listed below, particularly methanol and ethylene glycol, cause poisoning in extremely small quantities. The average swallow contains 0.27 ml/kg (19). Table 4 illustrates the expected toxicity of these solvents; several are toxic at less than one swallow.

TABLE 4. Summary comparison of ethanol and other lethal alcohols[a]

Alcohol (uses, formula, mol wt)	Metabolites	Findings	Dose (ml/kg)	(ml/70)	Level (mg/dl)	Treatment
Ethanol (solvent, beverages; $CH_3\text{-}CH_2\text{-}OH$; 46.07)	Acetaldehyde, acetic acid (ADh)	CNS depression, hypoglycemia, acidosis, ketosis, gastrointestinal bleed	0.8 3.4 5.7	53 239 400	100 Intox 450 LD50 780 MDS	Ipecac+charcoal+cathartic Same three plus support Same four ± dialysis
Isopropyl (rubbing alcohol; $CH_3\text{-}CHOH\text{-}CH_3$; 60.09)	Acetone	Ketosis, CNS depression, tracheobronchitis	0.4 1.1 3.3	26 77 230	50 Intox 150 MLD 440 MDS	Ipecac+charcoal+cathartic+support Same Same ± dialysis
Methanol (wood alcohol; $CH_3\text{-}OH$; 32.04)	Formic acid (ADh)	Acidosis, blindness, CNS depression, headache, nausea, vomiting, abdominal pain	0.15 0.4 1.0	11 27 70	20 50 135 MDS	Ipecac+charcoal+cathartic Same three plus ethanol Same four ± dialysis
Ethylene glycol (antifreeze, coolant; $HO\text{-}CH_2\text{-}CH_2\text{-}OH$; 62.07)	Oxalic acid (ADh)	CNS depression, cardiotoxicity, osmolar gap, renal failure, oxylate crystalluria	0.3 1.0 1.4	21 70 100	40 140 200 MLD	Ipecac+charcoal+cathartic Same three plus ethanol Same four ± dialysis
Chloral hydrate [knockout drops, hypnotic; $Cl_3\text{-}CH\text{-}(OH)_2$; 165.42]	Trichloroethanol, trichloroacetic acid (ADh)	CNS depression, nausea, vomiting, hepatotoxicity, arrhythmia, hypotension	(mg/kg) 14 57 143 429	(g/70) 1 4 10 30	(µg/ml[b]) 8 Usual 32 MLD 80 Toxic 240 MDS	Ipecac+charcoal+cathartic Same Same Same

[a] ml/70, Dose per standard adult (70 kg); Intox, intoxicating dose; LD50, 50% lethal dose; MLD, minimum lethal dose; MDS, maximum dose reportedly survived. Airway + breathing + cardiac support; ± dialysis, hemodialysis may be appropriate; EtOH, Treat with ethanol to block ADh. CNS, Central nervous system. ADh, Metabolism dependent on ADh.

[b] Levels of trichloroethanol.

(i) Ethanol. Ethanol, certainly the most common and economically important neurotoxin, has not been reported as neurotoxic by inhalation. Ethanol (ethyl alcohol, EtOH, alcohol) is the active ingredient in beer, wine, liquor, and moonshine. Death usually results from respiratory failure and has been reported at blood ethanol concentrations of 260 mg/dl. Levels of 780 mg/dl have been survived, but the 50% lethal dose corresponds to a blood ethanol concentration of 450 mg/dl. The overall impact of ethanol abuse on our society is difficult to estimate. Consumption in the United States amounts to 5.26×10^9 gallons of beverage per year, causing 4,161 accidental poisoning deaths per year, and costs an estimated $15 billion per year. Alcoholism of laboratory workers should be recognized and treated. Due to the easy availability of pure ethanol in most laboratories, acute poisonings could occur.

Ethanol is completely and rapidly adsorbed after ingestion and distributes in total body water (60% of weight). Elimination proceeds at a level-independant rate (zero order) via alcohol dehydrogenase (ADh). Ethanol acts as a central nervous system depressant and gastrointestinal irritant.

Treatment for ethanol intoxication is the same as for any other sedative hypnotic: basic life support, emesis (if early), and referral to an emergency department if the patient is symptomatic or has ingested more than 1 ml of absolute ethanol per kg.

(ii) Isopropanol. Isopropanol is the principal component in most rubbing alcohols, but some contain mainly ethanol. Many people regularly drink isopropanol and claim a "good high," but patients usually come to medical attention because of accidental or mistaken ingestion. Absorption and distribution are similar to those of ethanol. Isopropanol is metabolized to acetone (80%), but more slowly than is ethanol (28). Toxicity is about twice that of ethanol (22).

(iii) Methanol. Methanol (methyl alcohol, wood alcohol, acetone alcohol, Manhattan spirit) is a widely used industrial solvent which has caused many fatal poisonings or blindness after ingestion, inhalation, or skin absorption (6). The narcotic effects of parent methanol approximate those of ethanol. Toxicity results from the synthesis via ADh of formic acid (25). Initial nausea, vomiting, headache, and abdominal pain are followed by a latent period of 12 to 24 hours and then acidosis.

(iv) Ethylene glycol. Ethylene glycol is a colorless, odorless, sweet-tasting liquid with a high vapor pressure which is widely used as a solvent, coolant, and antifreeze. Permanent antifreeze, the most common source, contains 95% ethylene glycol. Toxicity results from synthesis of oxalic acid via ADh.

(v) Methyl-*n*-butyl ketone. Methyl-*n*-butyl ketone was implicated in an industrial epidemic of 86 cases of peripheral neuropathy. Sensory symptoms predominated and electromyographic changes were found in 38 other workers (26).

(vi) Methyl ethyl ketone. Methyl ethyl ketone has failed to cause neurotoxicity in animal studies, but may be synergistic with methyl-*n*-butyl ketone or with toluene. Methyl ethyl ketone has been implicated in a case of retrobulbar neuritis which resolved in 6 days. The methanol detected in this patient was believed to represent a metabolite of methyl-*n*-butyl ketone and the toxic principal (7).

Mixed solvents. (i) Gasoline. Gasoline has produced central nervous system depression, hallucinations, chorea, and abnormal electroencephalogram. Peripheral neuropathy and dementia have been reported with heavy abuse, but the contribution of known neurotoxins in the mixture, such as triorthocresyl phosphate, is difficult to assess (39). Tetraethyl lead is increasingly recognized as a source of chronic, and in some cases severe, lead poisoning in children abusing leaded gasoline.

(ii) Acrylic paints. Acrylic paints and spray paints are among the most common substances abused (16), and changes in the formulations have had disastrous effects on seven huffers (31).

Fluorocarbons. "Freon" refers to the series of fluorinated hydrocarbons widely used as refrigerant gases and cocktail glass chillers, but mainly as a propellant in the more than two million types of aerosols sold in the United States. The recreational inhalation of Freon grew predictably out of the widespread solvent abuse prevalent among U.S. teenagers in the 1960s. Bass called attention to the "sudden sniffing death" syndrome in 1970, describing 110 sudden deaths throughout the United States (4).

Freon 12 (dichlorodifluoromethane) was the first Freon synthesized and is probably the most widely used of the Freons. Freon 12 is used in most aerosols, whereas Freon 22 is most widely used in refrigerators and air conditioners. Rapid absorption from the lungs results in peak levels within seconds of inhalation. Freon disappears from the blood with a half-life of approximately 4 min and a whole-body elimination half-life of approximately 40 min.

Little doubt remains as to the cardiotoxic effects of Freon 12 and the other fluorocarbons. There are, however, two schools of thought as to the principal mechnisms of fatal arrythmias: cardiac sensitization versus direct toxic effects. Inhalation of 10 to 20% Freon 12 produced a direct, dose-related depression of left ventricular function in rabbits even in the absence of hypoxia and arrhythmia (37). Changes found in the liver at autopsy support a direct hepatotoxic effect in large-dose exposures (27). Freon 22 has been less well studied, but exhibits qualitatively similar toxicity. Pathology staff using Freon 22 to accelerate freezing of tissue samples experienced 3.6 times as many heart palpitations as a control group (34).

Caustics

The term "caustic" or "corrosive" used here refers to substances which produce severe tissue destruction generally without any toxicologic activity. Most common are the strong acids or alkalis, but any strong oxidizing or reducing agent may also be included. While both alkalis and acids cause tissue destruction, acids more frequently produce a coagulation necrosis which results in a superficial burn. Alkalis, in contrast, tend to produce a penetrating tissue destruction. Thus an ingested alkali more frequently causes esophageal burns leading to perforation and stricture formation. Acid ingestions, in contrast, more frequently cause burns in the stomach, particularly in the pyloric region.

Corrosives abound in the laboratory as metal and glassware cleaners, as well as reagents. For either an internal or external exposure to a caustic, dilution represents the primary intervention. The heat generated from any neutralization clearly contraindicates the use of this measure in management of caustic exposures (11).

Skin exposures
1. Wash the exposed area twice with copious amounts of water and any available soap.
2. Any persistent irritation or pain after the washing requires a physician's evaluation.

Ingestion
1. Immediate therapy consists of dilution with milk or water. Do not exceed 15 ml/kg in a child or 250 ml (½ pint) in an adult, as excessive fluids promote vomiting.
2. Patient should be kept NPO (no fluids by mouth) after initial dilution until they undergo medical evaluation.
3. Medical evaluation will require direct visual examination (esophagoscopy). Absence of oral burns does not preclude esophageal burns.
4. Call the nearest regional poison center or medical center.

Inhalation
1. Remove patient to fresh air.
2. Any symptomatic patient should receive oxygen and be transported for evaluation to the nearest medical facility.

Although the various techniques used in management of exposure to caustic agents are more alike than different, some specific comments about the following agents seem appropriate.

Hydrofluoric acid. Hydrofluoric acid may cause deep, penetrating, and painful necrotic lesions, depending on the concentration. When the concentration is greater than 50%, the burn is felt immediately and tissue destruction becomes rapidly apparent. Exposure to a 20 to 50% solution may cause burns which remain inapparent for 1 to 8 h after exposure, and burns by less than 20% hydrofluoric acid may not be noticed until up to 24 h after exposure.

In addition to immediate dilution, a solution of 25% magnesium sulfate or a high-molecular-weight quaternary ammonium compound (Hyamine no. 162) or benzalkonium chloride (Zephiran) may benefit the patient with protracted pain (33). This treatment should be administered under the direction of a physician. For more details on emergency and subsequent treatment of hydrofluoric acid exposure, see Chapter 30.

Chlorine-active compounds. Sodium hypochlorite, typically found as a 5.4% solution in the common household bleach Clorox, is a widely used antimicrobial agent. Clorox presents a relatively mild hazard in terms of ingestion. When Clorox is mixed with an acid toilet bowl cleaner, however, chlorine gas is produced. When mixed with ammonia, chloramine gas results. Either agent produces irritation and burning of the mucous membranes. Chlorine gas represents a greater hazard to the lungs, and medical evaluation is appropriate.

Mercury, lead, and other heavy metals

Of the 77 metals, 52 have industrial or environmental importance. Iron, copper, zinc, molybdenum, and many others are essential to normal cellular function. Others such as lead and mercury have no known physiologic function, but represent a major environmental and acute health hazard. Carcinogenic potential seems well established for nickel carbonyl, chromium, and arsenic and is suspected for several others. Hypersensitivity reactions, such as the well-recognized response to nickel, represent a principal hazard of the metal compounds.

Metals form stable complexes with many ligands, especially sulfur and nitrogen. Little generalization is possible regarding the critical organ or toxic effects of the numerous metal compounds available in the laboratory. Minimizing exposure to the metal compounds, clearly the best approach, should be assiduously followed for these and other toxic compounds. Any of several excellent references should be consulted for specific agents (12). Two specific agents are briefly discussed here, mercury because it is frequently encountered in metallic form, and lead because of its ubiquitous and insidious nature.

Mercury. The ancient Egyptians recognized the health risk posed by mercury and used slaves and convicts to work the mines. Although this element is particularly toxic in the ionized form, metallic mercury represents little toxic hazard. Ingestion of mercury from thermometers or spillage from Miller-Abbott tubes does not result in any evidence of systemic mercury poisoning if it stays within the gut. Mercuric salts and organomercurials react with sulfhydryl-rich macromolecules, but metallic mercury undergoes no such reactions (8).

A spill of metallic mercury, such as from a thermometer or weight, should therefore be cleaned up with an appropriate adsorbent. Heating the metallic mercury or cleaning up with a vacuum cleaner should be carefully avoided as the resulting vapors and oxides can produce serious poisoning.

Lead poisoning. Lead's stability under normal chemical exposures has resulted in its use in glass,

TABLE 5. Summary of special toxicology managements[a]

Name of poison	Target organ	Clinical effects	Indications for treatment	Special treatment
Acetaminophen	Liver	Nausea, vomiting, diarrhea, hepatitis	>100 mg/kg >140 mg/kg or >200 μg/ml over 2 h	Ipecac + charcoal + cathartic Same; to ED[b] + N-acetyl cystein (140 mg/kg loading dose; 70 mg/kg every 4 h for 17 doses)
Anticholinergics (Jimson weed, tricyclic antidepressants, antihistamines)	Heart, CNS[c]	Tachycardia (QRS > 100 ms), hallucinations	Convulsions, hallucinations, hypertension, serious arrhythmia	Alkalinization; physostigmine (1 to 2 mg, slow intravenously)
Barbiturates and ethanol	CNS	Somnolence, respiratory arrest, cellular hypoxia	Respiratory depression, hypotension	Ipecac + charcoal + cathartic; respiratory support (1 to 2 mg, slow IVπ)
Carbon monoxide	Hemoglobin	Cellular hypoxia	Suspicion	Maximum 100% oxygen ± hyperbaric oxygen
Cyanide (laetrile, fires)	Cytochromes	Headache, nausea, vomiting, cardiovascular collapse	History Symptoms	Amyl nitrate Sodium nitrate, sodium thiosulfate
Iron (300 mg of ferrous sulfate = 60 mg of elemental iron)	Gastrointestinal, cardiovascular coagulopathies	Gastroenteritis, shock	>20 mg/kg >60 mg/kg SeFe >500	Ipecac + cathartic Same + to ED Desferroxamine; lavage, 5% $NaHCO_3$
Lithium	CNS	Coma, hyperreflex, T-wave abnormalities	>1.5 mEq/liter >4.0 mEq/liter	Osmotic diuresis Hemodialysis
Narcotics (heroin, demerol, propoxyphene, pentazocine)	CNS	Miosis, respiratory depression	Suspicion	Naloxone (2 mg test intravenous dose; 0.4 mg/amp)
Organophosphate insecticides	Cholinergic	Salivation, lacrimation, urination, defecation (SLUD syndrome)	History Symptoms	Atropine (1 to 2 mg of PRN) Proladoxine (2-PAM)
Petroleum distillates (gasoline, kerosene)	Lungs (swallow = 0.3 ml/kg)	Aspiration pneumonia	>0.1 ml/kg >1 ml/kg	Cathartic only Ipecac + cathartic
Salicylates	Lungs; hemostasis	Pulmonary edema, bleeding	>100 mg/kg >240 mg/kg >360 mg/kg	Ipecac + charcoal + cathartic Same + ED Same + ED + hemodialysis

[a] The specific measures listed and their indications should be considered in addition to basic support and prevention of absorption.

[b] ED, Transport to emergency department.

[c] CNS, Central nervous system depression.

solder, storage batteries, paint, and ornaments and as an antiknock additive to automobile fuel. Lead remains toxic throughout all phases of its processing, and absorption by ingestion or inhalation at any stage may result in poisoning. Current lead levels in air, water, and food reflect a 500-fold increase compared to prehistoric times. A futher fivefold increase in our average daily intake results in clinically apparent lead toxicity (32).

Most adult intoxication results from industrial exposure. Breathing lead dust and fumes presents the greatest risk. Any known excessive exposure in the laboratory, with or without symptoms, should prompt an evaluation. Blood lead levels provide the best available measure of an individual's lead burden. Blood lead exceeding 50 μg/dl reflects an increased lead burden and should be treated. Levels between 40 and 50 μg/dl require further evaluation.

Summary of special treatments

Table 5 summarizes some of the specific measures commonly employed in acute poisoning in addition to basic life support and prevention of absorption. Thus, the table reflects both tests which the laboratory will likely be involved in running and poisonings for which immediate intervention is important.

Poison information resources

Recognition of acute poisoning as a major source of morbidity and mortality, particularly in the pediatric age group, gave rise to the concept of a Poison Control Center in 1953. Since that time approximately 600 Poison Control Centers have been officially designated by state health departments. For many of these centers at first, the volume of calls was low and the quality of responses was variable at best. Since 1970, the development of regional poison centers with full-time, specially trained poison information specialists answering a large number of calls represents a major advance in the standardization and quality of poison information service. Recommended regional Poison Control Centers for the United States can be found in Appendix 3 of this volume.

LITERATURE CITED

1. **Allen, N.** 1975. Chemical neurotoxins in industry and environment, p. 235–248. *In* D. B. Tower (ed.), The nervous system. Raven Press, New York.
2. **American Heart Association.** 1980. Standards and guidelines for cardiopulmonary resuscitation (CPR) and emergency cardiac care (ECC). J. Am. Med. Assoc. **244:**453–509.
3. **Austen, K. F.** 1974. Systematic anaphylaxis in the human being. N. Engl. J. Med. **13:**661–664.
4. **Barclay, W. R.** 1978. Emergency treatment of insect-sting allergy (commentary). J. Am. Med. Assoc. **240:**2735.
5. **Bass, M.** 1970. Sudden sniffing death. J. Am. Med. Assoc. **212:**2075–2079.
6. **Bennett, I. L., H. C. Freeman, G. L. Mitchell, and M. N. Cooper.** 1953. Acute methyl alcohol poisoning: a review based on experience in an outbreak of 323 cases. Medicine **32:**431–463.
7. **Berg, E. F.** 1971. Retrobulbar neuritis—a case report of presumed solvent toxicity. Ann. Opthalmol. **3:**1351–1353.
8. **Bidstrup, P. L.** 1964. Toxicity of mercury and its compounds, p. 2–87. *In* E. Browning (ed.), Elsevier monographs on toxic agents. Elsevier Publishing Co., New York.
9. **Cavalleri, A., and V. Cosi.** 1978. Polyneuritis incidence in shoe factory workers: case reports and etiological considerations. Arch. Environ. Health **33:**192–197.
10. **Cohr, K. H., and J. Stokholm.** 1979. Toluene: a toxicologic review. Scand. J. Work Environ Health **5:**71–90.
11. **Fow, M. I.** 1979. First aid recommendations changed for caustic ingestions. National Clearinghouse for Poison Control Centers Bull. **23:**1–2.
12. **Friberg, L., F. Nordberg, and V. B. Vouk.** 1979. Handbook on the toxicology of metals. Elsevier/North-Holland Biomedical Press, New York.
13. **Goldbert, T. M., R. Patterson, and J. J. Pruzansky.** 1969. Systematic allergic reactions to ingested antigens. J. Allergy **44:**96–107.
14. **Goldfrank, L., and A. Mayer.** 1978. Anaphylaxis—the IVP emergency. Hosp. Phys. **8:**28–32.
15. **Haddon, W.** 3. Energy damage and the ten countermeasure strategies. J. Trauma **13:**321–331.
16. **Hayden, J. W., E. G. Comstock, and B. S. Comstock.** 1976. The clinical toxicology of solvent abuse. Clin. Toxicol. **9:**169–184.
17. **Herskowitz, A., N. Ishii, and H. Schaumburg.** 1971. *n*-Hexane neuropathy: a syndrome occurring as a result of industrial exposure. N. Engl. J. Med. **285:**82–85.
18. **House, H. O.** 1980. Prudent practices for handling hazardous chemicals in laboratories, p. 285. National Academy Press, Washington, D.C.
19. **Jones, D. V., and C. E. Work.** 1961. Volume of a swallow. Am. J. Dis. Child. **102:**427.
20. **Key, M. M., and A. F. Herschel.** 1977. Occupational diseases: a guide to their recognition. U.S. Department of Health, Education and Welfare. Publication no. 017-033-00266-5. U. S. Government Printing Office, Washington, D.C.
21. **Knox, J. W., and J. R. Nelson.** 1966. Permanent encephalopathy from toluene inhalation. N. Engl. J. Med. **275:**1494–1496.
22. **Lehman, A. J., and H. F. Chase.** 1944. The acute and chronic toxicity of isopropyl alcohol. J. Lab. Clin. Med. **29:**561–567.
23. **Linden, C. H., K. Kulig, and B. H. Rumack.** 1985. Hydrocarbons, p. 849–857. *In* R. F. Edlich and D. A. Spyker (ed.), Current emergency therapy '85. Aspen, Rockville, Md.
24. **Lutzemberger, A. C., N. Rizzuto, A. Simonati, A. Rossi, and G. Toschi.** 1980. Neurotoxic effects of 2,5-hexanedione in rats: early morphological and func-

tional changes in nerve fibres and neuromuscular junctions. Neurotoxicology **2**:25–32.

25. **McMartin, K. E., J. J. Ambre, and T. R. Tephly.** 1980. Methanol poisoning in human subjects. Role for formic acid accumulation in the metabolic acidosis. Am. J. Med. **68**:414–418.

26. **Mendell, J. R., et al.** 1974. Toxic polyneuroapthy produced by methyl-*n*-butyl ketone. Science **185**:787–789.

27. **Morita, M., A. Miki, H. Kazama, and M. Sakata.** 1977. Case report of deaths caused by freon gas. Forensic Sci. **10**:253–260.

28. **Nordman, R., C. Ribieve, and H. O. Rouach.** 1973. Metabolic pathways involved in the oxidation of isopropanol into acetone by the intact rat. Life Sci. **13**:919–932.

29. **Pittelkow, R.** 1975. Occupational dermatoses, p. 191–203. *In* C. Zens (ed.), Occupational medicine, principles and practical applications. Yearbook Medical Publishers, Chicago.

30. **Prockop, L.** 1979. Neurotoxic volatile substances. Neurology **29**:862–865.

31. **Prockop, L. D., M. Alt, and J. Tison.** 1974. "Huffer's" neuropathy. J. Am. Med. Assoc. **229**:103–104.

32. **Settle, D. M., and C. C. Patterson.** 1980. Lead in albacore: guide to lead pollution in Americans. Science **207**:1167–1176.

33. **Shewmake, S. W., and B. G. Anderson.** 1979. Hydrofluoric acid burns: a report of a case and review of the literature. Arch. Dermatol. **115**:593–596.

34. **Speizer, F. E., D. H. Wegman, and A. Ramirez.** 1975. Palpitation rates associated with fluorocarbon exposure in a hospital setting. N. Engl. J. Med. **292**:624–626.

35. **Spyker, D. A.** 1985. Activated charcoal reborn: progress in poison managment. Arch. Intern. Med. **145**:43–44.

36. **Streicher, H. Z., P. A. Gabow, A. H. Moss, D. Kono, and W. D. Kaehny.** 1981. Syndromes of toluene sniffing in adults. Ann. Intern. Med. **94**:758–762.

37. **Taylor, G. J. and R. T. Drew.** 1975. Cardiovascular effects of acute and chronic inhalations of fluorocarbon 12 in rabbits. J. Pharm. Exp. Ther. **192**:129–135.

38. **Towfighe, J., N. K. Gonastas, D. Pleasure, H. S. Cooper, and L. McGee.** 1976. Glue sniffer's neuropathy. Neurology **26**:238–243.

39. **Valpey, R., S. M. Sumi, M. K. Copas, and G. J. Goble.** 1978. Acute and chronic progressive encephalopathy due to gasoline sniffing. Neurology **28**:507–510.

Chapter 30

Laboratory Personnel as the First Responders

RICHARD F. EDLICH, ELAINE LEVESQUE, RAYMOND F. MORGAN, JOHN G. KENNEY,
KIMBERLY A. SILLOWAY, AND JOHN G. THACKER

With the development of comprehensive emergency medical services systems in the United States, it has become clear that appropriate early action can save lives. The first individual (the "first responder") to aid an injured or ill patient is a prime link in the chain of an emergency medical system, providing immediate, urgent medical help until specially trained ambulance personnel (emergency medical technicians) arrive to continue the care of the patient.

Victims of heart attacks are among those who benefit most from immediate treatment. Cardiac arrest victims are more than twice as likely to survive if they are given cardiopulmonary resuscitation (CPR) by a first responder, even if a well-trained paramedic ambulance team is quickly available (46). Based on these statistics, it has been estimated that the first responder could save between 100,000 and 200,000 lives each year in the United States.

During the past decade, many first aid courses have been developed to train the general public to become first responders. The most notable of these is the CPR course prepared by the American Heart Association (4). Certification of competency in CPR is based on specially prepared written and performance tests. CPR and other first aid training programs are offered by the American Heart Association, the American Red Cross, and other agencies across the country. Judd and Ponsell (30) have recently designed an expanded training program, entitled "The First Responder," covering the emergency treatment of a broad spectrum of accidents and illnesses.

It is the purpose of this chapter to describe the role of laboratory personnel as first responders in the treatment of persons with cardiorespiratory emergencies or burn or cold injuries. As first responders, the laboratory personnel will provide immediate life support care until more highly trained medical help (paramedics, emergency medical technicians) arrives. Since first responders are only the first part of a comprehensive emergency medical system, we have also included information in this chapter regarding treatment of some of these illnesses or injuries in the hospital setting, to provide a broad perspective of emergency care.

CARDIAC AND BREATHING EMERGENCIES

CPR of the victim of cardiac arrest should be started immediately regardless of the cause. The time interval between when the heart stops and resuscitation begins is critical. Lund and Skulberg (33) found that when this interval was less than 1 min, 61% of heart victims survived to be discharged from the hospital. When this delay lasted 5 to 10 min, the survival rate was only 9%. Reports from Charlottesville, Va. (18), Seattle, Wash. (3), and Belfast, Ireland (1) indicate that mortality among cardiac victims is sharply reduced when a fast-responding emergency medical system is available.

CPR training of laymen has been proven to be feasible, successful, and economically practical. Alvarez and Cobb (3) reported that a CPR training program can be conducted at a cost of $1.25 per student. Training aids, which include resuscitation manikins, provide the student with actual practice in CPR. Twelve million persons plan to be trained (4). All laboratory personnel should be given this training, with specific reference to the emergency medical system available to the laboratory.

Specific resuscitation techniques for choking and unconscious victims are given in Appendix 4, First Aid Guide, of this volume. External cardiac compression for pulseless victims should be performed only by a person properly trained in the technique.

BURN INJURIES

Burns in the laboratory generally result from either a heat source, chemical agents, electricity, or UV radiation.

Thermal Burns

Heat damages tissue by protein denaturation. The magnitude of burn injury depends on the intensity of the heat, the length of time in contact with it, and the conductivity of the tissue involved. The specific heat conductivity of human tissue determines the rate and amount of heat transfer and, in turn, ultimate injury. The water content of tissues, local natural oils and secretions, dermal appendages, and natural insulating materials, such as the cornified layers of the skin, affect specific tissue conductivity. The uppermost layer of skin, the epidermis, is relatively uniform in thickness in all body regions (75 to 150 μm) except

TABLE 1. Specific heat of some common substances

Substance (at 60°C)	Specific heat (cal/g per °C)
Gold, lead, platinum	0.03
Tin	0.05
Silver	0.06
Copper, bronze, brass	0.09
Steel	0.11
Glass	0.20
Oxygen (gas)	0.22
Sugar	0.27
Leather	0.36
Wood	0.42
Rubber	0.45
Butane gas (liquified)	0.55
Paraffin	0.69
Water	1.00
Ammonia	1.12

TABLE 2. Flash points of some common liquids

Liquid	Flash point °C	Flash point °F
Corn oil	249	480
Lard	215	419
Cottonseed oil	305	581
Olive oil	275	527
Peanut oil	282	540
Oleo	232	450

for the soles of the feet and palms of the hands, where it attains a thickness of 0.4 to 0.6 mm. Full-thickness injury to palms and soles is rare. The dermal appendages of very young and very old persons are superficial, which renders such individuals susceptible to full-thickness burn injury. The presence of a red discoloration of burned skin in elderly and infant patients, as a result of hemoglobin fixed in the tissues, is a sign of deep thermal injury. By contrast, the epidermal appendages of the human scalp and male beard are very deep, making these sites more resistant to severe burn injury.

The specific heat of a substance is the amount of heat necessary to raise a certain volume of that substance a spécific number of degrees. It is usually expressed in calories per gram per degree centigrade. The higher a material's specific heat, the more heat it is capable of storing and releasing. The specific heat of water (1.00), the most common cause of scald injuries, is higher than that of all the gases and solids tested so far (Table 1). Only two liquids have a higher specific heat: ammonia, at 1.125, and ether, which achieves a specific heat of 1.041 only after its temperature has been increased to 180°C (356°F).

Many liquids, like water, cannot be heated beyond a certain temperature without changing state. Water can be heated only up to 100°C (212°F) at atmospher-

ic pressure before it vaporizes. When other liquids reach a certain temperature, they ignite or oxidize. The temperature at which the vapors of a volatile liquid spontaneously ignite is the flash point of the liquid. Any liquid having a flash point below 37.8°C (100°F) is termed ''flammable''; ''combustible'' liquids have a flash point above this temperature (Table 2).

Human skin can tolerate temperatures up to 44°C (111°F) for ca. 6 h before an apparent injury (35). At temperatures between 44 and 51°C (124°F), the rate of cellular destruction doubles with each degree of increase. Complete epidermal necrosis is noted 90 min after contact with 46°C (115°F). At 47°C (117°F), only 45 min is required to result in full destruction of all layers of the skin. When temperatures rise above 70°C (158°F), less than 2 s is needed for complete injury to the skin.

The length of time during which the human skin is exposed to the heat source depends on both the heat source and the victim. When spilled or splashed, viscous oils and greases will cling to the victim's skin, whereas water quickly runs to the floor. In immersion scalds, where the victim is immersed in a hot liquid, the duration of contact with heat will be considerably longer than in spill scalds. Clothing can prolong heat contact by slowing the descent of spilled hot liquids or otherwise absorbing heat from the source, serving as a temporary heat sink over the skin. If the clothing ignites, the injury will be that much more severe.

The depth of burn injury can be estimated by its appearance, biomechanical properties, and function (26) (Table 3). When the burned skin has a de-

TABLE 3. Clinical diagnosis of depth of burn injury

Depth of burn	Appearance	Hair	Biomechanical properties	Sensation	Pain
Superficial partial thickness	Pink with blisters	Present	Elevated, soft, and pliable	Light touch, pinprick, and pressure	Exquisitely painful
Deep partial thickness	White	Absent	Elevated, soft, and pliable	Pressure	Painful
Full thickness	Brown with thrombosed vein	Absent	Depressed, leathery	No sensation	Painless

pressed, white or brown, leathery appearance with underlying clotted vessels, it is usually a full-thickness ("third-degree") burn injury. Palpation of the burned skin demonstrates that it is inelastic. Because the nerve endings have been destroyed, full-thickness burns are painless and insensate.

A superficial partial-thickness ("second-degree") burn is characterized by pink skin with blisters. The burned skin is usually raised, soft and pliable, and exquisitely painful and sensitive to the touch. Stubbles of hair are often visible in these injuries. A deep partial-thickness burn exhibits whitened skin without blisters that is raised, soft, and pliable. Pinprick cannot be perceived, but gentle pressure on the burned skin can still be detected.

The extent of body surface involved in a burn injury is commonly estimated by the "rule of nines" (Fig. 1) (25). This is accomplished by dividing the major anatomic portions of the body into multiples of 9% of the major body surface area. This is not a precisely accurate method of measuring body surface area, but more exact measurements are so complicated that they have little practical value. The small differences between the surface areas of the adult and the infant reflect the surface area of the infant's head, which is proportionally larger than that of an adult.

Regardless of the type of heat source, the basic principles of treatment of the burn are to (i) stop the burning process, (ii) initiate CPR if necessary, (iii) care for the burn wound, and (iv) transfer the patient to professional medical care.

Obviously, the shorter the time the patient is in contact with the heat source, the less the injury will be. The burning patient himself by his actions has considerable control over his fate. If a victim's clothes have ignited, he must roll to the ground to

FIG. 2. Extinguish flaming clothing by rolling on the ground.

FIG. 3. Use a blanket to extinguish residual flames.

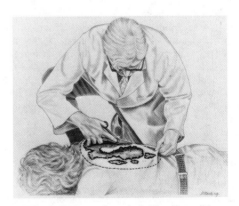

FIG. 4. Remove charred clothing if possible.

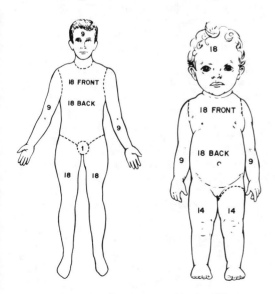

FIG. 1. Rule of nines for adults and children.

extinguish them (Fig. 2). Unfortunately, this response to burning is seldom encountered in our experience (26); in most cases the victim either runs for help or remains motionless. In the presence of flammable liquids, rolling on the ground does not always extinguish the flames. The first responder can extinguish residual flames by covering them with a wool blanket (Fig. 3). Once the flames are out, the rescuer should remove the burned clothing from contact with the skin (Fig. 4). Charred clothing may still be hot enough to increase the depth of the burn. Occasionally burned clothing will stick to the underlying burned skin. In such cases, it should not be removed, since this might cause bleeding which would demand immediate attention.

A new water gel fire blanket (Trilling Resources Ltd., Hartsdale, N.Y.) has been advocated to suppress flaming clothing. These blankets contain a gel that rapidly extinguishes flames. This benefit must be weighed against potential disadvantages. The fire safety blanket is stored in a container that is not easily accessible to the rescuer. Moreover, the effect of the gel on injured skin has not been adequately tested. Until these potential drawbacks have been resolved, we recommend that rescuers continue to use the wool blankets that are readily available in most laboratories. If a blanket is not at hand, the fire may be extinguished with a dry chemical multipurpose (ABC) fire extinguisher. This dry chemical can control a broad range of fires without damaging human skin or its underlying soft tissues.

As soon as the burning is stopped, the first responder must check whether the patient is conscious and breathing. If the patient is having difficulty, the first responder should begin resuscitation as described in Appendix 4. If the person is breathing, the first responder should look for signs of inhalation injury, such as burns of the mouth and nose, singed nasal hairs, sooty sputum, and a brassy cough. If one or more of these signs are present, it may be necessary for trained ambulance personnel to administer oxygen.

The best immediate treatment for the burn wound is cold water immersion, an accepted first-aid treatment for more than a century and one of the most effective measures of relieving pain from a recent burn. Experimental studies have demonstrated that local cooling will lower mortality and help prevent burn shock (31,37). Cooling of the burn is accomplished by immersing the wound in water at a temperature of 1 to 5°C (34 to 41°F) for ca. 30 min. To be effective, this treatment must begin as soon as possible during the first 30 min after burning; if cooling has been delayed more than 30 min, it has no therpeutic benefit. Local cooling, of less than 9% of the body surface, may be continued to relieve pain after this 30-min period. However, prolonged cooling of a larger body area may result in a dangerous drop in body core temperature, and cardiac arrest may occur. Ice should never be applied directly to the burn since this may result in frostbite.

The cardinal feature of a burn injury is an accumulation of fluid within the injured area and, to a lesser degree, in the unburned skin. This fluid is a consequence of a thermally induced defect in vascular permeability that allows intravascular fluids to extravasate into the tissue. Loss of intravascular fluids may result in burn shock, which is recognized by arterial hypotension and rapid pulse. The first responder should treat burn shock by elevating the patient's legs about 12 in.

A burn wound has been characterized as an open pan of water from which the water can evaporate. Healthy skin normally functions as a lid over this pan, limiting transepidermal water loss. The evaporation of 1 liter of water involves a heat loss of 560 kcal. As a result of the increased water evaporation from his burned skin, the patient will often complain of being chilled. This heat loss can be minimized by covering the patient with a clean sheet followed by clean blankets.

Chemical Burns

Accidental contact with hazardous chemicals is one of the potential dangers in a laboratory setting. More than 25,000 products marketed for use in the laboratory, industry, agriculture, or the home are capable of producing chemical burns. Emergency department physicians and poison control centers can attest to the rising number of such burns. It is estimated that more than 60,000 victims of chemical burn injury seek professional care, with more than 3,000 deaths related to skin or gastrointestinal chemical injuries (19).

Most chemical agents damage the skin by producing a chemical reaction rather than a hyperthermic injury. Although some chemicals produce considerable heat of hydration when they come in contact with water, their ability to produce direct chemical changes in the skin accounts for the most significant injury. Chemical agents implicated in skin burn injuries can be classified according to their chemical activity into five groups: oxidizing and reducing agents, corrosives, protoplasmic poisons, desiccants, and vesicants (29). The degree of skin destruction is determined mainly by the concentration of the toxic agent and the duration of its contact. When the skin is exposed to toxic chemicals, its keratinous covering is destroyed and its underlying dermal tissues are exposed to continued necrotizing action.

The absorption of some agents through the skin may cause systemic toxicity: dichromate poisoning produces liver failure, acute tubular necrosis, and death; oxalic acid and hydrofluoric acid injuries may result in hypocalcemia; tannic acid or phosphorus may cause hepatic necrosis (45); picric acid and phosphorus burns may be followed by nephrotoxicity; absorption of phenol may be associated with central nervous system depression and hypotension. Inhalation injury will result from exposure in a closed space to toxic fumes, such as ammonia or strong acids. Whenever patients develop signs of systemic toxicity, hemodialysis or exchange transfusions must be considered.

Treatment

Chemical burns continue to destroy tissue until the causative agent is inactivated or removed. Therefore, it is imperative to initiate treatment as soon as possible after contact. Speed is most important in treating chemical burns. When hydrotherapy is initated 1 min after contact with either an acid or alkali, the severity of the skin injury is far less than when treatment is delayed for 3 min (28). This early treatment is followed by a return of the pH of skin to normal. When the contact time is more than 1 h before hydrotherapy, the pH of a sodium hydroxide (NaOH) burn cannot be reversed. Similarly, a brief washing of a hydrochloric acid burn more than 15 min after exposure does not significantly alter the acidity of the damaged skin. Hydrotherapy is indicated in all cases *except* lithium and sodium metal burns, as discussed below.

Since contact time is a critical determinant of the severity of injury, hydrotherapy of skin exposed to a toxic liquid chemical must be initiated immediately by the victim or witness to the injury (39). Gentle irrigation, with large volumes of water under low pressure for a long time, dilutes the toxic agent and washes it out of the skin. When workers' clothes are immersed with such agents, valuable time can be lost in removing their clothing before copious washing is commenced. The patient's clothes should be removed during hydrotherapy by the rescuer, who should wear rubber gloves to protect his hands from contact with the chemical. In patients whose skin is exposed to strong alkalis, prolonged hydrotherapy is especially important to limit the severity of the injury. In experimental animals, the pH of chemically burned skin does not approach a normal level unless more than 1 h of continuous irrigation has been maintained (11). This aberration of pH in NaOH skin burns persists for 12 h despite hydrotherapy, in contrast to HCl skin burns, in which the pH returns to normal within 2 h of hydrotherapy. The mechanism by which NaOH maintains an alkaline pH despite treatment is related to the by-products of its chemical reaction to skin. Alkalis, including sodium hydroxide, combine with proteins or fats in tissue to form protein complexes or soaps which are soluble. These complexes permit passage of hydroxyl ions deep into the tissue, limiting their contact with the water dilutent on the skin surface. Acids do not form complexes, and the free hydrogen ions are easily neutralized. In either case, hydrotherapy must be continued after the patient is transported to an emergency department with advanced life support capability. If the chemical is localized to the patient's extremity, the injured part can be immersed in a sink under flowing tap water. For other anatomic sites, the patient is placed supine in a Hubbard tank in which the temperature of the water can be regulated. With alkali burns, the hydrotherapy is continued for 12 h or more. Hydrotherapy treatment of acid burns usually lasts 2 to 3 h.

When the victim's clothing comes in contact with a solid chemical like lye or white phosphorus, it is necessary to remove the contaminated clothing before beginning hydrotherapy (39). All visible particles must be removed from the patient's skin during copious irrigation with water. Particles of phosphorus should be submerged in cold water, since they ignite spontaneously in the air when dry. The water should be delivered to the wound at the lowest possible pressure, since high-pressure irrigation (shower) may disperse the liquid or solid chemical into the victim's or rescuer's eyes.

Do not attempt to neutralize the chemical. As early as 1927, the dangers of such attempts were demonstrated in experimental animals (22). In every instance, animals with alkali or acid burns that were washed with water survived longer than animals treated with chemical neutralizers. Davidson (22) attributed the striking difference between the results of these two treatment methods to the additional trauma of the heat of neutralization superimposed on the already existing burn.

In three specific circumstances, antidotes may be judiciously employed to limit the magnitude of the chemical injury. These antidotes must be viewed as adjuncts to hydrotherapy and are usually administered by physicians in the hospital. During transport of the patients, the emergency medical technician must identify the chemical agents involved in the burn injury to ensure that the specific antidotes are available when the patient arrives in the emergency department.

An important exception to the value of hydrotherapy as an emergency burn treatment is in the case of lithium and sodium metal burns. In these cases, water is definitely contraindicated since it combines with the metals to produce sodium or lithium hydroxide and releases hydrogen gas, often with explosive effects. Consequently, the emergency treatment of skin contacted by these metals is to remove all large particles from the skin and immerse them in mineral oil to prevent further combustion. The wound should be covered in mineral oil, and the patient should be rapidly transported to an emergency department. Any but the smallest, most superficial wounds should be excised and subsequently closed, once the margins of the burn wound have been established.

(i) Hydrofluoric acid. Hydrofluoric acid, one of the strongest acids, is used primarily in the semiconductor industry and is a primary component of rust-removing agents. It is also employed in germicides, dyes, plastics, and glass etching. Its toxicity has been attributed to the fluoride ion, which permeates the tissue as a result of the penetrating power of the acid. Even at low concentrations, the fluoride ion will cause liquefaction necrosis of the soft tissues and decalcification and corrosion of the bone. This destruction is accompanied by the formation in tissue of relatively insoluble calcium and magnesium fluoride, which are the only insoluble salt complexes in the tissue that will inactivate the fluoride ion. All other salts are soluble and fully dissociable and will release fluoride to continue tissue injury.

Prompt recognition and specific treatment are required if tissue damage by this strong acid is to be avoided. The skin must be copiously washed, beginning immediately after the injury and continuing until the patient is examined by a physician (16). If there is no evidence of injury (erythema, vesicles), the exposed skin can then be covered with a bulky dressing, either soaked in a solution containing a quaternary ammonia compound or calcium gluconate or impregnated with a magnesium oxide topical ointment (23). Each of these compounds combines with the fluoride ion to produce a nonionized fluoride complex. When there is any evidence of skin injury, usually associated with excruciating pain, it is necessary to inactivate the fluoride ions that have penetrated the skin. In these cases, a physician first performs a regional nerve block to permit debridement of the burn wound and injection of the antidote. Under aseptic conditions, all vesicles must be excised, and the injured skin is injected with a 10% solution of calcium gluconate (9). A rough guide to the usual effective dose of calcium gluconate is ca. 0.5 ml/cm^2 of burn surface area. The infiltration is extended 0.5 cm from the margin of the obviously injured tissue into the surrounding, apparently uninjured tissue. Outlining the painful skin before the regional nerve block is performed helps to localize the penetrating toxic fluoride ions.

After this procedure, the injected wound is covered by a dressing and reexamined 24 h later for evidence of new skin damage that was not noted at the initial examination. If a new injury is detected, 10% calcium gluconate is again injected. As with thermal injuries, ischemic injuries of exposed digits must be detected and treated. Monitoring the digital vessels with a flowmeter detects early changes in flow that signal the need for an escharotomy. On admission to the hospital, the injured extremity of the chemical burn victim should be immobilized and elevated above his heart.

In treating hydrofluoric acid burns of the finger, it is important to remember that this acid can easily penetrate and damage the nail matrix, which has no protective cover of stratum corneum (9). If the penetrating acid is not rapidly inactivated, it will damage the matrix soft tissue and underlying bond. The involved nail matrix will be greenish-gray. Development of necrosis can be prevented by excising the nail and injecting the matrix with calcium gluconate. Kohnlein and his associates favor primary excision and grafting for hydrogen fluoride burns. In their study, calcium gluconate injections were of little value (32).

(ii) Phenol and its derivatives. Phenol is an aromatic acidic alcohol. This compound and its derivatives are highly reactive corrosive contact poisons that damage cells by denaturing and precipitating cellular proteins. Their characteristic odor usually signals their presence. The industrial use of phenols has mushroomed; they are used as starting materials for many organic polymers and plastics and are used widely in agriculture, cosmetics, and medicine. Because of their antiseptic properties, they are employed frequently in commercially available germicides. A number of phenol derivatives (e.g., hexylresorcinol) are more bactericidal than phenol itself. Dilute solutions of phenol are used by plastic surgeons for chemical face peels.

After the skin has come into contact with phenol, treatment must be instituted immediately. Since dilute solutions of phenol are more rapidly absorbed through skin than concentrated ones, gentle swabbing of the surface of the skin with sponges soaked in water is not advised. Irrigation of the skin surface with large volumes of water delivered under low pressures is preferred. Since phenol may become trapped in the hair or beard, hair contacted by the chemical agent should be removed as soon as possible. Experimental studies indicate that water alone is effective in reducing the severity of burns and preventing death in animals with skin exposed to phenol and its derivatives (14). However, the most effective treatment is undiluted polyethylene glycol, either 300 or 400. A quick wipe of the skin with these solutions reduces mortality and burn severity in experimental animals. These solutions can be easily employed in phenol burns of the face, since they are not irritating to the eyes. These solutions increase the solubility of phenol, which explains their effectiveness. In the absence of polyethylene glycol, however, decontamination with water should not be delayed and must be performed until a polyethylene glycol solution is obtained. In most instances polyethylene glycol treatment will be utilized by a physician at the hospital. Removal of the phenol should be undertaken in a well-ventilated room so that hospital personnel are not exposed to high concentrations of phenol fumes.

In animal studies, exposure to as little as 0.625 mg of phenol per kg causes death (14,17). After phenol penetrates the dermis, it produces necrosis of the papillary dermis, and this necrotic tissue may temporarily delay its absorption. The absorbed phenol has a dramatic effect on the central nervous system. In humans, the prominent effect is depression, and death results from respiratory failure. As a result of central vasomotor depression, as well as its direct effect on myocardium and small blood vessels, phenol also results in a marked decrease in blood pressure. This agent is also a powerful antipyretic that produces a fall in body temperature. Metabolic acidosis may result from the direct effect of the acidic phenol. A number of substituted phenols, such as resorcinol and picric acid, have systemic actions distinct from that of phenol. For example, central nervous system stimulation is commonly encountered after the absorption of resorcinol. Picric acid hemolyzes erythrocytes and causes acute hemorrhagic glomerular nephritis and acute liver disease.

Treatment of the systemic symptoms is purely symptomatic. The patient should be kept warm. Therapy for shock should consist of the intravenous administration of 0.9% sodium chloride. Metabolic acidosis may require treatment with sodium bicarbonate administered intravenously. Alkalinization also prevents the precipitation of hemoglobin in urine which occurs with hemolysis. Hemochromogen excretion in the urine can also be enhanced by administering mannitol intravenously, which causes an osmotic diuresis. Anticonvulsants may be used to prevent seizures resulting from central nervous system stimulation.

(iii) White phosphorus. Elemental phosphorus exists in two forms: the red, granular nonabsorbable form is nontoxic; the other, a yellow, waxy, translucent solid (white phosphorus), ignites spontaneously in the air and must be preserved in water. When it ignites in air, it is oxidized to phosphorus pentoxide. With the addition of water, this compound forms metaphosphoric and orthophosphoric acids.

The ability of phosphorus to ignite spontaneously in air at temperatures over 34°C (94°F) has encouraged its use as an incendiary agent in military weapons and fireworks. In an explosion of a phosphorus munition, flaming droplets may become imbedded beneath the skin and will oxidize adjacent tissue unless they are removed (19). In nonmilitary industry, white phosphorus is used in the manufacture of insecticides, rodent poisons, and fertilizers. Its mechanism of injury appears to be its heat of reaction rather than the liberation of inorganic acids or cellular dehydration by the hydroscopic phosphorus pentoxide. This thermal injury often results in a painful partial- or full-thickness burn.

Prehospital care includes removing contaminated clothing and submerging the affected skin in cool water. Warm water should be avoided since white phosphorus melts at 44°C (111°F). The patient's skin should be searched for phosphorus particles, which should be removed and submerged in water. The burned skin should be covered with towels soaked in cool water during transport to the hospital.

Upon the patient's arrival in the emergency department, the burned skin is washed with a suspension of 5% sodium bicarbonate and 3% copper sulfate in 1% hydroxylethyl cellulose (8). Phosphorus particles will become coated with black cupric phosphide, allowing easy detection. Another benefit of this treatment is that the copper sulfate decreases the rate of oxidation of the phosphorus particles, limiting their damage to the tissue. These blackened particles can still cause injury and must be removed.

This antidote has its drawbacks: copper sulfate is toxic if systemically absorbed. Absorption of copper sulfate can be minimized by the surface-active agent in the antidote suspension, or by sodium lauryl sulfate. Although untested for this use, Pluronic F-68 may be an even safer suspending agent (40). Before the advent of these agents, prolonged treatment of phosphorus burns with copper sulfate solutions was followed by the development of systemic copper poisoning, which is manifested by vomiting, diarrhea, hemolysis, oliguria, hematuria, hepatic necrosis, and cardiopulmonary collapse. After the burned skin is treated with suspensions of copper sulfate, the antidote must be thoroughly washed from the skin.

Metabolic derangements have been identified in white phosphorus burns. Postburn serum electrolyte changes consisting of a depression of serum calcium and elevation of serum phosphorus have been reported (10). Also identified are postburn abnormalities of the electrocardiogram consisting of prolongation of the QT interval, bradycardia, and ST-T wave changes. These effects may explain early sudden death in patients with apparently inconsequential white phosphorus burns.

Ocular injuries

Ocular injuries, with the possibility of blindness, are among the most disastrous chemical burns. As with other chemical burns, immediate therapy is essential. Regardless of the nature of the chemical, copious irrigation is most important. With eyes submerged in a container of tap water, the patient should open and close his eyes continuously. In the absence of a container, the face and eyes should be held beneath a faucet and washed with running water. Irrigation should be continued while the patient is being transported to the hospital.

In the emergency department, the eye should receive further hydrotherapy in the form of a low-flow stream of 0.9% sodium chloride from an intravenous tube. The patient's response to the spilled chemical in the eye, including severe blepharospasm and tearing and forceful rubbing of the eye, can frustrate emergency treatment. Lid retractors may be necessary to evert the lids and ensure adequate irrigation of the conjuctival sac. Topical anesthetic agents may help to limit pain. Irrigation should continue until the pH of the conjunctival sac returns to its normal level. This can be monitored with a ninhydrin reagent strip. Any foreign material or solid chemical should be removed. After irrigation, the eye should be stained with fluorescein to detect corneal injury. Such injured eyes are examined with a slit lamp to determine the extent of damage to the anterior segment of the eye and anterior third of the vitreous chamber (43). Initial slit lamp examination of alkali burns of the eye reveals swelling and clouding of the anterior chamber. The intraocular pressure of these injured eyes should be measured to detect any increase. The injured eyes should be treated with a long-acting cycloplegic and mydriatic and possibly with a carbonic anhydrase inhibitor for 2 weeks, or until the pain disappears, to avoid pupillary constriction, increased intraocular pressure, and early glaucoma. All eyes with a corneal abasion should be treated with an emollient, broad-spectrum antibiotic ointment (e.g., chloramphenicol, gentamicin) instilled in the conjunctival sac. Neomycin-containing preparations should be avoided because of their toxicity to regenerating epithelium. Mobility of the globe should be encouraged to avoid the development of symblepharon. Conjunctival adhesions should be freed with a glass rod or spatula.

Alkali chemicals are the most toxic and anhydrous ammonia appears to be the worst offender. Alkali burns of an even seemingly mild degree can lead to devastating injury because of the propensity of the alkali to react with the lipid in the corneal epithelial cells to form a soluble soap, which then penetrates the corneal stroma. The alkali moves rapidly through the stroma and the endothelial cells to enter the anterior chamber. Anhydrous ammonia can penetrate into the anterior chamber in less than 1 min.

Alkali usually kills each tissue layer of the anterior segment of the eye that it contacts and results in occlusive vasculitis about the corneoscleral limbus which makes repair of these tissues difficult. As the tissues of the anterior segment of the eye degenerate, perforation follows, with the development of endophthalmitis and loss of the eyes. If perforation can be prevented, sight may eventually be restored by corneal transplantation. Recent experimental studies have concluded that the destruction of corneal stroma can be minimized by drug therapy (e.g., N-acetylcysteine, steroids, etc.) (12, 13, 36, 38, 41, 44). However, the necessary frequency of applications, the significant number of clinical failures, and the potential side effects of some of these treatments have limited their therapeutic usefulness.

Acid burns are better tolerated by the eyes because this organ, like other living tissue, has marked acid-buffering capacity. Acid is rapidly neutralized by the tear film, the proteins present in tears, and the conjunctival epithelial cells. Consequently, acid typically causes epithelial and basement membrane damage, but rarely damages deep endothelial cells. Acid burns that injure the periphery of the cornea and conjunctiva often heal uneventfully, leaving a clear corneal epithelium. In contrast, acid burns of the central part of the cornea may lead to corneal ulcer formation with neovascularization and scarring, requiring later surgical reconstruction.

When toxic chemicals are being used in a laboratory, protective gloves, goggles, and an apron must be worn to prevent skin contact. If the staff is educated and protected against these hazardous products, chemical burn injuries should become a subject only for forensic and military medicine.

Electrical Burns

Electricity is one of the most potentially dangerous commodities in our society. According to statistics, 0.8 to 1% of accidental deaths are caused by electrical injury, with approximately one quarter of these caused by natural lightning (21). Electrical burns are caused by either electric current or its arc. The standard wall outlet provides 120 V of alternating current at 60 cycles per s (Hertz [Hz]). High-tension voltage is defined as 1,000 V or more; low-tension power is less than 1,000 V (42).

Contact with a high-voltage current may be associated with an arc or a light flash. An electric arc is formed between two bodies of opposite charge (the power source and the body that is grounded). The arc is composed of ionized particles from both oppositely charged poles. The temperature of the particles and the immediately surrounding gases of the arc can be as high as 3,000°C (5,432°F) (5).

Physiologic effects

Low-voltage electric currents that pass through the body have well defined but reversible physiologic effects which are related to the amperage, frequency, and pathway of the current (Table 4). The threshold of human perception of 60 Hz (alternating) current is ca. 1.1 mA (21). The lowest perceptible flow of direct current is slightly higher (5.2 mA). Low currents give a sensation of tingling, which develops into muscular contractions and the sensation of heat which intensify as the current increases.

When a person's hands contact an electric current greater than 15 mA (for men; 10 mA for women), the person cannot release his grasp of the conductor and is said to "freeze" to the circuit despite struggles to free himself. Currents at this level are very frightening, painful, and hard to endure for even a short time. If the current is not interrupted promptly, the victim will become exhausted and collapse.

In the event of accident, the current must be switched off immediately. If this is not possible or will take too long, the rescuer must separate the victim from the contact, using a long piece of wood (e.g., a broom), a polydacron rope, or other dry, nonconductive object. Surgeon's rubber gloves do not provide adequate insulation. The rescuer must be certain he is standing on a dry surface.

Slightly higher alternating currents flowing across the chest are sufficient to cause contraction of the chest muscles, which stops breathing. After interruption of the current, the rescuer may initiate mouth-to-mouth ventilation on the victim. When a direct current in the range of 50 mA is interrupted, powerful muscular contractions result which throw the victim a considerable distance. Such violent muscular reactions may cause fractures (27).

TABLE 4. Physiological effects of electrical currents

Current (mA): 1-s contact	Effect
1	Threshold of perception
10–15	Sustained muscular contraction
50–100	Respiratory paralysis, ventricular fibrillation
1,000–5,000	Sustained myocardial contractions

Alternating current in the range of 50 to 100 mA may produce respiratory paralysis, ventricular fibrillation, or both. This paralysis appears to be due to the direct effect of the current passing through the respiratory center to the brain. The resultant apnea frequently lasts for several hours after the electrical power source is turned off; during this time, breathing must be restored by mouth-to-mouth ventilation. Ventricular fibrillation often accompanies respiratory paralysis. The heart's susceptibility to fibrillation first increases and then decreases with increasingly stronger currents. At relatively high currents (1 to 5 A), the likelihood of ventricular fibrillation is negligible, since the heart will be in sustained contraction. If this high current is terminated soon after the electrical shock, the heart will revert to normal sinus rhythm. In cardiac defibrillation, these very high currents are applied to the chest to depolarize the entire heart muscle.

When disconnecting the victim from the electrical circuit does not restore pulses, external cardiac compressions with ventilations must be started by the first responder to restore breathing and circulation (27). Ideally, this resuscitation will be continued by the paramedics, using advanced life support techniques. Other electrocardiographic changes besides ventricular fibrillation may be evident in the patient contacted by an electrical current. Tachycardia and ST segment alternations frequently appear and may persist for several weeks after electrical injury (7). Telemetered monitoring of these patients is recommended during transport to the hospital.

Pathologic effects

Burns due to contact with high-voltage electric circuits conform to two general types. In burns due to an elctric arc, the current courses from the contact point to the ground beneath the body. Circumscribed burns occur where portions of the arc contact the victim. These contact points may be multiple, single, or diffuse and may vary in depth. Entry points on flexor surfaces often produce "kissing" entry lesions resulting from severe tetanic muscle contractions. Baxter (7) indicated that the most common of these lesions is the circumscribed deep wound on the volar surface of the forearm in association with contact wounds of the palm. Arc burn injuries may be complicated by a flame burn if the arc ignites the victim's clothing.

The other type of burn injury, causing hidden destruction of deeper tissues, is due to an electric current that passes between the power source and the anatomical point of contact (entrance wound), through the victim, and then between him (via the exit wound) and the grounding mechanism. Such electricially conductive burns are simply thermal injuries, occurring when the electrical energy is converted to thermal energy. The extent of the electrical burn is related to the magnitude and frequency of the current and the resistance of the tissue. Skin represents an initial barrier to the flow of current and serves as insulation to the deeper tissues. The resistance offered by the calloused palm may reach 10^6 ohms, whereas the average resistance of normal dry and wet skin is 5,000 and 1,000 ohms, respectively.

The volume of tissue through which a current flows determines the seriousness of injury more than the resistance of the individual tissues. The current concentrates at its entrance to the body, diverges as it passes through, and finally converges again before exiting. Consequently, the most severe damage to the tissue occurs at the sites of contact. High-voltage entrance wounds are charred, centrally depressed, and leathery, whereas exit wounds are more likely to "explode" as the charge exits (7).

Electric currents follow the shortest path between the contact points and will involve any vital structures in the way. Fatalities are nearly 60% in hand-to-hand current passages, but considerably lower (20%) in hand-to-foot current passages (2). The anatomic location of the contact sites is a critical determinant of the degree of injury. The magnitude of the underlying tissue damage, especially to muscle, occurs at the time of the initial insult and does not appear to be progressive.

The heating of internal tissues by the electric current has irreversible, serious local and systemic consequences. The local injury may be as severe as necrosis of the entire limb; more commonly, the underlying tissue injury involves a portion of the superficial and deep muscles of one or more compartments. A wide spectrum of severe systemic complications have also been reported, including neurological complications, injury to abdominal viscera, severe potassium deficiency, and cataracts (27). Tetanus prophylaxis is mandatory in all burn patients, regardless of the mechanism of injury (25).

UV Radiation Burns

Injuries due to UV radiation are liable to occur in any occupation in which insufficiently screened sources of UV light are used (24). The original reports describe injuries resulting from using UV lamps in laboratories and photographic workshops. The medical employment of UV radiation has brought its toll of accidents to health professionals and patients, owing to carelessness in the use of protective equipment. The more recent and somewhat unjustified reliance on the "anti-germ" UV lamps sometimes used in operating theaters and laboratories may produce injuries. Although the UV radiation from such sources may be small, the direct radiation may be reinforced by reflection from hard-surfaced walls.

Consequently, many users of UV lamps are concerned about possible exposure. Often their fear is quite justified, as in the case of possible UV hazard from units producing 200 W per linear inch. This lamp assembly can radiate the equivalent of hours of sunlight in just minutes or even seconds. A set of guidelines has been established by the American Council of Governmental Industrial Hygienists regarding the safety and hazards of different kinds of UV sources. The threshold limit values for occupational exposure to UV radiation upon skin or eyes are as follows.

(i) For the near-UV spectral region (320 to 420 nm), total irradiation of the unprotected skin or eye should not exceed 1 mW/cm^2 for periods greater than 10^3 s (ca. 16 min). For exposure times less than 10^3 s, 1 J/cm^2 should not be exceeded.

(ii) For the actinic UV spectral region (220 to 315 nm), radiation exposure incident upon the unprotected eye or skin should not exceed the values given in Table 5 within an 8-h period. These threshold limit values do not, however, apply either to UV lasers or to individuals who are particularly photosensitive or whose medications may have altered their photosensitivity.

Injuries due to UV radiation occur when its energy is absorbed into the outer parts of the atoms of the tissue, producing a photochemical reaction. Proteins in tissue are the most susceptible to this reaction. Initially, irradiated proteins show physical and chemical changes involving a decrease in stability, which ultimately results in coagulation. UV injuries are usually confined to the skin and eyes. The adverse reactions of normal, healthy skin are classified into two types: the immediate type, due to acute exposure, and the delayed type, which occurs after chronic exposure to UV radiation (15).

Acute exposure

After acute exposure, the UV energy absorbed at different levels of the skin results in cell damage in the dyskerototic cells of the stratum malpighi and stratum corneum. An inflammatory infiltrate develops in the underlying papillary dermis and is thought to be mediated by histamine, serotonin, and kinins. Prostaglandins and related derivatives recently have been implicated in the development of erythema, and increased levels of these hormones have been found in human tissue exposed to UV energy. These substances of low molecular weight are synthesized by microsomal enzymes in all mammalian cells.

Depending on the skin type and duration of exposure, the injury can range in severity from a mild asymptomatic erythema to a more intense skin reaction that includes exquisite tenderness, pain, swelling, and blistering. The most serious injury includes systemic signs such as fever, chills, nausea, and prostration.

Treatment. Treatment of acute exposure may include (i) restoring any loss in intravascular volume, (ii) suppressing UV-induced erythema, and (iii) providing analgesia. When skin is burned by UV radiation, intravascular fluids extravasate into the burned tissue, thereby reducing intravascular volume. In severe UV radiation injuries, this reduction of vascular volume may be sufficient to produce hypotension. In such cases, the intravascular volume must be replaced with a crystalloid solution.

Most studies substantiate that synthesis and release of prostaglandins may be the primary mechanism in the production of erythema. Prostaglandin synthetase inhibitors administered systemically or topically are highly efficient suppressors of UV-caused erythema, but have no apparent preventive effect on the ultimate damage to the skin. Topical and systemic steriods have also been advocated in the treatment of UV-caused skin erythema. However, when steriod treatment was evaluated by means of random, double-blind studies, there were no significant differences in responses between the steroid-treated groups and the controls.

The basic form of benzocaine is bioactive; it penetrates the intact and damaged sunburned skin and limits the sensation of pain, burning, and itching (20). When properly formulated in concentrations between 5 and 20%, benzocaine is an effective and safe topical analgesic, anesthetic, and antipruritic agent on the intact skin. Among the benzocaine preparations that are commercially available, those consisting of 20% benzocaine base in propylene glycol are most effective. Relief is obtained for periods of 4 to 6 h. The lack of efficacy of some of the manufactured preparations is most commonly related to insufficient concentrations of the active ingredient (less than 5% benzocaine).

Epidemiologic data on allergy, irritancy, and other reactions to benzocaine do not support the contention that it is a potent sensitizer. It has been and is still one of the safest topical anesthetic agents. Because it has a low degree of water solubility, the quantities of benzocaine absorbed are relatively insignificant, and plasma levels that cause the systemic reactions that characterize the soluble ''caine''-type drugs do not occur with this drug. A 1% solution of lidocaine

TABLE 5. Relative spectral effectiveness of UV light

Wavelength (nm)	TLV[a] (mJ/cm^2)
200	100
210	40
220	25
230	16
240	10
250	7.0
260	4.6
270	3.0
280	3.4
290	4.7
300	10
305	50
310	200
315	1,000

[a]TLV, Threshold limit value.

hydrochloride exaggerates rather than relieves the pain associated with sunburn.

Delayed reaction

Repeated exposure to UV radiation can lead to aging of the skin and cutaneous neoplasia, usually basal- and squamous-cell carcinomas, that are not melanomatous. Premature aging of the skin is associated with damage to the elastic supporting tissues of the dermis. Skin moisturizers and toners cannot reverse these signs of premature aging.

Photophthalmia

UV radiation is absorbed almost entirely by the cornea and may result in an acute inflammatory reaction called photophthalmia (24). The clinical picture of acute photophthalmia is characteristic. There is always an asymptomatic latent period, the length of which varies with intensity and duration of exposure. For ordinary exposure, it lasts about 6 to 8 h, so that the unpleasant effects start during the following evening or night. With extreme intensities of exposure, the latent period can be reduced to 30 min. After the latent period, the earliest sign is a granular appearance of the cornea seen by a slit lamp. These granules probably represent swollen epithelial cells. Thereafter, irregularity of the corneal reflex appears, with areas of fluorescein staining accompanied by hyperemia of the conjunctiva. With the onset of the clinical findings, the patient usually complains of considerable discomfort as if his eye is full of sand, a sensation that increases to severe pain accompanied by marked photophobia, profuse lacrimation, and blepharospasm. Depending on the length of exposure, the symptoms may persist for 6 to 48 h. Symptoms are almost completely relieved by a topical anesthetic agent, which however, delays epithelization of the cornea. Bandaging the eyes, preferably with cold compresses, reduces the blepharospasm.

With repeated exposure, chronic photophthalmitis develops which is manifested by increased sensitivity to light and chronic blepharoconjunctivitis. The most appropriate treatment of either acute or chronic photophthalmitis is prevention. When exposed to UV radiation above the threshold limit value, laboratory personnel should wear protective glasses and clothing that block UV radiation to their eyes and skin.

LOCAL COLD INJURIES

Laboratory personnel who contact supercold surfaces (cryostats, freeze-fracture equipment, etc.) may sustain a local injury. The mechanisms of a local cold injury can be divided into phenomena that affect cells and extracellular fluids (direct effect) and those that disrupt the function of the organized tissue and the integrity of the circulation (indirect effects) (6). Generally, no serious damage is seen until tissue freezing (frostbite) results. Tissue begins to freeze at $-2.2°C$ ($-7.2°F$), but the supercooling phenomenon usually brings the point of ice crystal formation to ca. $-15°C$ ($-59°F$). During frostbite, ice crystals form from the available extracellular free water. As the ice crystals, many times the size of individual cells, form, solutes in the residual fluid become concentrated. The resulting increase in osmotic pressure results in withdrawal of water from the intracellular compartment, producing intracellular dehydration. The cell content becomes hyposmolar, and toxic concentrations of electrolytes may cause cell death. No gross rupture of the cell membrane is usually evident. A reversal of this process probably occurs during thawing of frozen tissues. The fulminating vascular reaction and stasis (indirect effect) which supervene in frostbite are associated with the release of prostaglandins, which have been implicated in progressive dermal ischemia. Increased tissue survival has been demonstrated by using antiprostaglandin agents and thromboxane inhibitors.

Frostbite can be classified into four degrees of severity, several days after the injury. In first degree, hyperemia and edema are evident. Second-degree frostbite is characterized by hyperemia and edema with large clear blisters that may extend the entire length of the digit. Third-degree frostbite has hyperemia, edema, and vesicles filled with hemorrhagic fluid that are usually smaller than in second-degree frostbite and do not extend to the tip of the involved digit. Fourth degree, the most severe type, involves complete necrosis with gangrene and loss of the affected part.

Initially, the patient experiences discomfort or pain in the tissues which are freezing. This progresses to numbness, and all sensation is soon lost. On examination, the frozen tissue is white and anesthetic because of blood loss due to the intense vasoconstriction. There are several factors that predispose to this cold injury (6). Clinical experience suggests that frostbite occurs at higher temperatures in patients with preexisting arterial disease. It has also been demonstrated repeatedly that a person who has already suffered frostbite is more prone to develop this cold injury than an individual with no frostbite history.

Treatment

The preferred initial treatment for frostbite is rapid rewarming in a water bath at a temperature of 39 to 42°C (102 to 108°F) (34). Rapid rewarming should not be done in the laboratory if there is a hospital within a short distance. The temperature of the water bath is monitored carefully as the bath cools. Additional hot water is added to the bath only once the injured part has been removed; the water is stirred and the temperature is retested before the injured part is replaced in the bath. Rewarming is continued until the frostbitten tissue has a flushed appearance, demonstrating that circulation is reestablished. This rewarming procedure usually lasts 30 to 45 min. Since rewarming is quite painful, narcotics are often required to relieve the pain.

After rewarming, the clear blisters should be debrided. To prevent further contact of prostaglandin and thromboxane in the blister fluid with the already damaged underlying vessels, topical treatment with aloe vera (Dermaide Aloe; Dermaide Research Corp., Chicago, Ill.) is then instituted every 6 h. This agent is a specific thromboxane inhibitor that prevents dermal ischemia in frostbite injuries. Unlike clear blisters, hemorrhagic blisters reflect damage to the underlying viable tissue; these blisters are left intact. Aloe vera is applied topically every 6 h to the surface of the hemorrhagic blisters. The injured tissue should be protected from further trauma and infection. Usually the frostbitten area is elevated to minimize formation of edema. Strict aseptic technique (mask, gloves, etc.) should be employed at each dressing change. Unless contraindicated by medical history, aspirin, an antiprostaglandin agent, is given orally every 6 h for 72 h (adults, 325 mg; children, 125 mg). Penicillin G (5×10^5 U every 6 h) is given intravenously until the edema resolves, to prevent secondary streptococcal infection. Tetanus prophylaxis is mandatory. Hydrotherapy and physical therapy are instituted during each dressing change. The injured part should be moved vigorously for at least several minutes every hour. Pain is relieved by analgesics.

DISCUSSION

When faced with an accident or emergency in the laboratory setting, laboratory personnel should serve as first responders in a comprehensive emergency medical system. By initiating immediate treatment, the first responder can often save lives and limit the consequences of the illness or injury. While the first responder attends to the victim, other laboratory personnel should immediately dial the 911 emergency telephone number to summon trained medical assistance.

Competence in CPR should be a prerequisite for employment for laboratory personnel. Ideally, this training in basic life support should be expanded to include training as a first responder. Degradation of these skills can occur at a surprisingly rapid rate (47). Periodic practice is essential, and refresher training should be required at least once a year.

The best treatment of accidental injuries in the laboratory will always be prevention, particularly in dealing with hazardous chemicals. Exposure to these agents can be significantly reduced by educational programs, labeling of toxic products, and protective clothing. Through promotional campaigns coordi-nated through regional poison control centers, laboratory personnel can be made aware of potentially hazardous chemicals. Many products once believed to be innocuous are now recognized as dangerous. When we identify products that are toxic to the skin, the information is added to our regional poison information system. We also communicate with the manufacturer and the Consumer Product Safety Commission so that the problem may be remedied nationally. The Consumer Product Safety Commission can be reached via their toll-free hotline at (800) 638-2772. This number also serves hearing-impaired persons with teletypewriters.

LITERATURE CITED

1. **Adgey, A. A. J., M. E. Scott, J. D. Allen, P. G. Nelson, J. S. Geddes, S. A. Zaidi, and J. F. Pantridge.** 1969. Management of ventricular fibrillation outside hospital. Lancet i:1169–1171.
2. **Alexander, L.** 1938. Electrical injuries to the central nervous system. Med. Clin. North Am. 22:633.
3. **Alvarez, H., III, and L. A. Cobb.** 1975. Experiences with CPR training of the general public, p. 33–37. In National Conference on Standards for Cardiopulmonary Resuscitation and Emergency Cardiac Care, 16–18 May 1973. American Heart Association, Dallas, Tex.
4. **Anonymous.** 1980. Standards and guidelines for cardiopulmonary resuscitation (CPR) and emergency cardiac care (EEC). J. Am. Med. Assoc. 244:453–509.
5. **Baldridge, R. R.** 1954. Electric burns: report of a case. N. Engl. J. Med. 250:46–49.
6. **Bangs, C., and M. P. Hamlet.** 1983. Hypothermia and cold injuries, p. 27–57. In P. S. Auerback and E. C. Greehr (ed.), Management of wilderness and environmental emergencies. Macmillan Publishing Co., New York.
7. **Baxter, C. R.** 1970. Present concepts in the management of major electrical injury. Surg. Clin. North Am. 50:1401–1418.
8. **Ben-Hur, N., and J. Appelbaum.** 1973. Biochemistry, histopathology and treatment of phosphorus burns. An experimental study. Isr. J. Med. Sci. 9:40–48.
9. **Blunt, C. P.** 1964. Treatment of hydrofluoric acid skin burns by injection with calcium gluconate. Ind. Med. Surg. 33:869–871.
10. **Bowen, T. E., T. J. Whelan, Jr., and T. G. Nelson.** 1971. Sudden death after phosphorus burns. Experimental observations of hypocalcemia, hyperphosphatemia and electrocardiographic abnormalities following production of a standard white phosphorus burn. Ann. Surg. 174:779–784.
11. **Bromberg, B. E., I. C. Song, and R. H. Walden.** 1965. Hydrotherapy of chemical burns. Plast. Reconstr. Surg. 35:85–95.
12. **Brown, S. I., M. P. Tragakis, and D. B. Pearce.** 1972. Treatment of the alkali-burned cornea. Am. J. Ophthalmol. 74:316–320.
13. **Brown, S. I., and C. A. Weller.** 1970. The pathogenesis and treatment of collagenase-induced diseases of the cornea. Trans. Am. Acad. Ophthalmol. 74:375–383.

14. **Brown, V. K. H., V. L. Box, and J. J. Simpson.** 1975. Decontamination procedures for skin exposed to phenolic substances. Arch. Environ. Health **30:**1–6.

15. **Bryant, C. A., J. G. Kenney, K. E. Greer, and R. F. Edlich.** 1984. Photosensitivity and other adverse reactions to sunlight, p. 241–246. *In* R. F. Edlich and D. A. Spyker (ed.), Current emergency therapy '84. Appleton-Century-Crofts, Norwalk, Conn.

16. **Carney, S. A., M. Hall, J. C. Lawrence, and C. R. Ricketts.** 1974. Rationale of the treatment of hydrofluoric acid burns. Br. J. Ind. Med. **31:**317–321.

17. **Conning, D. M., and M. J. Hayes.** 1970. The dermal toxicity of phenol: an investigation of the most effective first-aid measures. Br. J. Ind. Med. **27:**155–159.

18. **Crampton, R. S., R. F. Aldrich, J. A. Gascho, J. R. Miles, Jr., and R. Stillerman.** 1975. Reduction of prehospital, ambulance and community coronary death rates by the community-wide emergency cardiac care system. Am. J. Med. **58:**151–165.

19. **Curreri, P. W., M. J. Asch, and B. A. Pruitt, Jr.** 1970. The treatment of chemical burns: specialized diagnostic, therapeutic, and prognostic considerations. J. Trauma **10:**634–642.

20. **Dalili, H., and J. Adriani.** 1971. The efficacy of local anesthetic agents in blocking the sensation of itch, burning and pain in normal and unburned skin. Clin. Pharm. Ther. **12:**913–919.

21. **Dalziel, C. F.** 1956. Effects of electric shock on man. IRE Trans. Med. Electron. **5:**44–62.

22. **Davidson, E. C.** 1927. Treatment of acid and alkali burns: experimental study. Ann. Surg. **85:**481–489.

23. **Dibbell, D. G., R. E. Inverson, W. Jones, D. R. Laub, and M. S. Madison.** 1970. Hydrofluoric acid burns of the hand. J. Bone Joint Surg. **52A:**931–936.

24. **Duke-Elder, S., and P. A. MacFaul.** 1972. Injuries. Non-mechanical injuries, p. 912–933. *In* S. Duke-Elder (ed.), System of ophthalmology, vol. 14, part 2. C. V. Mosby, St. Louis.

25. **Edlich, R. F., B. W. Haynes, N. Larkham, M. S. Allen, W. Ruffin, Jr., J. M. Hiebert, and M. T. Edgerton.** 1978. Emergency Department treatment, triage and transfer protocols for the burn patient. J. Am. Coll. Emerg. Phys. **7:**152–158.

26. **Edlich, R. F., B. W. Haynes, W. Ruffin, Jr., M. S. Allen, D. Rockwell, J. M. Hiebert, G. T. Rodeheaver, and M. T. Edgerton.** 1978. Prehospital treatment of the burn patient. EMT J. **2:**42–48.

27. **Edlich, R. F., J. M. Hiebert, S. Halfacre, J. G. Thacker, and G. T. Rodeheaver.** 1982. Early assessment and management: commercial electric current injuries, p. 95–106. *In* B. W. Wolcott and D. A. Rund (ed.) Emergency medicine annual: 1982. Appleton-Century-Crofts, Norwalk, Conn.

28. **Gruber, R. P., D. R. Laub, and L. M. Vistnes.** 1975. The effect of hydrotherapy on the clinical course of pH of experimental cutaneous chemical burns. Plast. Reconstr. Surg. **55:**200–204.

29. **Jelenko, C., III.** 1974. Chemicals that "burn." J. Trauma **14:**65–72.

30. **Judd, R. L., and D. D. Ponsell.** 1982. The first responder: "the critical first minutes." The C. V. Mosby Co., St. Louis.

31. **King, T. C., and J. M. Zimmerman.** 1965. First-aid cooling of the fresh burn. Surg. Gynecol. Obstet. **120:**1271–1273.

32. **Kohnlein, H. E., P. Merkle, and H. W. Springorum.** 1973. Hydrogen fluoride burns: experiments and treatment. Surg. Forum **24:**50.

33. **Lund, I., and A. Skulberg.** 1976. Cardiopulmonary resuscitation by lay people. Lancet **ii:**702–704.

34. **McCauley, R. I., D. N. Hing, M. C. Robson, and J. P. Heggers.** 1983. Frostbite: a rational approached based on pathophysiology. J. Trauma **23:**143–147.

35. **Moritz, A. R., and F. C. Henriques, Jr.** 1947. Studies of thermal injury: relative importance of time and surface temperature in the causation of cutaneous burns. Am. J. Pathol. **23:**695–720.

36. **Newsome, N. A., and J. Gross.** 1977. Prevention by medroxyprogesterone of perforation in the alkali-burned rabbit cornea: inhibition of collagenolytic activity. Invest. Ophthalmol. Visual Sci. **16:**21–31.

37. **Ofeigsson, O. J.** 1965. Water cooling: first-aid treatment for scalds and burns. Surgery **57:**391–400.

38. **Pfister, R. R., and C. A. Paterson.** 1977. Additional clinical and morphological observations on the favorable effect of ascorbate in experimental ocular alkali burns. Invest. Ophthalmol. Visual Sci. **16:**478–487.

39. **Rodeheaver, G. T., J. M. Hiebert, and R. F. Edlich.** 1982. Initial treatment of chemical skin and eye burns. Comp. Ther. **8:**37–43.

40. **Rodeheaver, G. T., S. L. Smith, J. G. Thacker, M. T. Edgerton, and R. F. Edlich.** 1975. Mechanical cleansing of contaminated wounds with a surfactant. Am. J. Surg. **129:**241–245.

41. **Slansky, H. H., M. B. Berman, C. H. Dohlman, and J. Rose.** 1970. Cysteine and acetylcysteine in the prevention of corneal ulcerations. Ann. Ophthalmol. **2:**488–491.

42. **Sturim, H. S.** 1971. The treatment of electrical injuries. J. Trauma **11:**959–965.

43. **Sudarsky, R. D.** 1965. Ocular injury due to formic acid. Arch. Ophthalmol. **74:**805–806.

44. **Sugar, A., and S. R. Waltman.** 1973. Corneal toxicity of collagenase inhibitors. Invest. Ophthalmol. **12:**779–782.

45. **Summerlin, W. T., A. I. Walder, and J. A. Moncrief.** 1967. White phosphorus burns and massive hemolysis. J. Trauma **7:**476–484.

46. **Thompson, R. G., A. P. Hallstrom, and L. A. Cobb.** 1979. Bystander initiated cardiopulmonary resuscitation in the management of ventricular fibrillation. Ann. Intern. Med. **90:**737–740.

47. **Weaver, F. J., A. G. Ramirez, S. B. Dorfman, and A. E. Raizner.** 1979. Trainees' retention of cardiopulmonary resuscitation. How quickly they forget. J. Am. Med. Assoc. **241:**901–903.

Appendix 1
Biosafety in Microbiological and Biomedical Laboratories†

Contents

Section I

Introduction

Microbiology laboratories are special, often unique, work environments that may pose special infectious disease risks to persons in or near them. Personnel have contracted infections in the laboratory throughout the history of microbiology. Published reports around the turn of the century described laboratory-associated cases of typhoid, cholera, glanders, brucellosis, and tetanus (123). In 1941, Meyer and Eddie (75) published a survey of 74 laboratory-associated brucellosis infections that had occurred in the United States and concluded that the "handling of cultures or specimens or the inhalation of dust containing *Brucella* organisms is eminently dangerous to laboratory workers." A number of cases were attributed to carelessness or poor technique in the handling of infectious materials.

In 1949, Sulkin and Pike (113) published the first in a series of surveys of laboratory-associated infections summarizing 222 viral infections—21 of which were fatal. In at least a third of the cases the probable source of infection was considered to be associated with the handling of infected animals and tissues. Known accidents were recorded in 27 (12%) of the reported cases.

In 1951, Sulkin and Pike (114) published the second of a series of summaries of laboratory-associated infections based on a questionnaire sent to 5,000 laboratories. Only one-third of the 1,342 cases cited had been reported in the literature. Brucellosis outnumbered all other reported laboratory-acquired infections and together with tuberculosis, tularemia, typhoid, and streptococcal infection accounted for 72% of all bacteria infections and for 31% of infections caused by all agents. The overall case fatality rate was 3%. Only 16% of all infections reported were associated with a documented accident. The majority of these were related to mouth pipetting and the use of needle and syringe.

This survey was updated in 1965 (93), adding 641 new or previously unreported cases, and again in 1976 (90), summarizing a cumulative total of 3,921 cases. Brucellosis, typhoid, tularemia, tuberculosis, hepatitis, and Venezuelan equine encephalitis were

†Reprinted from U.S. Department of Health and Human Services, *Biosafety in Microbiological and Biomedical Laboratories,* 1st ed., HHS Publication no. (CDC) 84-8395, U.S. Government Printing Office, Washington, D.C., March 1984.

the most commonly reported. Fewer than 20% of all cases were associated with a known accident. Exposure to infectious aerosols was considered to be a plausible but unconfirmed source of infection for the more than 80% of the reported cases in which the infected person had "worked with the agent."

In 1967, Hanson et al. (53) reported 428 overt laboratory-associated infections with arboviruses. In some instances the ability of a given arbovirus to produce human disease was first confirmed as the result of unintentional infection of laboratory personnel. Exposure to infectious aerosols was considered the most common source of infection.

In 1974, Skinhoj (104) published the results of a survey which showed that personnel in Danish clinical chemistry laboratories had a reported incidence of hepatitis (2.3 cases per year per 1,000 employees) seven times higher than that of the general population. Similarly, a 1976 survey by Harrington and Shannon (55) indicated that medical laboratory workers in England had "a five times increased risk of acquiring tuberculosis compared with the general population." Hepatitis and shigellosis were also shown to be continuing occupational risks and together with tuberculosis were the three most commonly reported occupation-associated infections in Britain.

Although these reports suggest that laboratory personnel are at increased risk of being infected by the agents they handle, actual rates of infection are typically not available. However, the studies of Harrington and Shannon (55) and of Skinhoj (104) indicate that laboratory personnel have higher rates of tuberculosis, shigellosis, and hepatitis than the general population.

In contrast to the documented occurrence of laboratory-acquired infections in laboratory personnel, laboratories working with infectious agents have not been shown to represent a threat to the community. For example, although 109 laboratory-associated infections were recorded at the Center for Disease Control in 1947–1973 (97), no secondary cases were reported in family members or community contacts. The National Animal Disease Center has reported a similar experience (115), with no secondary cases occurring in laboratory and nonlaboratory contacts of 18 laboratory-associated cases occurring in 1960–1975. A secondary case of Marburg disease in the wife of a primary case was presumed to have been transmitted sexually two months after his dismissal from the hospital (70). Three secondary cases of smallpox were reported in two laboratory-associated outbreaks in England in 1973 (96) and 1978 (130). There were earlier reports of six cases of Q fever in employees of a commercial laundry which handled linens and uniforms from a laboratory where work with the agent was conducted (84), and two cases of Q fever in household contacts of a rickettsiologist (5). These cases are representative of the sporadic nature and infrequent association of community infections with laboratories working with infectious agents.

In his 1979 review (92), Pike concluded, "the knowledge, the techniques, and the equipment to prevent most laboratory infections are available." No single code of practice, standards, guidelines, or other publication, however, provides detailed descriptions of techniques, equipment, and other considerations or recommendations for the broad scope of laboratory activities conducted in the United States with a variety of indigenous and exotic infectious agents. The booklet *"Classification of Etiologic Agents on the Basis of* Hazard" (15) has, since 1969, served as a general reference for some laboratory activities utilizing infectious agents. That booklet and the concept of categorizing infectious agents and laboratory activities into four classes or levels served as a basic format for *Biosafety in Microbiological and Biomedical Laboratories*. This publication will provide specific descriptions of combinations of microbiological practices, laboratory facilities, and safety equipment and recommendations for use in four categories or biosafety levels of laboratory operation with selected infectious agents of man.

The descriptions of biosafety levels 1–4 parallel those of P1–4 in the *NIH Guidelines for Research Involving Recombinant DNA* (43) and are consistent with the general criteria used in assigning agents to Classes 1–4 in *Classification of Etiologic Agents on the Basis of Hazard* (15). Four biosafety levels are also described for infectious disease activities utilizing small laboratory animals. Recommendations for biosafety levels for specific agents are made on the basis of the potential hazard of the agent and of the laboratory function or activity.

Section II

Principles of Biosafety

The term "containment" is used in describing safe methods for managing infectious agents in the laboratory environment where they are being handled or maintained. Primary containment, the protection of personnel and the immediate laboratory environment from exposure to infectious agents, is provided by good microbiological technique and the use of appropriate safety equipment. The use of vaccines may provide an increased level of personal protection. Secondary containment, the protection of the environment external to the laboratory from exposure to infectious materials, is provided by a combination of facility design and operational practices. The purpose of containment is to reduce exposure of laboratory workers and other persons to, and to prevent escape into the outside environment of, potentially hazardous agents. The three elements of containment include laboratory practice and technique, safety equipment, and facility design.

Laboratory practice and technique. The most important element of containment is strict adherence to standard microbiological practices and techniques. Persons working with infectious agents or infected materials must be aware of potential hazards and must be trained and proficient in the practices and techniques required for safely handling such material. The director or person in charge of the laboratory is responsible for providing or arranging for appropriate training of personnel.

When standard laboratory practices are not sufficient to control the hazard associated with a particular agent or laboratory procedure, additional measures may be needed. The laboratory director is responsible for selecting additional safety practices, which must be in keeping with the hazard associated with the agent or procedure.

Each laboratory should develop or adopt a biosafety or operations manual which identifies the hazards that will or may be encountered and which specifies practices and procedures designed to minimize or eliminate risks. Personnel should be advised of special hazards and should be required to read and to follow the required practices and procedures. A scientist trained and knowledgeable in appropriate laboratory techniques, safety procedures, and hazards associated with handling infectious agents must direct laboratory activities.

Laboratory personnel, safety practices, and techniques must be supplemented by appropriate facility design and engineering features, safety equipment, and management practices.

Safety equipment (primary barriers). Safety equipment includes biological safety cabinets and a variety of enclosed containers. The biological safety cabinet is the principal device used to provide containment of infectious aerosols generated by many microbiological procedures. Three types of biological safety cabinets (Class, I, II, III) used in microbiological laboratories are illustrated in Fig. 1–3 and described in Appendix A. Open-fronted Class I and Class II biological safety cabinets are partial containment cabinets which offer significant levels of protection to laboratory personnel and to the environment when used with good microbiological techniques. The gas-tight Class III biological safety cabinet provides the highest attainable level of protection to personnel and the environment.

An example of an enclosed container is the safety centrifuge cup, which is designed to prevent aerosols from being released during centrifugation.

Safety equipment also includes items for personal protection such as gloves, coats, gowns, shoe covers, boots, respirators, face shields, and safety glasses. These personal protective devices are often used in combination with biological safety cabinets and other devices which contain the agents, animals, or materials being worked with. In some situations in which it is impractical to work in biological safety cabinets, personal protective devices may form the primary barrier between personnel and the infectious materials. Examples of such activities include certain animal studies, animal necropsy, production activities, and activities relating to maintenance, service, or support of the laboratory facility.

Facility design (secondary barriers). The design of the facility is important in providing a barrier to protect persons working in the facility but outside the laboratory and those in the community from infectious agents which may be accidentally released from the laboratory. Laboratory management is responsible for providing facilities commensurate with the laboratory's function. Three facility designs are described below, in ascending order by level of containment.

(1.) **The basic laboratory.** This laboratory provides general space in which work is done with viable agents which are not associated with disease in healthy adults. Basic laboratories include those facilities described in the following pages as Biosafety Levels 1 and 2 facilities.

This laboratory is also appropriate for work with infectious agents or potentially infectious materials when the hazard levels are low and laboratory personnel can be adequately protected by standard laboratory practice. While work is commonly conducted on the open bench, certain operations are confined to biological safety cabinets. Conventional laboratory designs are adequate. Areas known to be sources of general contamination, such as animal rooms and waste staging areas, should not be adjacent to patient care activities. Public areas and general offices to which nonlaboratory staff require frequent access should be separated from spaces which primarily support laboratory functions.

(2.) **The containment laboratory.** This laboratory has special engineering features which make it possible for laboratory workers to handle hazardous materials without endangering themselves, the community, or the environment. The containment laboratory is described in the following pages as a Biosafety Level 3 facility. The unique features which distinguish this laboratory from the basic laboratory are the provisions for access control and a specialized ventilation system. The containment laboratory may be an entire building or a single module or complex of modules within a building. In all cases, the laboratory is separated by a controlled access zone from areas open to the public.

(3.) **The maximum containment laboratory.** This laboratory has special engineering and containment features that allow activities involving infectious agents that are extremely hazardous to the laboratory worker or that may cause serious epidemic disease to be conducted safely. The maximum containment laboratory is described on the following pages as a Biosafety Level 4 facility. Although the maximum containment laboratory is generally a separate building, it can be constructed as an isolated area within a building. The laboratory's distinguishing characteristic is that it has secondary barriers to prevent hazardous materials from escaping into the environment. Such barriers include sealed openings into the laboratory, airlocks or liquid disinfectant barriers, a clothing-change and shower room contiguous to the laboratory ventilation system, and a treatment system to decontaminate exhaust air.

Biosafety levels. Four biosafety levels are described which consist of combinations of laboratory practices and techniques, safety equipment, and laboratory facilities appropriate for the operations performed and the hazard posed by the infectious agents and for the laboratory function or activity.

Biosafety Level 1. Biosafety level 1 practices, safety equipment, and facilities are appropriate for undergraduate and secondary educational training and teaching laboratories and for other facilities in which work is done with defined and characterized strains of viable microorganisms not known to cause disease in healthy adult humans. *Bacillus subtilis, Naegleria gruberi,* and infectious canine hepatitis virus are representative of those microorganisms meeting these criteria. Many agents not ordinarily associated with disease processes in humans are, however, opportunistic pathogens and may cause infection in the young, the aged, and immunodeficient or immunosuppressed individuals. Vaccine strains which have undergone multiple in vivo passages should not be considered avirulent simply because they are vaccine strains.

(ii) Biosafety Level 2. Biosafety Level 2 practices, equipment, and facilities are applicable to clinical, diagnostic, teaching, and other facilities in which work is done with the broad spectrum of indigenous moderate-risk agents present in the community and associated with human disease of varying severity. With good microbiological techniques, these agents can be used safely in activities conducted on the open bench, provided the potential for producing aerosols is low. Hepatitis B virus, the salmonellae, and *Toxoplasma* spp. are representative of microorganisms assigned to this containment level. Primary hazards to personnel working with these agents may include accidental autoinoculation, ingestion, and skin or mucous membrane exposure to infectious materials. Procedures with high aerosol potential that may increase the risk of exposure of personnel must be conducted in primary containment equipment or devices.

(iii) Biosafety Level 3. Biosafety Level 3 practices, safety equipment, and facilities are applicable to clinical, diagnostic, teaching, research, or production facilities in which work is done with indigenous or exotic agents where the potential for infection by aerosols is real and the disease may have serious or lethal consequences. Autoinoculation and ingestion also represent primary hazards to personnel working with these agents. Examples of such agents for which Biosafety Level 3 safeguards are generally recommended include *Mycobacterium tuberculosis*, St. Louis encephalitis virus, and *Coxiella burnetii*.

(iv) Biosafety Level 4. Biosafety Level 4 practices, safety equipment, and facilities are applicable to work with dangerous and exotic agents which pose a high individual risk of life-threatening disease. All manipulations of potentially infectious diagnostic materials, isolates, and naturally or experimentally infected animals pose a high risk of exposure and infection to laboratory personnel. Lassa fever virus is representative of the microorganisms assigned to Level 4.

Animal biosafety levels. Four biosafety levels are also described for activities involving infectious disease activities with experimental mammals. These four combinations of practices, safety equipment, and facilities are designated Animal Biosafety Levels 1, 2, 3, and 4 and provide increasing levels of protection to personnel and the environment.

The laboratory director is directly and primarily responsible for the safe operation of the laboratory. His/her knowledge and judgment are critical in assessing risks and appropriately applying these recommendations. The recommended biosafety level represents those conditions under which the agent can ordinarily be safely handled. Special characteristics of the agents used, the training and experience of personnel, and the nature or function of the laboratory may further influence the director in applying these recommendations.

Work with known agents should be conducted at the biosafety level recommended in Section V unless specific information is available to suggest that virulence, pathogenicity, antibiotic resistance patterns, and other factors are significantly altered to require more stringent or allow less stringent practices to be used.

Clinical laboratories, and especially those in health care facilities, receive clinical specimens with requests for a variety of diagnostic and clinical support services. Typically, clinical laboratories receive specimens without pertinent information such as patient history or clinical findings which may be suggestive of an infectious etiology. Furthermore, such specimens are often submitted with a broad request for microbiological examination for multiple agents (e.g., sputum samples submitted for "routine," acid-fast, and fungal cultures).

It is the responsibility of the laboratory director to establish standard procedures in the laboratory which realistically address the issue of the infective hazard of clinical specimens. Except in extraordinary circumstances (e.g., suspected hemorrhagic fever) the initial processing of clinical specimens and identification of isolates can be and are safely conducted using a combination of practices, facilities, and safety equipment described as Biosafety Level 2. Biological safety cabinets (Class I or II) should be used for the initial processing of clinical specimens when the nature of the test requested or other information is suggestive that an agent readily transmissible by infectious aerosols is likely to be present. Class II biological safety cabinets are also used to protect the integrity of the specimens or cultures by preventing contamination from the laboratory environment.

Segregating clinical laboratory functions and limiting or restricting access to laboratory areas are the responsibility of the laboratory director.

Importation and interstate shipment of certain biomedical materials. The importation of etiologic agents and vectors of human diseases is subject to the requirements of the Public Health Service Foreign Quarantine regulations. Companion regulations of the Public Health Service and the Department of Transportation specify packaging, labeling, and shipping requirements for etiologic agents and diagnostic specimens shipped in interstate commerce (see Appendix D).

The U.S. Department of Agriculture regulates the importation and interstate shipment of animal pathogens and prohibits the importation, possession, or use of certain exotic animal disease agents which pose a serious disease threat to domestic livestock and poultry (see Appendix E).

Section III

Laboratory Biosafety Level Criteria

The essential elements of the four biosafety levels for activities involving infectious microorganisms and laboratory animals are summarized in Tables 1 and 2. The levels are designated in ascending order by degree of protection provided to personnel, the environment, and the community.

Biosafety Level 1. Biosafety Level 1 is suitable for work involving agents of no known or minimal potential hazard to laboratory personnel and the environment. The laboratory is not separated from the general traffic patterns in the building. Work is generally conducted on open bench tops. Special containment equipment is not required or generally used. Laboratory personnel have specific training in the procedures conducted in the laboratory and are

supervised by a scientist with general training in microbiology or a related science.

The following standard and special practices, safety equipment, and facilities apply to agents assigned to Biosafety Level 1.

A. Standard microbiological practices
1. Access to the laboratory is limited or restricted at the discretion of the laboratory director when experiments are in progress.
2. Work surfaces are decontaminated once a day and after any spill of viable material.
3. All contaminated liquid or solid wastes are decontaminated before disposal.
4. Mechanical pipetting devices are used; mouth pipetting is prohibited.
5. Eating, drinking, smoking, and applying cosmetics are not permitted in the work area. Food may be stored in cabinets or refrigerators designated and used for this purpose only. Food storage cabinets or refrigerators should be located outside of the work area.
6. Persons wash their hands after they handle viable materials and animals and before leaving the laboratory.
7. All procedures are performed carefully to minimize the creation of aerosols.
8. It is recommended that laboratory coats, gowns, or uniforms be worn to prevent contamination or soiling of street clothes.

B. Special practices
1. Contaminated materials that are to be decontaminated at a site away from the laboratory are placed n a durable leakproof container which is closed before being removed from the laboratory.
2. An insect and rodent control program is in effect.

C. Containment equipment
Special containment equipment is generally not required for manipulations of agents assigned to Biosafety Level 1.

D. Laboratory facilities
1. The laboratory is designed so that it can be easily cleaned.
2. Bench tops are impervious to water and resistant to acids, alkalis, organic solvents, and moderate heat.
3. Laboratory furniture is sturdy. Spaces between benches, cabinets, and equipment are accessible for cleaning.
4. Each laboratory contains a sink for handwashing.
5. If the laboratory has windows that open, they are fitted with fly screens.

Biosafety Level 2. Biosafety Level 2 is similar to Level 1 and is suitable for work involving agents of moderate potential hazard to personnel and the environment. It differs in that (i) laboratory personnel have specific training in handling pathogenic agents and are directed by competent scientists, (ii) access

TABLE 1. Summary of recommended biosafety levels for infectious agents

Biosafety level	Practices and techniques	Safety equipment	Facilities
1	Standard microbiological practices	None: primary containment provided by adherence to standard laboratory practices during open bench operations	Basic
2	Level 1 practices plus: laboratory coats; decontamination of all infectious wastes; limited access; protective gloves and biohazard warning signs as indicated	Partial containment equipment (i.e., Class I or II Biological Safety Cabinets) used to conduct mechanical and manipulative procedures that have high aerosol potential which may increase the risk of exposure to personnel	Basic
3	Level 2 practices plus: special laboratory clothing; controlled access	Partial containment equipment used for all manipulations of infectious material	Containment
4	Level 3 practices plus: entrance through change room where street clothing is removed and laboratory clothing is put on; shower on exit; all wastes are decontaminated on exit from the facility	Maximum containment equipment (i.e., Class III biological safety cabinet or partial containment equipment in combination with full-body, air-supplied, positive-pressure personnel suit) used for all procedures and activities	Maximum containment

TABLE 2. Summary of recommended biosafety levels for activities in which experimentally or naturally infected vertebrate animals are used

Biosafety level	Practices and techniques	Safety equipment	Facilities
1	Standard animal care and management practices	None	Basic
2	Laboratory coats; decontamination of all infectious wastes and of animal cages prior to washing; limited access; protective gloves and hazard warning signs as indicated	Partial containment equipment and/or personal protective devices used for activities and manipulations of agents or infected animals that produce aerosols	Basic
3	Level 2 practices plus: special laboratory clothing; controlled access	Partial containment equipment and/or personal protective devices used for all activities and manipulations of agents or infected animals	Containment
4	Level 3 practices plus: entrance through clothes change room where street clothing is removed and laboratory clothing is put on shower on exit; all wastes are decontaminated before removal from the facility	Maximum containment equipment (i.e., Class III biological safety cabinet or partial containment equipment in combination with full-body, air supplied, positive-pressure personnel suit) used for all procedures and activities	Maximum containment

to the laboratory is limited when work is being conducted, and (iii) certain procedures in which infectious aerosols are created are conducted in biological safety cabinets or other physical containment equipment.

The following standard and special practices, safety equipment, and facilities apply to agents assigned to Biosafety Level 2.

A. **Standard microbiological practices**

1. Access to the laboratory is limited or restricted by the laboratory director when work with infectious agents is in progress.

2. Work surfaces are decontaminated at least once a day and after any spill of viable material.

3. All infectious liquid or solid wastes are decontaminated before disposal.

4. Mechanical pipetting devices are used; mouth pipetting is prohibited.

5. Eating, drinking, smoking, and applying

cosmetics are not permitted in the work area. Food may be stored in cabinets or refrigerators designated and used for this purpose only. Food storage cabinets or refrigerators should be located outside of the work area.

6. Persons wash their hands after handling infectious materials and animals and when they leave the laboratory.

7. All procedures are performed carefully to minimize the creation of aerosols.

B. Special practices

1. Contaminated materials that are to be decontaminated at a site away from the laboratory are placed in a durable leakproof container which is closed before being removed from the laboratory.

2. The laboratory director limits access to the laboratory. In general, persons who are at increased risk of acquiring infection or for whom infection may be unusually hazardous are not allowed in the laboratory or animal rooms. The director has the final responsibility for assessing each circumstance and determining who may enter or work in the laboratory.

3. The laboratory director establishes policies and procedures whereby only persons who have been advised of the potential hazard and meet any specific entry requirements (e.g., immunization) enter the laboratory or animal rooms.

4. When the infectious agent(s) in use in the laboratory require special provisions for entry (e.g., vaccination), a hazard warning sign, incorporating the universal biohazard symbol, is posted on the access door to the laboratory work area. The hazard warning sign identifies the infectious agent, lists the name and telephone number of the laboratory director or other responsible person(s), and indicates the special requirement(s) for entering the laboratory.

5. An insect and rodent control program is in effect.

6. Laboratory coats, gowns, smocks, or uniforms are worn while in the laboratory. Before leaving the laboratory for nonlaboratory areas (e.g., cafeteria, library, administrative offices), this protective clothing is removed and left in the laboratory or covered with a clean coat not used in the laboratory.

7. Animals not involved in the work being performed are not permitted in the laboratory.

8. Special care is taken to avoid skin contamination with infectious materials; gloves should be worn when handling infected animals and when skin contact with infectious materials is unavoidable.

9. All wastes from laboratories and animal rooms are appropriately decontaminated before disposal.

10. Hypodermic needles and syringes are used only for parenteral injection and aspiration of fluids from laboratory animals and diaphragm bottles. Only needle-locking syringes or disposable syringe-needle units (i.e., needle is integral to the syringe) are used for the injection or aspiration of infectious fluids. Extreme caution should be used when handling needles and syringes to avoid autoinoculation and the generation of aerosols during use and disposal. Needles should not be bent, sheared, replaced in the sheath or guard, or removed from the syringe following use. The needle and syringe should be promptly placed in a puncture-resistant container and decontaminated, preferably by autoclaving, before discard or reuse.

11. Spills and accidents which result in overt exposures to infectious materials are immediately reported to the laboratory director. Medical evaluation, surveillance, and treatment are provided as appropriate, and written records are maintained.

12. When appropriate, considering the agent(s) handled, base-line serum samples for laboratory and other at-risk personnel are collected periodically, depending on the agents handled or the function of the facility.

13. A biosafety manual is prepared or adopted. Personnel are advised of special hazards and are required to read instructions on practices and procedures and to follow them.

C. Containment equipment

Biological safety cabinets (Class I or II) (see Appendix A) or other appropriate personal protective or physical containment devices are used whenever:

1. Procedures with a high potential for creating infectious aerosols are conducted (82). These may include centrifuging, grinding, blending, vigorous shaking or mixing, sonic disruption, opening containers of infectious materials whose internal pressures may be different from ambient pressures, inoculating animals intranasally, and harvesting infected tissues from animals or eggs.

2. High concentrations or large volumes of infectious agents are used. Such materials may be centrifuged in the open laboratory if sealed heads or centrifuge safety cups are used and if they are opened only in a biological safety cabinet.

D. Laboratory facilities

1. The laboratory is designed so that it can be easily cleaned.

2. Bench tops are impervious to water and resistant to acids, alkalis, organic solvents, and moderate heat.

3. Laboratory furniture is sturdy, and spaces

between benches, cabinets, and equipment are accessible for cleaning.

4. Each laboratory contains a sink for handwashing.

5. If the laboratory has windows that open, they are fitted with fly screens.

6. An autoclave for decontaminating infectious laboratory wastes is available.

Biosafety Level 3. Biosafety Level 3 is applicable to clinical, diagnostic, teaching, research, or production facilities in which work is done with indigenous or exotic agents which may cause serious or potentially lethal disease as a result of exposure by the inhalation route. Laboratory personnel have specific training in handling pathogenic and potentially lethal agents and are supervised by competent scientists who are experienced in working with these agents. All procedures involving the manipulation of infectious material are conducted within biological safety cabinets or other physical containment devices or by personnel wearing appropriate personal protective clothing and devices. The laboratory has special engineering and design features. It is recognized, however, that many existing facilities may not have all the facility safeguards recommended for Biosafety Level 3 (e.g., access zone, sealed penetrations, directional airflow, etc.). In these circumstances, acceptable safety may be achieved for routine or repetitive operations (e.g., diagnostic procedures involving the propagation of an agent for identification, typing, and susceptibility testing) in laboratories where facility features satisfy Biosafety Level 2 recommendations, provided the recommended "Standard Microbiological Practices," "Special Practices," and "Containment Equipment" for Biosafety Level 3 are rigorously followed. The decision to implement this modification of Biosafety Level 3 recommendations should be made only by the laboratory director.

The following standard and special safety practices, equipment and facilities apply to agents assigned to Biosafety Level 3.

A. Standard microbiological practices

1. Work surfaces are decontaminated at least once a day and after any spill of viable material.

2. All infectious liquid or solid wastes are decontaminated before disposal.

3. Mechanical pipetting devices are used; mouth pipetting is prohibited.

4. Eating, drinking, smoking, storing food, and applying cosmetics are not permitted in the work area.

5. Persons wash their hands after handling infectious materials and animals and when they leave the laboratory.

6. All procedures are performed carefully to minimize the creation of aerosols.

B. Special practices

1. Laboratory doors are kept closed when experiments are in progress.

2. Contaminated materials that are to be decontaminated at a site away from the laboratory are placed in a durable leakproof container which is closed before being removed from the laboratory.

3. The laboratory director controls access to the laboratory and restricts access to persons whose presence is required for program or support purposes. Persons who are at increased risk of acquiring infection or for whom infection may be unusually hazardous are not allowed in the laboratory or animal rooms. The director has the final responsibility for assessing each circumstance and determining who may enter or work in the laboratory.

4. The laboratory director establishes policies and procedures whereby only persons who have been advised of the potential biohazard, who meet any specific entry requirements (e.g., immunization), and who comply with all entry and exit procedures enter the laboratory or animal rooms.

5. When infectious materials or infected animals are present in the laboratory or containment module, a hazard warning sign, incorporating the universal biohazard symbol, is posted on all laboratory and animal room access doors. The hazard warning sign identifies the agent, lists the name and telephone number of the laboratory director or other responsible person(s), and indicates any special requirements for entering the laboratory, such as the need for immunizations, respirators, or other personal protective measures.

6. All activities involving infectious materials are conducted in biological safety cabinets or other physical containment devices within the containment module. No work in open vessels is conducted on the open bench.

7. The work surfaces of biological safety cabinets and other containment equipment are decontaminated when work with infectious materials is finished. Plastic-backed paper toweling used on nonperforated work surfaces within biological safety cabinets facilitates cleanup.

8. An insect and rodent control program is in effect.

9. Laboratory clothing that protects street clothing (e.g., solid front or wrap-around gowns, scrub suits, coveralls) is worn in the laboratory. Laboratory clothing is not worn outside the laboratory, and it is decontaminated before being laundered.

10. Special care is taken to avoid skin contamination with infectious materials; gloves should

be worn when handling infected animals and when skin contact with infectious materials is unavoidable.

11. Molded surgical masks or respirators are worn in rooms containing infected animals.

12. Animals and plants not related to the work being conducted are not permitted in the laboratory.

13. All wastes from laboratories and animal rooms are appropriately decontaminated before disposal.

14. Vacuum lines are protected with high efficiency particulate air (HEPA) filters and liquid disinfectant traps.

15. Hypodermic needles and syringes are used only for parenteral injection and aspiration of fluids from laboratory animals and diaphragm bottles. Only needle-locking syringes or disposable syringe-needle units (i.e., needle is integral to the syringe) are used for the injection or aspiration of infectious fluids. Extreme caution should be used when handling needles and syringes to avoid autoinoculation and the generation of aerosols during use and disposal. Needles should not be bent, sheared, replaced in the sheath or guard, or removed from the syringe following use. The needle and syringe should be promptly placed in a puncture-resistant container and decontaminated, preferably by autoclaving, before discard or reuse.

16. Spills and accidents which result in overt or potential exposures to infectious materials are immediately reported to the laboratory director. Appropriate medical evaluation, surveillance, and treatment are provided and written records are maintained.

17. Base-line serum samples for all laboratory and other at-risk personnel should be collected and stored. Additional serum specimens may be collected periodically, depending on the agents handled or the function of the laboratory.

18. A biosafety manual is prepared or adopted. Personnel are advised of special hazards and are required to read instructions on practices and procedures and to follow them.

C. Containment equipment

Biological safety cabinets (Class I, II, or III) (see Appendix A) or other appropriate combinations of personal protective or physical containment devices (e.g., special protective clothing, masks, gloves, respirators, centrifuge safety cups, sealed centrifuge rotors, and containment caging for animals) are used for all activities with infectious materials which pose a threat of aerosol exposure. These include: manipulation of cultures and of those clinical or environmental materials which may be a source of infectious aerosols; the aerosol challenge of experimental animals; harvesting of tissues or fluids from infected

animals and embryonated eggs; and necropsy of infected animals.

D. Laboratory facilities

1. The laboratory is separated from areas which are open to unrestricted traffic flow within the building. Passage through two sets of doors is the basic requirement for entry into the laboratory from access corridors or other contiguous areas. Physical separation of the high containment laboratory from access corridors or other laboratories or activities may also be provided by a double-doored clothes change room (showers may be included), airlock, or other access facility which requires passage through two sets of doors before entering the laboratory.

2. The interior surfaces of walls, floors, and ceilings are water resistant so that they can be easily cleaned. Penetrations in these surfaces are sealed or capable of being sealed to facilitate decontaminating the area.

3. Bench tops are impervious to water and resistant to acids, alkalis, organic solvents, and moderate heat.

4. Laboratory furniture is sturdy, and spaces between benches, cabinets, and equipment are accessible for cleaning.

5. Each laboratory contains a sink for handwashing. The sink is foot, elbow, or automatically operated and is located near the laboratory exit door.

6. Windows in the laboratory are closed and sealed.

7. Access doors to the laboratory or containment module are self-closing.

8. An autoclave for decontaminating laboratory wastes is available, preferably within the laboratory.

9. A ducted exhaust air ventilation system is provided. This system creates directional airflow that draws air into the laboratory through the entry area. The exhaust air is not recirculated to any other area of the building, is discharged to the outside, and is dispersed away from occupied areas and air intakes. Personnel must verify that the direction of the airflow (into the laboratory) is proper. The exhaust air from the laboratory room can be discharged to the outside without being filtered or otherwise treated.

10. The HEPA-filtered exhaust air from Class I or Class II biological safety cabinets is discharged directly to the outside or through the building exhaust system. Exhaust air from Class I or II biological safety cabinets may be recirculated within the laboratory if the cabinet is tested and certified at least every twelve months. If the HEPA-filtered exhaust air from Class I or II biological safety cabinets is to be discharged to the outside through the

building exhaust air system, it is connected to this system in a manner (e.g., thimble unit connection [80]) that avoids any interference with the air balance of the cabinets or building exhaust system.

Biosafety Level 4. Biosafety Level 4 is required for work with dangerous and exotic agents which pose a high individual risk of life-threatening disease. Members of the laboratory staff have specific and thorough training in handling extremely hazardous infectious agents, and they understand the primary and secondary containment functions of the standard and special practices, the containment equipment, and the laboratory design characteristics. They are supervised by competent scientists who are trained and experienced in working with these agents. Access to the laboratory is strictly controlled by the laboratory director. The facility is either in a separate building or in a controlled area within a building, which is completely isolated from all other areas of the building. A specific facility operations manual is prepared or adopted.

Within work areas of the facility, all activities are confined to Class III biological safety cabinets or Class I or Class II biological safety cabinets used along with one-piece positive-pressure personnel suits ventilated by a life support system. The maximum containment laboratory has special engineering and design features to prevent microorganisms from being disseminated into the environment.

The following standard and special safety practices, equipment, and facilities apply to agents assigned to Biosafety Level 4.

A. Standard microbiological practices

1. Work surfaces are decontaminated at least once a day and immediately after any spill of viable material.
2. Only mechanical pipetting devices are used.
3. Eating, drinking, smoking, storing food, and applying cosmetics are not permitted in the laboratory.
4. All procedures are performed carefully to minimize the creation of aerosols.

B. Special practices

1. Biological materials to be removed from the Class III cabinet or from the maximum containment laboratory in a viable or intact state are transferred to a nonbreakable, sealed primary container and then enclosed in a nonbreakable, sealed secondary container which is removed from the facility through a disinfectant dunk tank, fumigation chamber, or an airlock designed for this purpose.
2. No materials, except for biological materials that are to remain in a viable or intact state, are removed from the maximum containment laboratory unless they have been autoclaved or decontaminated before they leave the facility. Equipment or material which might be damaged by high temperatures or steam is

decontaminated by gaseous or vapor methods in an airlock or chamber designed for this purpose.

3. Only persons whose presence in the facility or individual laboratory rooms is required for program or support purposes are authorized to enter. Persons who may be at increased risk of acquiring infection or for whom infection may be unusually hazardous are not allowed in the laboratory or animal rooms. The supervisor has the final responsibility for assessing each circumstance and determining who may enter or work in the laboratory. Access to the facility is limited by means of secure, locked doors; accessibility is managed by the laboratory director, biohazards control officer, or other person responsible for the physical security of the facility. Before entering, persons are advised of the potential biohazards and instructed as to appropriate safeguards for insuring their safety. Authorized persons comply with the instructions and all other applicable entry and exit procedures. A logbook, signed by all personnel, indicates the date and time of each entry and exit. Practical and effective protocols for emergency situations are established.
4. Personnel enter and leave the facility only through the clothing change and shower rooms. Personnel shower each time they leave the facility. Personnel use the airlocks to enter or leave the laboratory only in an emergency.
5. Street clothing is removed in the outer clothing change room and kept there. Complete laboratory clothing, including undergarments, pants and shirts or jumpsuits, shoes, and gloves, is provided and used by all personnel entering the facility. Head covers are provided for personnel who do not wash their hair during the exit shower. When leaving the laboratory and before proceeding into the shower area, personnel remove their laboratory clothing and store it in a locker or hamper in the inner change room.
6. When infectious materials or infected animals are present in the laboratory or animal rooms, a hazard warning sign, incorporating the universal biohazard symbol, is posted on all access doors. The sign identifies the agent, lists the name of the laboratory director or other responsible person(s), and indicates any special requirements for entering the area (e.g., the need for immunizations or respirators).
7. Supplies and materials needed in the facility are brought in by way of the double-doored autoclave, fumigation chamber, or airlock which is appropriately decontaminated between each use. After securing the outer doors, personnel within the facility retrieve

the materials by opening the interior doors of the autoclave, fumigation chamber, or airlock. These doors are secured after materials are brought into the facility.

8. An insect and rodent control program is in effect.

9. Materials (e.g., plants, animals, and clothing) not related to the experiment being conducted are not permitted in the facility.

10. Hypodermic needles and syringes are used only for parenteral injection and aspiration of fluids from laboratory animals and diaphragm bottles. Only needle-locking syringes or disposable syringe-needle units (i.e., needle is integral part of unit) are used for the injection or aspiration of infectious fluids. Needles should not be bent, sheared, replaced in the needle guard, or removed from the syringe following use. The needle and syringe should be placed in a puncture-resistant container and decontaminated, preferably by autoclaving before discard or reuse. Whenever possible, cannulas are used instead of sharp needles (e.g., gavage).

11. A system is set up for reporting laboratory accidents and exposures and employee absenteeism, and for the medical surveillance of potential laboratory-associated illnesses. Written records are prepared and maintained. An essential adjunct to such a reporting-surveillance system is the availability of a facility for the quarantine, isolation, and medical care of personnel with potential or known laboratory-associated illnesses.

C. Containment equipment

All procedures within the facility with agents assigned to Biosafety Level 4 are conducted in a Class III biological safety cabinet or in Class I or II biological safety cabinets used in conjunction with one-piece positive pressure personnel suits ventilated by a life support system. Activities with viral agents (e.g., Rift Valley fever virus) that require Biosafety Level 4 secondary containment capabilities and for which highly effective vaccines are available and used can be conducted within Class I or Class II biological safety cabinets within the facility without the one-piece positive-pressure personnel suit being used, if (i) the facility has been decontaminated; (ii) no work is being conducted in the facility with other agents assigned to Biosafety Level 4; and (iii) all other standards and special practices are followed.

D. Laboratory facilities

1. The maximum containment facility consists of either a separate building or a clearly demarcated and isolated zone within a building. Outer and inner change rooms separated by a shower are provided for personnel entering and leaving the facility. A double-doored autoclave, fumigation chamber, or ventilated airlock is provided for passage of those materials, supplies, or equipment which are not brought into the facility through the change room.

2. Walls, floors, and ceilings of the facility are constructed to form a sealed internal shell which facilitates fumigation and is animal and insect proof. The internal surfaces of this shell are resistant to liquids and chemicals, thus facilitating cleaning and decontamination of the area. All penetrations in these structures and surfaces are sealed. Any drains in the floors contain traps filled with a chemical disinfectant of demonstrated efficacy against the target agent, and they are connected directly to the liquid waste decontamination system. Sewer and other ventilation lines contain HEPA filters.

3. Internal facility appurtenances, such as light fixtures, air ducts, and utility pipes, are arranged to minimize the horizontal surface area on which dust can settle.

4. Bench tops have seamless surfaces which are impervious to water and resistant to acids, alkalis, organic solvents, and moderate heat.

5. Laboratory furniture is of simple and sturdy construction, and spaces between benches, cabinets, and equipment are accessible for cleaning.

6. A foot-, elbow-, or automatically operated handwashing sink is provided near the door of each laboratory room in the facility.

7. If there is a central vacuum system, it does not serve areas outside the facility. In-line HEPA filters are placed as near as practicable to each use point or service cock. Filters are installed to permit in-place decontamination and replacement. Other liquid and gas services to the facility are protected by devices that prevent backflow.

8. If water fountains are provided, they are foot operated and are located in the facility corridors outside the laboratory. The water service to the fountain is not connected to the backflow-protected distribution system supplying water to the laboratory areas.

9. Access doors to the laboratory are self-closing and lockable.

10. Any windows are breakage resistant.

11. A double-doored autoclave is provided for decontaminating materials passing out of the facility. The autoclave door which opens to the area external to the facility is sealed to the outer wall and automatically controlled so that the outside door can only be opened after the autoclave "sterilization" cycle has been completed.

12. A pass-through dunk tank, fumigation chamber, or an equivalent decontamination method is provided so that materials and equipment

that cannot be decontaminated in the auto-clave can be safely removed from the facility.

13. Liquid effluents from laboratory sinks, biological safety cabinets, floors, and autoclave chambers are decontaminated by heat treatment before being released from the maximum containment facility. Liquid wastes from shower rooms and toilets may be decontaminated with chemical disinfectants or by heat in the liquid waste decontamination system. The procedure used for heat decontamination of liquid wastes is evaluated mechanically and biologically by using a recording thermometer and an indicator microorganism with a defined heat susceptibility pattern. If liquid wastes from the shower rooms are decontaminated with chemical disinfectants, the chemical used is of demonstrated efficacy against the target or indicator microorganisms.

14. An individual supply and exhaust air ventilation system is provided. The system maintains pressure differentials and directional airflow as required to assure flows inward from areas outside of the facility toward areas of highest potential risk within the facility. Manometers are used to sense pressure differentials between adjacent areas maintained at different pressure levels. If a system malfunctions, the manometers sound an alarm. The supply and exhaust airflow is interlocked to assure inward (or zero) airflow at all times.

15. The exhaust air from the facility is filtered through HEPA filters and discharged to the outside so that it is dispersed away from occupied buildings and air intakes. Within the facility, the filters are located as near the laboratories as practicable in order to reduce the length of potentially contaminated air ducts. The filter chambers are designed to allow in situ decontamination before filters are removed and to facilitate certification testing after they are replaced. Coarse filters and HEPA filters are provided to treat air supplied to the facility in order to increase the lifetime of the exhaust HEPA filters and to protect the supply air system should air pressures become unbalanced in the laboratory.

16. The treated exhaust air from Class I and II biological safety cabinets can be discharged into the laboratory room environment or to the outside through the facility air exhaust system. If exhaust air from Class I or II biological safety cabinets is discharged into the laboratory, the cabinets are tested and certified at 6-month intervals. *The treated exhaust air from Class III biological safety cabinets is discharged, without recirculation through two sets of HEPA filters in series, via the facility exhaust air system.* If the treated exhaust air

from any of these cabinets is discharged to the outside through the facility exhaust air system, it is connected to this system in a manner (e.g., thimble unit connection [80]) that avoids any interference with the air balance of the cabinets or the facility exhaust air system.

17. A specially designed suit area may be provided in the facility. Personnel who enter this area wear a one-piece positive-pressure suit that is ventilated by a life support system. The life support system includes alarms and emergency backup breathing air tanks. Entry to this area is through an airlock fitted with airtight doors. A chemical shower is provided to decontaminate the surface of the suit before the worker leaves the area. The exhaust air from the suit area is filtered by two sets of HEPA filters installed in series. A duplicate filtration unit, an exhaust fan, and an automatically starting emergency power source are provided. The air pressure within the suit area is lower than that of any adjacent area. Emergency lighting and communications systems are provided. All penetrations into the internal shell of the suit area are sealed. A double-doored autoclave is provided for decontaminating waste materials to be removed from the suit area.

Section IV

Vertebrate Animal Biosafety Level Criteria

If experimental animals are used, institutional management must provide facilities and staff and establish practices which reasonably assure appropriate levels of environmental quality, safety, and care. Laboratory animal facilities are extensions of the laboratory and in some situations are integral to and inseparable from the laboratory. As a general principle, the Biosafety Level (facilities, practices, and operational requirements) recommended for working with infectious agents in vivo and in vitro are comparable.

These recommendations presuppose that laboratory animal facilities, operational practices, and quality of animal care meet applicable standards and regulations and that appropriate species are selected for animal experiments (e.g., *Guide for the Care and Use of Laboratory Animals,* HEW Publication no. [NIH] 78-23, Rev. 1978, and *Laboratory Animal Welfare Regulations,* 9 CFR, Subchapter A, Parts 1, 2 and 3).

Ideally, facilities for laboratory animals used for studies of infectious or noninfectious disease should be physically separate from other activities such as animal production and quarantine, clinical laboratories, and especially from facilities that provide patient care. Animal facilities should be designed and constructed to facilitate cleaning and housekeeping.

A "clean hall/dirty hall" layout is very useful in reducing cross contamination. Floor drains should be installed in animal facilities only on the basis of clearly defined needs. If floor drains are installed, the drain trap should always contain water.

These recommendations describe four combinations of practices, safety equipment, and facilities for experiments on animals infected with agents which are known or believed to produce infections in humans. These four combinations provide increasing levels of protection to personnel and to the environment and are recommended as minimal standards for activities involving infected laboratory animals. These four combinations, designated Animal Biosafety Levels 1–4, describe animal facilities and practices applicable to work on animals infected with agents assigned to corresponding Biosafety Levels 1–4.

Facility standards and practices for invertebrate vectors and hosts are not specifically addressed in standards written for commonly used laboratory animals. "Laboratory Safety for Arboviruses and Certain Other Viruses of Vertebrates" (112), prepared by the Subcommittee on Arbovirus Laboratory Safety of the American Committee on Arthropod-Borne Viruses, serves as a useful reference in the design and operation of facilities using arthropods.

Animal Biosafety Level 1
A. Standard practices
1. Doors to animal rooms open inward, are self-closing, and are kept closed when experimental animals are present.
2. Work surfaces are decontaminated after use or after any spill of viable materials.
3. Eating, drinking, smoking, and storing food for human use are not permitted in animal rooms.
4. Personnel wash their hands after handling cultures and animals and before leaving the animal room.
5. All procedures are carefully performed to minimize the creation of aerosols.
6. An insect and rodent control program is in effect.

B. Special practices
1. Bedding materials from animal cages are removed in such a manner as to minimize the creation of aerosols and disposed of in compliance with applicable institutional or local requirements.
2. Cages are washed manually or in a cagewasher. Temperature of final rinse water in a mechanical washer should be 180°F.
3. The wearing of laboratory coats, gowns, or uniforms in the animal room is recommended. It is further recommended that laboratory coats worn in the animal room not be worn in other areas.

C. Containment equipment
Special containment equipment is not required for animals infected with agents assigned to Biosafety Level 1.

D. Animal facilities
1. The animal facility is designed and constructed to facilitate cleaning and housekeeping.
2. A handwashing sink is available in the animal facility.
3. If the animal facility has windows that open, they are fitted with fly screens.
4. It is recommended, but not required, that the direction of airflow in the animal facility is inward and that exhaust air is discharged to the outside without being recirculated to other rooms.

Animal Biosafety Level 2
A. Standard practices
1. Doors to animal rooms open inward, are self-closing, and are kept closed when infected animals are present.
2. Work surfaces are decontaminated after use or spills of viable materials.
3. Eating, drinking, smoking, and storing of food for human use are not permitted in animal rooms.
4. Personnel wash their hands after handling cultures and animals and before leaving the animal room.
5. All procedures are carefully performed to minimize the creation of aerosols.
6. An insect and rodent control program is in effect.

B. Special practices
1. Cages are decontaminated, preferably by autoclaving, before they are cleaned and washed.
2. Surgical-type masks are worn by all personnel entering animal rooms housing nonhuman primates.
3. Laboratory coats, gowns, or uniforms are worn while in the animal room. This protective clothing is removed before leaving the animal facility.
4. The laboratory or animal facility director limits access to the animal room to personnel who have been advised of the potential hazard and who need to enter the room for program or service purposes when work is in progress. In general, persons who may be at increased risk of acquiring infection or for whom infection might be unusually hazardous are not allowed in the animal room.
5. The laboratory or animal facility director establishes policies and procedures whereby only persons who have been advised of the potential hazard and meet any specific requirements (e.g., for immunization) may enter the animal room.
6. When the infectious agent(s) in use in the animal room requires special entry provisions (e.g., vaccination), a hazard warning sign,

incorporating the universal biohazard symbol, is posted on the access door to the animal room. The hazard warning sign identifies the infectious agent, lists the name and telephone number of the animal facility supervisor or other responsible person(s), and indicates the special requirement(s) for entering the animal room.

7. Special care is taken to avoid skin contamination with infectious materials; gloves should be worn when handling infected animals and when skin contact with infectious materials is unavoidable.

8. All wastes from the animal room are appropriately decontaminated—preferably by autoclaving—before disposal. Infected animal carcasses are incinerated after being transported from the animal room in leakproof, covered containers.

9. Hypodermic needles and syringes are used only for the parenteral injection or aspiration of fluids from laboratory animals and diaphragm bottles. Only needle-locking syringes or disposable needle-syringe units (i.e., the needle is integral to the syringe) are used for the injection or aspiration of infectious fluids. Needles should not be bent, sheared, replaced in the sheath or guard, or removed from the syringe following use. The needle and syringe should be promptly placed in a puncture-resistant container and decontaminated, preferably by autoclaving, before discard or reuse.

10. If floor drains are provided, the drain traps are always filled with water or a suitable disinfectant.

11. When appropriate, considering the agents handled, base-line serum samples from animal care and other at-risk personnel are collected and stored. Additional serum samples may be collected periodically, depending on the agents handled or the function of the facility.

C. Containment equipment

Biological safety cabinets, other physical containment devices, and/or personal protective devices (e.g., respirators, face shields) are used whenever procedures with a high potential for creating aerosols are conducted (82). These include necropsy of infected animals, harvesting of infected tissues or fluids from animals or eggs, intranasal inoculation of animals, and manipulations of high concentrations or large volumes of infectious materials.

D. Animal facilities

1. The animal facility is designed and constructed to facilitate cleaning and housekeeping.

2. A handwashing sink is available in the room where infected animals are housed.

3. If the animal facility has windows that open, they are fitted with fly screens.

4. It is recommended, but not required, that the direction of airflow in the animal facility is inward and that exhaust air is discharged to the outside without being recirculated to other rooms.

5. An autoclave which can be used for decontaminating infectious laboratory waste is available in the building with the animal facility.

Animal Biosafety Level 3
A. Standard practices

1. Doors to animals rooms open inward, are self-closing, and are kept closed when work with infected animals is in progress.

2. Work surfaces are decontaminated after use or spills of viable materials.

3. Eating, drinking, smoking, and storing of food for human use are not permitted in the animal room.

4. Personnel wash their hands after handling cultures and animals and before leaving the laboratory.

5. All procedures are carefully performed to minimize the creation of aerosols.

6. An insect and rodent control program is in effect.

B. Special practices

1. Cages are autoclaved before bedding is removed and before they are cleaned and washed.

2. Surgical-type masks or other respiratory protection devices (e.g., respirators) are worn by personnel entering rooms housing animals infected with agents assigned to Biosafety Level 3.

3. Wrap-around or solid-front gowns or uniforms are worn by personnel entering the animal room. Front-button laboratory coats are unsuitable. Protective gowns must remain in the animal room and must be decontaminated before being laundered.

4. The laboratory director or other responsible person restricts access to the animal room to personnel who have been advised of the potential hazard and who need to enter the room for program or service purposes when infected animals are present. In general, persons who may be at increased risk of acquiring infection or for whom infection might be unusually hazardous are not allowed in the animal room.

5. The laboratory director or other responsible person establishes policies and procedures whereby only persons who have been advised of the potential hazard and meet any specific requirements (e.g., for immunization) may enter the animal room.

6. Hazard warning signs, incorporating the uni-

versal biohazard warning symbol, are posted on access doors to animal rooms containing animals infected with agents assigned to Biosafety Level 3. The hazard warning sign should identify the agent(s) in use, list the name and telephone number of the animal room supervisor or other responsible person(s), and indicate any special conditions of entry into the animal room (e.g., the need for immunizations or respirators).

7. Personnel wear gloves when handling infected animals. Gloves are removed aseptically and autoclaved with other animal room wastes before being disposed of or reused.

8. All wastes from the animal room are autoclaved before disposal. All animal carcasses are incinerated. Dead animals are tranported from the animal room to the incinerator in leakproof covered containers.

9. Hypodermic needles and syringes are used only for gavage or for parenteral injection or aspiration of fluids from laboratory animals and diaphragm bottles. Only needle-locking syringes or disposable needle-syringe units (e.g., the needle is integral to the syringe) are used. Needles should not be bent, sheared, replaced in the sheath or guard, or removed from the syring following use. The needle and syringe should be promptly placed in a puncture-resistant containers and decontaminated, preferably by autoclaving, before discard or reuse. Whenever possible, cannulas should be used instead of sharp needles (e.g., gavage).

10. If floor drains are provided, the drain traps are always filled with water or a suitable disinfectant.

11. If vacuum lines are provided, they are protected with HEPA filters and liquid disinfectant traps.

12. Boots, shoe covers, or other protective footwear and disinfectant footbaths are available and used when indicated.

C. Containment equipment

1. Personal protective clothing and equipment and/or other physical containment devices are used for all procedures and manipulations of infectious materials or infected animals.

2. The risk of infectious aerosols from infected animals or their bedding can be reduced if animals are housed in partial containment caging systems, such as open cages placed in ventilated enclosures (e.g., laminar flow cabinets), solid-wall and -bottom cages covered by filter bonnets, or other equivalent primary containment systems.

D. Animal facilities

1. The animal facility is designed and constructed to facilitate cleaning and housekeeping and is separated from areas which are open to unrestricted personnel traffic within the building. Passage through two sets of doors is the basic requirement for entry into the animal room from access corridors or other contiguous areas. Physical separation of the animal room from access corridors or other activities may also be provided by a double-doored clothes change room (showers may be included), airlock, or other access facility which requires passage through two sets of doors before entering the animal room.

2. The interior surfaces of walls, floors, and ceilings are water resistant so that they may be easily cleaned. Penetrations in these surfaces are sealed or capable of being sealed to facilitate fumigation or space decontamination.

3. A foot-, elbow-, or automatically operated handwashing sink is provided near each animal room exit door.

4. Windows in the animal room are closed and sealed.

5. Animal room doors are self-closing and are kept closed when infected animals are present.

6. An autoclave for decontaminating wastes is available, preferably within the animal room. Materials to be autoclaved outside the animal room are transported in a covered leakproof container.

7. An exhaust air ventilation system is provided. This system creates directional airflow that draws air into the animal room through the entry area. The building exhaust can be used for this purpose if the exhaust air is not recirculated to any other area of the building, is discharged to the outside, and is dispersed away from occupied areas and air intakes. Personnel must verify that the direction of the airflow (into the animal room) is proper. The exhaust air from the animal room that does not pass through biological safety cabinets or other primary containment equipment can be discharged to the outside without being filtered or otherwise treated.

8. The HEPA-filtered exhaust air from Class I or Class II biological safety cabinets or other primary containment devices is discharged directly to the outside or through the building exhaust system. Exhaust air from these primary containment devices may be recirculated within the animal room if the cabinet is tested and certified at least every 12 months. If the HEPA-filtered exhaust air from Class I or Class II biological safety cabinets is discharged to the outside through the building exhaust system, it is connected to this system in a manner (e.g., thimble unit connection [80]) that avoids any interference with the air balance of the cabinets or building exhaust system.

Animal Biosafety Level 4

A. Standard practices

1. Doors to animal rooms open inward and are self-closing.
2. Work surfaces are decontaminated after use or spills of viable materials.
3. Eating, drinking, smoking, and storing of food for human use is not permitted in the animal room.
4. All procedures are carefully performed to minimize the creation of aerosols.
5. An insect and rodent control program is in effect.
6. Cages are autoclaved before bedding is removed and before they are cleaned and washed.

B. Special practices

1. Only persons whose entry into the facility or individual animal rooms is required for program or support purposes are authorized to enter. Persons who may be at increased risk of acquiring infection or for whom infection might be unusually hazardous are not allowed in the animal facility. Persons at increased risk may include children, pregnant women, and persons who are immunodeficient or immunosuppressed. The supervisor has the final responsibility for assessing each circumstance and determining who may enter or work in the laboratory. Access to the facility is limited by secure, locked doors; accessibility is controlled by the animal facility supervisor, biohazards control officer, or other person responsible for the physical security of the facility. Before entering, persons are advised of the potential biohazards and instructed as to appropriate safeguards. Personnel comply with the instructions and all other applicable entry and exit procedures. Practical and effective protocols for emergency situations are established.
2. Personnel enter and leave the facility only through the clothing change and shower rooms. Personnel shower each time they leave the facility. Head covers are provided to personnel who do not wash their hair during the exit shower. Except in an emergency, personnel do not enter or leave the facility through the airlocks.
3. Street clothing is removed in the outer clothing change room and kept there. Complete laboratory clothing, including undergarments, pants and shirts or jumpsuits, shoes, and gloves, is provided and used by all personnel entering the facility. When exiting, personnel remove laboratory clothing and store it in a locker or hamper in the inner change room before entering the shower area.
4. When infectious materials or infected animals are present in the animal rooms, a hazard warning sign, incorporating the universal biohazard symbol, is posted on all access doors. The sign identifies the agent, lists the name and telephone number of the animal facility supervisor or other responsible person(s), and indicates any special conditions of entry into the area (e.g., the need for immunizations and respirators).
5. Supplies and materials to be taken into the facility enter by way of the double-door autoclave, fumigation chamber, or airlock, which is appropriately decontaminated between each use. After securing the outer doors, personnel inside the facility retrieve the materials by opening the interior doors of the autoclave, fumigation chamber, or airlock. This inner door is secured after materials are brought into the facility.
6. Materials (e.g., plants, animals, clothing) not related to the experiment are not permitted in the facility.
7. Hypodermic needles and syringes are used only for gavage or for parenteral injection and aspiration of fluids from laboratory animals and diaphragm bottles. Only needle-locking syringes or disposable syringe-needle units (i.e., needle is integral part of unit) are used. Needles should not be bent, sheared, replaced in the guard or sheath, or removed from the syringe following use. The needle and syringe should be promptly placed in a puncture-resistant container and decontaminated, preferably by autoclaving, before discard or reuse. Whenever possible, cannulas should be used instead of sharp needles (e.g., gavage).
8. A system is developed and is operational for the reporting of animal facility accidents and exposures, employee absenteeism, and for the medical surveillance of potential laboratory-associated illnesses. An essential adjunct to such a reporting-surveillance system is the availability of a facility for the quarantine, isolation, and medical care of persons with potential or known laboratory-associated illnesses.
9. Base-line serum samples are collected and stored for all laboratory and other at-risk personnel. Additional serum specimens may be collected periodically, depending on the agents handled or the function of the laboratory.

C. Containment equipment

Laboratory animals infected with agents assigned to Biosafety Level 4 are housed in a Class III biological safety cabinet or in partial-containment caging systems (such as open cages placed in ventilated enclosures, solid-wall and -bottom cages covered with filter bonnets, or other equivalent primary containment systems) in specially designed areas in which all personnel are required to wear one-piece positive pressure suits ventilated with a life support system. Animal work with viral agents that require Biosafety Level 4 secondary containment and for which highly effective vaccines are available and used may be conducted with partial-containment cages and without the one-piece positive pressure personnel suit if the facility has been decontaminated, if no concurrent experiments are being done in the facility which require Biosafety Level 4 primary and secondary containment, and if all other standard and special practices are followed.

D. Animal facility

1. The animal rooms are located in a separate building or in a clearly demarcated and isolated zone within a building. Outer and inner change rooms separated by a shower are provided for personnel entering and leaving the facility. A double-doored autoclave, fumigation chamber, or ventilated airlock is provided for passage of materials, supplies, or equipment which are not brought into the facility through the change room.

2. Walls, floors, and ceilings of the facility are constructed to form a sealed internal shell which facilitates fumigation and is animal and insect proof. The internal surfaces of this shell are resistant to liquids and chemicals, thus facilitating cleaning and decontamination of the area. All penetrations in these structures and surfaces are sealed.

3. Internal facility appurtenances, such as light fixtures, air ducts, and utility pipes, are arranged to minimize the horizontal surface area on which dust can settle.

4. A foot-, elbow-, or automatically operated handwashing sink is provided near the door of each animal room within the facility.

5. If there is a central vacuum system, it does not serve areas outside of the facility. The vacuum system has in-line HEPA filters placed as near as practicable to each use point or service cock. Filters are installed to permit in-place decontamination and replacement. Other liquid and gas services for the facility are protected by devices that prevent backflow.

6. External animal facility doors are self-closing and self-locking.

7. Any windows must be resistant to breakage and sealed.

8. A double-doored autoclave is provided for decontaminating materials that leave the facility. The autoclave door which opens to the area external to the facility is automatically controlled so that it can only be opened after the autoclave "sterilization" cycle is completed.

9. A pass-through dunk tank, fumigation chamber, or an equivalent decontamination method is provided so that materials and equipment that cannot be decontaminated in the autoclave can be safely removed from the facility.

10. Liquid effluents from laboratory sinks, cabinets, floors, and autoclave chambers are decontaminated by heat treatment before being discharged. Liquid wastes from shower rooms and toilets may be decontaminated with chemical disinfectants or by heat in the liquid waste decontamination system. The procedure used for heat decontamination of liquid wastes must be evaluated mechanically and biologically by using a recording thermometer and an indicator microorganism with a defined heat susceptibility pattern. If liquid wastes from the shower rooms are decontaminated with chemical disinfectants, the chemicals used must have documented efficacy against the target or indicator microorganisms.

11. An individual supply and exhaust air ventilation system is provided. The system maintains pressure differentials, and directional airflow is required to assure inflow from areas outside the facility toward areas of highest potential risk within the facility. Manometers are provided to sense pressure differentials between adjacent areas and are maintained at different pressure levels. The manometers sound an alarm when a system malfunctions. The supply and exhaust airflow is interlocked to assure inward (or zero) airflow at all times.

12. Air can be recirculated within an animal room if it is filtered through a HEPA filter.

13. The exhaust air from the facility is filtered by HEPA filters and discharged to the outside so that it is dispersed away from occupied buildings and air intakes. Within the facility, the filters are located as near to the laboratories as practicable in order to reduce the length of potentially contaminated air ducts. The filter chambers are designed to allow in situ decontamination before filters are removed and to facilitate certification testing after they are replaced. Coarse filters are provided for treatment of air supplied to the facility in order to increase the lifetime of the HEPA filters.

14. The treated exhaust air from Class I or Class II biological safety cabinets can be discharged into the animal room environment or to the outside through the facility air exhaust system. If exhaust air from Class I or II biological safety cabinets is discharged into the animal room, the cabinets are tested and certified at 6-month intervals. *The treated exhaust air from Class III biological safety cabinets is discharged without recirculation via the facility exhaust air system.* If the treated exhaust air from any of these cabinets is discharged to the outside through the facility exhaust air system, it is connected to this system in a

manner that avoids any interference with the air balance of the cabinets or the facility exhaust air system.

15. A specially designed suit area may be provided in the facility. Personnel who enter this area wear a one-piece positive-pressure suit that is ventilated by a life support system. The life support system is provided with alarms and emergency backup breathing air tanks. Entry to this area is through an airlock fitted with airtight doors. A chemical shower is provided to decontaminate the surface of the suit before the worker leaves the area. The exhaust air from the area in which the suit is used is filtered by two sets of HEPA filters installed in series. A duplicate filtration unit and exhaust fan are provided. An automatically starting emergency power source is provided. The air pressure within the suit area is lower than that of any adjacent area. Emergency lighting and communication systems are provided. All penetrations into the inner shell of the suit area are sealed. A double-doored autoclave is provided for decontaminating waste materials to be removed from the suit area.

Section V

Recommended Biosafety Levels for Infectious Agents and Infected Animals

Selection of an appropriate biosafety level for work with a particular agent or animal study depends upon a number of factors. Some of the most important are: the virulence, pathogenicity, biological stability, route of spread, and communicability of the agent; the nature or function of the laboratory; the procedures and manipulations involving the agent; the quantity and concentration of the agent; the endemicity of the agent; and the availability of effective vaccines or therapeutic measures.

Agent summary statements in this section provide guidance for the selection of appropriate biosafety levels. Specific information on laboratory hazards associated with a particular agent and recommendations regarding practical safeguards that can significantly reduce the risk of laboratory-associated diseases are included. Agent summary statements are presented for agents which meet one or more of the following criteria: the agent is a proven hazard to laboratory personnel working with infectious materials (e.g., hepatitis B virus, tubercle bacilli); the potential for laboratory-associated infection is high even in the absence of previously documented laboratory-associated infections (e.g., exotic arboviruses); or the consequences of infection are grave (e.g., Creutzfeldt-Jakob disease, botulism).

Recommendations for the use of vaccines and toxoids are included in agent summary statements when such products are available—either as licensed or Investigational New Drug (IND) products. When applicable, recommendations for the use of these products are based on current recommendations of the Public Health Service Advisory Committee on Immunization Practice and are specifically targeted to at-risk laboratory personnel and others who must work in or enter laboratory areas. These specific recommendations should in no way preclude the routine use of such products as diptheria-tetanus toxoids, poliovirus vaccine, influenza vaccine, and others because of the potential risk of community exposures irrespective of any laboratory risks. Appropriate precautions should be taken in the administration of live attenuated virus vaccines in individuals with altered immunocompetence.

Risk assessments and Biosafety Levels recommended in the agent summary statements presuppose a population of immunocompetent individuals. Those with altered immunocompetence may be at increased risk when exposed to infectious agents. Immunodeficiency may be hereditary, congenital, or induced by a number of neoplastic diseases, by therapy, or by radiation. The risk of becoming infected or the consequences of infection may also be influenced by such factors as age, sex, race, pregnancy, surgery (e.g., splenectomy, gastrectomy), predisposing diseases (e.g., diabetes, lupus erythematosus), or altered physiological function. These and other variables must be considered in individualizing the generic risk assessments of the agent summary statements for specific activities.

The basic biosafety level assigned to an agent is based on the activities typically associated with the growth and manipulation of quantities and concentrations of infectious agents required to accomplish identification of typing. If activities with clinical materials pose a lower risk to personnel than those activities associated with manipulation of cultures, a lower biosafety level is recommended. On the other hand, if the activities involve large volumes or highly concentrated preparations ("production quantities") or manipulations which are likely to produce aerosols or which are otherwise intrinsically hazardous, additional personnel precautions and increased levels of primary and secondary containment may be indicated. "Production quantities" refers to large volumes or concentrations of infectious agents considerably in excess of those typically used of identification and typing activites. Propagation and concentration of infectious agents, as occurs in large-scale fermentations, antigen and vaccine production, and a variety of other commercial and research activities, clearly deal with signficant masses of infectious agents that are reasonably considered "production quantities." However, in terms of potentially increased risk as a function of the mass of infectious agents, it is not possible to define "production quantities" in finite volumes or concentrations for any given agent. Therefore, the laboratory director must make a risk assessment of the activities conducted and select practices, containment equipment, and facilities appropriate to the risk, irrespective of the volume or concentration of agent involved.

Occasions will arise when the laboratory director should select a biosafety level higher than that recommended. For example, a higher biosafety level may be indicated by the unique nature of the proposed activity (e.g., the need for special containment for experimentally generated aerosols for inhalation studies) or by the proximity of the laboratory to areas of special concern (e.g., a diagnostic laboratory located near patient care areas). Similarly, a recommended biosafety level may be adapted to compensate for the absence of certain recommended safeguards. For example, in those situations where Biosafety Level 3 is recommended, acceptable safety may be achieved for routine or repetitive operations (e.g., diagnostic procedures involving the propagation of an agent for identification, typing, and susceptibility testing) in laboratories where facility features satisfy Biosafety Level 2 recommendations, provided the recommended "Standard Microbiological Practices," "Special Practices," and "Containment Equipment" for Biosafety Level 3 are rigorously followed. The decision to adapt Biosafety Level 3 recommendations in this manner should be made only by the laboratory director. This adaptation, however, is not suggested for agent production operations or activities where procedures are frequently changing. The laboratory director should also give special consideration to selecting appropriate safeguards for materials that may contain a suspected agent. For example, sera of human origin may contian hepatitis B virus and should be handled under conditions which reasonably preclude cutaneous, mucous membrane, or parenteral exposure of personnel, and sputa submitted to the laboratory for assay for tubercle bacilli should be handled under conditions which reasonably preclude the generation of aerosols or which contain any aerosols that may be generated during the manipulation of clinical materials or cultures.

The infectious agents which meet the previously stated criteria are listed by category of agent on the following pages. To use these summaries, first locate the agent in the listing under the appropriate category of agent. Second, utilize the practices, safety equipment, and type of facilities recommended for working with clinical materials, cultures of infectious agents, or infected animals recommended in the agent summary statement and described in Section V.

The laboratory director is also responsible for appropriate risk assessment of agents not included in the Agent Summary Statements and for utilization of appropriate practices, containment equipment, and facilities for the agent used.

Risk assessment. The risk assessment of laboratory activities involving the use of infectious micro-organisms is ultimately a subjective process. Those risks associated with the agent, as well as with the activity to be conducted, must be considered in the assessment. The characteristics of infectious agents and the primary laboratory hazards of working with the agent are described generically for agents in Biosafety Levels 1–4 and specifically for individual agents or groups of agents on page 295 and in Section V, respectively, of this Appendix.

Hepatitis B virus (HBv) is an appropriate model for illustrating the risk assessment process. HBv is among the most ubiquitous of human pathogens and most prevalent of laboratory-associated infections. The agent has been demonstrated in a variety of body secretions and excretions. Blood, saliva, and semen have been shown to be infectious. Natural transmission is associated with parenteral inoculation or with contamination of the broken skin or of mucous membranes with infectious body fluids. There is no evidence of airborne or interpersonal spread through casual contact. Prophylactic measures include the use of a licensed vaccine in high-risk groups and the use of hepatitis B immune globulin following overt exposure.

The primary risk of HBv infection in laboratory personnel is associated with accidental parenteral inoculation, exposure of the broken skin or mucous membranes of the eyes, nose, or mouth, or ingestion of infectious body fluids. These risks are typical of those described for Biosafety Level 2 agents and are addressed by using the recommended standard and special microbiological practices to minimize or eliminate these overt exposures.

Hepatitis non-A non-B and AIDS—acquired immune deficiency syndrome—pose similar infection risks to laboratory personnel. The prudent practices recommended for HBv are applicable to these two disease entities, as well as to the routine laboratory manipulation of clinical materials of domestic origin.

The described risk assessment process is also applicable to laboratory operations other than those involving the use of primary agents of human disease. Microbiological studies of animal host-specific pathogens, soil, water, food, feeds, and other natural or manufactured materials, by comparison, pose substantially lower risks of laboratory infection. Microbiologists and other scientists working with such materials may nevertheless find the practices, containment equipment, and facility recommendations described in this publication of value in developing operational standards to meet their own assessed needs.

Agent Summary Statements

Parasitic Agents

Agent: **Nematode parasites of humans**
Laboratory-associated infections with *Strongy-*

loides spp. and hookworms have been reported (90). Allergic reactions to various antigenic components of nematodes (e.g., aerosolized *Ascaris* antigens) may represent an individual risk to sensitized persons. Laboratory animal-associated infections (including arthropods) have not been reported, but infective larvae in the feces of nonhuman primates and of dogs infected with *Strongyloides* spp. are a potential infection hazard for laboratory and animal care personnel.

Laboratory hazards. Eggs and larvae in freshly passed feces of infected hosts are usually not infective; development of the infective stages may take periods of 1 day to several weeks. Ingestion of the infective eggs or skin penetration of infective larvae are the primary hazards to laboratory and animal care personnel. Arthropods infected with filarial parasites pose a potential hazard to laboratory personnel. In laboratory personnel with frequent exposure to aerosolized antigens of *Ascaris* spp., development of hypersensitivity is common.

Recommended precautions. Biosafety Level 2 practices, containment equipment, and facilities are recommended for activities with *infective stages* of the parasites listed. Exposure to aerosolized sensitizing antigens of *Ascaris* spp. should be avoided. Primary containment (e.g., biological safety cabinet) may be required for work with these materials by hypersensitive individuals.

Agent: **Protozoal parasites of humans**

Laboratory-associated infections with *Toxoplasma* spp., *Plasmodium* spp. (including *P. cynomologi*), *Trypanosoma* spp., and *Leishmania* spp. have been reported (21, 49, 90, 100). In addition, infections with *Entamoeba histolytica*, *Giardia* spp., and *Coccidia* spp. can result from ingestion of cysts in feces.

Accidental laboratory infections as well as human volunteer studies have proven the transmissibility of *P. cynomologi* from nonhuman primates to humans via infected mosquitoes (40). Although laboratory animal-associated infections have not been reported, contact with lesion material from rodents with cutaneous leishmaniasis and with feces or blood of experimentally or naturally infected animals may be a direct source of infection for laboratory personnel.

Laboratory hazards. Infective stages may be present in blood, feces, lesion exudates, and infected arthropods. Depending on the parasite, accidental parenteral inoculation, transmission by arthropod vectors, skin penetration, and ingestion are the primary laboratory hazards. Aerosol or droplet exposure of the mucous membranes of the eyes, nose, or mouth with trophozoites are potential hazards when working with cultures of *Naegleria fowleri*, *Leishmania* spp., *Trypanosoma cruzi*, or tissue homogenates or blood containing hemoflagellates. Because of the grave consequence of toxoplasmosis in the developing fetus, women of childbearing age should be discouraged from working with viable *Toxoplasma* spp.

Recommended precautions. Biosafety Level 2 practices, containment equipment, and facilities are recommended for activities with *infective stages* of the parasites listed. Infected arthropods should be maintained in facilities which reasonably preclude the exposure of personnel or their escape to the outside. Primary containment (e.g., biological safety cabinet) or personal protection (e.g., face shield) may be indicated when working with cultures of *T. cruzi*, *Leishmania*, *N. fowleri*, or tissue homogenates or blood containing hemoflagellates. Gloves are recommended for activities where there is the likelihood of direct skin contact with infective stages of the parasites listed.

Agent: **Trematode parasites of humans**

Laboratory-associated infections with *Schistosoma* spp. and *Fasciola* spp. have been reported—none associated directly with laboratory animals (90).

Laboratory hazards. Infective stages of *Schistosoma* spp. (cercariae) and *Fasciola* spp. (metacercariae) may be found, respectively, in the water or encysted on aquatic plants in laboratory aquaria used to maintain snail intermediate hosts. Skin penetration by schistosome cercariae and ingestion of fluke metacercariae are the primary laboratory hazards. Dissection or crushing of schistosome-infected snails may also result in exposure of skin or mucous membrane to cercariae-containing droplets. Additionally, metacercariae may be inadvertently transferred from hand to mouth by fingers or gloves following contact with contaminated aquatic vegetation or surfaces of aquaria. Most laboratory exposure to *Schistosoma* spp. would predictably result in low worm burdens with minimal disease potential. Safe and effective drugs are available for the treatment of schistosomiasis.

Recommended precautions. Biosafety Level 2 practices, containment equipment, and facilities are recommended for activities with *infective stages* of the parasites listed. Gloves should be worn when there may be direct contact with water containing cercariae or vegetation containing metacercariae from naturally or experimentally infected snail intermediate hosts. Snails and cercariae in the water of laboratory aquaria should be killed by chemicals (e.g., hypochlorites, iodine) or heat before discharge to sewers.

Agent: **Cestode parasites of humans**

Although laboratory-associated infections with *Echinococcus granulosus* or *Taenia solium* have not been reported, the consequences of such infections following the ingestion of infective eggs of *T. solium* or *E. granulosus* are potentially grave.

Laboratory hazards. Infective eggs may be present in the feces of dogs or other canids (the definitive hosts of *E. granulosus*) or in the feces of humans (the definitive host of *T. solium*). Ingestion of infective eggs from these sources is the primary laboratory hazard. Cysts and cyst fluids of *E. granulosus* are not infectious for humans.

Recommended precautions. Biosafety Level 2 practices, containment equipment, and facilities are recommended for work with infective stages of these parasites. Special attention should be given to personal hygiene practices (e.g., handwashing) and avoidance of ingestion of infective eggs. Gloves are recommended when there may be direct contact with feces or surfaces contaminated with fresh feces of dogs infected with *E. granulosus* or humans infected with *T. solium* adults.

Fungal Agents

Agent: *Blastomyces dermatitidis*

Laboratory-associated local infections following accidental parenteral inoculation with infected tissues or cultures containing yeast forms of *B. dermatitidis* (39, 54, 66, 103, 127) have been reported. A single pulmonary infection (asymptomatic) occurred following the presumed inhalation of conidia. Subsequently this individual developed an osteolytic lesion from which *B. dermatitidis* was cultured (30). Presumably, pulmonary infections are associated only with sporulating mold forms (conidia).

Laboratory hazards. Yeast forms may be present in the tissues of infected animals and in clinical specimens. Parenteral (subcutaneous) inoculation of these materials may cause local granulomas. Mold form cultures of *B. dermatitidis* containing infectious conidia may pose a hazard of aerosol exposure.

Recommended precautions. Biosafety Level 2 and Animal Biosafety Level 2 practices, containment equipment, and facilities are recommended for activities with clinical materials, animal tissues, and infected animals.

Biosafety Level 3 practices, containment equipment, and facilities are recommended for processing mold cultures, soil, and other environmental materials known or likely to contain infectious conidia.

Agent: *Coccidioides immitis*

Laboratory-associated coccidioidomycosis is a documented hazard (12, 28, 31, 32, 33, 64, 68, 79, 105, 106, 107). Smith reported that 28 of 31 (90%) laboratory-associated infections in his institution resulted in clinical disease, whereas more than half of infections acquired in nature were asymptomatic (128).

Laboratory hazards. Because of its size (2 to 5 μm), the arthrospore is conducive to ready dispersal in air and retention in the deep pulmonary spaces. The much larger size of the spherule (30 to 60 μm) considerably reduces the effectiveness of this form of the fungus as an airborne pathogen.

Spherules of the fungus may be present in clinical specimens and animal tissues, and infectious arthrospores may be present in mold cultures and soil samples. Inhalation of arthrospores from soil samples or mold cultures or following transformation from the spherule form in clinical materials is the primary laboratory hazard. Accidental percutaneous inoculation of the spherule form may result in local granuloma formation (118). Disseminated disease may occur at a greater frequency in pregnant women, blacks, and Filipinos than in whites.

Recommended precautions. Biosafety Level 2 practices, containment equipment, and facilities are recommended for handling and processing clinical specimens and animal tissues. Animal Biosafety Level 2 practices and facilities are recommended for experimental animal studies when the route of challenge is parenteral.

Biosafety Level 3 practices and facilities are recommended for all activities with sporulating mold form cultures of *C. immitis* and for processing soil or other environmental materials known or likely to contain infectious arthrospores.

Agent: *Cryptococcus neoformans*

A single account of a laboratory exposure to *C. neoformans* as a result of a laceration by a scalpel blade heavily contaminated with encapsulated cells is reported (50). This vigorous exposure, which did not result in local or systemic evidence of infection, suggests that the level of pathogenicity for normal immunocompetent adults is low. Respiratory infections as a consequence of laboratory exposure have not been recorded.

Laboratory hazards. Accidental parenteral inoculation of cultures or other infectious materials represents a potential hazard to laboratory personnel—particularly to those that may be immunocompromised. Bites by experimentally infected mice and manipulations of infectious environmental materials (e.g., pigeon droppings) may also represent a potential hazard to laboratory personnel.

Recommended precautions. Biosafety Level 2 and Animal Biosafety Level 2 practices, containment equipment, and facilities are recommended, respectively, for activities with known or potentially infectious clinical, environmental, or culture materials and with experimentally infected animals.

The processing of soil or other environmental materials known or likely to contain infectious yeast cells should be conducted in a Class I or Class II biological safety cabinet. This precaution is also indicated for culture of the perfect or sexual state of the agent.

Agent: *Histoplasma capsulatum*

Laboratory-associated histoplasmosis is a documented hazard in facilities conducting diagnostic or investigative work (90, 91). Pulmonary infections have resulted from handling mold form cultures (78). Local infection has resulted from skin puncture during autopsy of an infected human (119) and from accidental needle inoculation of a viable culture (116). Collecting and processing soil samples from endemic areas has caused pulmonary infections in laboratory workers. Spores are resistant to drying and may remain viable for long periods of time. The small size of the infective conidia (microconidia are less than 5 μm) is conducive to airborne dispersal and intrapulmonary retention. Furcolow reported that 10 spores were almost as effective as a lethal inoculum in mice as 10,000 to 100,000 spores. (45).

Laboratory hazards. The infective stage of this dimorphic fungus (conidia) is present in sporulating mold form cultures and in soil from endemic areas. The yeast form in tissues or fluids from infected animals may produce local infection following parenteral inoculation.

Recommended precautions. Biosafety Level 2 and Animal Biosafety Level 2 practices, containment equipment, and facilities are recommended for handling and processing clinical specimens and animal tissues and for experimental animal studies when the route of challenge is parenteral.

Biosafety Level 3 practices and facilities are recommended for processing mold cultures, soil, or other environmental materials known or likely to contain infectious conidia.

Agent: *Sporothrix schenckii*

S. schenckii has caused a substantial number of local skin or eye infections in laboratory personnel. Most cases have been associated with accidents and have involved splashing culture material into the eye (41, 125), scratching (13) or injecting (117) infected material into the skin, or being bitten by an experimentally infected animal (60, 61). Skin infections have resulted also from handling cultures (74, 81) or necropsy of animals (44) without a known break in technique. No pulmonary infections have been reported to result from laboratory exposure, although naturally occurring lung disease, albeit rare, is thought to result from inhalation.

Recommended precautions. Biosafety Level 2 and Animal Biosafety Level 2 practices, containment equipment, and facilities are recommended for all laboratory and experimental animal activities with *S. schenckii*.

Agents: **Pathogenic members of the genera** *Epidermophyton*, *Microsporum*, **and** *Trichophyton*

Although skin, hair, and nail infections by these dermatophytic molds are among the most prevalent of human infections, the processing of clinical material has not been associated with laboratory infections. Infections have been acquired through contacts with naturally or experimentally infected laboratory animals (mice, rabbits, guinea pigs, etc.) and, rarely, with handling cultures (71, 90, 51).

Laboratory hazards. Agents are present in the skin, hair, and nails of human and animal hosts. Contact with infected laboratory animals with inapparent or apparent infections is the primary hazard to laboratory personnel. Cultures and clinical materials are not an important source of human infection.

Recommended precautions. Biosafety Level 2

and Animal Biosafety Level 2 practices, containment equipment, and facilities are recommended for all laboratory and experimental animal activities with dermatophytes.

Bacterial Agents

Agent: *Bacillus anthracis*

Forty (40) cases of laboratory-associated anthrax, occurring primarily at facilities conducting anthrax research, have been reported (38, 90). No laboratory-associated cases of anthrax have been reported in the United States for more than 20 years.

Naturally and experimentally infected animals pose a potential risk to laboratory and animal care personnel.

Laboratory hazards. The agent may be present in blood, skin lesion exudates, and, rarely, in urine and feces. Direct and indirect contact of the intact and broken skin with cultures and contaminated laboratory surfaces, accidental parenteral inoculation, and, rarely, exposure to infectious aerosols are the primary hazards to laboratory personnel.

Recommended precautions. Biosafety Level 2 practices, containment equipment, and facilities are recommended for activities using clinical materials and diagnostic quantities of infectious cultures. Animal Biosafety Level 2 practices and facilities are recommended for studies utilizing experimentally infected laboratory rodents. A licensed vaccine is available through the Centers for Disease Control; however, vaccination of laboratory personnel is not recommended unless frequent work with clinical specimens or diagnostic cultures is anticipated (e.g., animal disease diagnostic laboratory). Biosafety Level 3 practices and facilities are recommended for work involving production volumes or concentrations of cultures and for activities which have a high potential for aerosol production. In these facilities vaccination is recommended for all persons working with the agent, all persons working in the same laboratory room where the cultures are handled, and persons working with infected animals.

Agent: *Brucella (B. abortus, B. canis, B. melitensis, B. suis)*

B. abortus, B. canis, B. melitensis, and *B. suis* have all caused illness in laboratory personnel (77, 90, 110). Brucellois is the most commonly reported laboratory-associated bacterial infection (90). Hypersensitivity to *Brucella* antigens is also a hazard to laboratory personnel.

Occasional cases have been attributed to exposure to experimentally and naturally infected animals or their tissues.

Laboratory hazards. The agent may be present in blood, cerebrospinal fluid, semen, and occasionally urine. Most laboratory-associated cases have occurred in research facilities and have involved exposure to *Brucella* organisms being grown in large quantities. Direct skin contact with cultures or with infectious clinical specimens from animals (e.g., blood, uterine discharges) are also commonly implicated. Aerosols generated during laboratory procedures have caused large outbreaks (59). Mouth pipetting, accidental parenteral inoculations, and sprays into eyes, nose, and mouth have also resulted in infection.

Recommended precautions. Biosafety Level 2 practices are recommended for activities with clinical materials of human or animal origin containing or potentially containing pathogenic *Brucella* spp. Biosafety Level 3 and Animal Biosafety Level 3 practices, containment equipment, and facilities are recommended, respectively, for all manipulations of cultures of the pathogenic *Brucella* spp. listed in this summary and for experimental animal studies. Vaccines are not available for use in humans.

Agent: *Chlamydia psittaci, C. trachomatis*

Infections with psittacosis, lymphogranuloma venereum (LGV), and trachoma are documented hazards and the fifth most commonly reported laboratory-associated bacterial infection. The majority of cases were of psittacosis, occurred before 1955, and had the highest case-fatality rate of all groups of infectious agents (90). Contact with and exposure to infectious aerosols in the handling, care, or necropsy of naturally or experimentally infected birds are the major sources of laboratory-associated psittacosis. Infected mice and eggs are less important sources of *C. psittaci*. Laboratory animals are not a reported source of human infection with *C. trachomatis*.

Laboratory hazards. *C. psittaci* may be present in the tissues, feces, nasal secretions, and blood of infected birds and in blood, sputum, and tissues of infected humans. *C. trachomatis* may be present in genital, bubo, and conjunctival fluids of infected humans. Exposure to infectious aerosols and droplets created during the handling of infected birds and tissues is the primary hazard to laboratory personnel working with psittacosis. The primary laboratory hazards of *C. trachomatis* are accidental parenteral inoculation and direct and indirect exposure of mucous membranes of the eyes, nose, and mouth to genital, bubo, or conjunctival fluids, cell culture materials, and fluids from infected eggs. Infectious aerosols may also pose a potential source of infection.

Recommended precautions. Biosafety Level 2 practices, containment equipment, and facilities are recommended for activities involving the necropsy of infected birds and the diagnostic examination of tissues of cultures known or potentially infected with *C. psittaci or C. trachomatis*. Wetting the feathers of infected birds with a detergent-disinfectant prior to necropsy can appreciably reduce the risk of aerosols of infected feces and nasal secretions on the feathers and external surfaces of the bird. Animal Biosafety Level 2 practices and facilities and respiratory pro-

tection are recommended for personnel working with caged birds naturally or experimentally infected. Gloves are recommended for the necropsy of birds and mice, the opening of inoculated eggs, and when there is the likelihood of direct skin contact with infected tissues, bubo fluids, and other clinical materials. Additional primary containment and personnel precautions, such as those recommended for Biosafety Level 3, may be indicated for activities with high potential for droplet or aerosol production and for activities involving production quantities or concentrations of infectious materials. Vaccines are not available for use in humans.

Agent: *Clostridium botulinum*

While there are no reported cases of botulism associated with the handling of the agent or toxin in the laboratory or working with naturally or experimentally infected animals, the consequences of such intoxications would be grave.

Laboratory hazards. *C. botulinum* or its toxin may be present in a variety of food products, clinical materials (serum, feces), and environmental samples (soil, surface water). Exposure to the toxin of *C. botulinum* is the primary laboratory hazard. The toxin may be absorbed after ingestion or following contact with the skin, eyes, or mucous membranes, including the respiratory tract. Accidental parenteral inoculation may also represent a significant exposure to toxin. Broth cultures grown under conditions of optimal toxin production may contain 2×10^6 mouse LD_{50} per ml (111).

Recommended precautions. Biosafety Level 2 practices, containment equipment, and facilities are recommended for all activities with materials known or potentially containing the toxin. A pentavalent (ABCDE) botulism toxoid is available through the Centers for Disease Control as an Investigational New Drug (IND). This toxoid is recommended for personnel working with cultures of *C. botulinum* or its toxins. Solutions of sodium hydroxide (0.1 N) readily inactivate the toxin and are recommended for decontaminating work surfaces and spills of cultures or toxin. Additional primary containment and personnel precautions, such as those recommended for Biosafety Level 3, may be indicated for activities with a high potential for aerosol or droplet production, those involving production quantities of toxin, and those involving purified toxins. Animal Biosafety Level 2 practices and facilities are recommended for diagnostic studies and titration of toxin.

Agent: *Clostridium tetani*

Although the risk of infection to laboratory personnel is negligible, Pike (90) has recorded five incidents related to exposure of personnel during manipulation of the toxin.

Laboratory hazards. Accidental parenteral inoculation and ingestion of the toxin are the primary hazards to laboratory personnel. Since tetanus toxin is poorly absorbed through mucous membranes, aerosols and droplets probably represent minimal hazards.

Recommended precautions. Biosafety Level 2 practices, containment equipment, and facilities are recommended for activities involving the manipulation of cultures or toxin. While the risk of laboratory-associated tetanus is low, the administration of an adult diphtheria-tetanus toxoid at 10-year intervals may further reduce the risk to laboratory and animal care personnel of toxin exposures and wound contamination (24).

Agent: *Corynebacterium diphtheriae*

Laboratory-associated infections with *C. diphtheriae* are documented. Pike (90) list 33 cases reported in the world literature.

Laboratory animal-associated infections have not been reported.

Laboratory hazards. The agent may be present in exudates or secretions of the nose, throat (tonsil), pharynx, larynx, and wounds, in blood, and on the skin. Inhalation, accidental parenteral inoculation, and ingestion are the primary laboratory hazards.

Recommended precautions. Biosafety Level 2 practices, containment equipment, and facilities are recommended for all activities utilizing known or potentially infected clinical materials or cultures. Animal Biosafety Level 2 facilities are recommended for studies utilizing infected laboratory animals. While the risk of laboratory-associated diphtheria is low, the administration of an adult diphtheria-tetanus toxoid at 10-year intervals may further reduce the risk to laboratory and animal care personnel of toxin exposures and work with infectious materials. (24).

Agent: *Francisella tularensis*

Tularemia is the third most commonly reported laboratory-associated bacterial infection (90). Almost all cases occurred at facilities involved in tularemia research. Occasional cases have been related to work with naturally or experimentally infected animals or their ectoparasites.

Laboratory hazards. The agent may be present in lesion exudate, respiratory secretions, cerebrospinal fluid, blood, urine, tissues from infected animals, and fluids from infected arthropods. Direct contact of skin or mucous membranes with infectious materials, accidental parenteral inoculation, ingestion, and exposure to aerosols and infectious droplets have resulted in infection. Cultures have been more commonly associated with infection than clinical materials and infected animals. The human 25–50% infectious dose is on the order of 10 organisms by the respiratory route (121).

Recommended precautions. Biosafety Level 2 practices, containment equipment, and facilities are recommended for activities with clinical materials of human or animal origin containing or potentially containing *F. tularensis*. Biosafety Level 3 and Animal Biosafety Level 3 practices and facilities are recommended, respectively, for all manipulations of

cultures and for experimental animal studies. An investigational live attenuated vaccine (10) is available through the Centers for Disease Control and is recommended for persons working with the agent or with infected animals and for persons working in or entering the laboratory or animal room where cultures or infected animals are maintained.

Agent: *Leptospira interrogans*—**all serovars**

Leptospirosis is a well-documented laboratory hazard. Sixty-seven laboratory-associated infections and 10 deaths have been reported (90).

An experimentally infected rabbit was identified as the source of an infection with *L. interrogans* serovar *icterohemorrhagiae* (97). Direct and indirect contact with fluids and tissues of experimentally or naturally infected mammals during handling, care, or necropsy are potential sources of infection. In animals with chronic kidney infections, the agent is shed in the urine in enormous numbers for long periods of time.

Laboratory hazards. The agent may be present in urine, blood, and tissues of infected animals and humans. Ingestion, accidental parenteral inoculation, and direct and indirect contact of skin or mucous membranes with cultures or infected tissues or body fluids—especially urine—are the primary laboratory hazards. The importance of aerosol exposure is not known.

Recommended precautions. Biosafety Level 2 practices, containment equipment, and facilities are recommended for all activities involving the use or manipulation of known or potentially infectious tissues, body fluids, and cultures and for the housing of infected animals. Gloves are recommended for the handling and necropsy of infected animals and when there is the likelihood of direct skin contact with infectious materials. Vaccines are not available for use in humans.

Agent: *Legionella pneumophila;* other *Legionella-like agents*

A single documented nonfatal laboratory-associated case of legionellosis due to presumed aerosol or droplet exposure during animal challenge studies with Pontiac fever agent (*L. pneumophila*) is recorded (16). Human-to-human spread has not been documented.

Experimental infections are readily produced in guinea pigs and embryonated chicken eggs (72). Challenged rabbits develop antibodies but not clinical disease. Mice are refractory to parenteral exposure. Unpublished studies by Kaufmann, Feeley, and others at the Centers for Disease Control have shown that animal-to-animal transmission did not occur in a variety of experimentally infected mammalian and avian species.

Laboratory hazards. The agent may be present in pleural fluids, tissue, sputa, and environmental sources (e.g., cooling tower water). Since the natural mode of transmission appears to be airborne, the greatest potential hazard is the generation of aerosols during the manipulation of cultures or of other concentrations of infectious materials (e.g., infected yolk sacs and tissues.)

Recommended precautions. Biosafety Level 2 practices, containment equipment, and facilities are recommended for all activities involving the use or manipulation of known or potentially infectious clinical materials or cultures and for the housing of infected animals. Primary containment devices and equipment (e.g., biological safety cabinets, centrifuge safety cups) should be used for activities likely to generate potentially infectious aerosols. Vaccines are not available for use in humans.

Agent: *Mycobacterium leprae*

Inadvertent parenteral human-to-human transmission of leprosy following an accidental needle stick in a surgeon (69) and the use of a presumably contaminated tattoo needle (87) have been reported. There are no cases reported as a result of working in a laboratory with biopsy or other clinical materials of human or animal origin. While naturally occurring leprosy or leprosy-like diseases have been reported in armadillos (120) and in nonhuman primates (35,76), humans are the only known important reservoir of this disease.

Laboratory hazards. The infectious agent may be present in tissues and exudates from lesions of infected humans and experimentally or naturally infected animals. Direct contact of the skin and mucous membranes with infectious materials and accidental parenteral inoculation are the primary laboratory hazards associated with handling infectious clinical materials.

Recommended precautions. Biosafety Level 2 practices, containment equipment, and facilities are recommended for all activities with known or potentially infectious clinical materials from infected humans and animals. Extraordinary care should be taken to avoid accidental parenteral inoculation with contaminated sharp instruments. Animal Biosafety Level 2 practices and facilities are recommended for animal studies utilizing rodents, armadillos, and nonhuman primates.

Agent: *Mycobacterium* spp. other than *M. tuberculosis, M. bovis,* or *M. leprae*

Pike reported 40 cases of nonpulmonary "tuberculosis" thought to be related to accidents or incidents in the laboratory or autopsy room (90). Presumably these infections were due to mycobacteria other than *M. tuberculosis* or *M. bovis.* A number of mycobacteria which are ubiquitous in nature are associated with diseases other than tuberculosis or leprosy in humans, domestic animals, and wildlife. Characteristically, these organisms are infectious but not contagious. Clinically, the diseases associated with infections by these "atypical" mycobacteria can be divided into three general categories:

1. Pulmonary diseases resembling tuberculosis which may be associated with infection with *M. kansasii, M. avium* complex, and rarely with *M. xenopi, M. malmoense, M. asiaticum, M. simiae,* and *M. szulgai*
2. Lymphadenitis which may be associated with infection with *M. scrofulaceum, M. avium* complex, and, rarely, with *M. fortuitum* and *M. kansasii*
3. Skin ulcers and soft tissue wound infections which may be associated with infection with *M. ulcerans, M. marinum, M. fortuitum,* and *M. chelonei*

Laboratory hazards. The agents may be present in sputa, exudates from lesions, tissues, and environmental samples (e.g., soil and water). Direct contact of skin or mucous membranes with infectious materials, ingestion, and accidental parenteral inoculation are the primary laboratory hazards associated with clinical materials and cultures. Infectious aerosols created during the manipulation of broth cultures or tissue homogenates of these organisms associated with pulmonary disease also pose a potential infection hazard to laboratory personnel.

Recommended precautions. Biosafety Level 2 practices, containment equipment, and facilities are recommended for activities with clinical materials and cultures of *Mycobacterium* spp. other than *M. tuberculosis* or *M. bovis.* Animal Biosafety Level 2 practices and facilities are recommended for animal studies with the mycobacteria other than *M. tuberculosis, M. bovis,* or *M. leprae.*

Agent: *Mycobacterium tuberculosis, M. bovis*

M. tuberculosis and *M. bovis* infections are a proven hazard to laboratory personnel as well as to others who may be exposed to infectious aerosols in the laboratory (90,93). The incidence of tuberculosis in laboratory workers working with *M. tuberculosis* is three times higher than that of laboratorians not working with the agent (95). Naturally or experimentally infected nonhuman primates are a proven source of human infection (e.g., the annual tuberculin conversion rate in personnel working with infected nonhuman primates is about 70/10,000 compared with less than 3/10,000 in the general population (62). Experimentally infected guinea pigs or mice do not pose the same problem, since droplet nuclei are not produced by coughing in these species; however, litter from infected animals may become contaminated and serve as a source of infectious aerosols.

Laboratory hazards. Tubercle bacilli may be present in sputum, gastric lavage fluids, cerebrospinal fluid, urine, and lesions from a variety of tissues (3). Exposure to laboratory-generated aerosols is the most important hazard encountered. Tubercle bacilli may survive in heat-fixed smears (1) and may be aerosolized in the preparation of frozen sections and during manipulation of liquid cultures. Because of the low infectious dose of *M. tuberculosis* for humans (i.e., 50% infectious dose < 10 bacilli)

(98,99) and in some laboratories a high rate of isolation of acid-fast organisms from clinical specimens (> 10%) (47), sputa and other clinical specimens from suspected or known cases of tuberculosis must be considered potentially infectious and handled with appropriate precautions.

Recommended precautions. Biosafety Level 2 practices, containment equipment, and facilities (see American Thoracic Society laboratory service levels I and II) (2, 65) are recommended for preparing acid-fast smears and for culturing sputa or other clinical specimens, provided that aerosol-generating manipulations of such specimens are conducted in a Class I or II biological safety cabinet. Liquification and concentration of sputa for acid-fast staining may also be conducted on the open bench at Biosafety Level 2 by first treating the specimen with an equal volume of 5% sodium hypochlorite solution (undiluted household bleach) and waiting 15 min before centrifugation (85,108).

Biosafety Level 3 practices, containment equipment, and facilities (see American Thoracic Society laboratory service level III) (2,65) are recommended for activities involving the propagation and manipulation of cultures of *M. tuberculosis* or *M. bovis* and for animal studies utilizing nonhuman primates experimentally or naturally infected with *M. tuberculosis* or *M. bovis*. Animal studies utilizing guinea pigs or mice can be conducted at Animal Biosafety Level 2. Skin testing with purified protein derivative (PPD) of previously skin-tested-negative laboratory personnel can be used as a surveillance procedure. A licensed attenuated live vaccine (BCG) is available but is not routinely used in laboratory personnel.

Agent: *Neisseria gonorrhoeae*
Four cases of laboratory-associated gonorrhea have been reported in the United States (34, 90).

Laboratory hazards. The agent may be present in conjunctival, urethral, and cervical exudates, synovial fluid, urine, feces, and cerebrospinal fluid. Accidental parenteral inoculation and direct or indirect contact of mucous membranes with infectious clinical materials are the primary latoratory hazards. The importance of aerosols is not determined.

Recommended precautions. Biosafety Level 2 practices, containment equipment, and facilities are recommended for all activities involving the use or manipulation of clinical materials or cultures. Gloves should be worn when handling infected laboratory animals and when there is the likelihood of direct skin contact with infectious materials. Additional primary containment and personnel precautions, such as those described for Biosafety Level 3, may be indicated for aerosol or droplet production and for activities involving production quantities or concentrations of infectious materials. Vaccines are not available for use in humans.

Agent: *Neisseria meningitidis*
Meningococcal meningitis is a demonstrated but rare hazard to laboratory workers (4, 92).

Laboratory hazards. The agent may be present in pharyngeal exudates, cerebrospinal fluid, blood, and saliva. Parenteral inoculation, droplet exposure of mucous membranes, and infectious aerosol and ingestion are the primary hazards to laboratory personnel.

Recommended precautions. Biosafety Level 2 practices, containment equipment, and facilities are recommended for all activities utilizing known or potentially infectious body fluids and tissues. Additional primary containment and personnel precautions, such as those described for Biosafety Level 3, may be indicated for activities with high potential for droplet or aerosol production and for activities involving production quantities or concentrations of infectious materials. The use of licensed polysaccharide vaccines (19) should be considered for personnel regularly working with large volumes or high concentrations of infectious materials.

Agent: *Pseudomonas pseudomallei*
Two laboratory-associated cases of melioidosis are reported, one associated with a massive aerosol and skin contact exposure (48), the second resulting from an aerosol created during the open-flask sonication of a culture presumed to be *Pseudomonas cepacia* (102).

Laboratory hazards. The agent may be present in sputa, blood, wound exudates, and various tissues, depending on site of localization of the infection. Direct contact with cultures and infectious materials from humans, animals, or the environment, ingestion, autoinoculation, and exposure to infectious aerosols and droplets are the primary laboratory hazards. The agent has been demonstrated in blood, sputum, and abscess materials and may be present in soil and water samples from endemic areas.

Recommended precautions. Biosafety Level 2 practices, containment equipment, and facilities are recommended for all activities utilizing known or potentially infectious body fluids and tissues. Gloves should be worn when handling, and during necropsy of, infected animals and when there is the likelihood of direct skin contact with infectious materials. Additional primary containment and personnel precautions, such as those described for Biosafety Level 3, may be indicated for activities with a high potential for aerosol or droplet production and the activities involving production quantities or concentrations of infectious materials.

Agent: *Salmonella cholerasuis, S. enteritidis—all serotypes*
Salmonellosis is a documented hazard to laboratory personnel (90). Primary reservoir hosts include a broad spectrum of domestic and wild animals including birds, mammals, and reptiles, all of which may serve as a source of infection to laboratory personnel.

Laboratory hazards. The agent may be present in feces, blood, and urine and in food, feed, and environmental materials. Ingestion or parenteral inoculation are the primary laboratory hazards. The importance of aerosol exposure is not known. Naturally or experimentally infected animals are a potential source of infection for laboratory and animal care personnel and for other animals.

Recommended precautions. Biosafety Level 2 practices, containment equipment, and facilities are recommended for activities with clinical materials known or potentially containing the agents. Animal Biosafety Level 2 practices and facilities are recommended for activities with experimentally or naturally infected animals.

Agent: *Salmonella typhi*

Typhoid fever is a demonstrated hazard to laboratory personnel (7, 92).

Laboratory hazards. The agent may be present in feces, blood, gallbladder (bile), and urine. Humans are the only known reservoir of infection. Ingestion and parenteral inoculation of the organism represent the primary laboratory hazards. The importance of aerosol exposure is not known.

Recommended precautions. Biosafety Level 2 practices, containment equipment, and facilities are recommended for all activities utilizing known or potentially infectious clinical materials and cultures.

Licensed vaccines, which have been shown to protect 70–90% of recipients, may be a valuable adjunct to good safety practices in personnel regularly working with cultures or clinical materials which may contain *S. typhi* (7).

Agent: *Shigella* spp.

Shigellosis is a demonstrated hazard to laboratory personnel, with 49 cases reported in the United States (90). While outbreaks have occurred in captive nonhuman primates, humans are the only significant reservoir of infection. Experimentally infected guinea pigs, other rodents, and nonhuman primates are a proven source of infection.

Laboratory hazards. The agent may be present in feces, and, rarely, in blood of infected humans or animals. Ingestion or parenteral inoculation of the agent are the primary laboratory hazards. The oral 25–50% infectious dose of *S. flexneri* for humans is on the order of 200 organisms (132). The importance of aerosol exposure is not known.

Recommended precautions. Biosafety Level 2 practices, containment equipment, and facilities are recommended for all activities utilizing known or potentially infectious clinical materials or cultures. Animal Biosafety Level 2 facilities and practices are recommended for activities with experimentally or naturally infected animals. Vaccines are not available for use in humans.

Agent: *Treponema pallidum*

Syphilis is a documented hazard to laboratory

personnel who handle or collect clinical material from cutaneous lesions. Pike lists 20 cases of laboratory-associated infection (90). Humans are the only known natural reservoir of the agent.

No cases of laboratory animal-associated infections are reported; however, rabbit-adapted strains of *T. pallidum* (Nichols and possibly others) retain their virulence for humans.

Laboratory hazards. The agent may be present in materials collected from primary and secondary cutaneous lesions and in blood. Accidental parenteral inoculation and contact of mucous membranes or broken skin with infectious clinical materials (and, perhaps, infectious aerosols) are the primary hazards to laboratory personnel.

Recommended precautions. Biosafety Level 2 practices, containment equipment, and facilities are recommended for all activities involving the use or manipulation of blood or lesion materials from humans or infected rabbits. Gloves should be worn when there is a likelihood of direct skin contact with lesion materials. Periodic serological monitoring should be considered in personnel regularly working with infectious materials. Vaccines are not available for use in humans.

Agent: **Vibrionic enteritis (*Campylobacter fetus* subsp. *jejuni, Vibrio cholerae, V. parahaemolyticus*)**

Vibrionic enteritis due to *C. fetus jejuni, V. cholerae,* or *V. parahaemolyticus* is a documented but rare cause of laboratory-associated illnesses (92). Naturally and experimentally infected animals are a potential source of infection (94).

Laboratory hazards. All pathogenic vibrios may occur in feces. *C. fetus* may also be present in blood, exudates from abscesses, tissue, and sputa. Ingestion of *V. cholerae* and ingestion or parenteral inoculation of other vibrios constitute the primary laboratory hazard. The human oral infecting dose of *V. cholerae* in healthy non-achlorhydric individuals is of the order of 10^8 organisms (94). The importance of aerosol exposure is not known. The risk of infection following oral exposure may be increased in achlorhydric individuals.

Recommended precautions. Biosafety Level 2 practices, containment equipment, and facilities are recommended for activities with cultures or potentially infectious clinical materials. Animal Biosafety Level 2 practices and facilities are recommended for activities with naturally or experimentally infected animals. Although vaccines have been shown to provide partial protection of short duration (3–6 months) to nonimmune individuals in highly endemic areas (7), the routine use of cholera vaccine in laboratory staff is not recommended.

Agent: *Yersinia pestis*

Plague is a proven but rare laboratory hazard. Four cases have been reported in the United States (11, 90).

Laboratory hazards. The agent may be present in bubo fluid, blood, sputum, cerebrospinal fluid, feces, and urine from humans, depending on the clinical form and stage of the disease. Direct contact with cultures and infectious materials from humans or rodents, infectious aerosols or droplets generated during the manipulation of cultures and infected tissues and in the necropsy of rodents, accidental autoinoculation, ingestion, and bites by infected fleas collected from rodents are the primary hazards to laboratory personnel.

Recommended precautions. Biosafety Level 2 practices, containment equipment, and facilities are recommended for all activities involving the handling of potentially infectious clinical materials and cultures. Special care should be taken to avoid the generation of aerosols of infectious materials and during the necropsy of naturally or experimentally infected rodents. Gloves should be worn when handling field-collected or infected laboratory rodents and when there is the likelihood of direct skin contact with infectious materials. Necropsy of rodents is ideally conducted in a biological safety cabinet. Although field trials have not been conducted to determine the efficacy of licensed inactivated vaccines, experience with these products has been favorable (14). Immunization is recommended for personnel working regularly with cultures of *Y. pestis* or infected rodents (26).

Additional primary containment and personnel precautions, such as those described for Biosafety Level 3, are recommended for activities with high potential for droplet or aerosol production, for work with antibiotic-restraint strains, and for activities involving production quantities or concentrations of infectious materials.

Rickettsial Agents

Agent: *Coxiella burnetii*

Pike's summary indicates that Q fever is the second most commonly reported laboratory-associated infection, with outbreaks involving 15 or more persons recorded in several institutions (90). A broad range of domestic and wild mammals are natural hosts for Q fever and may serve as potential sources of infection for laboratory and animal care personnel. Exposure to naturally infected and often asymptomatic sheep and to their birth products is a documented hazard to personnel (20,109). The agent is remarkably resistant to drying and is stable under a variety of environmental conditions (121).

Laboratory hazards. The agent may be present in infected arthropods and in the blood, urine, feces, milk, tissues of infected animal or human hosts. The placenta of infected sheep may contain as many as 10^9 organisms per g of tissue, and milk may contain 10^5 organisms per g. Parenteral inoculation and exposure to infectious aerosols and droplets are the most likely sources of infection to laboratory and animal care personnel. The estimated human 25–50% infectious dose (inhalation) for Q fever is 10 organisms (122).

Recommended precautions. Biosafety Level 2 practices, containment equipment, and facilities are recommended for nonpropagative laboratory procedures, including serological examinations and staining of impression smears. Biosafety Level 3 practices and facilities are recommended for activities involving the inoculation, incubation, and harvesting of embryonated eggs or tissue cultures, the necropsy of infected animals, and the manipulation of infected tissues. Since infected guinea pigs and other rodents may shed the organisms in urine or feces (90), experimentally infected rodents should be maintained under Animal Biosafety Level 3. Recommended precautions for facilities using sheep as experimental animals are described by Spinelli (109) and by Bernard (6). An investigational new phase 1 Q fever vaccine (IND) is available from the U.S. Army Medical Research Institute for Infectious Diseases, Fort Detrick, Md. The use of this vaccine should be limited to those at high risk of exposure who have no demonstrated sensitivity to Q fever antigen.

Agent: *Rickettsia akari, Rochalimaea quintana,* and *Rochalimaea vinsonii*

Based on the experience of laboratories actively working with *Rickettsia akari,* it is likely that the five cases of rickettsialpox recorded by Pike (90) were associated with exposure to bites of infected mites rather than aerosol or contact exposure to infected tissues. There are no recorded cases of laboratory-associated infections with trench fever (*Rochalimaea quintana*) or vole rickettsia (*Rochalimaea vinsonii*).

Laboratory hazards. The agent of rickettsialpox may be present in blood and other tissues of infected house mice or humans and in the mite vector *Liponyssoides sanguineus.* Exposure to naturally or experimentally infected mites and accidental parenteral inoculation are the most likely sources of human infection with rickettsialpox. The agent of trench fever may be present in the blood and tissues of infected humans and in the body fluids and feces of infected human body lice (*Pediculus h. humanus*).

Recommended precautions. Biosafety Level 2 practices, containment equipment, and facilities are recommended for propagation and animal studies with *Rickettsia akari, Rochalimaea vinsonii,* and *Rochalimaea quintana.* Appropriate precautions should be taken to avoid exposure of personnel to infected mites that are maintained in the laboratory or that may be present on naturally infected house mice.

Agent: *Rickettsia prowazekii, Rickettsia typhi (R. mooseri), Rickettsia tsutsugamushi, Rickettsia canada,* and Spotted Fever Group agents of human disease other than *Rickettsia rickettsii* and *Rickettsia akari*

Pike reported 57 cases of laboratory-associated

typhus (type not specified), 56 cases of epidemic typhus with three deaths, and 68 cases of murine typhus (90). More recently, three cases of murine typhus were reported from a research facility (18). Two of these three cases were associated with work with infectious materials on the open bench; the third case resulted from an accidental parenteral inoculation. These three cases represented an attack rate of 20% in personnel working with infectious materials.

Laboratory hazards. Accidental parenteral inoculation and exposure to infectious aerosols are the most likely sources of laboratory-associated infections. Naturally or experimentally infected lice, fleas, and flying squirrels (*Glaucomys* spp.) (9) may also be a direct source of infection to laboratory personnel. The organisms are relatively unstable under ambient environmental conditions.

Recommended precautions. Biosafety Level 2 practices, containment equipment, and facilities are recommended for nonpropagative laboratory procedures, including serological and fluorescent-antibody procedures, and for the staining of impression smears. Biosafety Level 3 practices and facilities are recommended for all other manipulations of known or potentially infectious materials, including necropsy of experimentally infected animals and trituration of their tissues and inoculation, incubation, and harvesting of embryonated eggs or tissue cultures. Animal Biosafety Level 2 practices and facilities are recommended for activities with infected mammals other than flying squirrels or arthropods. Vaccines are not currently available for use in humans. Because the mode of transmission of *R. prowazekii* from flying squirrels to humans is not defined, Animal Biosafety Level 3 practices and facilities are recommended for animal studies with flying squirrels naturally or experimentally infected with *R. prowazekii*.

Agent: *Rickettsia rickettsii*

Rocky Mountain spotted fever is a documented hazard to laboratory personnel. Pike (90) reported 63 laboratory-associated cases, 11 of which were fatal. Oster (86) reported nine cases occurring over a 6-year period in one laboratory which were believed to have been acquired as a result of exposure to infectious aerosols.

Laboratory hazards. Accidental parenteral inoculation and exposure to infectious aerosols are the most likely sources of laboratory-associated infection (57). Successful aerosol transmission has been experimentally documented in nonhuman primates (101). Naturally and experimentally infected mammals, their ectoparasites, and their infected tissues are sources of human infection. The organism is relatively unstable under ambient environmental conditions.

Recommended precautions. Biosafety Level 2 practices, containment equipment, and facilities are recommended for all nonpropagative laboratory procedures, including serological and fluorescent-

antibody tests, and staining of impression smears. Biosafety Level 3 practices and facilities are recommended for all other manipulations of known or potentially infectious materials, including necropsy of experimentally infected animals and trituration of their tissues and inoculation, incubation, and harvesting of embryonated eggs or tissue cultures. Animal Biosafety Level 2 practices and facilities are recommended for holding of experimentally infected rodents; however, necropsy and any subsequent manipulation of tissues from infected animals should be conducted at Biosafety Level 3.

Because of the proven value of antibiotic therapy in the early stages of infection, it is essential that laboratories working with *R. rickettsii* have an effective system for reporting febrile illnesses in laboratory personnel, medical evaluation of potential cases, and, when indicated, institution of appropriate antibiotic therapy. Vaccines are not currently available for use in humans (see Appendix C).

Viral Agents

Agent: **Hepatitis A virus**

Laboratory-associated infections with hepatitis A virus do not appear to be an important occupational risk among laboratory personnel. However, the disease is a documented hazard in animal handlers and others working with chimpanzees which are naturally or experimentally infected (92).

Laboratory hazards. The agent may be present in feces of infected humans and chimpanzees. Ingestion of feces, stool suspensions, and other contaminated materials is the primary hazard to laboratory personnel. The importance of aerosol exposure has not been demonstrated. Attenuated or avirulent strains have not been fully defined but appear to result from serial passage in tissue culture.

Recommended precautions. Biosafety Level 2 practices, containment equipment, and facilities are recommended for activities with known or potentially infected feces from humans or chimpanzees. Animal Biosafety Level 2 practices and facilities are recommended for activities using naturally or experimentally infected chimpanzees. Animal care personnel should wear gloves and take other appropriate precautions to avoid possible fecal-oral exposure. Vaccines are not available for use in humans, but are in the developmental stages.

Agent: **Hepatitis B, hepatitis non-A non-B**

Pike concluded that hepatitis B is currently the most frequently occurring laboratory-associated infection (90). The incidence in some categories of laboratory workers is seven times greater than that of the general population (104). Epidemiological evidence indicates that hepatitis non-A non-B is a blood-borne disease similar to hepatitis B.

Laboratory hazards. The agent of hepatitis B may be present in blood and blood products of human origin and in urine, semen, cerebrospinal fluid, and saliva. Parenteral inoculation, droplet exposure of mucous membranes, and contact exposure of broken skin are the primary laboratory hazards. The virus may be stable in dried blood or blood components for several days. Attenuated or avirulent strains are not defined.

Recommended precautions. Biosafety Level 2 practices, containment equipment, and facilities are recommended for all activities utilizing known or potentially infectious body fluids and tissues. Additional primary containment and personnel precautions, such as those described for Biosafety Level 3, may be indicated for activities with high potential for droplet or aerosol production and for activities involving production quantities or concentrations of infectious materials. Animal Biosafety Level 2 practices, containment equipment, and facilities are recommended for activities utilizing naturally or experimentally infected chimpanzees or other nonhuman primates. Gloves should be worn when working with infected animals and when there is the likelihood of skin contact with infectious materials. A licensed inactivated vaccine is available and is recommended for laboratory personnel, who are at substantially greater risk of acquiring infection than the general population (27).

Agent: *Herpesvirus simiae* (B-virus)

Although B-virus presents a potential hazard to laboratory personnel working with the agent, laboratory-associated human infections with B-virus have, with rare exceptions, been limited to personnel having direct contact with living Old World monkeys (29, 56, 89). Exposure to in vitro monkey tissues (i.e., primary rhesus monkey kidney) has been associated with a single documented case (29).

B-virus is an indigenous chronic and/or recurrent infection of macaques and possibly other Old World monkeys and is a frequent enzootic infection of captive *Macaca mulatta*.

Laboratory personnel handling Old World monkeys run the risk of acquiring B-virus from a bite or contamination of broken skin or mucous membranes by an infected monkey. Fifteen fatal cases of human infections with B-virus have been reported (29).

Laboratory hazards. The agent may be present in oral secretions, thoracic and abdominal viscera, and central nervous system tissues of naturally infected macaques. Bites from monkeys with oral herpes lesions are the greatest hazard to laboratory and animal care personnel. Exposures of broken skin or mucous membranes to oral secretions or to infectious culture fluids are also potential hazards. The importance of aerosol exposure is not known. Attenuated or avirulent strains have not been defined.

Recommended precautions. Biosafety Level 2 practices, containment equipment, and facilities are recommended for all activities involving the use or manipulation of tissues, body fluids, and primary tissue culture materials from macaques. Additional containment and personnel precautions, such as those recommended for Biosafety Level 3, are recommended for activities involving the use or manipulation of any material known to contain *H. simiae*.

Biosafety Level 4 practices, containment equipment, and facilities are recommended for activities involving the propagation of *H. simiae*, manipulations of production quantities or concentrations of *H. simiae*, and housing vertebrate animals with proven natural or induced infection with the agent.

The wearing of gloves, masks, and laboratory coats is recommended for all personnel working with nonhuman primates—especially macaques and other Old World species—and for all persons entering animal rooms where nonhuman primates are housed. Vaccines are not available for use in humans.

Agent: **Herpesviruses**

The herpesviruses are ubiquitous human pathogens and are commonly present in a variety of clinical materials submitted for virus isolation. While these viruses are not demonstrated causes of laboratory-associated infections, they are primary as well as opportunistic pathogens, especially in immunocompromised hosts. Nonpolio enteroviruses, adenoviruses, and cytomegalovirus pose similar low potential infection risks to laboratory personnel. Although this diverse group of indigenous viral agents does not meet the criteria for inclusion in agent specific summary statements (i.e., demonstrated or high potential hazard for laboratory-associated infection; grave consequences should infection occur), the frequency of their presence in clinical materials and their common use in research warrants their inclusion in this publication.

Laboratory hazards. Clinical materials and isolates of herpesviruses, nonpolio enteroviruses, and other indigenous pathogens may pose a risk of infection following ingestion, accidental parenteral inoculation, droplet exposure of the mucous membranes of the eyes, nose, or mouth, or inhalation of concentrated aerosolized materials.

Recommended precautions. Biosafety Level 2 practices, containment equipment, and facilities are recommended for activities utilizing known or potentially infectious clinical materials or cultures of indigenous viral agents which are associated or identified as a primary pathogen of human disease. Although there is no definitive evidence that infectious aerosols are a significant source of laboratory-associated infections, it is prudent to avoid the generation of aerosols during the handling of clinical materials and isolates or during the necropsy of animals. Primary containment devices (e.g., biological safety cabinets) constitute the basic barrier protecting personnel from exposure to infectious aerosols.

Agent: **Influenza virus**

Laboratory-associated infections with influenza are not normally documented in the literature, but are known to occur by informal accounts and published reports, particularly when new strains showing antigenic drift or shift are introduced into a laboratory for diagnostic/research purposes (36).

Laboratory animal-associated infections are not reported; however there is a high possibility of human infection from infected ferrets and vice versa.

Laboratory hazards. The agent may be present in respiratory tissues or secretions of man or most infected animals and in the cloaca of many infected avian species. The virus may be disseminated in multiple organs in some infected animal species.

Inhalation of virus from aerosols generated by aspirating, dispensing, or mixing virus-infected samples or by infected animals is the primary laboratory hazard. Genetic manipulation of virus has unknown potential for altering host range and pathogenicity or for introducing into man transmissible viruses with novel antigenic composition.

Recommended precautions. Biosafety Level 2 practices, containment equipment, and facilities are recommended when receiving and inoculating routine laboratory diagnostic specimens. Autopsy material should be handled in a biological safety cabinet using Biosafety Level 2 procedures.

Activities utilizing noncontemporary virus strains. Biosafety considerations should take into account the available information about infectiousness of the strains being used and the potential for harm to the individual or society in the event that laboratory-acquired infection and subsequent transmission occurs. Research or production activities utilizing contemporary strains may be safely performed using Biosafety Level 2 containment practices. Susceptibility to infection with older noncontemporary human strains, with recombinants, or with animal isolates warrants the use of Biosafety Level 2 containment procedures. Current experience suggests, however, there is no evidence for laboratory-acquired infection with reference strains A/PR/8/34 and A/WS/33 or its commonly used neurotropic variants.

Agent: **Lymphocytic choriomeningitis (LCM) virus**

Laboratory-associated infections with lymphocytic choriomeningitis virus are well documented in facilities where infections occur in laboratory rodents, especially mice and hamsters (8, 90). Tissue cultures which have inadvertently become infected represent a potential source of infection and dissemination of the agent. Natural infections are occasionally found in nonhuman primates, swine, and dogs.

Laboratory hazards. The agent may be present in blood, cerebrospinal fluid, urine, secretions of the nasopharynx, feces, and tissues of infected humans and other animal hosts. Parenteral inoculation, inhalation, and contamination of mucous membranes or broken skin with infectious tissues or fluids from infected animals are common hazards. Aerosol transmission is well documented (8).

Recommended precautions. Biosafety Level 2 practices, containment equipment, and facilities are recommended for all activities utilizing known or potentially infectious body fluids or tissues and for tissue culture passage of mouse brain-passaged strains. All manipulations of known or potentially infectious passage and clinical materials should be conducted in a biological safety cabinet. Additional primary containment and personnel precautions such as those described for Biosafety Level 3 may be indicated for activities with high potential for aerosol production and for activities involving production quantities or concentrations of infectious materials. Animal Biosafety Level 2 practices and facilities are recommended for studies in adult mice with mouse brain-passaged strains. Animal Biosafety Level 3 practices and facilities are recommended for work with infected hamsters. Vaccines are not available for use in humans.

Agent: **Poliovirus**

Laboratory-associated infections with polioviruses are uncommon and are generally limited to unvaccinated laboratory persons working directly with the agent. Twelve cases have been reported in the world literature (90).

Laboratory animal-associated infections have not been reported (73); however, naturally or experimentally infected nonhuman primates could provide a source of infection to exposed unvaccinated persons.

Laboratory hazards. The agent may be found in the feces and in throat secretions. Ingestion and parenteral inoculation of infectious tissues or fluids by unimmunized personnel are the primary hazards to laboratory personnel. The importance of aerosol exposure is not known. Laboratory exposures pose negligible risk to appropriately immunized persons.

Recommended precautions. Biosafety Level 2 practices, containment equipment, and facilities are recommended for all activities utilizing known or potentially infectious culture fluids and specimen materials. All laboratory personnel working directly with the agent must have documented polio vaccination or demonstrated serologic evidence of immunity to all three poliovirus types (25).

Agent: **Poxviruses**

Sporadic cases of laboratory-associated poxvirus infections have been reported. Pike lists 24 cases of yaba and tanapox virus infection and 18 vaccinia and smallpox infections (90). Epidemiological evidence suggests that transmission of monkeypox virus from nonhuman primates or rodents to humans may have occurred in nature but not in the laboratory setting. Naturally or experimentally infected laboratory animals are a potential source of infection to exposed unvaccinated laboratory personnel.

Laboratory hazards. The agents may be present

in lesion fluids or crusts, respiratory secretions, or tissues of infected hosts. Ingestion, parenteral inoculation, and droplet or aerosol exposure of mucous membranes or broken skin to infectious fluids or tissues are the primary hazards to laboratory and animal care personnel. Some poxviruses are stable at ambient temperature when dried and may be transmitted by fomites.

Recommended precautions. The possession and use of variola viruses is restricted to the World Health Organization Collaborating Center for Smallpox and Other Poxvirus Infections located at the Centers for Disease Control, Atlanta, Ga. Biosafety Level 2 practices, containment equipment, and facilities are recommended for all activities involving the use or manipulation of poxviruses other than variola that pose an infection hazard to humans. All persons working in or entering laboratory or animal care areas where activities with vaccinia, monkeypox, or cowpox viruses are being conducted should have documented evidence of satisfactory vaccination within the preceding 3 years (23). Activities with vaccinia, cowpox, or monkeypox viruses in quantities or concentrations greater than those present in diagnostic cultures may also be conducted by immunized personnel at Biosafety Level 2, provided that all manipulations of viable materials are conducted in Class I or II biological safety cabinets or other primary containment equipment.

Agent: Rabies virus

Laboratory-associated rabies infections are extremely rare. Two have been documented. Both resulted from presumed exposure to high-titered infectious aerosols generated in a vaccine production facility (129) and a research facility (17). Naturally or experimentally infected animals, their tissues, and their excretions are a potential source of exposure to laboratory and animal care personnel.

Laboratory hazards. The agent may be present in all tissues of infected animals. Highest titers are demonstrated in central nervous system tissue, salivary glands, and saliva. Accidental parenteral inoculation, cuts, or sticks with contaminated laboratory equipment, bites by infected animals, and exposure of mucous membranes or broken skin to infectious droplets of tissue or fluids are the most likely sources of exposure for laboratory and animal care personnel. Infectious aerosols have not been a demonstrated hazard to personnel working with clinical materials and conducting diagnostic examinations. Fixed and attenuated strains of virus are presumed to be less hazardous, but the only two recorded cases of laboratory-associated rabies resulted from exposure to a fixed challenge virus standard (CVS) and an attenuated strain derived from SAD (Street Alabama Dufferin) strain (17, 129).

Recommended precautions. Biosafety Level 2 practices, containment equipment, and facilities are recommended for all activities utilizing known or potentially infectious materials. Preexposure immunization is recommended for all individuals working with rabies virus or infected animals or engaged in diagnostic, production, or research activities with rabies virus. Preexposure immunization is also recommended for all individuals entering or working in the same room where rabies virus or infected animals are used. When it is not feasible to open the skull or remove the brain within a biological safety cabinet, it is pertinent to wear heavy protective gloves to avoid cuts or sticks from cutting instruments or bone fragments, and to wear a face shield to protect the mucous membranes of the eyes, nose, and mouth from exposure to infectious droplets or tissue fragments. If a Stryker saw is used to open the skull, avoid striking the brain with the blade of the saw. Additional primary containment and personnel precautions, such as those described for Biosafety Level 3, may be indicated for activities with a high potential for droplet or aerosol production and for activities involving production quantities or concentrations of infectious materials.

Agents: Transmissible spongiform encephalopathies (Creutzfeldt-Jakob and kuru agents)

Laboratory-associated infections with the transmissible spongiform encephalopathies have not been documented. The consequences of infection are grave, however, and there is evidence that Creutzfeldt-Jakob disease has been transmitted to patients by corneal transplant and by contaminated electroencephalographic electrodes. There is no known nonhuman reservoir for Creutzfeldt-Jakob disease or kuru. Nonhuman primates and other laboratory animals have been infected by inoculation, but there is no evidence of secondary transmission.

Laboratory hazards. High titers of a transmissible agent have been demonstrated in brain and spinal cord of persons with kuru. In persons with Creutzfeldt-Jakob disease, a transmissible agent has been demonstrated in the brain, spleen, liver, lymph nodes, lungs, spinal cord, kidneys, cornea, and lens. Accidental parenteral inoculation, especially neural tissues, and including Formalin-fixed specimens, is extremely hazardous. Although nonneural tissues are less often infective, all tissues of humans and animals infected with these agents should be considered potentially hazardous. The risk of infection from aerosols, droplets, and exposure to intact skin, gastric, and mucous membranes is not known; however, there is no evidence of contact or aerosol transmission. These agents are characterized by extreme resistance to conventional inactivation procedures, including irradiation, boiling, and chemicals (Formalin, betapropiolactone, alcohols).

Recommended precautions. Biosafety Level 2 practices, containment equipment, and facilities are recommended for all activities utilizing known or potentially infectious tissues and fluids from naturally infected humans and from experimentally infected animals. Extreme care must be taken to avoid accidental autoinoculation or other traumatic parenteral inoculations of infectious tissues and fluids (46). Although there is no evidence to suggest that aerosol transmission occurs in the natural disease, it is prudent to avoid the generation of aerosols or droplets during the manipulation of tissues and fluids and during the necropsy of experimental animals. It is further recommended that gloves should be worn for activities which provide the opportunity for skin contact with infectious tissues and fluids. Vaccines are not available for use in humans.

Agent: **Vesicular stomatitis virus (VSV)**

Forty-six laboratory-associated infections with indigenous strains of VSV have been reported (112). Laboratory activities with indigenous strains of VSV present two different levels of risk to laboratory personnel and are related, at least in part, to the passage history of the strains utilized. Activities utilizing infected livestock, their infected tissues, and virulent isolates from these sources are a demonstrated hazard to laboratory and animal care personnel (52, 88). Seroconversion and clinical illness rates in personnel working with these materials are high (88). Similar risks may be associated with exotic strains such as Piry (112).

In contrast, anecdotal information indicates that activities with less virulent laboratory-adapted strains (e.g., VSV-Indiana [San Juan and Glasgow]) are rarely associated with seroconversion or illness. Such strains are commonly used by molecular biologists, often in large volumes and high concentrations, under conditions of minimal or no primary containment. Experimentally infected mice have not been a documented source of human infection.

Laboratory hazards. The agent may be present in vesicular fluid, tissues, and blood of infected animals and in blood and throat secretions of infected humans. Exposure to infectious aerosols and infected droplets, direct skin and mucous membrane contact with infectious tissues and fluids, and accidental autoinoculation are the primary laboratory hazards associated with virulent isolates. Accidental parenteral inoculation and exposure to infectious aerosols represent potential risks to personnel working with less virulent laboratory-adapted strains.

Recommended precautions. Biosafety Level 3 practices, containment equipment, and facilities are recommended for activities involving the use or manipulation of infected tissues and of virulent isolates from naturally or experimentally infected livestock. Gloves and respiratory protection are recommended for the necropsy and handling of infected animals. Biosafety Level 2 practices and facilities are recommended for activities utilizing laboratory-adapted strains of demonstrated low virulence. Vaccines are not available for use in humans.

Arboviruses

Arboviruses Assigned to Biosafety Level 2

The American Committee on Arthropod-Borne Viruses (ACAV) registered 424 arboviruses as of December 31, 1979. The ACAV's Subcommittee on Arbovirus Laboratory Safety (SALS) has categorized each of these 424 agents into one of four recommended levels of practice and containment which parallel the recommended practices, safety equipment, and facilities described in this publication as Biosafety Levels 1–4 (112). It is the intent of SALS to periodically update the 1980 publication by providing a supplemental listing and recommended levels of practice and containment for arboviruses registered since 1979. SALS categorizations were based on risk assessments from information provided by a worldwide survey of 585 laboratories working with arboviruses. SALS recommended that work with the majority of these agents should be conducted at the equivalent of Biosafety Level 2. These viruses are listed alphabetically on pages 325 and 326 and include the following agents which are the reported cause of laboratory-associated infections (53,90,112). The list of arboviruses in Biosafety Level 2 includes yellow fever virus (17D strain) and Venezuelan equine encephalomyelitis (VEE) virus (TC83 strain), provided that personnel working with these vaccine strains are immunized.

Virus	Cases (SALS)
Vesicular stomatitis	46
Colorado tick fever	16
Dengue	11
Pichinde	17
Western equine encephalomyelitis	7 (2 deaths)
Rio Bravo	7
Kunjin	6
Catu	6
Caraparu	5
Ross River	5
Bunyamwera	4
Eastern equine encephalomyelitis	4
Zika	4
Apeu	2
Marituba	2
Tacaribe	2
Muructucu	1
O'nyong nyong	1
Modoc	1
Oriboca	1
Ossa	1
Keystone	1
Bebaru	1
Bluetongue	1

The results of the SALS survey clearly indicate that the suspected source of the laboratory-associated infections listed above was other than exposure to infectious aerosols. Recommendations that work with these 334 arboviruses should be conducted at Biosafety Level 2 was based on the existence of adequate historical laboratory experience to assess risks for the virus which indicated that: (i) no overt laboratory-associated infections are reported; or (ii) infections resulted from exposures other than to infectious aerosols; or (iii) if aerosol exposures are documented they represent an uncommon route of exposure.

Laboratory hazards. Agents listed in this group may be present in blood, cerebrospinal fluid, central nervous system and other tissues, and infected arthropods, depending on the agent and the stage of infection. While the primary laboratory hazards are accidental parenteral inoculation, contact of the virus with broken skin or mucous membranes, and bites by infected laboratory rodents or arthropods, infectious aerosols may also be a potential source of infection.

Recommended precautions. Biosafety Level 2 practices, safety equipment, and facilities are recommended for activities with potentially infectious clinical materials and arthropods and for manipulations of infected tissue cultures, embryonated eggs, and rodents. Infection of newly hatched chickens with eastern and western equine encephalomyelitis viruses is especially hazardous and should be undertaken under Biosafety Level 3 conditions by immunized personnel. Investigational vaccines (IND) against eastern equine encephalomyelitis and western equine encephalomyelitis viruses are available through the Centers for Disease Control and the U.S. Army Medical Research Institute for Infectious Diseases, respectively. The use of these vaccines is recommended for personnel who work directly and regularly with these two agents in the laboratory. Western equine encephalomyelitis immune globulin (human) is also available from the Centers for Disease Control. The efficacy of this product has not been established.

Arboviruses Assigned to Biosafety Level 2

Abu Hammad	Icoaraci	Parana
Acado	Ieri	Pata
Acara	Ilesha	Pathum Thani
Aguacate	Iheus	Patois
Alfuy	Ingwavuma	Phnom-Penh bat
Almpiwar	Inkoo	Pichinde
Amapari	Ippy	Pixuna
Anhanga	Irituia	Pongola
Anhembi	Isfahan	Pretoria
Anopheles A	Itaporanga	Puchong
Anopheles B	Itaqui	Punta Salinas
Apeu	Jamestown Canyon	Punta Toro
Apoi	Japanaut	Qalyub
Aride	Jerry Slough	Quaranfil
Arkonam	Johnston Atoll	Restan
Aruac	Joinjakaka	Rio Bravo
Arumowot	Juan Diaz	Rio Grande
Aura	Jugra	Ross River
Avalon	Jurona	Royal Farm
Bagaza	Jutiapa	Sabo
Bahig	Kadam	Saboya
Bakau	Kaeng khoi	Saint Floris
Baku	Kaikalur	Sakhalin
Bandia	Kaisodi	Salehabad
Bangoran	Kamese	San Angelo
Bangui	Kammavanpettai	Sandfly F. (Naples)
Banzi	Kannamangalam	Sandfly F. (Sicilian)
Barur	Kao shuan	Sandjimba
Batai	Karimabad	Sathuperi
Batu	Karshi	Sawgrass
Bauline	Kasba	Sebokele
Bebaru	Kemervo	Seletar
Belmont	Kern Canyon	Sembalam
Bertioga	Ketapang	Shamonda
Bimiti	Keterah	Shark River
Birao	Keuraliba	Shuni
Bluetongue	Keystone	Silverwater
(indigenous)	Klamath	Simbu
Boraceia	Kokobera	Simian hemorrhagic
Botambi	Kolongo	fever
Boteke	Koongol	Sindbis
Bouboui	Kowanyama	Sixgun City
Bujaru	Kunjim	Snowshoe hare
Bunyamwera	Kununurra	Sokuluk
Burg el Arab	Kwatta	Soldado
Bushbush	La Crosse	Sororoca
Bussuquara	Lagos bat	Stratford
Buttonwillow	La Joya	Sunday Canyon
Bwamba	Landjia	Tacaiuma
Cacao	Langat	Tacaribe
Cache Valley	Lanjan	Taggert
Caimito	Latino	Tahyna
California	Lebombo	Tamiami
encephalitis	Le Dantec	Tanga
Calovo	Lipovnik	Tanjong Rabok
Candiru	Lokern	Tataguine
Cape Wrath	Lone Star	Tembe
Capim	Lukuni	Tembusu
Caraparu	M'poko	Tensaw
Carey Island	Madrid	Tete
Catu	Maguari	Tettnang
Chaco	Mahogany	Thimiri
Chagres	Hammock	Thottapalayam
Chandipura	Main Drain	Timbo
Changuinola	Malakal	Toure
Charleville	Manawa	Tribec
Chenuda	Manzanilla	Triniti
Chilibre	Mapputa	Trivittatus
Chobar	Maprik	Trubanaman
Gorge	Marco	Tsuruse
Clo Mor	Marituba	Turlock
Colorado Tick	Matariya	Tyuleniy
Fever	Matruh	Uganda S
Corriparta	Matucare	Umatilla
Cotia	Melao	Umbre
Cowbone Ridge	Mermet	Una
D'Aguilar	Minatitlan	Upolu
Dakar Bat	Minnal	Urucuri
Dengue-1	Mirim	Usutu
Dengue-2	Mitchell River	Uukuniemi
Dengue-3	Modoc	Vellore
Dengue-4	Moju	Venezuelan equine
Dera Ghazi Khan	Mono Lake	encephalomyelitis
Eastern equine	Mont. myotis leuk.	(TC-83)
encephalomyelitis	Moriche	Venkatapuram
Edge Hill	Mossuril	Vesicular stomatitis
Entebbe Bat	Mount Elgon bat	(see p. 324)
Epizootic	Murutucu	Wad Medani
hemorrhagic	Navarro	Wallal
disease	Nepuyo	Wanowrie
Eubenangee	Ngaingan	Warrego
Eyach	Nique	Western equine
Flanders	Nkolbisson	encephalomyelitis
Fort Morgan	Nola	Whataro
Frijoles	Ntaya	Witwatersrand

Gamboa	Nugget	Wongal	Mayaro	5
Gomoká	Nyamanini	Wongorr	Spondweni	4
Gossas	Nyando	Wyeomyia	St. Louis encephalitis	4
Grand Arbaud	O'nyong-nyong	Yaquina Head	Murray Valley	
Great Island	Okhotskiy	Yata	encephalitis	3
Guajara	Okola	Yellow fever (17D)	Semliki Forest	3 (1 death)
Guama	Olifantsvlei	Yogue	Powassan	2
Guaroa	Oriboca	Zaliv Terpeniya	Dugbe	2
Gumbo limbo	Ossa	Zegla	Issyk-kul	1
Hart Park	Pacora	Zika	Koutango	1
Hazara	Pacui	Zingilamo		
Huacho	Pahayokee	Zirqa		
Hughes	Palyam			

Arboviruses and Arenaviruses Assigned to Biosafety Level 3

SALS has recommended that work with the arboviruses included in the alphabetical listing on page 324 should be conducted at the equivalent of Biosafety Level 3 practices, safety equipment, and facilities. These recommendations are based on one of the following criteria: overt laboratory-associated infections with these agents have occurred by aerosol route if protective vaccines are not used or are unavailable, or laboratory experience with the agent is inadequate to assess risk and the natural disease in humans is potentially severe or life threatening or causes residual damage. Hantaan virus, which was not included in the SALS publication, has been placed at Biosafety Level 3 based on documented laboratory-associated infections. Rift Valley fever virus, which was classified by SALS at Containment Level 3 (i.e., HEPA filtration required for all air exhausted from the laboratory), was placed in Biosafety Level 3 provided that all personnel entering the laboratory or animal care area where work with this virus is being conducted are vaccinated. Laboratory or laboratory animal-associated infections have been reported with the following agents (53,90,112,124):

Virus	Cases (SALS)
Venezuelan equine encephalitis	150 (1 death)
Rift Valley fever	47 (1 death)
Chikungunya	39
Yellow fever	38 (8 deaths)
Japanese encephalitis	22
Louping ill	22
West Nile	18
Lymphocytic choriomeningitis	15
Orungo	13
Piry	13
Wesselsbron	13
Mucambo	10
Oropouche	7
Germiston	6
Bhanja	6
Hantaan (Korean hemorrhagic fever)	6

Large quantities and high concentrations of Semliki Forest virus are commonly used or manipulated by molecular biologists under conditions of moderate or low containment. Although antibodies have been demonstrated in individuals working with this virus, the first overt (and fatal) laboratory-associated infection with this virus was reported in 1979 (126). Because this infection may have been influenced by a compromised host, an unusual route of exposure or high dosage, or a mutated strain of the virus, this case and its outcome may not be typical. Since exposure to an infectious aerosol was not indicated as the probable mode of transmission in this case, it is suggested that most activities with Semliki Forest disease virus can be safely conducted at Biosafety Level 2.

Some viruses (e.g., Ibaraki, Israel turkey meningoencephalitis) are listed by SALS in Level 3, not because they pose a threat to human health, but because they are exotic diseases of domestic livestock or poultry.

Laboratory hazards. The agents listed in this group may be present in blood, cerebrospinal fluid, urine, and exudates, depending on the specific agent and stage of disease. The primary laboratory hazards are exposure to aerosols of infectious solutions and animal bedding, accidental parenteral inoculation, and broken skin contact. Some of these agents (e.g., Venezuelan equine encephalitis [VEE]) may be relatively stable in dried blood or exudates. Attenuated strains are identified in a number of the agents listed (e.g., yellow fever 17D strain and VEE TC83 strain).

Recommended precautions. Biosafety Level 3 practices, containment equipment, and facilities are recommended for activities using potentially infectious clinical materials and infected tissue cultures, animals, or arthropods.

A licensed attenuated live virus is available for immunization against yellow fever and is recommended for all personnel who work with this agent or with infected animals and for those who enter rooms where the agents or infected animals are present. An investigational vaccine (IND) is available for immunization against VEE and is recommended for all personnel working with VEE (and the related Everglades, Mucambo, Tonate, and Cabassou viruses) or infected animals or entering rooms where these agents or infected animals are present. Work with Hantaan (Korean hemorrhagic fever) virus in rats, voles, and other laboratory rodents should be conducted with special caution (Biosafety Level 4). An inactivated, investigational new Rift Valley fever vaccine (IND) is available from the U.S. Army Medical Research Institute for Infectious Diseases and recommended for all laboratory and animal care personnel working with the agent or infected animals and for all personnel entering laboratories or animal rooms when the agent is in use.

Arboviruses assigned to Biosafety Level 3

Aino	Murray Valley
Akabane	encephalitis
Araguari	Nariva
Batama	Ndumu
Batken	Negishi
Bhanja	New Minto
Bimbo	Nodamura
Bluetongue (exotic)[a]	Northway
Bobaya	Oropouche[c]
Bobia	Orungo
Buenaventura	Ouango
Cabassou[c]	Oubangui
Chikungunya[c]	Paramushir
Chim	Piry
Cocal	Ponteves
Dhori	Powassan
Dugbe	Razdan
Everglades[c]	Rift Valley fever[a,b,c]
Garba	Rochambeau
Germiston[c]	Rocio[c]
Getah	Sagiyama
Gordil	Sakpa
Guaratuba	Salanga
Ibaraki	Santa Rosa
Ihangapi	Saumarez Reef
Inini	Semliki Forest
Israel turkey	Sepik
meningo.	Serra do Navio
Issyk-kul	Slovakia
Itaituba	Spondweni
Japanese encephalitis	St. Louis
Kairi	encephalitis
Khasan	Tamdy
Korean hemorrhagic	Telok Forest
fever (Hantaan)	Thogoto
Koutango	Tlacotalpan
Kyzlagach	Tonate[c]
Louping ill[a]	VSV-Alagoas
Lymphocytic	Venezuelan equine
choriomeningitis	encephalomyelitis[c]
Mayaro	Wesselsbron[a,c]
Middelburg	West Nile
Mosqueiro	Yellow fever[c]
Mucambo[c]	Zinga[a,b,c]

[a]The importation, possession, or use of this agent is restricted by USDA regulation or administrative policy. See Appendix E.

[b]Zinga virus is now recognized as being identical to Rift Valley fever virus.

[c]SALS recommends that work with this agent should be conducted only in Biosafety Level 3 facilities which provide for HEPA filtration of all exhaust air prior to discharge from the laboratory. All persons working with agents for which a vaccine is available should be immunized.

Arboviruses, Arenaviruses, or Filoviruses Assigned to Biosafety Level 4

SALS has recommended that work with the arboviruses, arenaviruses, or filoviruses (63) included in the listing that follows should be conducted at the equivalent of Biosafety Level 4 practices, safety equipment, and facilities. These recommendations are based on documented cases of severe and frequently fatal naturally occurring human infections and aerosol-transmitted laboratory infections. Additionally, SALS recommended that certain agents with a close or identical antigenic relationship to the Biosafety Level 4 agents (e.g., Absettarov and Kumlinge viruses) also be handled at this level until sufficient laboratory experience is obtained to retain these agents at this level or to work with them at a lower level. Laboratory or laboratory animal-associated infections have been reported with the following agents (37,53,58,67,90,112,12).

Virus	Cases (SALS)
Kyasanur Forest disease	133
Hypr	37 (2 deaths)
Junin	21 (1 death)
Marburg	25 (5 deaths)
Russian spring-summer encephalitis	8
Congo-Crimean hemorrhagic fever	8 (1 death)
Omsk hemmorrhagic fever	5
Lassa	2 (1 death)
Machupo	1 (1 death)
Ebola	1

Rodents are natural reservoirs of Lassa fever virus (*Mastomys natalensis*), Junin and Machupo viruses (*Calomys* spp.), and perhaps other viruses assigned to Biosafety Level 4. Nonhuman primates were associated with the initial outbreaks of Kyasanur Forest disease (*Presbytis* spp.). and Marburg disease (*Cercopithecus* spp.), and arthropods are the natural vectors of the tick-borne encephalitis complex agents. Work with or exposure to rodents, nonhuman primates, or vectors naturally or experimentally infected with these agents represents a potential source of human infection.

Laboratory hazards. The infectious agents may be present in blood, urine, respiratory and throat secretions, semen, and tissues from human or animal hosts and in arthropods, rodents, and nonhuman primates. Respiratory exposure to infectious aerosols, mucous membrane exposure to infectious droplets, and accidental parenteral inoculation are the primary hazards to laboratory or animal care personnel (67,124).

Recommended precautions. Biosafety Level 4 practices, containment equipment, and facilities are recommended for all activities utilizing known or potentially infectious materials of human, animal, or arthropod origin. Clinical specimens from persons suspected of being infected with one of the agents listed in this summary should be submitted to a laboratory with a Biosafety Level 4 maximum containment facility (22).

Arboviruses, arenaviruses, and filoviruses assigned to Biosafety Level 4

Congo-Crimean hemorrhagic fever	Marburg
Tick-borne encephalitis virus complex	Ebola
(Absettarov, Hanzalova, Hypr,	Junin
Kumlinge, Kyasanur Forest disease,	Lassa
Omsk hemorrhagic fever, and	Machupo
Russian spring-summer encephalitis)	

Appendix A

Biological Safety Cabinets

Biological safety cabinets are among the most effective, as well as the most commonly used, primary containment devices in laboratories working with infectious agents. Each of the three types—Class I, II, and III—has performance characteristics which are described below. In addition to the design, construction, and performance standards for vertical laminar flow biological safety cabinets (Class III), the National Sanitation Foundation has also developed a list of such products which meet the reference standard. Utilization of this standard (80) and list should be the first step in selection and procurement of a biological safety cabinet.

Class I and II biological safety cabinets, when used in conjunction with good microbiological techniques, provide an effective partial containment system for safe manipulation of moderate and high-risk microorganisms (i.e., Biosafety Level 2 and 3 agents). Both Class I and II biological safety cabinets have comparable inward face velocities (75 linear ft/min) and provide comparable levels of containment in protecting the laboratory worker and the immediate laboratory environment from infectious aerosols generated within the cabinet.

It is imperative that Class I and II biological safety cabinets are tested and certified in situ at the time of installation within the laboratory, at any time the cabinet is moved, and at least annually thereafter. Certification at locations other than the final site may attest to the performance capability of the individual cabinet or model but does not supersede the critical certification prior to use in the laboratory.

As with any other piece of laboratory equipment, personnel must be trained in the proper use of the biological safety cabinets. The slide-sound training film developed by NIH (Effective Use of the Laminar Flow Biological Safety Cabinet) provides a thorough training and orientation guide. Of particular note are those activities which may disrupt the inward directional airflow through the work opening of Class I and II cabinets. Repeated insertion and withdrawal of the workers' arms in and from the work chamber, opening and closing doors to the laboratory or isolation cubicle, improper placement or operation of

materials or equipment within the work chamber, or brisk walking past the biological safety cabinet while it is in use are demonstrated causes of the escape of aerosolized particles from within the cabinet. Strict adherence to recommended practices for the use of biological safety cabinets is as important in attaining the maximum containment capability of the equipment as is the mechanical performance of the equipment itself.

Horizontal laminar flow "clean benches" are present in a number of clinical, pharmacy, and laboratory facilities. These "clean benches" provide a high-quality environment within the work chamber for manipulation of nonhazardous materials. Caution: Since the operator sits in the immediate downstream exhaust from the "clean bench," this equipment must never be used for the handling of toxic, infectious, or sensitizing materials.

Class I. The Class I biological safety cabinet (Fig. 1) is an open-fronted, negative-pressure, ventilated cabinet with a minimum inward face velocity at the work opening of at least 75 ft/min. The exhaust air from the cabinet may be used in three operational modes: with a full-width open front, with an installed front closure panel not equipped with gloves, and with an installed front closure panel equipped with arm-length rubber gloves.

Class II. The Class II vertical laminar-flow biological cabinet (Fig. 2) is an open-fronted, ventilated cabinet with an average inward face velocity at the work opening of at least 75 ft/per min. This cabinet provides a HEPA-filtered, recirculated mass airflow within the work space. The exhaust air from the cabinet is also filtered by HEPA filters. Design, construction, and performance standards for Class II cabinets have been developed by and are available from the National Sanitation Foundation, Ann Arbor, Mich. (80).

Class III. The Class III cabinet (Fig. 3) is a totally enclosed ventilated cabinet of gas-tight construction. Operations within the Class III cabinet are conducted through attached rubber gloves. When in use, the Class III cabinet is maintained under negative air pressure of at least 0.5 in. water gauge. Supply air is drawn into the cabinet through HEPA filters. The cabinet exhaust air is filtered by two HEPA filters, installed in series, before discharge outside of the facility. The exhaust fan for the Class III cabinet is generally separate from the exhaust fans of the facility's ventilation system.

Use of cabinets. Personnel protection provided by Class I and Class II cabinets is dependent on the inward airflow. Since the face velocities are similar, they generally provide an equivalent level of personnel protection. The use of these cabinets alone, however, is not appropriate for containment of highest-risk infectious agents because aerosols may accidentally escape through the open front.

The use of a Class II cabinet in the microbiological laboratory offers the additional capability and advan-

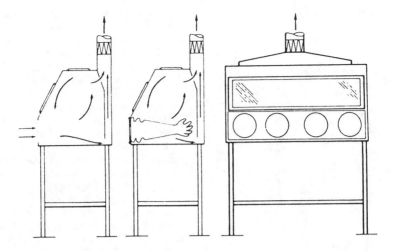

FIG. 1. Class I biological safety cabinet.

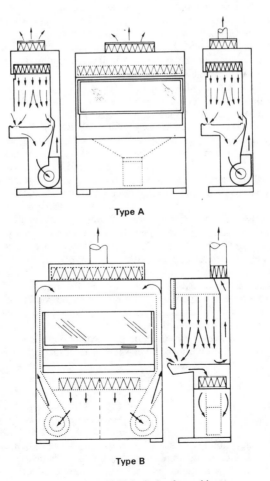

Type A

Type B

FIG. 2. Class II biological safety cabinets.

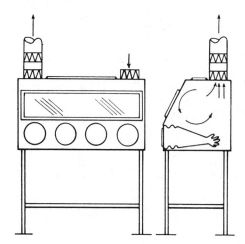

FIG. 3. Class III biological safety cabinet.

tage of protecting materials contained within it from extraneous airborne contaminants. This capability is provided by the HEPA-filtered, recirculated mass airflow within the work space.

The Class III cabinet provides the highest level of personnel and product protection. This protection is provided by the physical isolation of the space in which the infectious agent is maintained. When these cabinets are required, all procedures involving infectious agents are contained within them. Several Class III cabinets are therefore typically set up as an interconnected system. All equipment required by the laboratory activity, such as incubators, refrigerators, and centrifuges, must be an integral part of the cabinet system. Doubled-doored autoclaves and chemical dunk tanks are also attached to the cabinet system to allow supplies and equipment to be safely introduced and removed.

Personnel protection equivalent to that provided by Class III cabinets can also be obtained with a personnel suit area and Class I or Class II cabinets. This is one in which the laboratory worker is protected from a potentially contaminated environment by a one-piece positive-pressure suit ventilated by a life-support system. This area is entered through an airlock fitted with airtight doors. A chemical shower is provided to decontaminate the surfaces of the suit as the worker leaves the area. The exhaust air from the suit area is filtered by two HEPA units installed in series.

Appendix B

Immunoprophylaxis

An additional level of protection for at-risk personnel may be achieved with appropriate prophylactic vaccinations. A written organizational policy which defines at-risk personnel, which specifies risks

as well as benefits of specific vaccines, and which distinguishes between required and recommended vaccines is essential. In developing such an organizational policy, these recommendations and requirements should be specifically targeted at infectious diseases known or likely to be encountered in a particular facility.

Vaccines for which the benefits (levels of antibody considered to be protective) clearly exceed the risks (local or systemic reactions) should be required for all clearly identified at-risk personnel. Examples of such preparations include vaccines against yellow fever, rabies, and poliomyelitis. Recommendations for giving less efficacious vaccines, those associated with high rates of local or systemic reactions, or those that produce increasingly severe reactions with repeated use should be carefully considered. Products with these characteristics (e.g., cholera, tularemia, and typhoid vaccines) may be recommended but should not ordinarily be required for employment. A complete record of vaccines received on the basis of occupational requirements or recommendations should be maintained in the employee's permanent medical file.

Recommendations for the use of vaccines, adapted from those of the Public Health Service Advisory Committee on Immunization Practices, are included in the agent summary statements in Section V.

Appendix C

Surveillance of Personnel for Laboratory-Associated Rickettsial Infections

Under natural circumstances, the severity of disease caused by rickettsial agents varies considerably. In the laboratory, very large inocula which might produce unusual and perhaps very serious responses are possible. Surveillance of personnel for laboratory-associated infections with rickettsial agents can dramatically reduce the risk of serious consequences of disease.

Recent experience indicates that infections treated adequately with specific anti-rickettsial chemotherapy on the first day of disease do not generally present serious problems. Delay in instituting appropriate chemotherapy, however, may result in debilitating or severe acute disease ranging from increased periods of convalescence in typhus and scrub typhus to death in *R. rickettsii* infections. The key to reducing the severity of disease from laboratory-associated infections is a reliable surveillance system which includes (i) round-the-clock availability of an experienced medical officer, (ii) indoctrination of all personnel into the potential hazards of working with rickettsial agents and advantages of early therapy, (iii) a reporting system for all recognized overt exposures and accidents, (iv) the reporting of all febrile illnesses, especially those associated with headache, malaise, prostration, when no other certain cause

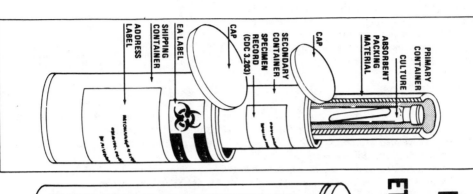

PACKAGING AND LABELING OF ETIOLOGIC AGENTS

FIGURE 1

FIGURE 2

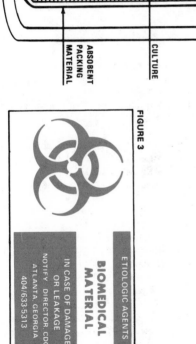

CROSS SECTION
OF PROPER PACKING

FIGURE 3

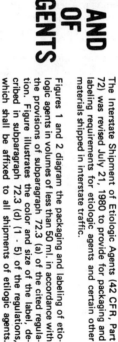

ETIOLOGIC AGENTS

**BIOMEDICAL
MATERIAL**

IN CASE OF DAMAGE
OR LEAKAGE
NOTIFY DIRECTOR CDC
ATLANTA, GEORGIA
404/633-5313

The Interstate Shipment of Etiologic Agents (42 CFR, Part 72) was revised July 21, 1980 to provide for packaging and labeling requirements for etiologic agents and certain other materials shipped in interstate traffic.

Figures 1 and 2 diagram the packaging and labeling of etiologic agents in volumes of less than 50 ml. in accordance with the provisions of subparagraph 72.3 (a) of the cited regulation. Figure illustrates the color and size of the label, described in subparagraph 72.3 (d) (1 - 5) of the regulations, which shall be affixed to all shipments of etiologic agents.

For further information on any provision of this regulation contact:

Centers for Disease Control
Attn: Biohazards Control Office
1600 Clifton Road
Atlanta, Georgia 30333

Telephone: 404-329-3883
FTS-236-3883

FIG. 4. Packaging and labeling requirements for interstate shipment of etiologic agents.

exists, and (v) a nonpunitive atmosphere that encourages reporting of any febrile illness.

Rickettsial agents can be handled in the laboratory with minimal real danger to life when an adequate surveillance system complements a staff who are knowledgeable about the hazards of rickettsial infections and who put to use the safeguards recommended in the agent summary statements.

Appendix D

Importation and Interstate Shipment of Human Pathogens and Related Materials

The importation or subsequent receipt of etiologic agents and vectors of human disease is subject to the Public Health Service Foreign Quarantine Regulations (42 CFR, Section 71.56). Permits authorizing the importation or receipt of regulated materials and specifying conditions under which the agent or vector is shipped, handled, and used are issued by the Centers for Disease Control.

The interstate shipment of indigenous etiologic agents, diagnostic specimens, and biological products is subject to applicable packaging, labeling, and shipping requirements of the Interstate Shipment of Etiologic Agents (42 CFR, Part 72). Packaging and labeling requirements for interstate shipment of etiologic agents are summarized and illustrated in Fig. 4.

Additional information on the importation and interstate shipment of etiologic agents of human disease and other related materials may be obtained by writing to:

Centers for Disease Control
Attention: Office of Biosafety
1600 Clifton Road, N.E.
Atlanta, GA 30333
Telephone: (404) 329-3883
FTS: 236-3883

Appendix E

Restricted Animal Pathogens

Nonindigenous pathogens of domestic livestock and poultry may require special laboratory design, operation, and containment features not generally addressed in this publication. The importation, possession, or use of the following agents is prohibited or restricted by law or by U.S. Department of Agriculture regulations or administrative policies:

African horse sickness virus	Nairobi sheep disease virus
African swine fever virus	(Ganjam virus)
Besnoitia besnoiti	Newcastle disease virus
Borna disease virus	(velogenic strains)
Bovine ephemeral fever	*Pseudomonas mallei*
Bovine infectious petechial	*Rickettsia ruminantium*
fever agent	Rift Valley fever virus

Camelpox virus	Rinderpest virus
Foot and mouth disease virus	Swine vesicular disease virus
Fowl plague virus	Teschen disease virus
Histoplasma (Zymonema)	*Theileria annulata*
farciminosum	*Theileria bovis*
Hog cholera virus	*Theileria hirci*
Louping ill virus	*Theileria lawrencei*
Lumpy skin disease virus	*Trypanosoma evansi*
Mycoplasma agalactiae	*Trypanosoma vivax*
Mycoplasma mycoides	Vesicular exanthema virus
	Wesselsbron disease virus

The importation, possession, use, or interstate shipment of animal pathogens other than those listed above may also be subject to regulations of the U.S. Department of Agriculture.

Additional information may be obtained by writing to:

Chief Staff Veterinarian
Organisms and Vectors
Veterinary Services
Animal and Plant Health Inspection Service
U.S. Department of Agriculture
Hyattsville, MD 20782
Telephone: (301) 436-8017
FTS: 436-8017

Appendix F

Resources for Information

Resources for information, consultation, and advice on biohazard control, decontamination procedures, and other aspects of laboratory safety management include:

Centers for Disease Control
Attention: Office of Biosafety
Atlanta, GA 30333
Telephone: (404) 329-3883
FTS 236-3883

National Institutes of Health
Attention: Division of Safety
Bethesda, MD 20205
Telephone: (301) 496-1357
FTS 496-1357

National Animal Disease Center
U.S. Department of Agriculture
Ames, IA 50010
Telephone: (515) 862-8258
FTS 862-8258

LITERATURE CITED

1. **Allen, B. W.** 1981. Survival and tubercle bacilli in heat-fixed sputum smears. J. Clin. Pathol. **34:**719–722.
2. **American Thoracic Society Policy Statement.**

1974. Quality of laboratory services for mycobacterial disease. Am. Rev. Respir. Dis. **110:**376–377.

3. **Anon.** 1980. Tuberculosis infection associated with tissue processing. Calif. Morbid. 30.

4. **Anon.** 1936. Bacteriologist dies of meningitis. J. Am. Med. Assoc. **106:**129.

5. **Beeman, E. A.** 1950. Q fever—an epidemiological note. 1950. Public Health Rep. **65:**88–92.

6. **Bernard, K. W., G. L. Parham, W. G. Winkler, and C. G. Helmick.** 1982. Q fever control measures: recommendations for research of facilities using sheep. Infect. Control **3**(6).

7. **Blaser, M. J., F. W. Hickman, J. J. Farmer III, D. J. Brenner, A. Balows, and R. A. Feldman.** 1980. *Salmonella typhi.* the laboratory as a reservoir of infection. J. Infect. Dis. **142:**934–938.

8. **Bowen, G. S., C. H. Calisher, W. G. Winkler, A. L. Kraus, E. H. Fowler, R. H. Garman, D. W. Fraser, and A. R. Hinman.** 1975. Laboratory studies of a lymphocytic choriomeningitis virus outbreak in man and laboratory animals. Am. J. Epidemiol. **102:**233–240.

9. **Bozeman, F. M., S. A. Masiello, M. S. Williams, and B. L. Elisberg.** 1975. Epidemic typhus rickettsiae isolated from flying squirrels. Nature (London) **255:**545–547.

10. **Burke, D. S.** 1977. Immunization against tularemia: analysis of the effectiveness of live *Francisella tularensis* vaccine in prevention of laboratory-acquired tularemia. J. Infect. Dis. **135:**55–60.

11. **Burmeister, R. W., W. D. Tigertt, and E. L. Overholt.** 1962. Laboratory-acquired pneumonic plague. Ann. Intern. Med. **56:**789–800.

12. **Bush, J. D.** 1943. Coccidioidomycosis. J. Med. Assoc. Alabama **13:**159–166.

13. **Carougeau, M.** 1909. Premier cas Africain de sporotrichose de deBeurmann: transmission de la sporotrichose du mulet a l'homme. Bull. Mem. Soc. Med. Hop. (Paris) **28:**507–510.

14. **Cavenaugh, D. C., B. L. Elisberg, C. H. Llewellyn, J. D. Marshall, Jr., J. H. Rust, J. E. Williams, and K. F. Meyer.** 1974. Plague immunization. IV. Indirect evidence for the efficacy of plague vaccine. J. Infect. Dis. **129** (Suppl):S37–S40.

15. **Center for Disease Control, Office of Biosafety.** 1974. Classification of etiologic agents on the basis of hazard, 4th ed. U.S. Department of Health, Education and Welfare, Public Health Service.

16. **Center for Disease Control.** 1976. Unpublished data. Center for infectious diseases. U.S. Department of Health, Education and Welfare, Public Health Service.

17. **Center for Disease Control.** 1977. Rabies in a laboratory worker, New York. Morbid. Mortal. Weekly Rep. **26:**183–184.

18. **Center for Disease Control.** 1978. Laboratory-acquired endemic typhus. Morbid. Mortal. Weekly Rep. **27:**215–216.

19. **Center for Disease Control.** 1978. Meningococcal polysaccharide vaccines. Recommendations of the Immunization Practices Advisory Committee (ACIP). Morbid. Mortal. Weekly Rep. **27:**327–328.

20. **Centers for Disease Control.** 1979. Q fever at a university research center—California. Morbid. Mortal. Weekly Rep. **28:**333–334.

21. **Centers for Disease Control.** 1980. Chagas' disease, Kalamazoo, Michigan. Morbid. Mortal. Weekly Rep. **20:**147–148.

22. **Centers for Disease Control.** 1983. Viral hemorrhagic fever: initial management of suspected and confirmed cases. Morbid. Mortal. Weekly Rep. **32**(Suppl.):25S–39S.

23. **Centers for Disease Control.** 1980. Smallpox vaccines. Recommendation of the Immunization Practices Advisory Committee (ACIP). Morbid. Mortal. Weekly Rep. **29:**417–420.

24. **Centers for Disease Control.** 1981. Recommendations of the Immunization Practices Advisory Committee (ACIP). Diphtheria, tetanus, and pertussis. Morbid. Mortal. Weekly Rep. **30:**392–396.

25. **Centers for Disease Control.** 1982. Recommendations of the Immunization Practices Advisory Committee (ACIP). Poliomyelitis prevention. Morbid. Mortal. Weekly Rep. **29:**22–26, 31–34.

26. **Centers for Disease Control.** 1982. Plague vaccine. Selected recommendations of the Public Health Service Advisory Committee on Immunization Practices (ACIP). Morbid. Mortal. Weekly Rep. **31:**301–304.

27. **Centers for Disease Control.** 1982. Recommendations of the Immunizations Practices Advisory Committee (ACIP). Inactivated hepatitis B virus vaccine. Morbid. Mortal. Weekly Rep. **31:**317–328.

28. **Conant, N. F.** 1955. Development of a method for immunizing man against coccidioidomycosis, Third Quarterly Progress Report. Contract DA-18-064-CML-2563, Duke University, Durham, NC. Available from Defense Documents Center, AD 121-600.

29. **Davidson, W. L., and K. Hummeler.** 1960. B-virus infection in man. Ann. N.Y. Acad. Sci. **85:**970–979.

30. **Denton, J. F., A. F. DiSalvo, and M. L. Hirsch.** 1967. Laboratory-acquired North American blastomycosis. J. Am. Med. Assoc. **199:**935–936.

31. **Dickson, E. C.** 1937. Coccidioides infection: part I. Arch. Intern. Med. **59:** 1029–1044.

32. **Dickson, E. C.** 1937. "Valley fever" of the San Joaquin Valley and fungus coccidioides. Calif. Western Med. **47:**151–155.

33. **Dickson, E. C., and M. A. Gifford.** 1938. Coccidioides infection (coccidioidomycosis). II. The primary type of infection. Arch. Intern. Med. **62:**853–871.

34. **Diena, B. B., R. Wallace, F. E. Ashton, W. Johnson, and B. Patenaude.** 1976. Gonococcal conjunctivitis: accidental infection. Can. Med. Assoc. J. **115:**609.

35. **Donham, K. J., and J. R. Leninger.** 1977. Spontaneous leprosy-like disease in a chimpanzee. J. Infect. Dis. **136:**132–136.

36. **Dowdle, W. R., and M. A. W. Hattwick.** 1977. Swine influenza virus infections in humans. J. Infect. Dis. **136**(Suppl.):S386–S389.

37. **Edmond, R. T. D., B. Evans, E. T. W. Bowen, and G. Lloyd.** 1977. A case of Ebola virus infection. Br. Med. J. **2:**541–544.

38. **Ellingson, H. V., P. J. Kadull, H. L. Bookwalter, and C. Howe.** 1946. Cutaneous anthrax: report of twenty-five cases J. Am. Med. Assoc. **131:**1105–1108.

39. **Evans, N.** 1903. A clinical report of a case of blastomycosis of the skin from accidental inoculation J. Am. Med. Assoc. **40:**1172–1175.

40. **Eyles, D. E., G. R. Coatney, and M. E. Getz.**

1960. Vivax-type malaria parasite of macaques transmissible to man. Science **131**:1812–1813.

41. **Fava, A.** 1909. Un cas de sporotrichose conjonctivale et palpebrale primitives. Ann. Ocul. (Paris) **141**:338–343.

42. **Federal Register.** 1976. Recombinant DNA research guidelines. **41**:27902–27943.

43. **Federal Register.** 1982. Guidelines for research involving recombinant DNA molecules. **47**:38048–38068.

44. **Fielitz, H.** 1910. Ueber eine Laboratoriumsinfektion mit dem *Sporotrichum* de Beurmanni. Zentralbl. Bakteriol. Parasitenkd. Abt. 1 Orig. **55**:361–370.

45. **Furcolow, M. L.** 1961. Airborne histoplasmosis. Bacteriol. Rev. **25**:301–309.

46. **Gajdusek, D. C., C. J. Gibbs, D. M. Asher, P. Brown, A. Diwan, P. Hoffman, G. Nemo, R. Rohwer, and L. White.** 1977. Precautions in the medical care and in handling materials from patients with transmissible virus dementia (Creutzfeldt-Jakob Disease). N. Engl. J. Med. **297**:1253–1258.

47. **Good, R. C., and E. E. Snider, Jr.** 1982. Isolation of nontuberculosis mycobacteria in the U.S. J. Infect. Dis. **146**(6).

48. **Green, R. N., and P. G. Tuffnell.** 1968. Laboratory-acquired melioidosis. Am. J. Med. **44**:599–605.

49. **Gutteridge, W. E., B. Cover, and A. J. D. Cooke.** 1974. Safety precautions for working with *Trypanosoma cruzi*. Trans. R. Soc. Trop. Med. Hyg. **68**:161.

50. **Halde, C.** 1964. Percutaneous *Cryptococcus neoformans* inoculation without infection. Arch. Dermatol. **89**:545.

51. **Hanel, E., Jr., and R. H. Kruse.** 1967. Laboratory-acquired mycoses. Miscellaneous Publication 28. Department of the Army.

52. **Hanson, R. P., et al.** 1950. Human infections with the virus of vesicular stomatitis. J. Lab. Clin. Med. **36**:754–758.

53. **Hanson, R. P., S. E. Sulkin, E. L. Buescher, W. McD. Hammond, R. W. McKinney, and T. E. Work.** 1967. Arbovirus infections of laboratory workers. Science **158**:1283–1286.

54. **Harrell, E. R.** 1964. The known and the unknown of the occupational mycoses, p. 176–178. *In* Occupational diseases acquired from animals. Continued Education Series no. 124. University of Michigan School of Public Health, Ann Arbor.

55. **Harrington, J. M., and H. S. Shannon.** 1976. Incidence of tuberculosis, hepatitis, brucellosis and shigellosis in British medical laboratory workers. Br. Med. J. **1**:759–762.

56. **Hartley, E. G.** 1966. "B" virus disease in monkeys and man. Br. Vet. J. **122**:46–50.

57. **Hattwick, M. A. W., R. J. O'Brien, and B. F. Hanson.** 1976. Rocky Mountain spotted fever: epidemiology of an increasing problem. Ann. Intern. Med. **84**:732–739.

58. **Hennessen, W.** 1971. Epidemiology of "Marburg Virus" disease, p. 161–165. *In* G. A. Martini and R. Siegert (ed)., Marburg virus disease. Springer-Verlag, New York.

59. **Huddleson, I. F., and M. Munger.** 1940. A study of an epidemic of brucellosis due to *Brucella melitensis*. Am. J. Public Health **30**:944–954.

60. **Jeanselme, E., and P. Chevallier.** 1910. Chancres sporotrichosiques des doigts produits par la morsure d'un rat inocule de sporotrichose. Bull. Mem. Soc. Med. Hop. (Paris) **30**:176–178.

61. **Jeanselme, E., and P. Chevallier.** 1911. Transmission de la sporotrichose a l'homme par les morsures d'un rat blanc inocule avec une nouvelle variété de *Sporotrichum*: lymphangite gommeuse ascendante. Bull. Mem. Soc. Med. Hop. (Paris) **31**:287–301.

62. **Kaufmann, A. F., and D. C. Anderson.** 1978. Tuberculosis control in nonhuman primates, p. 227–234. *In* R. J. Montali (ed.), Mycobacterial infections of zoo animals. Smithsonian Institution Press, Washington, D.C.

63. **Kiley, M. P., E. T. W. Bowel, G. A. Eddy, M. Isaacson, K. M. Johnson, J. B. McCormick, F. A. Murphy, S. R. Pattyn, D. Peters, W. Prozesky, R. Regnery, D. I. H. Simpson, W. Slenczka, P. Sureau, G. van der Groen, P. A. Webb, and H. Sulff.** 1982. Filoviridae: taxonomic home for Marburg and Ebola viruses? Intervirology **18**:24–32.

64. **Klutsch, K., N. Hummer, H. Braun, A. Heidland.** 1965. Zur Klinik der Coccidioidomykose. Dtsch. Med. Wochenschr. **90**:1498–1501.

65. **Kubica, G. P., W. Gross, J. E. Hawkins, H. M. Sommers, A. L. Vestal, and L. G. Wayne.** 1975. Laboratory services for mycobacterial diseases. Am. Rev. Respir. Dis. **112**:773–787.5.

66. **Larsh, H. W., and J. Schwartz.** 1977. Accidental inoculation-blastomycosis. Cutis **19**:334–336.

67. **Leifer, E., D. J. Gocke, and H. Bourne.** 1970. Lassa fever, a new virus disease of man from West Africa. II. Report of a laboratory-acquired infection treated with plasma from a person recently recovered from the disease. Am. J. Trop. Med. Hyg. **19**:677–679.

68. **Looney, J. M., and T. Stein.** 1950. Coccidioidomycosis. N. Engl. J. Med. **242**:77–82.

69. **Marchoux, P. E.** 1934. Un cas d'inoculation accidentelle du bacille de Hanson en pays non lepreux. Int. J. Leprosy **2**:1–7.

70. **Martini, G. A., and H. A. Schmidt.** 1968. Spermatogenic transmission of Marburg virus. Klin. Wochenschr. **46**:398–400.

71. **McAleer, R.** 1980. An epizootic in laboratory guinea pigs due to *Trichophyton mentagrophytes*. Aust. Vet. J. **56**:234–236.

72. **McDade, J. E., and C. C. Shepard.** 1979. Virulent to avirulent conversion of Legionnaire's disease bacterium (*Legionella pneumophila*). Its effect on isolation techniques. J. Infect. Dis. **139**:707–711.

73. **Melnick, J. L., H. A. Wenner, and C. A. Phillips.** 1979. Enteroviruses, p. 471–534. *In* E. H. Lennette and N. J. Schmidt (ed.), Diagnostic procedures for viral, rickettsial and chlamydial infections, 5th ed. American Public Health Association, Washington, D.C.

74. **Meyer, K. F.** 1915. The relationship of animal to human sporotrichosis: studies on American sporotrichosis III. J. Am. Med. Assoc. **65**:579–585.

75. **Meyer, K. F., and B. Eddie.** 1941. Laboratory infections due to *Brucella*. J. Infect. Dis. **68**:24–32.

76. **Meyers, W. M., G. P. Walsh, H. L. Brown, Y. Fukunishi, C. H. Binford, P. J. Gerone, and R. H. Wolf.** 1980. Naturally acquired leprosy in a mangabey monkey (*Cercocebus* sp.). Int. J. Leprosy **48**:495–496.

77. **Morisset, R., and W. W. Spink.** 1969. Epidemic canine brucellosis due to a new species, *Brucella canis.* Lancet ii:1000–1002.

78. **Murray, J. F., and D. H. Howard.** 1964. Laboratory-acquired histoplasmosis. Am. Rev. Respir. Dis. **89:**631–640.

79. **Nabarro, J. D. N.** 1948. Primary pulmonary coccidioidomycosis: case of laboratory infection in England. Lancet i:982–984.

80. **National Sanitation Foundation.** 1983. Standard 49. Class II (laminar flow) biohazard cabinetry. National Sanitation Foundation, Ann Arbor, Mich.

81. **Norden, A.** 1951. Sporotrichosis: clinical and laboratory features and a serologic study in experimental animals and humans. Acta Pathol. Microbiol. Scand. Suppl. **89:**3–119.

82. **Office of Research Safety, National Cancer Institute, and the Special Committee of Safety and Health Experts.** 1978. Laboratory safety monograph: a supplement to the NIH Guidelines for Recombinant DNA Research. National Institutes of Health, Bethesda, Md.

83. **Oliphant, J. W., and R. R. Parker.** 1948. Q fever: three cases of laboratory infection. Public Health Rep. **63:**1364–1370.

84. **Oliphant, J. W., D. A. Gordon, A. Meis, and R. R. Parker.** 1949. Q fever in laundry workers, presumably transmitted from contaminated clothing. Am. J. Hyg. **49:**76–82.

85. **Oliver, J., and T. R. Reusser.** 1942. Rapid method for the concentration of tubercle bacilli. Am. Rev. Tuberc. **45:**450–452.

86. **Oster, C. N., et al.** 1977. Laboratory-acquired Rocky Mountain spotted fever. The hazard of aerosol transmission. N. Engl. J. Med. **297:**859–862.

87. **Parritt, R. J., and R. E. Olsen.** 1947. Two simultaneous cases of leprosy developing in tattoos. Am. J. Pathol. **23:**805–817.

88. **Patterson, W. C., L. O. Mott, and E. W. Jenney.** 1958. A study of vesicular stomatitis in man. J. Am. Vet. Med. Assoc. **133:**57–62.

89. **Perkins, F. T., and E. G. Hartley.** 1966. Precautions against B virus infection. Br. Med. J. **1:**899–901.

90. **Pike, R. M.** 1976. Laboratory-associated infections: summary and analysis of 3,921 cases. Heath Lab. Sci. **13:**105–114.

91. **Pike, R. M.** 1978. Past and present hazards of working with infectious agents. Arch. Pathol. Lab. Med. **102:**333–336.

92. **Pike, R. M.** 1979. Laboratory-associated infections: incidence, fatalities, causes and prevention. Annu. Rev. Microbiol. **33:**41–66.

93. **Pike, R. M., S. E. Sulkin, and M. L. Schulze.** 1965. Continuing importance of laboratory-acquired infections. Am. J. Public Health **55:**190–199.

94. **Prescott, J. F., and M. A. Karmali.** 1978. Attempts to transmit *Campylobacter enteritis* to dogs and cats. Can. Med. Assoc. J. **119:**1001–1002.

95. **Reid, D. D.** 1957. Incidence of tuberculosis among workers in medical laboratories. Br. Med. J. **2:**10–14.

96. **Report of the Committee of Inquiry into the Smallpox Outbreak in London in March and April 1973.** 1974. Her Majesty's Stationery Office, London.

97. **Richardson, J. H.** 1973. Provisional summary of 109 laboratory-associated infections at the Center for Disease Control, 1947–1973. Presented at the 16th Annual Biosafety Conference, Ames, Iowa.

98. **Riley, R. L.** 1957. Aerial dissemination of pulmonary tuberculosis. Am. Rev. Tuberc. **76:**931–941.

99. **Riley, R. L.** 1961. Airborne pulmonary tuberculosis. Bacteriol. Rev. **25:**243–248.

100. **Robertson, D. H. H., S. Pickens, J. H. Lawson, and B. Lennex.** 1980. An accidental laboratory infection with African trypanosomes of a defined stock. I and II. J. Infect. Dis. **2:**105–112, 113–124.

101. **Sastaw, S., and H. N. Carlisle.** 1966. Aerosol infection of monkeys with *Rickettsia rickettsii.* Bacteriol. Rev. **30:**636–645.

102. **Schlech, W. F., J. B. Turchik, R. E. Westlake, G. C. Klein, J. D. Band, and R. E. Wever.** 1981. Laboratory-acquired infection with *Pseudomonas pseudomallei* (melioidosis). N. Engl. J. Med. **305:**1133–1135.

103. **Schwarz, J., and G. L. Baum.** 1951. Blastomycosis. Am. J. Clin. Pathol. **21:**999–1029.

104. **Skinhoj, P.** 1974. Occupational risks in Danish clinical chemical laboratories. II. Infections. Scand. J. Clin. Lab. Invest. **33:**27–29.

105. **Smith, C. E.** 1950. The hazard of acquiring mycotic infections in the laboratory. Presented at 78th Ann. Meeting. American Public Health Association, St. Louis, Mo.

106. **Smith, C. E., D. Pappagianis, H. B. Levine, and M. Saito.** 1961. Human coccidioidomycosis. Bacteriol. Rev. **25:**310–320.

107. **Smith, D. T., and E. R. Harrell, Jr.** 1948. Fatal coccidioidomycosis: a case of laboratory infection. Am. Rev. Tuberc. **57:**368–374.

108. **Smithwick, R. W., and C. B. Stratigos.** 1978. Preparation of acid-fast microscopy smears for proficiency testing and quality control. J. Clin. Microbiol. **8:**110–111.

109. **Spinelli, J. S., et al.** 1981. Q fever crisis in San Francisco: controlling a sheep zoonosis in a lab animal facility. Lab. Anim. **10:**24–27.

110. **Spink, W. W.** 1956. The nature of brucellosis, p. 106–108. University of Minnesota Press, Minneapolis.

111. **Sterne, M., and L. M. Wertzel.** 1950. A new method of large-scale production of high-titer botulinum formol-toxoid types C and D. J. Immunol. **65:**175–183.

112. **Subcommittee on Arbovirus Laboratory Safety of the American Committee on Arthropod-borne Viruses.** 1980. Laboratory safety for arboviruses and certain other viruses of vertebrates. Am. J. Trop. Med. Hyg. **29:**1359–1381.

113. **Sulkin, S. E., and R. M. Pike.** 1949. Viral infections contracted in the laboratory. N. Engl. J. Med. **241:**205–213.

114. **Sulkin, S. E. and R. M. Pike.** 1951. Survey of laboratory-acquired infections. Am. J. Public Health **41:**769–781.

115. **Sullivan, J. F., J. R. Songer, and I. E. Estrem.** 1978. Laboratory-acquired infections at the National Animal Disease Center, 1960–1976. Health Lab. Sci. **15:**58–64.

116. **Tesh, R. B., and J. D. Schneidau, Jr.** 1966. Primary cutaneous histoplasmosis. N. Engl. J. Med. **275:**597–599.

117. **Thompson, D. W., and W. Kaplan.** 1977.

Laboratory-acquired sporotrichosis. Sabouraudia **15:**167–170.

118. **Tomlinson, C. C., and P. Bancroft.** 1928. Granuloma coccidioides: report of a case responding favorably to antimony and potassium tartrate. J. Am. Med. Assoc. **91:**947–951.

119. **Tosh, F. E., J. Balhuizen, J. L. Yates, and C. A. Brasher.** 1964. Primary cutaneous histoplasmosis: report of a case. Arch. Intern. Med. **114:**118–119.

120. **Walsh, G. P., E. E. Storrs, H. P. Burchfield, E. H. Cottrel, M. F. Vidrine, and C. H. Binford.** 1975. Leprosy-like disease occurring naturally in armadillos. J. Reticuloendothel. Soc. **18:**347–351.

121. **Wedum, A. G., and R. H. Kruse.** 1969. Assessment of risk of human infection in the microbiology laboratory. Miscellaneous Publication no. 30, Industrial Health and Safety Directorate, Fort Detrick, Federick, Md.

122. **Wedum, A. G., W. E. Barkley, and A. Hellman.** 1972. Handling of infectious agents. J. Am. Vet. Med. Assoc. **161:**1557–1567.

123. **Wedum, A. G.** 1975. History of microbiological safety. 18th Biological Safety Conference. Lexington, Ky.

124. **Weissenbacher, M. C., M. E. Grela, M. S. Sabattini, J. I. Maiztegui, C. E. Coto, M. J. Frigerio, P. M. Cossio, A. S. Rabinovich, and J. G. B. Oro.** 1978. Inapparent infections with Junin virus among laboratory workers. J. Infect. Dis. **137:**309–313.

125. **Wilder, W. H., and C. P. McCullough.** 1914. Sporotrichosis of the eye. J. Am. Med. Assoc. **62:**1156–1160.

126. **Willems, W. R., G. Kaluza, C. B. Boschek, and H. Bauer.** 1979. Semliki Forest virus: cause of a fatal case of human encephalitis. Science **203:**1127–1129.

127. **Wilson, J. W., E. P. Cawley, F. D. Weidman, and W. S. Gilmer.** 1955. Primary cutaneous North American blastomycosis. Arch. Dermatol. **71:**39–45.

128. **Wilson, J. W., C. E. Smith, and O. A. Plunkett.** 1953. Primary cutaneous coccidioidomycosis; the criteria for diagnosis and a report of a case. Calif. Med. **79:**233–239.

129. **Winkler, W. G.** 1973. Airborne rabies transmission in a laboratory worker. J. Am. Med. Assoc. **226:**1219–1221.

130. **World Health Organization.** 1978. Smallpox surveillance. Weekly Epidemiol. Rec. **53:**265–266.

Appendix 2

State and Territorial Public Health Laboratory Directors

Alabama
Dr. James L. Holston, Jr.
Director, Clinical Laboratory Administration
State Dept. of Public Health
University Drive
Montgomery, AL 36130
FTS Direct & Commercial:
(205) 277-8660, ext. 215

Alaska
Dr. Harry J. Colvin
Chief, Section of Laboratories
Alaska Div. of Public Health
Dept. of Health and Social Services
Pouch H-06-D
Juneau, AK 99811
FTS Direct & Commercial:
(907) 465-3140

Arizona
Dr. Jon M. Counts, Chief
Bureau of Laboratory Services
Arizona Dept. of Health Services
1520 W. Adams St.
Phoenix, AZ 85007
FTS Direct & Commercial:
(602) 255-1188

Arkansas
Mr. Robert L. Horn, Director
Div. of Public Health Laboratories
4815 West Markham St.
Little Rock, AR 72201
FTS Operator: 740-5011
Commercial: (501) 661-2217

California
Dr. G. W. Fuhs, Chief
Laboratory Services Branch
State Dept. of Health Services
2151 Berkeley Way
Berkeley, CA 94704
FTS Direct & Commercial:
(415) 540-2408

Colorado
Dr. Ronald L. Cada, Director
Div. of Laboratories
Dept. of Public Health
4210 East 11th Ave.

Denver, CO 80220
FTS Direct & Commercial:
(303) 320-1166

Connecticut
Dr. Jesse Tucker
Director of Laboratories
State Dept. of Health
P.O. Box 1689
Hartford, CT 06101
FTS Direct & Commercial:
(203) 566-5063

Delaware
Dr. Mahadeo P. Verma, Director
Div. of Public Health Laboratories
Jesse S. Cooper Memorial Bldg.
Capitol Square
Dover, DE 19901
FTS Direct & Commercial:
(302) 736-4734

District of Columbia
Dr. James B. Thomas, Acting Director
Bureau of Laboratories
Dept. of Human Services
300 Indiana Ave., N.W., Room 6154
Washington, DC 20001
FTS Direct & Commercial:
(202) 727-0557

Florida
Dr. Eldert C. Hartwig, Director
Office of Laboratory Services
Dept. of Health & Rehabilitative Services
P.O. Box 210 (1217 Pearl St.)
Jacksonville, FL 32231
FTS Direct & Commercial:
(904) 354-3961

Georgia
Dr. Frank M. Rumph
Director of Laboratories
Georgia Dept. of Human Resources
47 Trinity Ave.
Atlanta, GA 30334
Commercial: (404) 656-4852

Guam
Mr. Luis P. Flores

Laboratory Director
Public Health & Social Services
P.O. Box 2816
Agana, Guam 96910

Hawaii

Dr. Robert Katasse
Chief, Laboratories Branch
State Dept. of Health
P.O. Box 3378
Honolulu, HI 96801
FTS Direct & Commercial:
(808) 548-6324

Idaho

Dr. D. W. Brock
Chief, Bureau of Laboratories
Dept. of Health and Welfare
2220 Old Penitentiary Road
Boise, ID 83712
FTS Direct: 554-2236
Commercial: (208) 334-2236
Evaluations and Specimens:
Box 640
Boise, ID 83701

Illinois

Mr. Harry C. Bostick, Chief
Div. of Laboratories
Illinois Dept. of Public Health
535 W. Jefferson, 4th Floor
Springfield, IL 62761
FTS Direct & Commercial:
(217) 782-4977

Indiana

Mr. T. L. Eddleman, Director
Bureau of Laboratories
State Board of Health
1330 W. Michigan St.
Indianapolis, IN 46206
FTS Direct & Commercial:
(317) 633-0376

Iowa

Dr. W. J. Hausler, Jr., Director
University Hygienic Laboratory
University of Iowa
Iowa City, IA 52242
FTS Direct & Commercial
(319) 353-5990

Kansas

Dr. Roger H. Carlson, Director
Office of Laboratories and Research
Dept. of Health and Environment
Forbes Bldg., no. 740
Topeka, KS 66620
FTS Direct & Commercial:
(913) 862-9360

Kentucky

Dr. B. F. Brown, Director
Div. of Laboratory Services
Dept. for Health Services
Cabinet for Human Resources
275 E. Main St.
Frankfort, KY 40621
FTS Direct & Commercial:
(502) 564-4446

Louisiana

Dr. Henry Bradford, Director
Bureau of Laboratories
Office of Health Services and Environmental
   Quality
Louisiana State Dept. of Health
325 Loyola Ave., 7th Floor
New Orleans, LA 70112
FTS Direct & Commercial:
(504) 568-5373

Maine

Dr. Philip W. Haines, Director
Public Health Laboratory
Dept. of Human Services
State House—Station no. 12
Augusta, ME 04333
FTS Direct & Commercial:
(207) 289-2727

Maryland

Dr. J. Mehsen Joseph, Director
Laboratories Administration
State Dept. of Health & Mental Hygiene
P.O. Box 2355
Baltimore, MD 21203
FTS Direct: 932-2880
Commercial: (301) 383-2880

Massachusetts

Dr. George F. Grady, Director
State Laboratory Institute
Dept. of Public Health
305 South Street
Jamaica Plain, MA 02130
FTS Direct & Commercial:
(617) 522-3700

Michigan

Dr. George R. Anderson, Laboratory Director
Laboratory and Epidemiological Services
   Administration
Michigan Dept. of Public Health
P.O. Box 30035—3500 N. Logan
Lansing, MI 48909
FTS Direct & Commercial:
(517) 373-1381

Minnesota

Dr. C. Dwayne Morse, Director

Div. of Medical Laboratories
Minnesota Dept. of Health
717 S.E. Delaware St.
Minneapolis, MN 55440
FTS Direct & Commercial:
(612) 623-5210

Mississippi
Mr. R. H. Andrews, Director
Public Health Laboratories
State Board of Health
P.O. Box 1700
Jackson, MS 39205
FTS Direct & Commercial:
(601) 354-6672

Missouri
Dr. Elmer R. Spurrier, Director
Bureau of Laboratory Services
Missouri Div. of Health
307 W. McCarty
Jefferson City, MO 65101
FTS Direct & Commercial:
(314) 751-3334

Montana
Dr. Douglas Abbott, Chief
State Microbiology Laboratory
State Dept. of Health and Environmental Sciences
Cogswell Bldg.
Helena, MT 59620
FTS Operator: 585-5011
Commercial: (406) 444-2642

Nebraska
Mr. John Blosser, Director of Laboratories
State Dept. of Health
P.O. Box 2755
Lincoln, NE 68502
FTS Direct & Commercial:
(402) 471-2122

Nevada
Dr. George Reynolds, Administrator
Nevada State Health Laboratory
Dept. of Human Resources
1660 N. Virginia St.
Reno, NV 89503
FTS Direct & Commercial:
(702) 885-4475

New Hampshire
Mrs. Veronica C. Malmberg, Director
Diagnostic Laboratories
Div. of Public Health
State Laboratory Bldg.
Hazen Drive
Concord, NH 03301
FTS Direct & Commercial:
(603) 271-4657

New Jersey
Dr. Bernard F. Taylor, Director
Public Health and Environmental Laboratories
State Dept. of Health
CN 360
Trenton, NJ 08625
FTS Direct & Commercial:
(609) 292-5605

New Mexico
Dr. Loris W. Hughes, Director
Scientific Laboratory Div.
700 Camino de Salud, N.E.
Albuquerque, NM 87106
FTS Direct & Commercial:
(505) 841-2500

New York
Dr. David O. Carpenter, Director
Div. of Laboratories and Research
State Dept. of Health
Tower Bldg., Empire State Plaza
Albany, NY 12201
FTS Direct & Commercial:
(518) 474-4170

North Carolina
Mrs. Mildred A. Kerbaugh, Director
Public Health Laboratory
State Board of Health
P.O. Box 28047
Raleigh, NC 27611
FTS Direct & Commercial:
(919) 733-7834

North Dakota
Mr. A. A. Gustafson, Chief
Laboratory Services Section
State Dept. of Health
Box 1618
Bismarck, ND 58502-1618
FTS Operator: 783-4011
Commercial: (701) 224-2384

Ohio
Dr. Gary D. Davidson, Chief
Div. of Public Health Laboratories
State Dept. of Health
P.O. Box 2568
Columbus, OH 43216
FTS Direct & Commercial:
(614) 421-1078

Oklahoma
Dr. Garry L. McKee, Chief
Public Health Laboratory Service
State Dept. of Health
P.O. Box 24106
Oklahoma City, OK 73124

FTS Operator: 736-4011
Commercial: (405) 271-5070

Oregon
Dr. Michael R. Skeels, Manager-Director
Public Health Laboratory
Dept. of Human Resources
1717 S.W. 10th Ave.
Portland, OR 97201
FTS Direct & Commercial:
(503) 229-5884

Pennsylvania
Dr. Vern Pidcoe, Director
Bureau of Laboratories
Pennsylvania Dept. of Health
Pickering Way and Welsh Pool Road
Lionville, PA 19353
FTS Direct & Commercial:
(215) 363-8500

Puerto Rico
Dr. Jose L. Villamil, Director
Institute of Health Laboratories
Dept. of Health
Bldg. A—Call Box 70184
San Juan, PR 00922
FTS Direct & Commercial:
(809) 767-2014

Rhode Island
Dr. Raymond G. Lundgren, Jr.,
    Associate Director
Div. of Laboratories
Health Laboratory Bldg.
50 Orms St.
Providence, RI 02904
FTS Direct & Commercial:
(401) 274-1011

South Carolina
Dr. Arthur DiSalvo, Chief
Bureau of Laboratories
Dept. of Health & Environmental Control
P.O. Box 2202
Columbia, SC 29202
FTS Direct & Commercial:
(803) 758-4491

South Dakota
Dr. A. Richard Melton, Director
State Health Laboratory
Laboratory Bldg.
Pierre, SD 57501
FTS Operator: 782-7000
Commercial: (605) 773-3368

Tennessee
Dr. Michael W. Kimberly, Director
Div. of Laboratory Services

Tennessee Dept. of Public Health
Cordell Hull Bldg., Room 425
Nashville, TN 37219
FTS Direct & Commercial:
(615) 741-3596

Texas
Dr. Charles E. Sweet, Chief
Bureau of Laboratories
Texas Dept. of Health
1100 W. 49th St.
Austin, TX 78756
FTS Direct & Commercial:
(512) 458-7318

Utah
Dr. Francis M. Urry, Director
Utah State Health Laboratory
44 Medical Dr., Room 207
Salt Lake City, UT 84113
FTS Direct & Commercial:
(801) 533-6131

Vermont
Dr. Katherine A. Kelley
State Public Health Laboratory
State Dept. of Health
115 Colchester Ave.
Burlington, VT 05401
FTS Direct & Commercial:
(802) 863-7200

Virginia
Dr. Frank W. Lambert, Jr., Director
Bureau of Microbiological Science
Div. of Consolidated Laboratory Services
Dept. of General Services, Commonwealth of
    Virginia
Box 1877
Richmond, VA 23215
FTS Direct: 936-3756
Commercial: (804) 786-3756

Virgin Islands
Dr. Norbert Mantor, Director
Public Health Laboratory
P.O. Box 8585
St. Thomas, VI 00801
FTS Direct & Commercial:
(809) 776-8311

Washington
Dr. Jack Allard, Chief
Laboratory Section
State Dept. of Social & Health Services
1409 Smith Tower
Seattle, WA 98104
FTS Direct & Commercial:
(206) 464-6461

West Virginia
Dr. John W. Brough, Director
State Hygienic Laboratory
167 11th Ave.
South Charleston, WV 25303
FTS Direct: 885-3530
Commercial: (304) 348-3530

Wisconsin
Dr. Ronald H. Laessig, Director
State Laboratory of Hygiene
William D. Stovall Bldg.
465 Henry Mall

Madison, WI 53706
FTS Direct & Commercial:
(608) 262-1293

Wyoming
Dr. Donald T. Lee, Director
Public Health Laboratory Services
Div. of Health & Medical Services
State Office Bldg.
Cheyenne, WY 82001
FTS Operator: 328-1110
Commercial: (307) 777-7431

Appendix 3

Poison Information Centers by State†

Alabama
* Alabama Poison Center
  809 University Blvd., E.
  Tuscaloosa, AL 35401
  (205) 345-0600
  (800) 462-0800 (in Alabama)

  Poison Information Center
  Children's Hospital
  1601 Sixth Ave., South
  Birmingham, AL 35233
  (205) 933-4050
  (800) 292-6678

Alaska
  Anchorage Poison Center
  Providence Hospital
  3200 Providence Dr.
  Anchorage, AK 99504
  (907) 274-6535

  Fairbanks Poison Center
  Fairbanks Memorial Hospital
  1650 Cowles
  Fairbanks, AK 99701
  (907) 452-8181

Arizona
* Arizona Poison and Drug Information Center
  Arizona Health Sciences Center
  University of Arizona
  Tucson, AZ 85724
  (602) 626-6016
  (800) 362-0101 (in Arizona)

  Central Arizona Regional Poison Management
    Center
  St. Luke's Hospital
  1800 E. Van Buren
  Phoenix, AZ 85006
  (602) 251-8186

Arkansas
  Arkansas Poison and Drug Info
  University of Arkansas Medical Center

4301 Markham St.
Little Rock, AR 72205
(501) 666-5532

California
  Central Valley Poison Control Center
  Fresno Community Hospital and Medical Center
  Fresno and R Sts.
  Fresno, CA 93715
  (209) 442-1222

  Los Angeles County Medical Association
  Poison Information Center, Regional
  1925 Wilshire
  Los Angeles, CA 90057
  (213) 664-2121 (professional consultation)
  (213) 484-5151 (public consultation)

* Sacramento Medical Center
  University of California
  2315 Stockton Blvd.
  Sacramento, CA 95817
  (916) 453-3414
  (800) 852-7221

* San Diego Regional Poison Center
  University of California Medical Center
  225 W. Dickinson St.
  San Diego, CA 92103
  (619) 294-6000

* San Francisco Bay Area Regional Poison Center
  San Francisco General Hospital
  1001 Potrero Ave.
  San Francisco, CA 94102
  (415) 666-2845

Colorado
* Rocky Mountain Poison Center
  Denver General Hospital
  West 8th & Cherokee
  Denver, CO 80204
  (303) 629-1123 (24 h)

Connecticut
  Connecticut Poison Information Center
  University of Connecticut Health Center
  Farmington, CT 06032
  (203) 674-3456

†The toll-free information number (800) 555-1212 will provide telephone numbers for local poison control centers.

*Certified as Regional Center by the American Association of Poison Control Centers.

District of Columbia
* National Capital Poison Center
  Georgetown University Hospital
  3800 Reservoir Rd., N.W.
  Washington, DC 20007
  (202) 625-3333

Florida
  Gulf Region Poison Center
  Baptist Hospital
  1000 West Moreno St.
  Pensacola, FL 32501
  (904) 434-4811

* Tampa Bay Region Poison Control Center
  P.O. Box 18582
  Tampa, FL 33679
  (813) 251-6995
  (800) 282-3171

Georgia
* Georgia Poison Control Center
  Grady Memorial Hospital
  80 Butler St., SE
  Atlanta, GA 30303
  (404) 589-4400
  (404) 525-3323 (deaf TTD)
  (800) 282-5846

  Savannah Regional Poison Center
  Memorial Medical Center
  P.O. Box 23089
  Savannah, GA 31403
  (912) 355-5228

Hawaii
  Hawaii Poison Center
  Kapiolani Children's Medical Center
  1319 Panahou St.
  Honolulu, HI 96826
  (808) 541-4411

Idaho
  Idaho Drug Information and Poison Control Center
  Pocatello Region Medical Center
  755 Hospital Way Suite F-2
  Pocatello, ID 83201
  (208) 234-0777 ext. 5019
  (800) 632-9490 (in Idaho)

  Idaho Poison Center System
  Department of Health and Welfare
  Boise, ID 83702
  (208) 376-1211
  (800) 632-8000

Illinois
* Chicago Area Poison Resource Center
  Rush-Presbyterian-St. Luke's Medical Center
  1753 W. Congress Parkway

Chicago, IL 60612
(312) 942-5969
(800) 942-5969

Brokaw Hospital Poison Center
Virginia at Franklin
Normal, IL 61761
(309) 454-1400

Northern and Central Poison Resource Center
St. Francis Hospital
530 N.E. Glen Oak
Peoria, IL 61637
(309) 672-2334
(800) 322-5330 (in Illinois)

* Central & Southern Poison Resource Center
  St. Johns Hospital
  800 E. Carpenter
  Springfield, IL 62702
  (217) 753-3330
  (800) 252-2022 (in Illinois)

Indiana
* Indiana Poison Center
  Wishard Memorial Hospital
  1001 West 10th St.
  Indianapolis, IN 46202
  (317) 630-7351
  (800) 382-9097

Iowa
* Iowa Poison Information Center
  University of Iowa Hospital
  Iowa City, IA 52242
  (319) 356-2922
  (800) 272-6477 (in Iowa)

  Marion Health Center Poison Center
  801 5th St.
  Sioux City, IA 51103
  (712) 279-2066

Kansas
  Mid America Poison Center
  University of Kansas Medical Center
  39th & Rainbow Blvd.
  Kansas City, KS 66103
  (913) 588-6633

Kentucky
* Kentucky Poison Control Center
  Norton Children's Hospital
  P.O. Box 35070
  Louisville, KY 40232
  (502) 589-8222
  (800) 722-5725 (in Kentucky)

Louisiana
* Louisiana Poison Center
Louisiana State University Medical Center
P.O. Box 33932
Shreveport, LA 71130
(318) 425-1524
(800) 535-0525 (in Louisiana)

Maine
Maine Poison Control Center
Maine Medical Center
22 Bramhill St.
Portland, ME 04102
(207) 871-2950
(800) 442-6305 (in Maine)

Maryland
* Maryland Poison Information Center
University of Maryland
20 N. Pine St.
Baltimore, MD 21201
(301) 528-7701
(800) 492-2414 (in Maryland)

Massachusetts
* Massachusetts Poison Control System
Children's Hospital
300 Longwood Ave.
Boston, MA 02115
(617) 232-2120
(800) 682-9211 (in Massachusetts)

Michigan
Amway Poison Control
Ada, MI 49355
(616) 676-6307

* Southeast Regional Poison Center
Children's Hospital of Michigan
3901 Beaubien Blvd.
Detroit, MI 48201
(313) 494-5711 (24 h)

* Western Michigan Regional Poison Center
1840 Wealthy St., S.E.
Grand Rapids, MI 49506
(616) 774-7854
(800) 442-4571 (Michigan; from area code 616)
(800) 632-2727 (Michigan; other)

Great Lakes Poison Center
Bronson Methodist Hospital
252 E. Lovell St.
Kalamazoo, MI 49007
(616) 383-6409
(800) 442-4112 (from area code 616)

Midwestern Poison Center
1521 Gull Rd.

Kalamazoo, MI 49001
(616) 383-7070
(800) 632-4177

Saginaw Region Poison Center
Saginaw General Hospital
1447 North Harrison St.
Saginaw, MI 48602
(517) 755-1111

Minnesota
Hennepin Poison Center
701 Park Ave.
Minneapolis, MN 55415
(612) 347-3141

* Minnesota Poison Control System
St. Paul-Ramsey Medical Center
640 Jackson St.
St. Paul, MN 55101
(612) 221-2113
(800) 222-1222 (in Minnesota)

Mississippi
University of Mississippi Medical Center Poison
Service
University Medical Center
2500 W. State St.
Jackson, MS 39216
(601) 354-7660

Missouri
Poison Control Center
Children's Mercy Hospital
24th St. & Gilham Rd.
Kansas City, MO 64108
(816) 234-3000

* CCMH St. Louis Poison Center
Cardinal Glennon Memorial Hospital
1465 South Grand Blvd.
St. Louis, MO 63104
(314) 772-5200
(314) 865-4000 (24 h)
(800) 392-9111 (in Missouri)

Montana
Montana Poison Control System
Department of Health & Environmental Science
Helena, MT 59601
(800) 525-5042

Nebraska
* Mid Plains Poison Control Center
Children's Memorial Hospital
8301 Dodge St.
Omaha, NE 68114
(402) 390-5400
(800) 642-9999 (in Nebraska)
(800) 228-9515 (out of state)

Nevada
St. Mary's Hospital
235 W. 6th St.
Reno, NV 39503
(702) 323-2041

New Hampshire
New Hampshire Poison Center
Dartmouth-Hitchcock Medical Center
Hanover, NH 03756
(603) 646-5000

New Jersey
* New Jersey Poison Information
201 Lyons Ave.
Newark, NJ 10016
(201) 926-8005
(800) 962-1253 (in New Jersey)

Middlesex General University Hospital
180 Somerset St.
New Brunswick, NJ 08901
(201) 937-8583

New Mexico
* New Mexico Poison and Drug Information Center
Bernalillo County Medical Center
221 Lomas Blvd., N.E.
Albuquerque, NM 87131
(505) 843-2551
(800) 432-6866 (in New Mexico)

New York
Western New York Poison Center
Children's Hospital
219 Bryant St.
Buffalo, NY 14222
(716) 878-7000, 878-7654

* Long Island Poison Center
Nassau County Medical Center
2201 Hempstead Turnpike
East Meadow (Long Island), NY 11554
(516) 542-2323

New York Poison Center
Department of Health
455 First Ave., Room 123
New York, NY 10016
(212) 340-4494

Hudson Valley Poison Center
Nyack Hospital
North Midland Ave.
Nyack, NY 10960
(914) 353-1000

* Finger Lakes Poison Control Center
LIFELINE
973 East Ave.

University of Rochester Medical Center
Rochester, NY 14607
(716) 275-5151

North Carolina
Mercy Hospital Poison Control Center
2001 Vail Ave.
Charlotte, NC 28207
(704) 379-5827

* Duke University Poison Control Center
Box 3007
Durham, NC 27710
(919) 684-8111
(800) 672-1697 (in North Carolina)

North Dakota
St. Luke's Poison Center
St. Luke's Hospitals
Fifth St. at Mills Ave.
Fargo, ND 58122
(701) 280-5575

Ohio
Akron Regional Poison Center
Children's Hospital Medical Center
281 Locust St.
Akron, OH 44308
(216) 379-8562

Drug and Poison Information Center
University of Cincinnati Medical Center
Room 7701 Bridge
Cincinnati, OH 45267
(513) 872-5111

Southwest Ohio Regional Poison Center
231 Bethesda Ave.
Cincinnati, OH 45267-0144
(513) 872-5111

Greater Cleveland Poison Control Center
2119 Abington Rd.
Cleveland, OH 44106
(216) 844-1573

* Central Ohio Poison Center
700 E. Children's Dr.
Columbus, OH 43205
(614) 228-1323

Mahoning Valley Poison Center
1044 Belmont Ave.
Youngstown, OH 44501
(216) 746-2222

Oklahoma
Oklahoma Poison Control Center
Oklahoma Children's Memorial Hospital
P.O. Box 26307

Oklahoma City, OK 73126
(405) 271-5454
(800) 522-4611 (in Oklahoma)

Oregon
Oregon Poison Control & Drug Information
   Center
University of Oregon Health Sciences Center
3181 S.W. Sam Jackson Park Road
Portland, OR 97201
(503) 225-8968
(800) 452-7165 (in Oregon)

Pennsylvania
Lehigh Valley Poison Center
17th and Chew St.
Allentown, PA 18102
(215) 433-2311

Keystone Region Poison Center
Mercy Hospital
2500 Seventh Ave.
Altoona, PA 16603

Susquehanna Poison Center
Geisinger Medical Center
North Academy Ave.
Danville, PA 17821
(717) 275-6116

Hamot Medical Center
Poison Information Center
Erie, PA 16550
(814) 452-4242

Capital Area Poison Center
Milton Hershey Medical Center
Pennsylvania State University
Hershey, PA 17033
(717) 534-6111

* Pittsburgh Poison Center
Children's Hospital of Pittsburgh
125 De Soto St.
Pittsburgh, PA 15213
(412) 681-6669

National Poison Center Network (NPCN)
(412) 647-5600

Rhode Island
Rhode Island Hospital
593 Eddy St.
Providence, RI 02902
(401) 277-4000

South Carolina
Palmetto Poison Center
College of Pharmacy
University of South Carolina

Columbia, SC 29208
(803) 765-7359
(800) 922-1117

South Dakota
West River Poison Center
Rapid City Regional Hospital East
Rapid City, SD 57709
(605) 339-7874
(800) 952-0123 (in South Dakota)
(800) 843-0505 (in Iowa, Minnesota, Nebraska)

McKennan Poison Center
800 E. 21 Street
Sioux Falls, SD 57101
(605) 336-3894

Tennessee
Johnson City Medical Center
Poison Control Center
400 State of Franklin Rd.
Johnson City, TN 37601
(615) 461-6572

Southern Poison Center
College of Pharmacy
The University of Tennessee
874 Union
Memphis, TN 38163
(901) 528-6048

Texas
Montgomery County Poison Center
P.O. Box 1538
Conroe, TX 77305
(409) 539-7700

North Central Texas Poison Center
P.O. Box 35926
Dallas, TX 75235
(214) 920-2400

El Paso Poison Center
Thomason General Hospital
4815 Alameda Ave.
El Paso, TX 79905
(915) 533-1244

Cook Poison Center
W.I. Cook Children's Hospital
1212 Lancaster
Fort Worth, TX 76102
(817) 336-5521, ext. 17; 336-6611

Southeast Texas Poison Control Center
University of Texas Medical Branch
8th St. and Mechanic St.
Galveston, TX 77550
(713) 765-1420

East Texas Poison Center
Medical Center Hospital
Box 6400
Tyler, TX 75711
(214) 597-0351

Utah
Humana Hospital Davis NTH
1600 W. Antelope Dr.
Layton, UT 84041
(801) 825-4357

* Intermountain Regional Poison Control Center
University of Utah Medical Center
50 North Medical Dr.
Salt Lake City, UT 84132
(801) 581-2151
(800) 662-0062 (in Utah)

Vermont
Vermont Poison Center
Medical Center Hospital of Vermont
Burlington, VT 05401
(802) 658-3456

Virginia
Blue Ridge Poison Center
University of Virginia
Box 484, UVA Hospital
Charlottesville, VA 22908
(804) 924-5543
(800) 552-3723 (in Virginia)
(800) 446-9876 (outside Virginia)

Central Virginia Poison Center
Virginia Commonwealth University
Box 763, MCV Station
Richmond, VA 23298
(804) 786-9123; 911

Washington
* COHMC Poison Control Center
Children's Orthopedic Hospital & Medical Center
4800 Sand Point Way, N.E.
Seattle, WA 98105
(206) 526-2121
(800) 732-6985

Mary Bridge Poison Center
Mary Bridge Children's Hospital
311 South L St.

Tacoma, WA 98405
(206) 272-1281, ext. 259

Central Washington Poison Center
Yakima Valley Memorial Hospital
2811 Tieton Dr.
Yakima, WA 98902
(509) 248-4400
(800) 572-9176 (in Washington)

West Virginia
West Virginia Poison Center
3110 McCorkle Ave. SE
Charleston, WV 25304
(304) 348-4211

Wisconsin
Eau Claire Poison Center
Luther Hospital
1221 Whipple St.
Eau Claire, WI 54702
(715) 835-1515

Green Bay Poison Center
St. Vincent Hospital
P.O. Box 13508
Green Bay, WI 54307
(414) 433-8100

La Crosse Area Poison Center
700 West Ave. South
La Crosse, WI 54601
(608) 784-3971

Madison Poison Control Center
1300 University Ave.
Madison, WI 53706
(608) 262-3702

Milwaukee Poison Center
1700 West Wisconsin Ave.
Milwaukee, WI 53233
(414) 931-4114

Ontario, Canada
Ontario Poison Information Centre
Hospital for Sick Children
555 University Ave.
Toronto, Ontario, Canada
(416) 929-1900

Appendix 4

Emergency First Aid Guide[†]

In an emergency, proper treatment should be instituted by you as soon as possible. If care of the victim who is not breathing is delayed more than 4 min, death may occur. Patients with profuse bleeding which persists longer than 15 min may die.

CALL FOR HELP

1. If an injured person needs help and is breathing, telephone for help immediately.
2. If the victim is not breathing, start life-saving treatment and then ask someone to telephone your local rescue squad.
3. When you are talking to the emergency personnel:
   (i) Give the telephone number from which you are calling.
   (ii) Indicate the address of the victim's location, relating any special directions that would be helpful.
   (iii) Describe the victim's condition as clearly as possible—unconscious, burned, bleeding, etc. If poison is involved, name the substance if possible.
   (iv) Give your name.
   (v) Do not hang up until the emergency personnel end the conversation. Additional information may be needed, or they may wish to give you instructions on what to do until help arrives.

BLEEDING

External

1. Immediately apply continuous, firm pressure over the wound, using the cleanest material that is available to you. Even your hand can be used, if necessary.
2. Elevate the bleeding site above the victim's heart.
3. If bleeding continues, apply more forceful, direct pressure on the wound.
4. If extensive bleeding has occurred, have the victim lie down and elevate his legs 12 in.
5. Do not give the victim anything to eat or drink. If part of the person's body has been cut off,

[†]From the office of Richard Edlich.

wrap the part in the cleanest material you have and give it to the rescue squad when they arrive.

Internal

Even though you may not be able to see the bleeding, internal bleeding is a serious problem. Coughing or vomiting blood and/or passing blood in the urine or stool (a black, tarlike stool) are signs of internal bleeding. Victims of accidental injury may have bleeding which you cannot see.

1. Have the victim lie down and elevate his legs 12 in.
2. DO NOT GIVE THE VICTIM ANYTHING TO EAT OR DRINK WHILE WAITING FOR THE RESCUE SQUAD.

BURNS

Flame

1. Stop the burning by having the victim drop to the floor or ground and roll.
2. Smother any residual flames with a blanket.
3. If the victim is unconscious:
   (i) Open the airway by tilting the head back.
   (ii) Check for breathing.
   (iii) If the victim is breathless, start mouth-to-mouth ventilation (see Unconscious Person, below).
4. Put a cool, moist compress on the burned area, or immerse it in cold water (1 to 5°C; 34 to 41°F) until the rescue squad arrives.
5. Cover the breathing victim with a clean sheet.

Chemical

Liquid chemical

1. Pour large volumes of tap water over the burned area.
2. Remove clothes from burned area.
3. Continue flushing the area with water until the rescue squad arrives.

Dry chemical

1. Brush chemical off clothing before removal of clothing.

2. After removal of clothing, visible solid particles should be brushed from the skin.
3. Pour large volumes of water, from any source, over the site contacted by the chemical until the rescue squad arrives.

DO NOT USE ANTIDOTES.

Electrical

1. Turn off the electrical circuit.
2. If the circuit cannot be disconnected, disengage the victim by using a long piece of wood (e.g., broom) or another nonconductive object. During the rescue, avoid contact with the electrical source.
3. If the victim is unconscious:
   (i) Open the airway by tilting the head back.
   (ii) Check for breathing.
   (iii) If the victim is breathless, start mouth-to-mouth ventilation (see Unconscious Person, below).

CHOKING

If the person is choking and can speak or is coughing, do not interfere.

If the choking person cannot speak or cough, he may die withing a few minutes unless the object stuck in his throat is dislodged.

1. Support the victim's chest with one hand. With the heel of your other hand, administer four sharp blows between the victim's shoulder blades.
2. Stand behind the victim with your arms directly under the victim's armpits and encircling his or her chest. Place the thumb side of one fist on the middle of the victim's breastbone. You should then grasp your fist with your other hand and exert four backward thrusts with the intent of relieving the obstruction (Fig. 1).

If the person still cannot speak or cough:

3. Alternate between the above two maneuvers until the victim either is no longer choking or loses consciousness.
4. If the person loses consciousness, help him to floor, tilt his head back, and attempt to perform mouth-to-mouth ventilation (see Unconscious Person, below).
5. If the chest does not rise when you perform mouth-to-mouth ventilation, roll the person onto his side and give four sharp blows between his shoulder blades.
6. Roll the victim onto his back and kneel close to the side of his body. Place the heel of one hand against the lower half of the breastbone. Cover this hand with your other hand. Quickly press down on the breastbone four times (Fig. 2).
7. You should attempt to remove the foreign body by using the tongue-jaw lift maneuver (Fig. 3). Open the victim's mouth by grasping both the tongue and lower jaw between your thumb and index finger of one hand and lift the tongue and jaw forward. This maneuver draws the tongue and lower jaw from the back of the throat and allows the index finger of your other hand to remove and dislodge the foreign body.
8. Attempt to give mouth-to-mouth ventilation again. If the victim's chest still does not rise, repeat steps 5 through 8 until the rescue squad arrives.

FIG. 2. Choking, step 6.

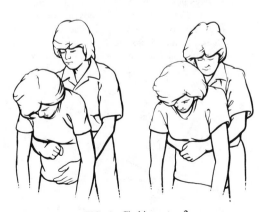

FIG. 1. Choking, step 2.

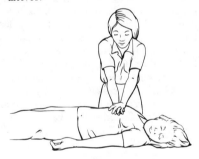

FIG. 3. Choking, step 7.

FROSTBITE

1. Gently remove all clothing from the affected area.
2. The skin of the affected area will appear gray white, or waxy.
3. Immerse the affected part in warm water (102 to 108°F; 39 to 42°C) until the skin becomes flushed or until the rescue squad arrives.
4. Cover the rewarmed part with dry, sterile dressings.
5. Elevate and immobilize the rewarmed part.

HEART ATTACK

Heart attacks are a leading cause of death. Signs and symptoms of a heart atack include one or more of the following:

1. Squeezing chest pain that may radiate to the arms, stomach, jaw, or shoulders
2. Shortness of breath
3. Profuse sweating
4. Nausea or vomiting
5. Dizziness, faint feeling, or weakness
6. Anxiety

If any of these signs and symptoms are detected, call the rescue squad immediately. While waiting for the rescue squad to arrive:

1. Reassure the patient that help is on the way. If possible, do not leave him by himself.
2. Loosen tight clothing (e.g., collars, belts).
3. Place the patient in a semi-sitting position to allow him to breathe more easily.
4. If the patient becomes unconscious:
   (i) Open the airway by tilting the head back.
   (ii) Check for breathing.
   (iii) If the victim is breathless, start mouth-to-mouth ventilation (see Unconscious Person, below).

HEAT STROKE

Signs and symptoms: hot/dry skin, altered state of consciousness, body temperature greater than 105°F (40.6°C).

Heat stroke occurs most frequently in the inactive aged and the overly active young and healthy.

1. Remove the victim's outer clothing.
2. If possible, wrap the patient in a sheet soaked with cool water and fan him vigorously until the rescue squad arrives or until his temperature is less than 103°F (39.4°C).

INJURIES

If possible, an injured person should remain still until the rescue squad arrives.

Broken bones

If bone ends are sticking out of the skin, do not push them back in. Cover the exposed ends with the cleanest material that is available to you.

Head injury

A person who has an injury to his head may also have a neck injury and must not move or be moved.

1. Stabilize the victim's head with your hands.
2. If the victim is unconscious, open his airway by using the modified jaw thrust maneuver (Fig. 4) (see Unconscious Person, below).
   (i) Kneel beside the victim's head.
   (ii) Place the index and long fingers of your hands at the angles of the jaw (below each ear lobe). After lifting upward on his jaw, open the victim's mouth by pushing with your thumbs on his chin.
3. Look, listen, and feel for breathing.
4. If the victim is not breathing, start mouth-to-mouth ventilation.

Impaled object

1. Do not remove any object that is impaled or imbedded in a person.
2. Keep the victim still.
3. If there is external bleeding resulting from the impaled object, control the bleeding by applying direct pressure on the person's skin *around* the object. NEVER PRESS ON THE IMPALED OBJECT.

POISONING

If you suspect a person has swallowed a potentially dangerous substance:

1. Call the nearest Poison Control Center (see Appendix 3 of this volume), whose staff will give you instructions on what to do.
2. Always have syrup of ipecac on hand since the Poison Control Center may advise you to give it to the victim. Syrup of ipecac can be purchased at any drugstore.
3. If the person is unconscious and starts to vomit,

FIG. 4. Modified jaw thrust maneuver.

turn him onto his side and clear the vomitus out of his mouth and throat.

4. If the victim becomes unconscious:
   (i) Open the airway by tilting the head back.
   (ii) Check for breathing.
   (iii) If the victim is breathless, start mouth-to-mouth ventilation (see Unconscious Person, below).
5. Do not throw away anything that might be associated with the poisonous substance (e.g., bottles, pills, vomitus, leaves, etc.). Give them to the rescue squad when they arrive.

SEIZURES

A seizure can result from epilepsy, head injuries, high fever, poisoning, or numerous other problems. During a seizure, the victim must be protected from injury:

1. If the victim's mouth can be opened easily, place something (e.g., a spoon wrapped with a handkerchief) between his teeth. Do not force his mouth open.
2. Loosen any tight clothing.
3. Move surrounding objects (e.g., furniture) away from the victim or place yourself between the objects and the victim.

The victim may stop breathing during the seizure but will usually resume breathing when the seizure activity stops. If the victim does not resume breathing after the seizure:

1. Open the airway by tilting the head back.
2. Check for breathing.
3. If the victim is breathless, start mouth-to-mouth ventilation (see Unconscious Person, below).

UNCONSCIOUS PERSON

Unconsciousness can be the result of many different illnesses or injuries. Regardless of the cause, you should initiate the following specific treatments:

1. Establish unresponsiveness by shaking the patient's shoulder and shouting at him. If the person does not respond, have someone call the rescue squad.
2. Kneel beside the victim. Before moving him, check for possible spinal injury (see Injuries, above.).
3. Place one of your hands beneath the victim's neck and the other on his forehead. Open the victim's airway by tilting his head back (Fig. 5). This maneuver will lift the victim's tongue (the most common cause of airway obstruction in the unconscious person) off the back of his throat. If this does not open the airway, use the modified jaw thrust maneuver (see Injuries, above).

4. Check for breathing by placing your ear as close as possible to the victim's mouth. Look for movement of the victim's chest. Feel for movement of air against your cheek. Listen for breath sounds. If breathing is detected, keep the person's head tilted back until the rescue squad arrives.
5. If the victim is breathless, start mouth-to-mouth ventiltion (Fig. 6.). While pinching his nostrils, cover his mouth completely with your mouth. Blow into the victim's mouth until his chest moves. This should be repeated three more times. After each ventilation, take your mouth off the victim's mouth to allow exhalation.
6. Check for pulse by placing two or three fingers (not your thumb) gently on the person's neck, to one side of the Adam's apple (Fig. 7).
7. If the pulse is present, continue mouth-to-mouth ventilation at the rate of one breath every five seconds until the rescue squad arrives.
8. If the pulse is absent, external cardiac compression should be started by a person properly trained in the technique. Mouth-to-mouth

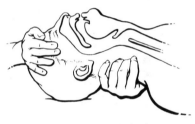

FIG. 5. Open the victim's airway.

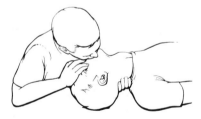

FIG. 6. Mouth-to-mouth ventilation.

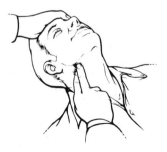

FIG. 7. Check for pulse.

ventilation and external cardiac compression should be continued until the rescue squad arrives.

If an infant or small child is unconscious, do not tilt his head as far back as you would for an adult. If you need to administer mouth-to-mouth ventilation, small puffs of air blown into both his mouth and nose should be sufficient to make the chest rise.

Author Index

Subject Index